Lecture Notes in Computer Science 1084

T0217114

Springer
Berlin
Heidelberg
New York
Barcelona
Budapest
Hong Kong
London
Milan
Paris
Santa Clara
Singapore
Tokyo

William H. Cunningham
S. Thomas McCormick
Maurice Queyranne (Eds.)

Integer Programming and Combinatorial Optimization

5th International IPCO Conference
Vancouver, British Columbia, Canada
June 3-5, 1996
Proceedings

 Springer

Series Editors

Gerhard Goos, Karlsruhe University, Germany

Juris Hartmanis, Cornell University, NY, USA

Jan van Leeuwen, Utrecht University, The Netherlands

Volume Editors

William H. Cunningham
University of Waterloo, Department of Combinatorics and Optimization
Waterloo, Ontario, N2L 3G1, Canada

S.Thomas McCormick
Maurice Queyranne
University of British Columbia
Faculty of Commerce and Business Administration
Vancouver, British Columbia, V6T 1Z2 Canada

Cataloging-in-Publication data applied for

Die Deutsche Bibliothek - CIP-Einheitsaufnahme

Integer programming and combinatorial optimization :
proceedings / 5th International IPCO Conference, Vancouver,
British Columbia, Canada, June 1996. William H. Cunningham
... (ed.). - Berlin ; Heidelberg ; New York ; Barcelona ;
Budapest ; Hong Kong ; London ; Milan ; Paris ; Santa Clara ;
Singapore ; Tokyo : Springer, 1996
(Lecture notes in computer science ; Vol. 1084)

ISBN 3-540-61310-2
NE: Cunningham, William H. [Hrsg.]; International IPCO Conference
 <5, 1996, Vancouver, British Columbia>; GT

CR Subject Classification (1991): G.1.6, G.2.1-2, F.2.2

1991 Mathematics Subject Classification: 90Cxx, 65Kxx, 05-06, 90-06

ISBN 3-540-61310-2 Springer-Verlag Berlin Heidelberg New York

© Springer-Verlag Berlin Heidelberg 1996
Printed in Germany

Typesetting: Camera-ready by author
SPIN 10513071   06/3142 – 5 4 3 2 1 0   Printed on acid-free paper

Preface

This volume contains the papers selected for presentation at IPCO V, the Fifth Integer Programming and Combinatorial Optimization Conference, Vancouver, June 3-5, 1996. The IPCO series of conferences highlights recent developments in theory, computation, and applications of integer programming and combinatorial optimization.

These conferences are sponsored by the Mathematical Programming Society, and are held in the years in which no International Symposium on Mathematical Programming takes place. Earlier IPCO conferences were held in Waterloo (Ontario) in May 1990; Pittsburgh (Pennsylvania) in May 1992; Erice (Sicily) in April 1993; and Copenhagen (Denmark) in May 1995. See the article "History and Scope of IPCO", by Karen Aardal, Ravi Kannan, and William R. Pulleyblank, published in 1994 in *OPTIMA* **43** (the newsletter of The Mathematical Programming Society) for further details.

The proceedings of the first three IPCO conferences were published by organizing institutions. Since then they have been published by Springer in the series, Lecture Notes in Computer Science. The proceedings of IPCO IV, *Integer Programming and Combinatorial Optimization,* edited by Egon Balas and Jens Clausen, were published in 1995 as Volume 920 of this series.

The 36 papers presented at IPCO V were selected from a total of 99 extended abstracts. The overall quality of these submissions was extremely high. As a result, many excellent papers could not be chosen.

The papers included in this volume in most cases represent expanded versions of the submitted abstracts, within a stated 15-page limit. They have not been refereed. We expect more detailed versions of most of these works to appear eventually in scientific journals.

The Program Committee thanks all the authors of submitted abstracts and papers for their support of the IPCO conferences.

March 1996

William H. Cunningham
S. Thomas McCormick
Maurice Queyranne

IPCO V Program Committee

William J. Cook, Rice University
Gérard Cornuéjols, Carnegie Mellon University
William H. Cunningham, University of Waterloo (chair)
Jan Karel Lenstra, Eindhoven University of Technology
László Lovász, Yale University
Thomas L. Magnanti, Massachusetts Institute of Technology
Maurice Queyranne, University of British Columbia
Giovanni Rinaldi, Istituto de Analisi dei Sistemi ed Informatica, Rome

Table of Contents

Session 7: Set Systems and Submodularity

Session 8: Scheduling I

Session 9: Probabilistic Methods

Colourful Linear Programming

Imre Bárány[1] * and Shmuel Onn[2] **

[1] Mathematical Institute of the Hungarian Academy of Sciences, P.O.Box 127,
Budapest, 1364 Hungary
barany@math-inst.hu
[2] Department of Operations Research, School of Industrial Engineering and
Management, Technion - Israel Institute of Technology, 32000 Haifa, Israel
onn@ie.technion.ac.il

Abstract. We consider the following *Colourful* generalisation of Linear Programming: given sets of points $S_1, \cdots, S_k \subset \mathbb{R}^d$, referred to as *colours*, and a point $b \in \mathbb{R}^d$, decide whether there is a *colourful* $T = \{s_1, \ldots, s_k\}$ such that $b \in \text{conv}(T)$, and if there is, find one. Linear Programming is obtained by taking $k = d + 1$ and $S_1 = \ldots = S_{d+1}$. If $k = d + 1$ and $b \in \bigcap_{i=1}^{d+1} \text{conv}(S_i)$ then a solution always exists, but finding it is still hard. We describe an iterative approximation algorithm for this problem, that finds a colourful T whose convex hull contains a point ϵ-close to b, and analyse its Real Arithmetic and Turing complexities. We then consider a class of linear algebraic relatives of this problem, and give a computational complexity classification for the related decision and counting problems that arise. In particular, Colourful Linear Programming is strongly $\mathcal{NP}$-complete. We also introduce and discuss the complexity of a hierarchy of (w_1, w_2)-Matroid-Basis-Nonbasis problems, and give an application of Colourful Linear Programming to the algorithmic problem of Tverberg's theorem in combinatorial geometry.

1 Introduction

The so-called *Carathéodory's Theorem* allows to pose the problem of Linear Programming as follows.

Linear Programming Problem. Given a finite set $S \subset \mathbb{Q}^d$ and a point $b \in \mathbb{Q}^d$, *Decide* whether there is a subset $T \subseteq S$ of size at most $d + 1$ such that $b \in \text{conv}(T)$, and if there is, *Find* one.

Carathéodory's Theorem admits a *colourful* generalisation, due to the first author [3]. To state it, we use the following terminology: given a family of sets $S_1 \cdots, S_k \subset \mathbb{R}^d$, referred to as *colours*, a *colourful set* is a set $T = \{s_1, \ldots, s_k\}$ where $s_i \in S_i$ for all i.

* Partially supported by Hungarian National Science Foundation no. 4296 and 016937.
** Partially supported by the Alexander von Humboldt Stiftung, the Fund for the Promotion of Research at the Technion, and Technion VPR Fund no. 191-198.

Theorem 1. Colourful Carathéodory's Theorem. *If each of $d+1$ given colours $S_0, \ldots, S_d \subset \mathbb{R}^d$ in d-space contains the point b in its convex hull, then $b \in conv(T)$ for some colourful set $T = \{s_0, \ldots, s_d\}$.*

A proof will be given in the next section.

The following algorithmic problem suggested by Theorem 1 is a natural generalisation of Linear Programming.

Colourful Linear Programming Problem. Given colours $S_1, \cdots, S_k \subset \mathbb{Q}^d$ and a point $b \in \mathbb{Q}^d$, *Decide* whether there is a colourful $T = \{s_1, \ldots, s_k\}$ such that $b \in conv(T)$, and if there is, *Find* one.

The specialization of this problem to Linear Programming is obtained by taking $S = S_1 = \ldots = S_{d+1}$.

In this article we study the complexity of this problem and its relatives in linear algebra, matroids, and combinatorial geometry. We provide a rather efficient approximation algorithm for the problem, and give a computational complexity classification for a hierarchy of related decision and counting problems in linear algebra. In particular, Colourful Linear Programming is strongly $\mathcal{NP}$-complete. We also introduce and discuss the complexity of a hierarchy of (w_1, w_2)-Matroid-Basis-Nonbasis problems, and give an application of Colourful Linear Programming to the algorithmic problem of Tverberg's theorem in combinatorial geometry.

The article is organized as follows. In Sections 2 and 3 we study an approximation algorithm for Colourful Linear Programming. We concentrate on the case $k = d + 1$ and $b = 0 \in \bigcap_{i=1}^{d+1} conv(S_i)$ where a solution is guaranteed to exist, but needs to be found. Given an $\epsilon > 0$, the algorithm finds a colourful T which is ϵ-close to 0, that is, whose convex hull contains a point which is ϵ-close to 0. Interestingly, our algorithm specializes, in the case $S_1 = \ldots = S_{d+1}$, to an algorithm of von Neumann for Linear Programming. Assuming that each S_i contains at most n points, that the points are normalized, and that a ball $B(0, \rho)$ is contained in $\bigcap_{i=1}^{d+1} conv(S_i)$, we obtain the following results on the Real arithmetic and Turing complexities of the algorithm, respectively (see Section 3 for the precise statements):

- **Theorem 4:** There is a positive constant c such that the number of real arithmetic operations taken by the algorithm to find a colourful T which is ϵ-close to 0 is $O\left(\frac{n \cdot d^c}{\rho^2} \log \frac{1}{\epsilon}\right)$.
- **Theorem 5:** There is a positive constant c such that the running time of the algorithm to find a colourful T which is ϵ-close to 0, when applied to rational data of bit size L, is $O\left(\frac{n \cdot (dL)^c}{\rho^2} \log \frac{1}{\epsilon}\right)$.

In Sections 4 and 5 we give a computational complexity classification of a hierarchy of related problems. Colourful Linear Programming is equivalent to deciding if $S_1, \ldots, S_k$ admit a colourful T which is *positively dependent*. Replacing "positively" by "linearly" and "dependent" by "independent", we get four decision and related counting problems. We have the following results, the hardness ones holding even if the number k of sets equals the dimension d.

- **Theorem 8:** All four counting problems are $\#P$-complete.
- **Theorems 6 and 7:** The complexity of deciding the existence of a colourful set of one of the types above is given in the following table:

	linearly	positively
dependnet	$\mathcal{NP}$-complete	$\mathcal{NP}$-complete
independent	polynomial time	

- **Theorem 10:** Colourful Linear Programming is strongly $\mathcal{NP}$-complete.

We also discuss the complexity of the following hierarchy of Basis-Nonbasis matroid problems, one for each pair (w_1, w_2) of positive numbers: it is the problem of deciding, given $w_1 + w_2$ matroids, whether there is a subset which is a basis in the first w_1 matroids but not in the others. Counting problems related to this hierarchy will be studied elsewhere, in an extension of [9] and [10].

In Section 6 we turn to an application to a problem from combinatorial geometry motivated by Tverberg's theorem: given a set of points $S \subset \mathbb{Q}^d$ and a positive integer k, a colouring (partition) $S = \biguplus_{i=1}^k S_i$ such that $\bigcap_{i=1}^k \operatorname{conv}(S_i) \neq \emptyset$ is sought. Based on a recent result by Sarkaria [13], the following statement is derived (see Section 6 for the precise statement).

- **Theorem 12:** The decision, counting, and search Tverberg colouring problems are polynomial time reducible to the corresponding suitable problems of Colourful Linear Programming.

We conclude with a brief discussion of another application.

2   Approximating a Colourful Point

As we shall see in Sections 4 and 5, the Colourful Linear Programming Problem is computationally hard. We now consider the perhaps more tractable problem of *approximating* the colourful point. This means finding a colourful set T such that $\operatorname{conv}(T)$ contains a point which is ϵ-close to 0. In this section and the next we describe and analyse a certain iterative pivoting algorithm for such an approximation. Interestingly, when all the sets are the same $S_0 = \ldots = S_d$, our algorithms essentially specialises to an algorithm of von Neumann for Linear Programming [6]. So our algorithm can be regarded as "colourful" refinement of von Neumann's algorithm.

Without loss of generality we shall assume that the input is *normalized*, that is, the norm of each input point $s \in \bigcup_{i=0}^d S_i$ satisfies $1 \leq |s| \leq 2$. The reason that we do not simply assume $|s| = 1$ is that we want to deal with rational data also. Note that normalizing an arbitrary input is easy to do: with real arithmetic computation simply divide s by its norm $|s|$; with a Turing machine on rational data, first scale $s \in \mathbb{Q}^d$ to have integer components, and then divide it by the largest integer n satisfying $n^2 \leq |s|^2$. It will also be useful to extend the analysis to the situation where we know that a Euclidean ball $B(0, \rho)$ of radius ρ (allowing still $\rho = 0$) about 0 is contained in the convex hull of each monochromatic set S_i. So, we address the following problem.

Colourful Point Approximation Problem. Given $0 < \epsilon < 1$ and normalized sets $S_0, \ldots, S_d$ of points in $\mathbb{R}^d$, such that $B(0, \rho) \subset \bigcap_{i=0}^{d} \text{conv}(S_i)$ for some $0 \leq \rho \leq 1$, find a colourful set containing a point x satisfying $|x| \leq \epsilon$.

In this section we describe our iterative algorithm, and bound the number of iterations required. In the next section we describe a variant of this algorithm, discuss the details of its efficient implementation, and give a complexity analysis for both real arithmetic and Turing computations.

Algorithm 1:

- *Initialization.* Put $k = 1$. Pick an arbitrary colourful set $T_1 = \{s_0, \ldots, s_d\}$. Let $x_1 = s_0$.
- *Iteration.* If $|x_k| \leq \epsilon$ then stop and output T_k and x_k. Otherwise update the colourful set as follows: choose a colour i such that $x_k \in \text{conv}(T_k \setminus \{s_i\})$; choose a point $t_i \in S_i$ minimizing the inner product $\langle x_k, t_i \rangle$; and let $T_{k+1} = T_k \cup \{t_i\} \setminus \{s_i\}$. Let x_{k+1} be the point of minimum norm in $\text{conv}(T_{k+1})$. Increment k and proceed to the next iteration.

Note that in the kth iteration either $x_k = 0$ and the algorithm stops, or else it is indeed possible to choose a colour i to be exchanged: either $\text{conv}(T_k)$ is full dimensional and x_k lies on its boundary, or $\text{conv}(T_k)$ has affine dimension less than d; in both cases x_k can be expressed as a convex combination of at most d points from T_k by the usual Carathéodory theorem. We now derive an upper bound on the number of iterations (Lemma 3 below), which turns out to be no worse than for von Neumann's noncolourful algorithm [6], [7].

Proposition 2. *Let $0 < \epsilon < 1$ and normalized $S_0, \ldots, S_d \subset \mathbb{R}^d$ be given. Let $0 \leq \rho \leq 1$ be a real number such that $B(0, \rho) \subset \bigcap_{i=0}^{d} \text{conv}(S_i)$. Then, when Algorithm 1 is applied, the following recursions hold while $x_k \neq 0$:*

$$\text{If } \rho = 0: \quad \frac{1}{|x_{k+1}|^2} \geq \frac{1}{4} + \frac{1}{|x_k|^2}; \qquad \text{If } \rho > 0: \quad |x_{k+1}|^2 \leq \left(1 - \frac{\rho^2}{4}\right) |x_k|^2 \quad (1)$$

Proof. Consider the kth iteration. The point $q = -\rho \frac{x_k}{|x_k|}$ lies in the ball $B(0, \rho)$ hence in $\text{conv}(S_i)$, and satisfies $\langle x_k, q \rangle = -\rho|x_k|$. Therefore, there must be a point in S_i, in particular the point t_i chosen by the algorithm, which also satisfies $\langle x_k, t_i \rangle \leq -\rho|x_k|$. Let $a := x_k$ and $b := t_i$, and let p be the point closest to the origin on the line spanned by a and b. So p is the projection of 0 onto that line, hence

$$p = \frac{\langle b - a, b \rangle a + \langle a - b, a \rangle b}{\langle b - a, b - a \rangle} \quad \text{and} \quad |p|^2 = \frac{|a|^2|b|^2 - \langle a, b \rangle^2}{|a|^2 + |b|^2 - 2\langle a, b \rangle}.$$

Since $\langle a, b \rangle \leq -\rho|a|$ this gives

$$|p|^2 \leq \frac{(|b|^2 - \rho^2)|a|^2}{|b|^2 + 2\rho|a| + |a|^2}.$$

Since the input is normalised we have $|b| \leq 2$, and so we get for $\rho = 0$

$$\frac{1}{|p|^2} \geq \frac{1}{|b|^2} + \frac{1}{|a|^2} \geq \frac{1}{4} + \frac{1}{|a|^2},$$

and for $\rho > 0$

$$|p|^2 \leq \frac{(|b|^2 - \rho^2)|a|^2}{|b|^2} \leq \left(1 - \frac{\rho^2}{4}\right)|a|^2.$$

Now $\langle a, b \rangle \leq 0$ and p is on the line spanned by a and b, so p in fact lies on the line segment $[a, b] = [x_k, t_i] \subset \text{conv}(T_{k+1})$. Since x_{k+1} is defined as the point in $\text{conv}(T_{k+1})$ of smallest norm, we have $|x_{k+1}| \leq |p|$ and the proposition follows. $\square$

One consequence of the proposition is the following proof of the Colourful Carathéodory Theorem:

Proof of Theorem 1. Let $S_0 \ldots, S_d$ be sets of points in $\mathbb{R}^d$ such that $0 \in \bigcap_{i=0}^{d} \text{conv}(S_i)$. Normalize the points by scaling: for any subset $T \subseteq \bigcup_{i=0}^{d} \text{conv}(S_i)$ this does not affect whether or not $0 \in \text{conv}(T)$. Now apply Algorithm 1. By Proposition 2 with $\rho = 0$, as long as $x_k \neq 0$ we have $|x_{k+1}| < |x_k|$ hence $x_{k+1} \neq x_k$. Since x_k is uniquely defined from T_k and there are only finitely many colourful sets, we will eventually have $0 = x_k \in \text{conv}(T_k)$ proving that the origin is in the convex hull of some colourful set. $\square$

Another consequence is the following estimate on the number of iterations of Algorithm 1.

Lemma 3. *Let $0 < \epsilon < 1$ and normalized $S_0, \ldots, S_d \subset \mathbb{R}^d$ be given, and let $0 \leq \rho \leq 1$ be such that $B(0, \rho) \subset \bigcap_{i=0}^{d} \text{conv}(S_i)$. Then we have the following upper bounds on the number of iterations performed by Algorithm 1 to find a colourful point which is within distance ϵ from the origin:*

$$\text{If } \rho = 0: \quad \left\lceil \frac{4}{\epsilon^2} \right\rceil = O\left(\frac{1}{\epsilon^2}\right); \quad \text{If } \rho > 0: \quad 1 + \left\lceil \frac{16}{\rho^2} \log \frac{2}{\epsilon} \right\rceil = O\left(\frac{1}{\rho^2} \log \frac{1}{\epsilon}\right) \quad (2)$$

Proof. For $\rho = 0$, let $I = \lceil \frac{4}{\epsilon^2} \rceil$. Summation of the expressions in (1) for $k = 1, \ldots, I - 1$ gives

$$\frac{1}{|x_I|^2} \geq \frac{1}{4}\left(\left\lceil \frac{4}{\epsilon^2} \right\rceil - 1\right) + \frac{1}{|x_1|^2} \geq \frac{1}{4}\left\lceil \frac{4}{\epsilon^2} \right\rceil \geq \frac{1}{\epsilon^2},$$

which proves the claim in this case.

For $\rho > 0$, let $t = \frac{4}{4 - \rho^2} > 1$ and let $I = 1 + \left\lceil \frac{2}{\log t} \log \frac{2}{\epsilon} \right\rceil$. Then $(I - 1) \log t \geq 2 \log \frac{2}{\epsilon}$, so

$$\left(\frac{1}{t}\right)^{I-1} \leq \left(\frac{\epsilon}{2}\right)^2, \quad \text{and so by Proposition 2,} \quad |x_I|^2 \leq \left(\frac{1}{t}\right)^{I-1} \cdot |x_1|^2 \leq \epsilon^2.$$

Since $\rho \leq 1$ we have

$$\log t = \log \left(\frac{4}{4 - \rho^2} \right) = \log \left(1 + \frac{\rho^2}{4 - \rho^2} \right) \geq \left(\frac{1}{2} \right) \frac{\rho^2}{4 - \rho^2} > \frac{\rho^2}{8},$$

so $\frac{2}{\log t} < \frac{16}{\rho^2}$. This proves the claim in this case as well. $\square$

3 A Variant of the Algorithm and its Complexity

Finding a point x_{k+1} of minimum norm in $\mathrm{conv}(T_{k+1})$ in the kth iteration of Algorithm 1 is a fairly heavy task which involves the minimization of a quadratic form, and can be solved only approximately. To avoid this, we present a variant of the algorithm in which only linear algebraic computations (such as solutions of systems of linear equations and computations of determinants) will be required. Such computations can be easily carried out in strongly polynomial time (i.e., polynomial number of arithmetic operations and polynomial time in Turing computations on rational data) via Gaussian elimination. To arrive at such a variant of the algorithm, we reconsider the choice of the new point x_{k+1} in the k-th iteration step of the algorithm. If x_{k+1} is chosen to be any point of $\mathrm{conv}(T_{k+1})$ which is expressible as a convex combination of at most d elements, then the next iteration can be carried out - namely a colour can be found in which a new point can replace the old one. If x_{k+1} is chosen to be any point of $\mathrm{conv}(T_{k+1})$ satisfying $|x_{k+1}| \leq |p|$, where p is the projection of 0 to the line segment $[x_k, t_i]$ as in the proof of Proposition 2, then the proposition remains true. So the bound on the number of iterations in Lemma 3 will hold for any variant of Algorithm 1 in which, at each iteration k, the point x_{k+1} is chosen so as to satisfy these two properties. This suggests the following variant of Algorithm 1. In the k-th iteration, the set T_{k+1} is constructed as before. Then the point p is computed. If $\mathrm{conv}(T_{k+1})$ happens to contain p on its boundary or is not full dimensional, then p is expressible as a convex combination of d or fewer points from T_{k+1}, so p itself can be taken as the next point x_{k+1}. Otherwise, $\mathrm{conv}(T_{k+1})$ is a d-simplex containing p in its interior. In this case it is easy to compute the at most $d + 1$ intersection points $\alpha_j p$ of the line spanned by p and 0 with the hyperplanes $\mathrm{aff}(T_{k+1} \setminus \{t_j\})$ spanned by the facets of $\mathrm{conv}(T_{k+1})$. It is then possible to either conclude that $0 \in \mathrm{conv}(T_{k+1})$ or to identify the boundary point αp of $\mathrm{conv}(T_{k+1})$ where it is stabbed by the ray spanned by p. This boundary point can then be taken to be the next point x_{k+1}. Here is a more formal description of this algorithm.

Algorithm 2:

- *Initialization.* Put $k = 1$. Pick an arbitrary colourful set $T_1 = \{s_0, \ldots, s_d\}$. Let $\lambda^1 = (1, 0, \ldots, 0)$ and $x_1 = \sum_{i=0}^{d} \lambda_i^1 s_i = s_0$.
- *Iteration.* If $|x_k| \leq \epsilon$ then stop and output T_k, x_k, and λ^k. Otherwise proceed as follows:

1. Update the colourful set: choose i such that $\lambda_i^k = 0$; choose a point $t_i \in S_i$ minimising the inner product $\langle x_k, t_i \rangle$; and let $T_{k+1} = \{t_0, \ldots, t_d\}$ where $t_j = s_j$ for all $j \neq i$.

2. Let $x = \sum_{j=0}^{d} \lambda_j t_j$ where $\lambda = (\lambda_0, \ldots \lambda_d)$ is the nonnegative vector of coefficients given by

$$\lambda_i = \frac{\langle x_k - t_i, x_k \rangle}{|t_i - x_k|^2} \quad \text{and} \quad \lambda_j = \frac{\langle t_i - x_k, t_i \rangle}{|t_i - x_k|^2} \lambda_j^k \quad (j \neq i).$$

Update λ by subtracting suitable affine relations on T_{k+1} while maintaining its nonnegativity, until $\{t_j : \lambda_j > 0\}$ becomes affinely independent.

3. If $x \neq 0$ and $\lambda_j > 0$ for all j, Then do the following for $j = 0, \ldots, d$: if the line $\text{lin}(x)$ and the hyperplane $\text{aff}(T_{k+1} \setminus \{t_j\})$ intersect, then let α_j be the real number such that $\alpha_j x$ is that intersection point, whereas if they do not intersect, then put $\alpha_j = \infty$. Put

$$\alpha = \max\{\alpha_j : 0 \leq j \leq d, \ \alpha_j < 1\}.$$

If $\alpha \leq 0$ then put $x := 0$, whereas if $\alpha > 0$ put $x := \alpha x$. Let λ be the unique vector of coefficients satsifying $\sum_{j=0}^{d} \lambda_j = 1$ and $x = \sum_{j=0}^{d} \lambda_j t_j$.

4. Put $\lambda^{k+1} := \lambda$ and $x_{k+1} := x = \sum_{j=0}^{d} \lambda_j^{k+1} t_j$.

Increment k and proceed to the next iteration.

The next two theorems discuss the Real Arithmetic and Turing complexities of this algorithm, respectively. The complexity is stated for the case of positive ρ, but can easily be adopted to get bounds for the case $\rho = 0$ as well. The first theorem gives a bound, which depends nicely on ϵ, on the number of arithmetic operations performed on normalised data.

Theorem 4. *Let $A(d, n, \epsilon, \rho)$ be the largest number of real arithmetic operations taken by Algorithm 2 to find a colourful point ϵ-close to 0, when applied to normalised sets $S_0, \ldots, S_d \subset \mathbb{R}^d$, each containing at most n points, and satisfying $B(0, \rho) \subset \bigcap_{i=0}^{d} \text{conv}(S_i)$. Then a positive constant c_1 exists such that, in the range $0 < \epsilon < 1$ and $\rho > 0$, there holds*

$$A(d, n, \epsilon, \rho) = O\left(\frac{n \cdot d^{c_1}}{\rho^2} \log \frac{1}{\epsilon}\right).$$

Proof. We give only an outline of the proof of correctness and complexity analysis of the algorithm. The complete details can be found in [4].

The algorithm maintains, in each iteration, a colourful set $T_k = \{s_0, \ldots, s_d\}$, an approximating point $x_k \in \text{conv}(T_k)$, and a coefficient vector λ^k expressing x_k as a convex combination $x_k = \sum_{j=0}^{d} \lambda_j^k t_j$ of an affinely independent subset $\{t_j : \lambda_j^k > 0\}$ of T_k of at most d points (except possibly for the last iteration). Consider the kth iteration loop. In step 1, since at least one $\lambda_i^k = 0$, it is possible to construct the new colourful set T_{k+1}. This step involves checking the inner product $\langle x_k, t \rangle$ for at most $|S_i| \leq n$ points t. Next, consider step 2: the point x computed here is the projection point p of 0 onto the segment $[x_k, t_i]$, as explained earlier in this section. The update of λ so that $\{t_j : \lambda_j > 0\}$ becomes affinely independent can be done by subtracting suitable affine relations, which can be found by Gaussian elimination in a number of arithmetic operations bounded by a polynomial in d. When step 2 is completed, λ expresses x as a convex combination of affinely independent points of T_{k+1}.

If now $\lambda_j = 0$ for some j, or if $x = 0$, then x_{k+1} can be taken to be x, and λ^{k+1} can be taken to be λ. In this case step 3 will be justfully skipped. Otherwise, all $\lambda_j > 0$ so conv(T_{k+1}) is a full dimensional simplex containing $x \neq 0$ in its interior. For $j = 0, \ldots, d$, it is possible, by using Cramer's rule and computing at most two determinants, to check if the line lin(x) through x and 0 intersects the hyperplane aff$(T_{k+1} \setminus \{t_j\})$, and if it does, determine the intersection point $\alpha_j x$. Since $x = 1 \cdot x$ is in the interior of conv(T_{k+1}), the line $lin(x)$, directed from 0 to x, enters conv(T_{k+1}) at the point αx where α is the largest α_j less than 1, as computed in step 3. Now, if $\alpha \leq 0$ then 0 is in the line segment $[\alpha x, x]$ hence in conv(T_{k+1}), so x_{k+1} can be taken to be 0. If $\alpha > 0$, then αx is in $[0, x]$ hence closer to 0 then x, so x_{k+1} can be taken to be αx. So in either case, the updated point x computed in step 3 is a good point for the next iteration. Since this point lies in the convex hull of T_{k+1}, which is affinely independent in this situation, there is a unique coefficient vector λ satisfying $\sum_{j=0}^{d} \lambda_j = 1$ and $x = \sum_{j=0}^{d} \lambda_j t_j$, and it is nonnegative and can be computed by a single matrix inversion. Moreover, if $x \neq 0$ then x is a boundary point of conv(T_{k+1}), so at least one λ_i is zero. If $x = 0$, then the algorithm will terminate in the following iteration.

The work in each iteration involves scanning the set S_i in step 1 and doing linear algebraic calculations, which involve a number of arithmetic operations polynomial in d and linear in n. Since, for positive ρ, the number of iterations is $O\left(\frac{1}{\rho^2} \log \frac{1}{\epsilon}\right)$ as guaranteed by Lemma 3, the theorem follows. $\square$

For rational data, we need to maintain the bit size of the numbers involved in the computations polynomially bounded in the size of the input throughout the algorithm. This is achieved by replacing step 4 in the iteration loop of Algorithm 2 by the following *rounding step*:

Rounding step for rational data: let $D = \left\lceil \frac{32d}{\epsilon \rho^2} \right\rceil$. Put $x_{k+1} := \sum_{j=0}^{d} \lambda_j^{k+1} t_j$, where

$$\lambda_j^{k+1} := \frac{\lceil D\lambda_j \rceil}{\sum_{r=0}^{d} \lceil D\lambda_r \rceil} \qquad (j = 0, \cdots, d).$$

Let the *bit size* L of the input point sets S_i denote the largest number of bits needed to encode in binary any coordinate of any point of $\cup S_i$. We now show that with the rounding step, the Turing running time of Algorithm 2 on normalized rational input, is pseudopolynomial in ρ and polynomial in the rest of the data. For each fixed $\rho > 0$, this is a polynomial time algorithm for normalized data, even when ϵ is encoded in binary.

Theorem 5. *Let $T(d, n, \epsilon, \rho, L)$ be the largest running time taken by a Turing machine executing Algorithm 2 with the rounding step, to find a colourful point ϵ-close to 0, when applied to normalized rational sets $S_0, \ldots, S_d \subset \mathbb{Q}^d$ of bit size L, each containing at most n points, and satisfying $B(0, \rho) \subset \bigcap_{i=0}^d conv(S_i)$. Then a positive constant c_2 exists such that, for rational ϵ, ρ in the range $0 < \epsilon < 1$ and $\rho > 0$, there holds*

$$T(d, n, \epsilon, \rho, L) = O\left(\frac{n \cdot (dL)^{c_2}}{\rho^2} \log\frac{1}{\epsilon}\right).$$

Proof. First, note that $1 - \frac{\rho^2}{8} \geq \sqrt{1 - \frac{\rho^2}{4}}$, so the number D used in the rounding step satisfies

$$D = \left\lceil \frac{32d}{\epsilon\rho^2} \right\rceil \geq \frac{4d}{\epsilon\left(1 - \sqrt{1 - \frac{\rho^2}{4}}\right)}.$$

Now consider the kth iteration of the algorithm, and let λ and $x = \sum_{j=0}^d \lambda_j t_j$ be as computed in this iteration before entering the rounding step. The number of arithmetic operations needed for that is bounded as in the proof of Theorem 4. Now apply the rounding step and obtain x_{k+1} and λ^{k+1}. First, note that

$$|x_{k+1}| \leq \left| \frac{1}{D} \sum_{j=0}^d \lceil D\lambda_j \rceil t_j \right| = \left| \sum_{j=0}^d \lambda_j t_j + \sum_{j=0}^d \frac{1}{D}(\lceil D\lambda_j \rceil - D\lambda_j) t_j \right|$$

$$\leq |x| + \frac{1}{D}\sum_{j=0}^d |t_j| \leq \left(\sqrt{1 - \frac{\rho^2}{4}}\right)|x_k| + \frac{1}{D}d \cdot 2$$

$$\leq \left(\sqrt{1 - \frac{\rho^2}{4}}\right)|x_k| + \frac{1 - \sqrt{1 - \frac{\rho^2}{4}}}{2}\epsilon \leq \left(\frac{1 + \sqrt{1 - \frac{\rho^2}{4}}}{2}\right)|x_k|.$$

Therefore, $|x_{k+1}|$ is a constant fraction of $|x_k|$. So the recursion (1) of Proposition 2 is replaced by the recursion $|x_{k+1}| \leq \left(\frac{1+\sqrt{1-\frac{\rho^2}{4}}}{2}\right)|x_k|$, and an analysis similar to that of Lemma 3 gives again the bound $O\left(\frac{1}{\rho^2}\log\frac{1}{\epsilon}\right)$ on the number of iterations.

Second, note that the numerator of each λ_j^{k+1} is a positive integer less than or equal to D, and its denominator is a positive integer less than or equal to dD, so the bit size of λ_j^{k+1} satisfies $size(\lambda_j^{k+1}) = O(\log d + \log\frac{1}{\epsilon} + \log\frac{1}{\rho})$, hence is polynomially bounded in the data. Therefore also $x_{k+1} = \sum_{r=0}^d \lambda_j^{k+1} t_j$ is polynomially bounded in d, $\log\frac{1}{\epsilon}$, $\log\frac{1}{\rho}$, and dL. So the numbers involved in computations throughout the algorithm remain polynomially bounded in the size of the input, and the theorem follows. $\square$

4 Colourful Linear Algebra and its Complexity

A set $S = \{s_1, \cdots, s_n\}$ of vectors in $\mathbb{R}^d$ is *positively dependent* if it has a non-trivial linear dependency $\sum_{i=1}^n \mu_i s_i = 0$ with all μ_i nonnegative. Since this is equivalent to $0 \in \text{conv}(S)$, Carathéodory's theorem now says that if $S \subset \mathbb{R}^d$ is positively dependent then it has a subset of size at most $d+1$ which also is. This is in analogy with the trivial statement for a linearly dependent set. Given now a family of colours $S_0 \cdots, S_d \subset \mathbb{R}^d$, the Colourful Carathéodory Theorem says the following: if each member S_i of the family is positively dependent, then the family admits a positively dependent colourful set $T = \{s_0, \ldots, s_d\}$. Again, this is in analogy with the trivial statement for linear dependence.

We now consider several algorithmic problems that arise, and that turn out to be hard for the linear-dependency case as well. For each problem, we shall distinguish a **Decision** problem, a **Search** problem, and a **Counting** problem. Here we shall restrict attention to the Turing machine computation model, and to the field $\mathbb{Q}$ of rational integers. However, these problems have obvious analogs over any field $\mathcal{F}$ and any type of $\mathcal{P}$-dependency (i.e., when the coefficients in the linear relation are required to come from a fixed subset $\mathcal{P} \subset \mathcal{F}$) and could be studied under other models of computation such as by Blum-Shub-Smale [5].

Colourful Set Problem. Given are k, d, and a family of nonempty colours $S_1, \cdots, S_k \subset \mathbb{Q}^d$.

- **Decide** if the family has a linearly (respectively, positively) dependent (respectively, independent) colourful set.
- **Find** such a colourful set if one exists.
- **Count** the number of such colourful sets.

Note that for Positively dependent sets, this is the same as the Colourful Linear Programming Problem stated in Section 1. Also, note that if $k = d + 1$ then the answer to the linear version of the decision problem becomes trivial. If in addition each S_i is positively dependent, the positive version of the decision problem also becomes trivial by Theorem 1. So the following statement is sharp.

Theorem 6. *Deciding linearly (respectively, positively) dependent colourful sets is $\mathcal{NP}$-complete, even if $k = d$ and each S_i is itself positively dependent.*

Proof. We prove the statements for the linear and positive versions simultaneously. Clearly both problems are in the complexity class $\mathcal{NP}$. We reduce the $\mathcal{NP}$-complete problem of *Partition* [8] to each. Given positive integers $a_1, \ldots, a_d$, let for $i = 2, \cdots, d$

$$u_1 = a_1 e_1 + \sum_{i=2}^d e_i, \quad v_1 = -a_1 e_1 + \sum_{i=2}^d e_i, \quad \text{and} \quad u_i = a_i e_1 - e_i, \quad v_i = -a_i e_1 - e_i,$$

where $e_1, \cdots, e_d$ denote the standard unit vectors in $\mathbb{Q}^d$, and let $S_i = \{u_i, v_i\}$ for all i. It is now easy to verify that there is a linearly dependent colourful set if and only if there is a positively dependent one if and only if there is a partition

$I \uplus J$ of $[d] = \{1, \ldots, d\}$ such that $\sum_{i \in I} a_i = \sum_{j \in J} a_j$: first, if $I \uplus J$ is such a partition then $\sum_{i \in I} u_i + \sum_{j \in J} v_j = 0$, so $\{u_i : i \in I\} \cup \{v_j : j \in J\}$ is a linearly and positively dependent colourful set. Conversely, let $T = \{s_1, \cdots, s_d\}$ be a colourful set admitting a nontrivial linear relation $\sum_{j=1}^{d} \lambda_j s_j = 0$. For $i = 2, \cdots, d$, by considering the ith coordinate of the equation $0 = \sum_{j=1}^{d} \lambda_j s_j$ we find that $0 = \lambda_1 - \lambda_i$. Therefore $\lambda_1 = \ldots = \lambda_d$ and so $\sum_{i=1}^{d} s_i = 0$ as well. So T is also positively dependent, and the sets $I = \{i : s_i = u_i\}$ and $J = \{j : s_j = v_j\}$ form a partition of $[d]$ with the desired property. To prove that the problem remains $\mathcal{N}P$-complete if each S_i is itself positively dependent, simply take $S_i = \{u_i, v_i, -u_i, -v_i\}$. Now, if $\{s_1, \cdots, s_d\}$ is a linearly dependent colourful set with $\sum_{i=1}^{d} \lambda_i s_i = 0$ then, flipping the sign of both s_i and λ_i if necessary, we find another linearly dependent colourful set with $s_i \in \{u_i, v_i\}$ for all i, and proceed as above to show that it is also positively dependent and to construct a suitable partition of $[d]$. $\square$

Searching for a colourful dependent set is no easier than deciding if one exists, and so is generally hard. However, if $k = d + 1$, then any colourful set is linearly dependent, and there always exists a positively dependent colourful set. The search problem becomes trivial for linearly dependent colourful sets, but not for positively dependent ones. The question of whether there is a polynomial time algorithm for the problem remains open and is central to this area.

In partial contrast to Theorem 6, we have the following statement, observed together with M. Loebl.

Theorem 7. *Deciding linearly independent colourful sets can be done in polynomial time.*

Proof. Let V be the disjoint union of the S_i (so a point of $\mathbb{Q}^d$ appearing in several S_i will have several copies in V). Define two matroids M_1, M_2 on V as follows: M_1 will be the matroid of linear dependencies on V, while M_2 will be the matroid whose bases are all colourful sets. Then a k-subset of V is a linearly independent colourful set if and only if it is independent in both M_1 and M_2, and so the decision (and search) problem reduce to 2-matroid intersection, which can be done in polynomial time. $\square$

This proof raises some questions about matroids, which we discuss later on. But first we show that counting is hard for all four variants of the problem.

Theorem 8. *Counting linearly (respectively, positively) dependent (respectively, independent) colourful sets is $\#P$-complete.*

Proof. All problems are clearly in $\#P$. We now reduce the $\#P$-complete problem of computing the *permanent* of a $\{0, 1\}$-matrix to each. Let $A = (A_{i,j})$ be such a matrix of size $d \times d$, and for $i = 1, \ldots, d$ define $S_i = \{e_j : A_{i,j} = 1\}$. Then

$$\text{perm}(A) := \sum \{\Pi_{i=1}^{d} A_{i, \pi(i)} : \pi \text{ permutation on } [d]\}$$

$$= \sum \{1 : \; \pi \text{ permutation}, \; e_{\pi(1)} \in S_1, \; \ldots, \; e_{\pi(d)} \in S_d\}$$
$$= \#\{\text{linearly independent colourful sets of } S_1, \ldots, S_d\}$$
$$= \Pi_{i=1}^{d}|S_i| - \#\{\text{linearly dependent colourful sets of } S_1, \ldots, S_d\}.$$

This proves the claim for linearly independent (respectively, dependent) colourful sets. The modifications for the positive analogs can be found in [4]. □

The linear variants of the colourful set problem form a special case of a general hierarchy of *Matroid Basis-Nonbasis* problems which we now introduce. Given are a pair $w = (w_1, w_2)$ of nonnegative integers and matroids $M_1, \ldots, M_{w_1+w_2}$ of the same rank, defined on the same set. A subset of the elements is a *w-set* if it is a basis in each of the first w_1 matroids and is not a basis in each of the last w_2 matroids. Note that, for complexity considerations, one has to specify the way in which the matroids are presented - e.g., by an independent set oracle, or by a matrix when the matroids are linear.

Matroid Basis-Nonbasis Problem. Given are a pair $w = (w_1, w_2)$ of nonnegative integers, positive integers d and n, and $w_1 + w_2$ matroids of rank d defined on the same set of n elements.

- **Decide if there exists a w-set.**
- **Find a w-set if one exists.**
- **Count the number of w-sets.**

The decision and search problems are polynomial time solvable for $w = (1, 0)$ by the so-called greedy algorithm and for $w = (2, 0)$ by the 2-matroid intersection algorithm mentioned before, even if the matroids are given by oracles. For $w = (3, 0)$ the decision problem is known to be $\mathcal{NP}$-complete. The complexity for $w = (0, 1)$, which includes as a special case the problem of checking if n given points in d-space are in general position, is unknown. Theorem 6 shows that for $w = (1, 1)$ the decision problem is $\mathcal{NP}$-complete. It would be interesting to settle the complexity of these problems restricted to special classes of matroids.

Another interesting problem related to the colourful set problem concerns common zeros of systems of quadratic forms. Identify $\mathbb{Q}^d$ with the vector subspace of linear forms in the algebra of polynomials $\mathbb{Q}[x_1, \cdots, x_d]$ in the obvious way. Let $\mathcal{F}$ be any field extension of $\mathbb{Q}$ (possibly $\mathcal{F} = \mathbb{Q}$). A quadratic form is *simple* (of rank 1) if it is the product $u \cdot v$ of two linear forms $u, v \in \mathbb{Q}^d$.

Theorem 9. *Deciding if a system of rational quadratic forms have a common nontrivial zero over $\mathcal{F}$ is NP-complete, even if the number of forms equals the number of indeterminates and each form is simple.*

Proof. By reduction from *Partition*: construct d sets $S_i = \{u_i, v_i\}$, where $u_i, v_i \in \mathbb{Q}^d$, as in the proof of Theorem 6. Now, the system $u_1 \cdot v_1, \ldots, u_d \cdot v_d$ of simple quadratic forms in $\mathbb{Q}[x_1, \cdots, x_d]$ admit a common nontrivial zero $\lambda \in \mathcal{F}^d$ if and only if λ is in the orthogonal complement in $\mathbb{Q}^d$ of some colourful set of the S_i, i.e., if and only if the S_i admit a linearly dependent colourful set. □

5   Strong Hardness of Colourful Linear Programming

Here we show that the Colourful Linear Programming Problem is *strongly $\mathcal{NP}$-complete* [8]. Thus, even a pseudo-polynomial time algorithm (of running time polynomial in the unary representation of the data - see [8] for the exact definition) does not exist unless $\mathcal{P} = \mathcal{NP}$.

Theorem 10. *Given colours $S_1, \cdots, S_d \subset \mathbb{Q}^d$ such that $0 \in \bigcap_{i=1}^{d} conv(S_i)$, it is strongly $\mathcal{NP}$-complete to decide whether there is a colourful $T = \{s_1, \ldots, s_d\}$ such that $0 \in conv(T)$.*

Proof. The proof is by reduction from 3-satisfiability. The details are quite technical and lengthy and can be found in [4]. □

6   Tverberg Colourings

A *Tverberg k-colouring* of a set S of points in $\mathbb{R}^d$ is a k-colouring of S such that the convex hulls of the monochromatic subsets have a point in common, i.e., a partition $S = \biguplus_{i=1}^{k} S_i$ such that the intersection $\bigcap_{i=1}^{k} conv(S_i)$ is nonempty. Tverberg's theorem [14] asserts the following.

Theorem 11. *Any set of more than $(k-1)(d+1)$ points in $\mathbb{R}^d$ has a Tverberg k-colouring.*

This theorem will follow from the proof of Theorem 12 below. Note that for $k = 2$, it reduces to the simple fact that any set of more than $d+1$ points in $\mathbb{R}^d$ is affinely dependent (*Radon's theorem*).

A remark in place here is that this theorem has an analogue over the integers. It says that for every k and d there is a finite number $t(k, d)$ such that any set of more than $t(k, d)$ points in $\mathbb{Z}^d \subset \mathbb{R}^d$ has an *integer* k-colouring, namely a k-colouring such that $\bigcap_{i=1}^{k} conv(S_i)$ contains an integer point. The determination of the numbers $t(k, d)$ is an outstanding open problem, even for $k = 2$. See [11] for the bounds $2^d < t(2, d) < d2^d$ and for related results.

Tverberg's theorem raises the following algorithmic problems.

Tverberg Colouring Problem. Given are k, d, and a set S of points in $\mathbb{Q}^d$.

- Decide if the set admits a Tverberg k-colouring.
- Find a Tverberg k-colouring if one exists.
- Count the number of Tverberg k-colourings.

Recently, Sarkaria [13] discovered that the Colourful Carathéodory Theorem 1 implies Tverberg's Theorem 11. Here we give a simplified variant of Sarkaria's argument, and use it to reduce the algorithmic Tverberg Colouring Problem to the Colourful Linear Programming Problem.

Theorem 12. *The Decision (respectively, Search, Counting) variant of the Tverberg Colouring Problem is polynomial time reducible to the Decision (respectively, Search, Counting) variant of the Colourful Linear Programming Problem.*

Proof. Let k and $S = \{v_1, \ldots, v_n\} \subset \mathbb{Q}^d$ be given. For $i = 1, \ldots, k-1$, let $f_i = e_i$ be the ith unit vector in $\mathbb{Q}^{k-1}$, and let $f_k = -\sum_{i=1}^{k-1} f_i$. Embed S in an open halfspace in $\mathbb{Q}^{d+1}$, say replace each v_i by $\bar{v}_i = \begin{bmatrix} v_i \\ 1 \end{bmatrix} \in \mathbb{Q}^{d+1}$. For $i = 1, \ldots, n$ construct the following set T_i of k matrices in $\mathbb{Q}^{(d+1) \times (k-1)}$:

$$T_i = \{\bar{v}_i f_1^T, \ldots, \bar{v}_i f_k^T\} \subset \mathbb{Q}^{(d+1) \times (k-1)}, \qquad i = 1, \ldots, n.$$

Now there is a natural bijection between colourful sets $\{t_1, \ldots, t_n\}$ of $T_1, \ldots, T_n$ and k-colourings $S_1, \ldots, S_k$ of S: given a colourful set $\{t_1, \ldots, t_n\}$, let

$$S_j = \{v_i : t_i = \bar{v}_i f_j^T\}, \qquad j = 1, \ldots, k.$$

We now show that a colourful set is positively dependent if and only if the corresponding colouring is Tverberg. Assume first that $\{t_1, \ldots, t_n\}$ has a nontrivial positive dependency $\sum_{i=1}^{n} \mu_i t_i = 0$. We have

$$0 = \sum_{j=1}^{k} \sum_{v_i \in S_j} \mu_i \bar{v}_i f_j^T = \sum_{j=1}^{k-1} \left(\sum_{v_i \in S_j} \mu_i \bar{v}_i - \sum_{v_l \in S_k} \mu_l \bar{v}_l \right) e_j^T. \tag{3}$$

Therefore

$$\sum_{v_i \in S_j} \mu_i \bar{v}_i = \sum_{v_l \in S_k} \mu_l \bar{v}_l \qquad j = 1, \ldots, k-1, \tag{4}$$

and in particular, by considering the $(d+1)$th coordinate of the $\bar{v}_i$, we have

$$M := \sum_{v_l \in S_k} \mu_l = \sum_{v_i \in S_1} \mu_i = \ldots = \sum_{v_i \in S_{k-1}} \mu_i, \qquad M = \frac{1}{k} \sum_{i=1}^{n} \mu_i > 0.$$

We conclude that $S_1, \ldots, S_k$ is Tverberg, since

$$\frac{1}{M} \sum_{v_l \in S_k} \mu_l v_l \in \bigcap_{j=1}^{k} \text{conv}(S_j).$$

Conversely, if $S_1, \ldots, S_k$ is Tverberg with

$$\sum_{v_i \in S_1} \mu_i v_i = \ldots = \sum_{v_i \in S_k} \mu_i v_i, \qquad \mu_i \geq 0, \qquad \sum_{v_i \in S_1} \mu_i = \ldots = \sum_{v_i \in S_k} \mu_i = 1,$$

then (4) hence (3) hold, so $\{t_1, \ldots, t_n\}$ is positively dependent. Thus, the Decision, Search, and Counting variants of the Tverberg Colouring Problem for S reduce to the corresponding variants of the Colourful Linear Programming Problem for $T_1, \ldots, T_n$.

In particular, since the T_i lie in $(k-1)(d+1)$-space and each is positively dependent by construction, the Colourful Carathéodory Theorem 1 implies that if $n > (k-1)(d+1)$ then $T_1, \ldots, T_n$ have a positively dependent colourful set, hence S has a Tverberg k-colouring, which proves Tverberg's Theorem 11. $\square$

7 Another Application Briefly

Tverberg's Theorem 11 and the related algorithmic problems reduce to the Colourful Theorem 1 and to Colourful Linear Programming. Similar reductions are possible for other Theorems and related algorithmic problems. One example is that of finding a *weak ϵ-Net*, a notion which has various applications in computational geometry: a set $T \subset \mathbb{R}^d$ is called a weak ϵ-net for a given set $S \subset \mathbb{R}^d$ if T intersects the convex hull of every ϵ-fraction of points of S. In [1] it is shown that for every finite set $S \subset \mathbb{R}^d$ and $\epsilon > 0$, there exists a weak ϵ-net T of size at most $(d+1)^{-(d+1)}\epsilon^{-(d+1)}$. This theorem was crucial in the solution of the Hadwiger–Debrunner (p, q)-problem by Alon and Kleitman [2]. It is possible to reduce this theorem and the algorithmic problem of finding a small ϵ-net to the Colourful Carathèodory Theorem and to Colourful Linear Programming.

Acknowledgements
The second author is grateful to A. Nemirovski for very helpful discussions and suggestions.

References

1. N. Alon, I. Bárány, Z. Füredi, and D. Kleitman, Point selections and weak ϵ-nets for convex hulls, Combinatorics, Prob., Comp. 1:189–200, 1992.
2. N. Alon, D. Kleitman, Piercing convex sets and the Hadwiger–Debrunner (p, q)-problem, Adv. Math. 96:103–112, 1992.
3. I. Bárány, A generalisation of Carathéodory's theorem, Discr. Math. 40:141–152, 1982.
4. I. Bárány and S. Onn, Colorful Linear Programming, full version preprint, 1996.
5. L. Blum, M. Shub, and S. Smale, On a theory of computation and complexity over the real numbers, Bull. AMS. 21:1–46, 1989.
6. G.B. Dantsig, An ϵ-precise feasible solution to a linear program in ϵ^{-2} iterations, Tech. Rep. SOL 92-5, Stanford University, October 1992.
7. R. Freund, personal communication.
8. M.R. Garey and D.S. Johnson, Computers and Intractability, W.H. Freeman, San Francisco, Cal., 1979.
9. P. Kleinschmidt and S. Onn, Oriented matroid polytopes and polyhedral cone fans are signable, in: 4th IPCO Proceedings, Lecture Notes in Computer Science, 920:198-211, 1995, Springer-Verlag.
10. P. Kleinschmidt and S. Onn, Signable posets and partitionable simplicial complexes, Discr. Comp. Geom., to appear.
11. S. Onn, On the geometry and computational complexity of Radon partitions in the integer lattice, SIAM J. Discr. Math. 4:436–447, 1991.
12. L. Lovász, personal communication.
13. K.S. Sarkaria, Tverberg's theorem via number fields, Israel J. Math. 79:317–320, 1992.
14. H. Tverberg, A generalisation of Radon's theorem, J. London Math. Society 41:123–128, 1966.

Test Sets and Inequalities for Integer Programs

Rekha R. Thomas[1] and Robert Weismantel[2]

[1] Texas A & M University, College Station, TX 77843-3368, rekha@math.tamu.edu
[2] Konrad–Zuse–Zentrum für Informationstechnik Berlin, Heilbronner Straße 10,
D–10711 Berlin, Germany, weismantel@zib–berlin.de

Abstract. This paper studies the passage from a linear to an integer program using tools provided by test sets and cutting planes. The first half of the paper examines the process by which the secondary polytope $\Sigma(A)$ associated with a matrix A refines to the state polytope $St(A)$. In the second half of the paper, we show how certain elements in a test set can be used to provide inequalities that are valid for the optimal solutions of a 0/1 integer program. We close with complexity results for certain integer programs, apparent from their test sets.

1 Introduction

In this paper we investigate connections between integer programs associated with a fixed matrix A and their linear relaxations. A set of integral vectors T_c is called a *test set* for the family of integer programs with coefficient matrix A for which a cost vector c is to be minimized if, for each non-optimal solution x to a program in the family, there exists an element $t \in T_c$ such that $ct > 0$ and $x - t$ is feasible for the same program. A *universal test set* for A contains a test set for all programs associated with A with respect to all cost vectors.

The cutting plane approach to an integer program studies valid inequalities for the polyhedron $P_b^I =$ convex hull $\{x \in \mathbf{N}^n : Ax = b\}$ starting with the LP feasible region $P_b = \{x \in R_+^n : Ax = b\}$. In section 3 we show that certain elements in a test set for 0/1 integer programs can be used to derive inequalities satisfied by optimal solutions to the program. Section 3 also presents complexity results for certain integer programs obtained by examining their test sets.

The *secondary polytope* of A [GKZ94], [BiS92], [BFS90] is defined to be the infinite *Minkowski integral* $\int_b P_b db$ and is denoted $\Sigma(A)$. Similarly, the *state polytope* $St(A)$ [BM88], [MR88], [St91], [ST94] is defined to be $\int_b P_b^I db$. We assume that A is such that P_b and P_b^I are polytopes for all b. The polytope $\Sigma(A)$ reflects properties of all polytopes P_b as b varies, and similarly $St(A)$ reflects properties of all polytopes P_b^I as b varies. The secondary and state polytope are closely related and in particular, $\Sigma(A)$ is a *Minkowski summand* of $St(A)$. In section 2 we examine the process by which $\Sigma(A)$ refines to $St(A)$. This sheds light on the passage from a linear program to an integer program in the generalized sense specified by the definitions of $\Sigma(A)$ and $St(A)$.

2 On secondary and state polytopes

For a fixed matrix $A \in \mathbb{Z}^{d \times n}$ of rank d and $b \in pos_{\mathbb{Z}}(A) = \{Ap : p \in \mathbb{N}^n\}$, let $P_b = \{x \in \mathbb{R}^n_+ : Ax = b\}$ and $P_b^I = $ convex hull $\{x \in \mathbb{N}^n : Ax = b\}$. Throughout this paper $\mathbb{N}$ stands for the set of natural numbers including zero. We will assume that $\{x \in \mathbb{R}^n_+ : Ax = 0\} = \{0\}$ which guarantees that both P_b and P_b^I are polytopes for all $b \in pos_{\mathbb{Z}}(A)$. For a fixed cost vector c (refined to be a linear order on $\mathbb{N}^n$ if necessary) and right hand side vector $b \in pos_{\mathbb{Z}}(A)$, let $IP_{A,c}(b)$ denote the integer program : $min \ cx : x \in P_b^I \cap \mathbb{N}^n$ and $LP_{A,c}(b)$ denote its linear relaxation : $min \ cx : x \in P_b$. A number of recent papers ([CT91], [Th94], [ST94]) have established the existence and construction of the *reduced Gröbner basis* $\mathcal{G}_c \subset \{x \in \mathbb{Z}^n : Ax = 0\} = ker_{\mathbb{Z}}(A)$ which is a uniquely defined minimal test set for the family of integer programs $IP_{A,c}(\cdot)$. The union of all reduced Gröbner bases obtained by varying the cost function c while keeping A fixed is a finite set UGB_A, called the *universal Gröbner basis* of A. The set UGB_A is a minimal *universal test set* for the family $IP_A = \{IP_{A,c}(b) : b \in pos_{\mathbb{Z}}(A), c \in \mathbb{R}^n\}$.

It can be established from the mechanics of the simplex method that the *circuits* of A which are the primitive vectors of minimal support in $ker_{\mathbb{Z}}(A)$, constitute a minimal universal test set for LP_A. Further, the circuits of A are contained in UGB_A. Associated with A, one can construct two $(n - d)$ dimensional polytopes: the *secondary polytope* $\Sigma(A)$ which is the *Minkowski integral* $\int_b P_b \, db$ and the *state polytope* $St(A) = \int_b P_b^I \, db$ where db is a suitable probability measure in both cases. In fact, any polytope *normally equivalent* to $\Sigma(A)$ (respectively, $St(A)$) is called a secondary polytope of A (respectively, state polytope of A). Two polytopes are normally equivalent if they have the same normal fan. The inner normal fan of $\Sigma(A)$ is an n-dimensional complete fan called the *secondary fan* of A, denoted $\mathcal{N}(\Sigma(A))$ and similarly, the inner normal fan of $St(A)$ is a complete polyhedral fan in $\mathbb{R}^n$, called the *Gröbner fan* of A, denoted as $\mathcal{N}(St(A))$. Construction methods for these fans are given in [BFS90] and [ST94]. The following relationships between $\Sigma(A)$ and $St(A)$ are collected from [St91] and [ST94].

Proposition 1. (i) The state polytope $St(A)$ is a *Minkowski summand* of the secondary polytope $\Sigma(A)$ and hence the Gröbner fan $\mathcal{N}(St(A))$ is a *refinement* of the secondary fan $\mathcal{N}(\Sigma(A))$. In particular, every full dimensional cone in $\mathcal{N}(\Sigma(A))$ either remains the same or partitions (refines) into subcones while passing to the Gröbner fan.
(ii) The edge directions of $\Sigma(A)$ and hence of all polytopes $P_b, b \in pos(A) = \{Ax : x \in \mathbb{R}^n_+\}$ are the circuits of A while the edge directions of $St(A)$ and hence of all polytopes $P_b^I, b \in pos_{\mathbb{Z}}(A)$ are the elements in UGB_A.
(iii) Let $c \in \mathbb{R}^n$ be a *generic* cost function for IP_A, namely, c is optimized at a unique vertex in each polytope P_b^I, $b \in pos_{\mathbb{Z}}(A)$. Then c lies in the interior of an n-dimensional cell K_c in $\mathcal{N}(St(A))$ and hence by (i), in the interior of an n-dimensional cell S_c in $\mathcal{N}(\Sigma(A))$. We say that two cost functions c_1 and c_2 are *equivalent* with respect to IP_A (respectively LP_A) if $IP_{A,c_1}(b)$ and $IP_{A,c_2}(b)$ (respectively $LP_{A,c_1}(b)$ and $LP_{A,c_2}(b)$) have the same set of optimal solutions

for all $b \in pos_{\mathbb{Z}}(A)$ (respectively, for all $b \in pos(A)$). Then the interior of K_c is precisely the equivalence class of c with respect to IP_A and the interior of S_c is the equivalence class of c with respect to LP_A.

(iv) For an element $v := v^+ - v^- \in ker_{\mathbb{Z}}(A)$, the line segment $[v^+, v^-]$ lies in $P^I_{Av^+}$ which is called the IP-fiber of v. If $v \in UGB_A$, $P^I_{Av^+}$ is called an IP-Gröbner fiber of A and $[v^+, v^-]$ is a primitive edge in $P^I_{Av^+}$. In this case, the linear relaxation P_{Av^+} is called a LP-Gröbner fiber of A. It was shown in [ST94] that the (finite) Minkowski sum of all IP-Gröbner fibers of A is a state polytope of A. If $v \in UGB_A$ is also a circuit of A then $[v^+, v^-]$ is a primitive edge of both P_{Av^+} (an LP-circuit fiber of A) and $P^I_{Av^+}$ (an IP-circuit fiber of A). The Minkowski sum of all LP-circuit fibers of A is a secondary polytope of A. The state (secondary) polytope constructed via this explicit finite Minkowski sum recipe will be referred to as **the** state (secondary) polytope of A.

2.1 Refining the secondary fan to the Gröbner fan

Let $B \in \mathbb{Z}^{(n-d) \times n}$ be a matrix whose rows form a basis (over $\mathbb{Z}$) of the lattice $ker_{\mathbb{Z}}(A)$. The vector configuration $\mathcal{B} \in \mathbb{Z}^{n-d}$ consisting of the columns of B is called a *Gale transform* of A. For the rest of this section, we fix the Gale transform $\mathcal{B}$. Each secondary cell S_c can be written as $S_c = S'_c \oplus ker(B)$ where S'_c is a pointed $n-d$ dimensional cone and $ker(B)$ is the d-dimensional nullspace of B. The collection of all pointed secondary cones is a complete polyhedral fan in $\mathbb{R}^{n-d}$ called the *pointed secondary fan* of A, denoted as $\mathcal{N}(\Sigma(A))'$. Similarly, each full dimensional Gröbner cone $K_c = K'_c \oplus ker(B)$ and the collection of all pointed Gröbner cones is the *pointed Gröbner fan* in $\mathbb{R}^{n-d}$ denoted as $\mathcal{N}(St(A))'$. The fan $\mathcal{N}(St(A))'$ is a refinement of $\mathcal{N}(\Sigma(A))'$.

For each basis μ of $\mathcal{B}$ define the cone $C_\mu = pos\{\mu_1, \mu_2, \ldots, \mu_{n-d}\}$. Then the following proposition gives the geometric construction of $\mathcal{N}(\Sigma(A))'$ in [BFS90].

Proposition 2. *The pointed secondary fan $\mathcal{N}(\Sigma(A))'$ is the multi-intersection in $\mathbb{R}^{n-d}$ of all cones C_μ, as μ ranges over all bases of $\mathcal{B}$.*

Corollary 3. *If A is of corank two, then the pointed secondary fan of A is the fan generated by the Gale transform of A.*

We first outline a geometric rule for refining the pointed secondary fan of a matrix of corank two, to its pointed Gröbner fan. We may assume that the vectors in $\mathcal{B}$ are sorted in a cyclic manner, possibly after relabeling. Therefore, if A is of corank two, then the full dimensional cells in $\mathcal{N}(\Sigma(A))'$ are the cones $S'_i = pos(b_i, b_{i+1})$ for $i = 1, \ldots, n$ where $B = [b_1, \ldots, b_n] \subset \mathbb{Z}^{2 \times n}$ and $b_{n+1} = b_1$.

Theorem 4. *If A is of corank two, then a vector u in a two dimensional pointed secondary cone S'_i generates the facet of a pointed Gröbner cone if and only if u is in the minimal Hilbert basis of S'_i.*

A full dimensional simplicial cone in $\mathbf{R}^n$ generated by the primitive vectors $v_1, \ldots, v_n$ is said to be *unimodular* if the determinant of the $n \times n$ matrix $[v_1 \ldots v_n]$ has absolute value one.

Corollary 5. *If A is of corank two, then*
(i) Each pointed Gröbner cone is unimodular.
(ii) A pointed secondary cone subdivides while passing to the Gröbner fan if and only if it is not unimodular.
(iii) Let $\Delta = max_{i,j}\{|det[b_i, b_j]|\}$. Then each secondary cell can split into at most Δ Gröbner cells.
(iv) There exist at most $n\Delta$ distinct facet normals among the polytopes P_b^I as b varies over $pos_{\mathbb{Z}}(A)$. More precisely, $u \in \mathbf{R}^n$ is a facet (inner) normal in some P_b^I if and only if Bu is in the Hilbert basis of some pointed secondary cone S_i'.

Property (i) in Corollary 5 fails for matrices of higher corank since typically, the pointed Gröbner fan in higher dimensions is not simplicial. To see that (ii) fails as well when the corank is increased, consider $A = \begin{bmatrix} 2 & 1 & 1 & 0 & 0 & 0 \\ 0 & 1 & 0 & 2 & 1 & 0 \\ 0 & 0 & 1 & 0 & 1 & 2 \end{bmatrix}$ of corank three, with $B = \begin{bmatrix} 1 & 0 & -2 & 0 & 0 & 1 \\ 1 & -2 & 0 & 1 & 0 & 0 \\ 1 & -1 & -1 & 0 & 1 & 0 \end{bmatrix}$. In this case, the unimodular pointed secondary cone $pos((0,1,1),(1,1,1),(0,0,1))$ splits into the pointed Gröbner cones $pos((0,1,1),(1,1,1),(1,1,2))$ and $pos((0,1,1),(0,0,1),(1,1,2))$.

By Proposition 1 (ii), the edge directions of $St(A)$ and hence the facet normals of full dimensional cells in $\mathcal{N}(St(A))$ are the elements in UGB_A. We represent an element $v \in UGB_A \subset ker_{\mathbb{Z}}(A)$, by the unique vector $\gamma_v \in \mathbb{Z}^{n-d}$ such that $\gamma_v B = v$. Then the elements in the set $\{\gamma_v : v \in UGB_A\}$ are precisely the facet normals in the pointed Gröbner fan $\mathcal{N}(St(A))'$. As seen earlier, a vector $a \in \mathbf{N}^n$ lies in the unique IP-fiber $P_{Aa}^I = $ convex hull $\{(a - ker_{\mathbb{Z}}(A)) \cap \mathbf{N}^n\} = $ convex hull $\{a - uB : a - uB \geq 0, u \in \mathbb{Z}^{n-d}\}$. Therefore, there exists a bijection between the lattice points in P_{Aa}^I and those in the polytope $F_{Ap}^I \subseteq \mathbf{R}^{n-d}$ defined as $F_{Aa}^I = $ convex hull $\{u \in \mathbb{Z}^{n-d} : a - uB \geq 0\}$. Further, minimizing over P_{Aa}^I is equivalent to maximizing over F_{Aa}^I. Consider an element $p - q \in UGB_A$. Then $p, q \in \mathbf{N}^n$ and $support(p) \cap support(q) = \emptyset$ and let $\gamma B = p - q$. The edge $[p, q]$ in the IP-Gröbner fiber P_{Ap}^I corresponds to the edge $[0, \gamma]$ in F_{Ap}^I. Then we have the following relationship bewteen the union of all facets in $\mathcal{N}(St(A))'$ with facet normal $\pm\gamma$ and the polytope F_{Ap}^I.

Theorem 6. *The union of all $(n-d-1)$-dimensional cells (facets) in $\mathcal{N}(St(A))'$ with normal vector $\pm\gamma$ is the outer normal cone of F_{Ap}^I at the edge $[0, \gamma]$.*

The construction of $\mathcal{N}(\Sigma(A))'$ given in [BFS90] can be augmented by the above theorem to provide a geometric rule for constructing $\mathcal{N}(St(A))'$ by first constructing $\mathcal{N}(\Sigma(A))'$ and then refining each pointed secondary cell to its associated Gröbner cells. We leave all proofs and details of results to the final version of this paper.

2.2 When do secondary and state polytopes coincide ?

Definition 7. A matrix $A \in \mathbb{Z}^{d \times n}$ of rank d is called c-unimodular if each of its maximal minors is one of $-c$, 0 or c, where c is a positive integral constant.

Theorem 8. *A matrix A is c-unimodular if and only if $P_b = P_b^I$ for all $b \in pos_{\mathbb{Z}}(A)$.*

The above theorem follows directly from Definition 7 and is well known in the case $c = 1$. When A is 1-unimodular, $pos_{\mathbb{Z}}(A) = pos(A) \cap \mathbb{Z}^d$ and hence $P_b = P_b^I$ for all $b \in pos(A) \cap \mathbb{Z}^d$. This result is precisely Theorem 19.2 in [Schr86].

Remark. If A is c-unimodular, $c > 1$, then $pos(A) \cap c \cdot \mathbb{Z}^d \subseteq pos_{\mathbb{Z}}(A)$ where the containment is often strict. For example, the incidence matrix A_4 of the complete graph K_4 is 2-unimodular. However, $A_4 \cdot (1,1,1,1,0,0)^t = (3,2,2,1) \notin 2 \cdot \mathbb{Z}^4$.

Theorem 8 along with Proposition 1 (ii), (iv) imply the following corollary.

Corollary 9. *If A is c-unimodular, then the circuits of A constitute UGB_A and $St(A) = \Sigma(A)$.*

However, there are matrices that are not c-unimodular for which the state and secondary polytopes coincide. For such matrices, it is no longer true that $P_b = P_b^I$ for all $b \in pos_{\mathbb{Z}}(A)$ and yet $\int_b P_b db = \int_b P_b^I db$. We establish a necessary and sufficient condition on A for the state and secondary polytopes to coincide and exhibit a family of matrices (possibly non-unimodular) with this property.

Theorem 10. *The state and secondary polytopes of a matrix A coincide if and only if (i) the circuits of A constitute UGB_A and (ii) each LP-circuit fiber of A is integral.*

Proof. The *if* direction of the proof follows easily from Proposition 1. We outline the *only if* direction of the proof. Clearly, if $St(A) = \Sigma(A)$, they have the same set of edge directions and hence the circuits of A constitute UGB_A and all Gröbner fibers are circuit fibers. Suppose the LP-circuit fiber P_{Ap} of the circuit $p - q$ is not integral. Since the polytopes P_{Ap} and P_{Ap}^I share the edge $[p, q]$, they cannot be normally equivalent. This implies that there is some edge $[e, f]$, in P_{Ap}^I possibly not primitive such that the normal cone of P_{Ap}^I at this edge is not equal to the normal cone of the primitive edge $[(e-f)^+, (e-f)^-]$ in its fiber. The proof of the theorem now follows from Theorem 6 and the construction of the polytopes $\Sigma(A)$ and $St(A)$ in Proposition 1 (iv). $\qquad\Box$

Clearly, c-unimodular matrices satisfy the conditions in Theorem 10 and hence have $\Sigma(A) = St(A)$. We now exhibit other, possibly non-unimodular, matrices that satisfy conditions (i) and (ii) in Theorem 10. The *Lawrence lifting* of a matrix $A \in \mathbb{Z}^{d \times n}$, is defined to be the enlarged matrix $\Lambda(A) = \begin{pmatrix} A & 0 \\ 1 & 1 \end{pmatrix}$ where 0 is a $d \times n$ matrix of zeroes and 1 denotes an identity matrix of size n. Lawrence matrices have many special properties and can be used to compute UGB_A. It was shown in [ST94] that the IP-Gröbner fibers of $\Lambda(A)$ are one dimensional. The following proposition establishes the LP analogue.

Proposition 11. *Every LP-circuit fiber of $\Lambda(A)$ is one dimensional and integral.*

Proof: Note that A and $\Lambda(A)$ have isomorphic kernels: $ker_{\mathbb{Z}}(\Lambda(A)) = \{(u, -u) : u \in ker_{\mathbb{Z}}(A)\}$. In particular, a vector u is a circuit of A if and only if the vector $(u, -u)$ is a circuit of $\Lambda(A)$. Proposition 11 follows from the following two observations:
(a) Each point x in the LP-circuit fiber of $(u, -u)$, is in the affine span of $[u^+, u^-]$ which implies that the fiber is one dimensional.
(b) If $(u, -u)$ is a circuit of $\Lambda(A)$, then the line segment $[u^+, u^-]$ is an edge in the LP-circuit fiber of $(u, -u)$. $\qquad\Box$

Using Theorem 10 and Proposition 11 we obtain the following.

Proposition 12. *If the circuits of $\Lambda(A)$ constitute $UGB_{\Lambda(A)}$ then the state and secondary polytopes of $\Lambda(A)$ coincide.*

The above result can also be inferred from [ST94] using the notion of *circuit and Gröbner arrangements.*

Example 1. Consider $A = \begin{bmatrix} 1 & 1 & 1 & 1 & 1 \\ 0 & 1 & 2 & 1 & 0 \\ 0 & 0 & 1 & 2 & 1 \end{bmatrix}$ an integral matrix of corank two. The maximal minors of A have absolute values one, two and three. The vectors $(0, 1, -1, 1, -1), (2, -1, 0, 1, -2), (2, 0, -1, 2, -3), (2, -2, 1, 0, -1)$ and $(2, -3, 2, -1, 0)$ are the circuits of A. The lifted circuits $(u, -u) \in \mathbb{Z}^{10}$ of $\Lambda(A)$ form $UGB_{\Lambda(A)}$ and hence $\Sigma(\Lambda(A)) = St(\Lambda(A))$. In this example, the circuits of A also form UGB_A but $\Sigma(A) \neq St(A)$. Since A is of corank two, the pointed secondary fan is given by the Gale transform $\mathcal{B} = \{(-2, -2), (3, 2), (-2, -1), (1, 0), (0, 1)\}$ of A. The pointed secondary cells $pos((1, 0), (3, 2)), pos((3, 2), (0, 1))$ and $pos((0, 1), (-2, -1))$ are not unimodular and hence split into two Gröbner cells each. Also, the LP-circuit fibers of $(0, 1, -1, 1, -1), (2, -1, 0, 1, -2)$ and $(2, -2, 1, 0, -1)$ are not integral. This example illustrates the necessity of both conditions in Theorem 10.

2.3 Total dual integrality and the Gröbner fan

We saw that for certain matrices A, $St(A) = \Sigma(A)$ which is equivalent to saying that no cell in the secondary fan partitions into subcones while passing to the Gröbner fan. For an arbitrary matrix A, typically, some cells in the secondary fan partition into subcones while others do not. We examine below properties of those secondary cells that do not subdivide. This is related to the classical concept of TDI-ness.

Definition 13. [Schr86] A rational system $yA \leq c$, $A \in \mathbb{Z}^{d \times n}$ is TDI if the minimum in the LP duality equation $max\{yb : yA \leq c\} = min\{cx : Ax = b, x \geq 0\}$ has an integral optimal solution for every $b \in \mathbf{Z}^d$ for which the minimum is finite.

In the terminology used so far, $yA \leq c$ is TDI if $LP_{A,c}(b)$ has an integral optimal solution for all $b \in pos_\mathbf{Z}(A)$. We now use a few facts from the algebraic version of the study of integer programs using Gröbner bases. The reader is referred to [St96] for the algebraic background. The monomial ideal generated by the *leading terms* of the elements in the reduced Gröbner basis $\mathcal{G}_c$ is called the initial ideal of I_A with respect to c and is denoted $init_c(I_A)$. Here I_A is the *toric ideal of* A, a central entity in the algebraic version of this theory. Also, a generic cost vector c induces a *regular triangulation* Δ_c, of the point configuration $\mathcal{A}$ formed by the columns of A. See [GKZ94], [BFS90] for details on regular triangulations. Regular triangulations of $\mathcal{A}$ and initial ideals of I_A are related as follows [St91]: the *radical* of $init_c(I_A)$ is the *Stanley-Reisner ideal* of Δ_c. Further, the regular triangulations of $\mathcal{A}$ are in bijection with the vertices of $\Sigma(A)$ while the initial ideals of I_A are in bijection with the vertices of $St(A)$. Recall that the secondary cell of a generic cost vector c is denoted S_c and the Gröbner cell is denoted K_c. If c is generic, then c lies in the interior of K_c and hence of S_c. In summary, the subcones into which a fixed secondary cell S_c subdivides while passing to the Gröbner fan, are in bijection with those distinct initial ideals of I_A whose radicals coincide with the Stanley-Reisner ideal of Δ_c. A regular triangulation Δ_c is said to be *unimodular* if all its maximal simplices have unit normalized volume. The following algebraic result can be inferred from Theorem 5.3 in [KSZ92].

Proposition 14. *The monomial ideal* $init_c(I_A)$ *is square-free if and only if the regular triangulation* Δ_c *is unimodular.*

Using Proposition 14 we can prove the following theorem. Again we omit proofs of theorems in this section since they often rely on commutative algebra.

Theorem 15. *For $A \in \mathbb{Z}^{d \times n}$ and a generic cost vector $c \in \mathbf{R}^n$, the following are equivalent:*
(i) The system $yA \leq c$ is TDI.
(ii) The optimal solution of $LP_{A,c}(b)$ is integral for all $b \in pos_\mathbf{Z}(A)$.
(iii) The initial ideal $init_c(I_A)$ is square free.

Theorem 15 gives an algorithmic criterion for $yA \le c$ to be TDI. The reduced Gröbner basis of I_A and hence $init_c(I_A)$ can be computed using a computer algebra package like MACAULAY [Mac] or MAPLE. We close this section with a theorem that relates TDI-ness of $yA \le c$ to properties of the cells S_c and K_c.

Theorem 16. *Consider the following properties of $A \in \mathbf{Z}^{d \times n}$ and a generic cost vector $c \in \mathbf{R}^n$.*
(i) The secondary cell S_c coincides with the Gröbner cell K_c.
(ii) The system $yA \le c$ is TDI.
(iii) The pointed secondary cell S_c is unimodular. Then
a) (ii) $\Rightarrow$ (i) but (i) $\not\Rightarrow$ (ii),
b) (i) $\not\Rightarrow$ (iii) and (iii) $\not\Rightarrow$ (i),
c) (ii) $\not\Rightarrow$ (iii) and (iii) $\not\Rightarrow$ (ii).

Therefore the TDI-ness of $yA \le c$ is a sufficient condition for the secondary cell S_c to not subdivide while passing to the Gröbner fan although the converse is false. Geometrically, the interior of S_c is the set of all cost functions that select the same LP optimum as c in the polytopes P_b for all $b \in pos_(A)$ while the interior of K_c is the set of all cost vectors that select the same optimal vertex as c in the polytopes P_b^I for all $b \in pos_Z(A)$. However since (i) $\not\Rightarrow$ (ii), it follows that all cost vectors in $S_c = K_c$ can pick the same LP optimum for all $b \in pos(A)$ and the same IP optimum for all $b \in pos_Z(A)$, although for a given b, the LP and IP optima may be different - the LP optimum being fractional. This further supports the fact that unimodularity of A is not a necessary condition for $St(A)$ and $\Sigma(A)$ to coincide.

3 Augmentation vectors

In this section we consider integer programs of the form

$$(IP) \qquad \max cx : Ax \le b, \; x \in \mathbf{N}^n$$

where $A = [a_1, \ldots, a_n] \in \mathbf{Z}^{m \times n}, b \in \mathbf{Z}^m$ and $c \in \mathbf{R}^n$.

In polyhedral combinatorics one often tries to find classes of inequalities that are valid for the integer polyhedron. Once such a class is known, one derives a separation algorithm that, for a given fractional point, finds an inequality that is violated by the fractional point, or proves that no inequality in this class is violated by the fractional point. This dual approach is justified, as separation and optimization are equivalent in terms of computational complexity [GLS88].

The primal counterpart of these questions is to ask for a set of vectors and an algorithm that either finds a vector (augmentation vector) in this set such that a current feasible point can be improved, or asserts that there is no vector in this set that yields an improvement of the current feasible solution. These questions are reasonable to ask (at least for 0/1 integer programs), since in this case, the augmentation problem, the *irreducible* augmentation problem and the optimization problem are strongly polynomial time equivalent (see [SWZ95]).

Below, we show how certain elements in a test set can be used to derive inequalities that are satisfied by all optimal solutions to 0/1 programs of the form (IP). We call such inequalities, *objective based cutting planes*. The second subsection demonstrates that, for every pair of columns a^i, a^j of a matrix $A \in \mathbf{N}^{m \times n}$ and for every feasible point x of (IP), one can decide efficiently whether x can be improved by a vector of the form $ta^i - t'a^j$ with $t, t' \in \mathbf{N}$.

The following definitions are needed in this section.

Definition 17. A vector $t \in \mathbf{Z}^N$, $ct > 0$ is *reducible* by $v \in \mathbf{Z}^N$, $cv > 0$ if $v^+ \leq t^+$, $v^- \leq t^-$ and $(Av)^+ \leq (At)^+$. If no such v exists, then t is *irreducible*.

Note that, if $t \in \mathbf{Z}^N$ is reducible by $v \in \mathbf{Z}^N$, then, whenever $x + t$ is feasible for (IP), $x + v$ is also feasible for (IP).

Definition 18. A subset T of a set $G \subseteq \mathbf{Z}^n$ is a *generating set* for G if every element in G can be written as a non-negative integer combination of elements in T. A generating set is *minimal* if it is minimal with respect to inclusion.

3.1 Minimum cover vectors

We now consider 0/1 programs of the form (IP) and introduce a subclass of irreducible elements called *minimum cover vectors*, found in every test set of (IP). With each such element we can associate an inequality that is satisfied by all optimal solutions of the given instance. If $cv > cw$, we write $v \succ_c w$.

Definition 19. For disjoint subsets $K, K' \subseteq N$ such that $\sum_{j \in K} a^j \leq \sum_{j \in K'} a^j$ and $\sum_{j \in K \cup \{i\}} a^j \not\leq \sum_{j \in K'} a^j$ for all $i \notin K$, the vector $\sum_{j \in K} e_j - \sum_{j \in K'} e_j$ is called a minimum cover vector if $\sum_{j \in K} e_j \succ_c \sum_{j \in K'} e_j$.

The name "minimum cover" reflects that K' is a minimal subset with respect to inclusion such that $\sum_{j \in K'} a^j$ covers $\sum_{j \in K} a^j$. From the definition it follows that if a minimum cover vector is reducible with respect to an objective function $c \in \mathbf{N}^n$, then the reduction vector is also a minimum cover vector. Hence, for every objective function $c \in \mathbf{N}^n$ there exists a subset of the set of all minimum cover vectors that must be contained in every test set of (IP). Minimum cover vectors give rise to inequalities that must be satisfied by every optimal solution of the given integer programming problem.

Lemma 20. *If $v = \sum_{j \in K} e_j - \sum_{j \in K'} e_j$ is a minimum cover vector, then every optimal solution x of (IP) satisfies the constraint*

$$\sum_{j \in K} x_j \geq \sum_{j \in K'} x_j - (|K'| - 1).$$

Proof. Clearly, x satisfies $\sum_{j \in K} x_j \geq 0$. Hence, if $\sum_{j \in K'} x_j \leq |K'| - 1$ then we are done. If $\sum_{j \in K'} x_j = |K'|$ then $\sum_{j \in K'} x_j - (|K'| - 1) = 1$ and the inequality states that $\sum_{j \in K} x_j \geq 1$. Since x has full support in K', if $\sum_{j \in K} x_j = 0$, x can be improved by $\sum_{j \in K} e_j - \sum_{j \in K'} e_j$. This contradicts that x is optimal. $\square$

Hence, minimum cover vectors can be used to derive inequalities satisfied by all optimal solutions of (IP). It can further be shown that the property of a minimum cover vector being reducible is reflected by the property that the associated inequality is dominated by an inequality that we derive from an irreducible vector. Suppose that the minimum cover vector $v := \sum_{j \in K} e_j - \sum_{j \in K'} e_j$ is reducible by $w := \sum_{j \in \overline{K}} e_j - \sum_{j \in \overline{K'}} e_j \succ_c 0$. Then w is a minimum cover vector and every feasible point x that satisfies the inequality associated with w will also satisfy the inequality associated with v for the following reason. Consider

$$\sum_{j \in \overline{K}} x_j \geq \sum_{j \in \overline{K'}} x_j - (|\overline{K'}| - 1)$$

the inequality associated with w. Adding to the left hand side of this inequality, the expression

$$\sum_{j \in K \setminus \overline{K}} x_j$$

maintains the validity of the inequality for all optimal solutions, because all variables are non-negative. Adding, in a second step, the expression

$$\sum_{j \in K' \setminus \overline{K'}} x_j - |K' \setminus \overline{K'}|$$

to the right hand side of the current inequality maintains the validity of the inequality for all optimal solutions, because all variables are bounded by 1. This new inequality coincides with the inequality that we associated with the minimum cover vector v. In this sense, reducibility of augmentation vectors and domination of inequalities are related, at least in this special case.

3.2 The augmentation problem for pairs of column vectors

In this subsection we first study the augmentation problem for a special knapsack problem $IP(\mu, \lambda, b, c)$ defined as

$$\max c_1 x_1 + c_2 x_2 : \mu x_1 + \lambda x_2 \leq b, \; x_1, x_2 \in \mathbb{N}$$

where $b \in \mathbf{N}$, $\mu, \lambda \in \mathbf{N}$, $gcd(\mu, \lambda) = 1$ and $c_1, c_2 \in \mathbf{N}$.

For a given integer $r \in \{0, 1, \ldots, \min\{\lambda, \mu\} - 1\}$, we define a set $G \subseteq \mathbf{N}^2$ and for every $v \in G$ a *remainder* $R(v)$ as follows:

$$G := \{(\lfloor \frac{y\lambda + r}{\mu} \rfloor, y) : y \in \mathbf{N}\} \text{ and for } v = (v_1, v_2) \in G, \ R(v) := v_2\lambda - v_1\mu.$$

The set G is related to a solution of the augmentation problem of $IP(\mu, \lambda, b, c)$ in the following way.

Let $z = (z_1, z_2) \in \mathbf{N}^2$ be a feasible point of $IP(\mu, \lambda, b, c)$ and define the number r as the rest capacity at z, i.e., $r := b - \mu z_1 - \lambda z_2$. If z is not optimal, then there exists an augmenting vector $v \in \mathbf{Z}^2$ such that $z + v$ is feasible. Let $v = (v_1, v_2)$. If one of v_1 or v_2 is zero, z can be improved by adding a unit vector to z. This case is easy to check. Otherwise we either have the possibility that $v_1 > 0$ and $v_2 < 0$ or $v_1 < 0$ and $v_2 > 0$. Without loss of generality, assume $v_1 > 0$ and $v_2 < 0$. Then $\lfloor \frac{(-v_2)\lambda + r}{\mu} \rfloor$ is the maximum amount of μ's that can be added to z after having taken out $-v_2$ units of weight λ. Therefore, $v_1 \leq \lfloor \frac{(-v_2)\lambda + r}{v\mu} \rfloor$ and further, z can be improved by an integral vector v with $v_1 > 0$ and $v_2 < 0$ if and only if there exists an integral vector $g \in G$ improving z.

Next we want to demonstrate that it is sufficient to inspect at most $log_2^2(\lambda)$ special elements in G in order to decide whether the feasible point z of $IP(\mu, \lambda, b, c)$ can be improved by an element in G. Note that $r := b - \mu z_1 - \lambda z_2$. In [W94] an algorithm was presented to compute a minimal generating set of G whose remainders are a unique sequence of decreasing numbers. We will now show (i) how to compute difference vectors of two consecutive elements in this generating set and (ii) that the number of different difference vectors is at most $log_2(\lambda)$. These difference vectors that we denote by $D[1], \ldots, D[\tau]$ and associated step lengths $S[1], \ldots, S[\tau]$ are computed via Algorithm 22. The generating set of G is defined in Definition 24. Lemmas 25 and 26 show that it is sufficient to test the vectors in Definition 24 in order to decide whether a feasible point of $IP(\mu, \lambda, b, c)$ can be improved by any vector in G. The step of testing the vectors of Definition 24 can be performed efficiently by resorting to the two arrays D and S that we compute with Algorithm 22. More precisely, we will show

Theorem 21. *After inspecting at most* $\sum_{i=0}^{\tau} log_2(S[i]) \leq log_2^2(\lambda)$ *elements in the set*

$$\{h[i, j] : i = 0, \ldots, \tau, \ j = 0, \ldots S[i]\}, \ \textit{see Definition 24,}$$

one can decide whether a feasible point of $IP(\mu, \lambda, b, c)$ *can be improved by an element in* G.

Algorithm 22 to compute arrays D and S:

Set $h := (\lfloor\frac{\lambda+r}{\mu}\rfloor, 1)$, $a = h$, $i = 0$, $S[0] = 0$.

If $R(h) \leq 0$, set $\tau := 0$ and exit.

WHILE $(R(h) > 0)$ perform Steps (0) - (5).

(0) Increment i.

(1) Compute a step length $\sigma := \max\{v \in \mathbf{N} : vR(h) + R(a) < \mu - r\}$.

(2) Set $d := \sigma h + a - (1, 0)$.

(3) Compute a step length $\epsilon := \max\{v \in \mathbf{N} : R(h) + vR(d) \geq -r\}$.

(4) Update $a := a + \sigma h$ and $h := h + \epsilon d$.

(5) Set $D[i] = d$ and $S[i] = \epsilon$.

Define $\tau := i$.

Set $S[\tau] := \max\{v \in \mathbf{N} : R(h) + vR(d) > 0\} + 1$.

An easy analysis of Algorithm 22 shows that each performance of the While-loop reduces the remainder of the vector h by a factor of at least $\frac{1}{2}$. Hence, the sizes of the two arrays D and S can be bounded by $\log_2(\lambda)$ and we obtain

Lemma 23. *$\tau \leq \log_2(\lambda)$.*

Definition 24. Set $h[0, 0] := (\lfloor\frac{\lambda+r}{\mu}\rfloor, 1)$. For i from 1 to τ and j from 0 to $S[i]$ define

$$h[i, j] := h[i - 1, S[i - 1]] + jD[i].$$

By inspecting this subset of vectors in G we can decide whether a feasible point of $IP(\mu, \lambda, b, c)$ can be improved by any vector in G. This is made precise in Lemmas 25 and 26.

Lemma 25. *The set $\bigcup_{i=0}^{\tau} \bigcup_{j=0}^{S[i]} \{h[i, j]\}$ is a minimal generating set for $\{g \in G : g_2 \leq h_2[\tau, S[\tau]]\}$.*

Proof. The proof follows from Algorithm 2.3, Lemma 2.4 and Theorem 2.5 in [W94] that show how to compute the unique minimal generating set for $\{(g_1, g_2) \in G : g_2 \leq h_2[\tau, S[\tau]]\}$ that consists of the sequence of vectors in $\{(g_1, g_2) \in G : g_2 \leq h_2[\tau, S[\tau]]\}$ with decreasing remainders. □

Lemma 26. *A feasible point z of $IP(\mu, \lambda, b, c)$ can be improved by a vector g such that $(g_1, -g_2) \in G$ if and only if there exists $i \in \{0, \ldots, \tau\}$ and $j \in \{0, \ldots S[i]\}$ such that $c_1 h_1[i, j] - c_2 h_2[i, j] > 0$ and $z + (h_1[i, j], -h_2[i, j])$ is feasible for $IP(\mu, \lambda, b, c)$.*

Proof. The first if-statement is obvious. We analyze the only if statement. Let $z = (z_1, z_2)$ be the feasible point of $IP(\mu, \lambda, b, c)$ and $r = b - \mu z_1 - \lambda z_2$. Suppose that z can be improved by some $g \in \mathbb{Z}^n$ with $(g_1, -g_2) = (\lfloor \frac{-g_2 \lambda + r}{\mu} \rfloor, -g_2) \in G$. We write $(g_1, -g_2)$ in the following form:

$$(g_1, -g_2) = th[\tau, S[\tau]] + g',$$

where $t := \max\{i \in \mathbb{N} : -g_2 \geq i h_2[\tau, S[\tau]]\}$. We obtain

$$R(g') = R((g_1, -g_2)) - tR(h[\tau, S[\tau]]) \geq R((g_1, -g_2)) > -r$$

because $(g_1, -g_2) \in G$. It follows now that $g'_1 \leq \lfloor \frac{g'_2 \lambda + r}{\mu} \rfloor$. Since $g'_2 \leq h_2[\tau, S[\tau]]$ we conclude that $(\lfloor \frac{g'_2 \lambda + r}{\mu} \rfloor, g'_2) \in \{v \in G : v_2 \leq h_2[\tau, S[\tau]]\}$. By Lemma 25 there is a representation

$$(\lfloor \frac{g'_2 \lambda + r}{\mu} \rfloor, g'_2) = \sum_{i=0}^{\tau} \sum_{j=0}^{S[i]} \theta_{i,j} h[i,j]$$

with $\theta_{i,j} \in \mathbb{N}$ for all $i \in \{0, \ldots, \tau\}$ and $j \in \{0, \ldots S[i]\}$. The lemma now follows from the subsequent inequalities that imply that one of the summands of the right hand side must be positive.

$$0 < c_1 g_1 + c_2 g_2 = tc_1 h_1[\tau, S[\tau]] - tc_2 h_2[\tau, S[\tau]] + c_1 g'_1 - c_2 g'_2$$

$$\leq tc(h_1[\tau, S[\tau]], -h_2[\tau, S[\tau]]) + \sum_{i=0}^{\tau} \sum_{j=0}^{S[i]} \theta_{i,j} c(h_1[i,j], -h_2[i,j]).$$

$\square$

The following algorithm proves Theorem 21.

We start inspecting whether $h := h[0,0]$ is an augmentation vector that is applicable at z. If the answer is no we proceed for $i = 1$ up to τ as follows:

(a) Check whether $c_1 D_1[i] - c_2 D_2[i] > 0$. If not, Goto Step (c).

(b) Use binary search to assert whether there exists an augmenting vector of the form $h + v(D_1[i], -D_2[i])$ with $v \in \{1, \ldots, S[i]\}$ that is applicable at z.

(c) Set $i := i + 1$ and $h := h + S[i]D[i]$. Goto Step (a).

The correctness of the algorithm follows easily from the following arguments: Whenever we perform Step (a) of the algorithm, $c_1 h_1 - c_2 h_2 \leq 0$. By Lemma 25 and Lemma 26 we only need to check the vectors $h[i,j]$ to decide whether or not z can be improved by an element in G. For a given index $i \in \{1, \ldots, \tau\}$ we can incorporate a binary search to check whether there exists a vector of the form $h[i,v]$ with $v \in \{1, \ldots, S[i]\}$ that improves z. For, if we know that for $v \in \{1, \ldots, S[i]\}$, $z + h + vD[i]$ is not feasible, then $z + h + wD[i]$ is not feasible for all $S[i] \geq w \geq v$. Similarly, if $c(h + v(D_1[i], -D_2[i])) \leq 0$ for some $v \in \{1, \ldots, S[i]\}$, then $c(h + w(D_1[i], -D_2[i])) \leq 0$ for all $1 \leq w \leq v$ as $c_1 h_1 - c_2 h_2 \leq 0$. This completes the proof. □

Taking Algorithm 22, Lemmas 23, 25, 26, Theorem 21 and our initial discussions on the relationship between the set G and the augmentation problem for $IP(\mu, \lambda, b, c)$ into account, we get the following theorem.

Theorem 27. *The augmentation problem for the knapsack problem $IP(\mu, \lambda, b, c)$ can be solved in time $O(log_2^2(\lambda))$.*

It is not too difficult to extend all the results in this section to the case when one considers several two-dimensional knapsack constraints simultaneously. Suppose we consider the program

$$(IP) \quad \max c_1 x_1 + c_2 x_2 : \mu_i x_1 + \lambda_i x_2 \leq b_i, \text{ for } i = 1, \ldots, m, \; x_1, x_2 \in \mathbf{N}$$

with $b \in \mathbf{N}^m$, $\mu_1, \lambda_1, \ldots \mu_m, \lambda_m \in \mathbf{N}$ and $c_1, c_2 \in \mathbf{N}$.

Let $z = (z_1, z_2) \in \mathbf{N}^2$ be a feasible point of IP and define the vector r as the vector of rest capacities at z, i.e., $r_i := b_i - \mu_i z_1 - \lambda_i z_2$. If z is not optimal, then there exists an augmenting vector $v = (v_1, v_2) \in \mathbf{Z}^2$ such that $z + v$ is feasible. If one of v_1 or v_2 is zero, z can be improved by adding a unit vector to z. Otherwise we assume without loss of generality, that $v_1 > 0$ and $v_2 < 0$. Then $\lfloor \frac{(-v_2)\lambda_i + r_i}{\mu_i} \rfloor$ is the maximum amount of μ_i's that can be added to z after having taken out v_2 units of weight λ_i. Therefore, $v_1 \leq \lfloor \frac{(-v_2)\lambda_i + r_i}{\mu_i} \rfloor$ for all $i = 1, \ldots, m$. Next we determine for each constraint $\mu_i x_1 \leq \lambda_i x_2 + r_i$ the range of x_2-values where this constraint defines the upper bound on x_1. More formally, we determine for every $i \in \{1, \ldots, m\}$ the interval $S_i \subseteq \mathbf{N}$ such that a vector $(v_1, v_2) \in \mathbf{N}^2$ with $v_2 \in S_i$ satisfies $v_1 \leq \lfloor \frac{(-v_2)\lambda_i + r_i}{\mu_j} \rfloor$ for all $j = 1, \ldots, m$ if and only if already $v_1 \leq \lfloor \frac{(-v_2)\lambda_i + r_i}{\mu_i} \rfloor$ holds. Such intervals can be computed by determining for every pair of constraints the point intersecting the two corresponding half spaces and ordering these intersection points with respect to increasing values of x_2.

For every interval S_i, $i = 1, \ldots, m$ we are left with one knapsack constraint $\mu_i x_1 \leq \lambda_i x_2 + r_i$ subject to $x_2 \in \mathbf{N} \cap S_i$ since all other constraints are not binding. Algorithm 22, Lemmas 23 ,25 and 26 can be adapted to take into account the lower and upper bound on x_2 given by the interval S_i. As a consequence, we obtain the following theorem.

Theorem 28. *Let (IP) $\max c^T x : Ax \leq b$, $x \in \mathbf{N}^n$ be an integer program with $A \in \mathbf{N}^{m \times n}$ of rank m and $b \in \mathbf{N}^m$, $c \in \mathbf{N}^n$. For every feasible point z and every pair of columns i, j one can decide in time $O(m^2 log_2(m) < A >^2)$ whether z can be improved by a vector $g \in \mathbf{Z}^n$ such that $g_u = 0$ for all $u \in \{1, \ldots, n\} \setminus \{i, j\}$.*

References

[BM88] D. Bayer and I. Morrison, Gröbner bases and geometric invariant theory I, *Journal of Symbolic Computation*, 6, 1988, 209–217.

[Mac] MACAULAY : a computer algebra system for algebraic geometry, available by anonymous ftp from *zariski.harvard.edu*.

[BiS92] L.J. Billera and B. Sturmfels, Fiber polytopes, *Annals of Mathematics*, 135, 1992, 527–549.

[BFS90] L.J. Billera, P. Filliman and B. Sturmfels, Constructions and complexity of secondary polytopes, *Advances in Mathematics*, 83, 1990, 155–179.

[CT91] P. Conti and C. Traverso, Buchberger algorithm and integer programming, *Proceedings AAECC-9 (New Orleans)*, Springer Verlag LNCS 539, 1991, 130–139.

[GKZ94] I.M. Gel'fand, M. Kapranov and A. Zelevinsky, *Multidimensional Determinants, Discriminants and Resultants*, Birkhäuser, Boston, 1994.

[GLS88] M. Grötschel, L. Lovász, and A. Schrijver: *Geometric Algorithms and Combinatorial Optimization*, Algorithms and Combinatorics 2, Springer, Berlin, 1988; Second edition 1993.

[KSZ92] M.M. Kapranov, B. Sturmfels and A.V.Zelevinsky, Chow polytopes and general resultants, *Duke Math. Journal*, Vol 67, 1992, 189–218.

[MR88] T. Mora and L. Robbiano, The Gröbner fan of an ideal, *Journal of Symbolic Computation*, 6, 1988, 183–208.

[Schr86] A. Schrijver, *Theory of Linear and Integer Programming*, Wiley-Interscience Series in Discrete Mathematics and Optimization, New York, 1986.

[St91] B. Sturmfels, Gröbner bases of toric varieties, *Tôhoku Math. Journal*, 43, 1991, 249–261.

[St96] B. Sturmfels, *Gröbner Bases and Convex Polytopes*, AMS University Lecture Series 8, Providence, RI, 1996.

[ST94] B.Sturmfels and R.Thomas, Variation of cost functions in integer programming, *Technical Report*, 1994, Cornell University.

[Th94] R.Thomas, A geometric Buchberger algorithm for integer programming, *Mathematics of Operations Research*, 20, 1995, 864–884.

[SWZ95] A. Schulz, R. Weismantel and G. Ziegler, 0/1 integer programming: optimization and augmentation are equivalent, to appear in *Proceedings of the European Symposium on Algorithms 95*, "Lecture Notes in Computer Science", Springer-Verlag, 1995.

[W94] R. Weismantel, "Hilbert bases and the facets of special knapsack polytopes", *Preprint* SC 94-19, Konrad-Zuse-Zentrum Berlin, 1994.

An Optimal, Stable Continued Fraction Algorithm for Arbitrary Dimension

Carsten Rössner and Claus P. Schnorr

Fachbereich Mathematik/Informatik
Universität Frankfurt
PSF 11 19 32
D–60054 Frankfurt am Main, Germany
{roessner,schnorr}@cs.uni-frankfurt.de

Abstract. We analyse a continued fraction algorithm (abbreviated CFA) for arbitrary dimension n showing that it produces simultaneous diophantine approximations which are up to the factor $2^{(n+2)/4}$ best possible. Given a real vector $x = (x_1, \ldots, x_{n-1}, 1) \in \mathbb{R}^n$ this CFA generates a sequence of vectors $(p_1^{(k)}, \ldots, p_{n-1}^{(k)}, q^{(k)}) \in \mathbb{Z}^n$, $k = 1, 2, \ldots$ with increasing integers $|q^{(k)}|$ satisfying for $i = 1, \ldots, n-1$

$$|x_i - p_i^{(k)}/q^{(k)}| \leq 2^{(n+2)/4} \sqrt{1 + x_i^2} / |q^{(k)}|^{1 + \frac{1}{n-1}}.$$

By a theorem of Dirichlet this bound is best possible in that the exponent $1 + \frac{1}{n-1}$ can in general not be increased.

1 Introduction

We analyse a CFA which computes for real vectors $x \in \mathbb{R}^n$ diophantine approximations to x that are up to the factor $2^{(n+2)/4}$ best possible. Given $x \in \mathbb{R}^n$ this CFA constructs a sequence of lattice bases of the lattice $\mathbb{Z}^n$ consisting of vectors that approximate the line $x\mathbb{R}$. For given $\epsilon > 0$, this CFA either finds an integer relation $m \in \mathbb{Z}^n - 0$ for x, i.e. $< m, x >= 0$, of Euclidean length at most $2^{n/2} \epsilon^{-1}$ or it proves that no integer relation of length $\leq \epsilon^{-1}$ exists. For this the algorithm uses $O(n^4 (n + |\log \epsilon|))$ arithmetic operations on real numbers with exact arithmetic. For a rational input vector $x := (q_1, \ldots, q_n)/q_n$, with $q_1, \ldots, q_n \in \mathbb{Z}$ the algorithm has polynomial bit complexity in the input size $\sum_{i=1}^n \lceil \log |q_i| \rceil + |\log \epsilon|$. Our analysis relies on the *dual* lattice basis which we show to consist of very short vectors, see Theorems 1, 2. From this we greatly improve the known bounds for the primary lattice basis and for diophantine approximation. The crucial role of the dual basis escaped in all previous studies.

Our algorithm is a variant of the HJLS–algorithm of Hastad, Just, Lagarias, Schnorr [HJLS89] for finding integer relations for a real vector x which in turn relies on the algorithms of Bergman [Berg80], Ferguson, Forcade [FF79] and Lenstra, Lenstra, Lovász [LLL82]. It also incorporates ideas of Just [Ju92], Ferguson and Bailey [FB92] and Rössner, Schnorr [RS95]. We present a stable floating point version of this algorithm, prove stability in Theorem 6 and demonstrate its stability by experimental data.

The problem of higher dimensional CFA has been widely studied by Jacobi [Ja1868], Perron [Pe1907], Bernstein [Bern71], Szekeres [Sz70], Ferguson, Forcade [FF79], Bergman [Berg80] and Lenstra, Lenstra, Lovász [LLL82]. The HJLS–algorithm of [HJLS89] is a variant of the algorithms in [FF79], [Berg80] and [LLL82]. It finds short integer relations for x in polynomial time using exact arithmetic on real numbers. Just [Ju92] showed that a variant of this algorithm provides diophantine approximations satisfying $|x_i - p_i^{(k)}/q^{(k)}| \leq 2^{(n+2)/4} \sqrt{1+x_i^2}/|q^{(k)}|^{1+\frac{1}{2n(n-1)}}$. We improve the analysis of [Ju92].

Ferguson and Bailey [FB92] have implemented a close variant of the HJLS–algorithm which they call the PSLQ–algorithm. Their experimental results show that this CFA produces simultaneous diophantine approximations that are far better than for any other known algorithm. Recenty Bailey, Borwein, Plouffe [BBP96] found surprising new approximation algorithms for π, $\ln(2)$ using this CFA. While this CFA could so far not be analyzed we prove for the first time the superiority of this CFA.

2 Preliminaries

Let $\mathbb{R}^n$ be the n–dimensional real vector space equipped with the ordinary inner product $< . , . >$ and Euclidean length $\|y\| := < y, y >^{1/2}$. We let $[y_1, \ldots, y_m]$ denote the matrix with column vectors $y_1, \ldots, y_m$ and $\lceil . \rfloor$ is the nearest integer function to a real number r, $\lceil r \rfloor = \lfloor r + 0.5 \rfloor$.

A non–zero vector $m \in \mathbb{Z}^n$ is called an *integer relation* for $x \in \mathbb{R}^n$ if $< x, m > = 0$. We let $\lambda(x)$ denote the length $\|m\|$ of the shortest integer relation m for x, $\lambda(x) = \infty$ if no relation exists.

Throughout this paper, $b_1, \ldots, b_n$ is an ordered basis of the integer lattice $\mathbb{Z}^n$ and its *dual* basis $a_1, \ldots, a_n$ is defined by $[a_1, \ldots, a_n]^\top := [b_1, \ldots, b_n]^{-1}$. Let $x \in \mathbb{R}^n$ be a non–zero vector, set $b_0 := x$. We associate with the basis $b_1, \ldots, b_n$ the orthogonal projections

$$\pi_{i,x} : \mathbb{R}^n \longrightarrow span(x, b_1, , \ldots, b_{i-1})^\perp \quad \text{and}$$
$$\pi_i : \mathbb{R}^n \longrightarrow span(b_1, , \ldots, b_{i-1})^\perp \quad \text{for } i = 1, \ldots, n,$$

where $span(b_j, \ldots, , b_{i-1})$ denotes the linear space generated by $b_j, \ldots, b_{i-1}$ and $span(b_j, \ldots, b_{i-1})^\perp$ its orthogonal complement in $\mathbb{R}^n$. We abbreviate $\hat{b}_{i,x} := \pi_{i,x}(b_i)$ and $\hat{b}_i := \pi_i(b_i)$. The vectors $\hat{b}_{1,x}, \ldots, \hat{b}_{n,x}$ (resp. $\hat{b}_1, \ldots, \hat{b}_n$) are pairwise orthogonal. They are called the *Gram–Schmidt orthogonalization* of $x, b_1, \ldots, b_n$ (resp. $b_1, \ldots, b_n$). The *Gram–Schmidt coefficients* $\mu_{i,j}$ of the factorization $[x, b_1, \ldots, b_n] = [x, \hat{b}_{1,x}, \ldots, \hat{b}_{n,x}]$ $(\mu_{i,j})^\top_{0 \leq i,j \leq n}$ are defined as $\mu_{i,j} := < b_i, \hat{b}_{j,x} > /\|\hat{b}_{j,x}\|^2$. If $\hat{b}_{j,x} = 0$, we set $\mu_{i,j} = 0$ for $i \neq j$ and $\mu_{j,j} = 1$. The matrix $(\mu_{i,j})_{0 \leq i,j \leq n}$ is lower triangular with all diagonal elements 1. Finally we note that $a_n = \hat{b}_n/\|\hat{b}_n\|^2$ since both a_n and $\hat{b}_n$ are orthogonal to $b_1, \ldots, b_{n-1}$.

The (ordered) vectors $x, b_1, \ldots, b_n$ are *size–reduced* if $|\mu_{k,j}| \leq \frac{1}{2}$ holds for $1 \leq j < k \leq n$ and L^3–*reduced* if they are size–reduced and the inequality

$\frac{3}{4} \|\pi_{k-1,x}(b_{k-1})\|^2 \leq \|\pi_{k-1,x}(b_k)\|^2$ holds for $k = 2, \ldots, n$. If L^3-reduced the vectors satisfy $\|\widehat{b}_{i,x}\|^2 \leq 2 \|\widehat{b}_{i+1,x}\|^2$ for $i = 1, \ldots, n-1$.

Models of Computation. We distinguish three models of computation for the CFA.

Exact Real Arithmetic. For real input $x \in \mathbb{R}^n$ we use exact arithmetic over real numbers. This version of the CFA can use either Gram–Schmidt orthogonalization or Givens Rotation with square roots. The analysis of the HJLS–algorithm applies.

Exact Integer Arithmetic. For rational input $x \in \mathbb{Q}^n$ we can use exact arithmetic over the integers. The rational numbers $\mu_{i,j}$, $\|\widehat{b}_{j,x}\|^2$ are represented by their numerator and denominator. This version of the CFA uses Gram–Schmidt orthogonalization. The analysis of the L^3-algorithm [LLL82] for lattice basis reduction applies.

Floating Point Arithmetic. For rational input x we can speed up the CFA in that we replace the exact arithmetic on the rationals $\mu_{i,j}$, $\|\widehat{b}_{j,x}\|$ by floating point arithmetic. The vectors $x, b_1, \ldots, b_n$, $a_1, \ldots, a_n$ are kept in exact representation. In order to minimize floating point errors we use, instead of the $\mu_{i,j}$, the normalized coefficients $\tau_{i,j} := \mu_{i,j} \|\widehat{b}_{j,x}\|$. We call the entities $\tau_{i,j}$ for $0 \leq i, j \leq n$ the *orthonormalization* of $x, b_1, \ldots, b_n$. Note that $\tau_{i,i} = \|\widehat{b}_{i,x}\|$. The L^3-property $\frac{3}{4} \|\pi_{k-1,x}(b_{k-1})\|^2 \leq \|\pi_{k-1,x}(b_k)\|^2$ is expressed by $\frac{3}{4} \tau_{k-1,k-1}^2 \leq \tau_{k,k}^2 + \tau_{k,k-1}^2$. The $\tau_{i,j}$ are not rational but require square roots, we compute them in floating point arithmetic using Givens Rotation.

We present our algorithm in its floating point version. From this description the details for the other models of computation are straightforward and left to the reader.

3 The Algorithm Description

This algorithm improves the HJLS–algorithm [HJLS89] towards numerical stability. Given a real vector $x \in \mathbb{R}^n$ and $\epsilon > 0$, the HJLS–algorithm either finds an integer relation m for x with $\|m\| \leq 2^{n/2-1} \min\{\lambda(x), \epsilon^{-1}\}$ or it proves $\lambda(x) \geq \epsilon^{-1}$. The HJLS–algorithm performs reduction and exchange steps on the linearly dependent system of vectors of the matrix $x, b_1, \ldots, b_n$ where initially $b_1, \ldots, b_n$ are set to the unit vectors in $\mathbb{R}^n$. The vector x remains unchanged and the vectors $b_1, \ldots, b_n$ remain a basis of the lattice $\mathbb{Z}^n$. The HJLS–algorithm uses exact arithmetic on real numbers. Its reduction and exchange steps minimize $\max_{1 \leq i \leq n} \|\widehat{b}_{i,x}\|$.

The HJLS–algorithm terminates if either $x \in span(b_1, \ldots, b_{n-1})$, i.e. if a swap $b_{n-1} \leftrightarrow b_n$ results in $\widehat{b}_{n-1,x} = 0$, or if $\max_{1 \leq i \leq n} \|\widehat{b}_{i,x}\| \leq \epsilon$. In the first case, the last vector a_n of the dual basis is an integer relation for x. In the latter case, we have $\lambda(x) \geq \epsilon^{-1}$ which follows from

[HJLS89] Proposition 3.1. *Every basis $b_1, \ldots, b_n$ of $\mathbb{Z}^n$ satisfies*

$$\lambda(x) \geq 1/ \max_{1 \leq i \leq n} \|\widehat{b}_{i,x}\|. \tag{1}$$

Our main modifications of the HJLS–algorithm are as follows:

1. We iteratively swap vectors b_{k-1}, b_k with $2 \leq k \leq n-1$ that do not satisfy the L^3–condition $\frac{3}{4} \|\pi_{k-1,x}(b_{k-1})\|^2 \leq \|\pi_{k-1,x}(b_k)\|^2$. The selection of k, either minimal as in the L^3–algorithm or so that $i := k$ maximizes $2^i \|\widehat{b}_{i,x}\|^2$ as proposed by Bergman, is irrelevant.

2. Before swapping the last two vectors b_{n-1} and b_n the basis $\pi_{1,x}(b_1), \ldots, \pi_{1,x}(b_n)$ is L^3–reduced. Here we follow Just [Ju92]. In the model of exact real arithmetic our algorithm essentially coincides with the algorithm of Just [Ju92] and her analysis of diophantine approximation properties applies.

3. We apply reduction in size, i.e. we reduce b_k so that $|\mu_{k,i}| \leq 1/2$ for $i = 1, \ldots, k-1$. Reduction in size has been neglected in [HJLS89] since it is pointless for the exact real arithmetic.

4. In the floating point version orthonormalization of the vectors $x, b_1, \ldots, b_n$ is done by Givens Rotation with a floating point error that is linear in n and $\max_{0 \leq i \leq n} \|b_i\|$, see [Ge75, GL89]. Givens Rotations has already been used in the parallel L^3–algorithms of Heckler, Thiele [HT93] and Joux [Jo93]. Ferguson and Bailey[FB92] essentially use Givens Rotation in connection with the HJLS–algorithm.

The test on $\tau_{n,n} \neq 0$ actually checks whether $\tau_{n,n} > 2^{-r}$ where r is the number of precision bits of the floating point arithmetic.

Stable Continued Fraction Algorithm (SCFA)

INPUT $x \in \mathbb{R}^n\text{-}0, \epsilon > 0$.

1. *Initiation.* Let $b_i \in \mathbf{Z}^n$ be the i-th unit vector. Compute the orthonormalization $\tau_{i,j}$ for $0 \leq i,j \leq n$ of $x, b_1, \ldots, b_n$ using Givens Rotation (see Section 5). $s := 1$.

2. L^3*–reduction of* $\pi_{1,x}(b_1), \ldots, \pi_{1,x}(b_{n-1})$.

While there exists k with $1 < k < n$ and $\frac{3}{4}\tau_{k-1,k-1}^2 > \tau_{k,k}^2 + \tau_{k,k-1}^2$ size–reduce b_k with respect to b_{k-1} by setting $b_k := b_k - \lceil \tau_{k,k-1}/\tau_{k-1,k-1} \rfloor b_{k-1}$, swap b_{k-1} and b_k and update the orthonormalization using Givens Rotation.

Reduce $b_1, \ldots, b_n$ in size. While $|\tau_{s,s}| \leq \epsilon$ increment s to $s+1$.

Output $(p_1, \ldots, p_{n-1}, q) := b_1$, the next approximation for x, see Theorem 3.

3. *Swap the last vectors.* Swap b_{n-1} and b_n, and update the orthonormalization using Givens Rotation. If $\tau_{n,n} = 0$ and $s < n$ then goto 2.

4. *Termination.* Compute $[a_1, \ldots, a_n]^\top := [b_1, \ldots, b_n]^{-1}$.

If $\tau_{n,n} > 0$ a relation for x is found. Output the nearby point $x' := x$ and the relation a_n for x.

If $s = n$ then $\tau_{i,i} \leq \epsilon$ holds for $i = 1, \ldots, n$.
Compute $\pi_n(x) \in span(b_1, \ldots, b_{n-1})^\perp$, output the nearby point $x' := x - \pi_n(x)$, the relation a_n for x' and "$\lambda(x) \geq \epsilon^{-1}$".

If $\epsilon = 0$ then SCFA produces a possibly infinite sequence of vectors b_1, occuring after L^3–reduction, that are good diophantine approximations to x.

Correctness Properties. 1. Upon termination of step 2 we have $\tau_{i,i} \leq \epsilon$ for $i = 1, \ldots, s-1$ and the projected basis $\pi_{1,x}(b_1), \ldots, \pi_{1,x}(b_{n-1})$ is L^3-reduced.
2. Before swapping b_{n-1} and b_n we have $s < n$ (note that $s \neq n$ since $\tau_{n,n} = 0$) and $\tau_{s,s}^{-1} < \epsilon^{-1}$. Therefore the L^3-reducedness of $\pi_{1,x}(b_1), \ldots, \pi_{1,x}(b_{n-1})$ implies that $\|\widehat{b}_{n-1,x}\|^{-1} < 2^{(n-1-s)/2} \epsilon^{-1}$.
3. We recall from [RS95], Theorem 6 that SCFA computes a nearby point x' and a non-zero vector $a_n \in \mathbf{Z}^n$ so that $< a_n, x' > = 0$ and $\lambda(\overline{x}) \geq \epsilon^{-1}/2$ holds for all $\overline{x} \in \mathbb{R}^n$ with $\|x - \overline{x}\| < \|x - x'\|/2$. If $x' \neq x$ we have $\lambda(x) \geq \epsilon^{-1}$.

4 Analysis of SCFA in Exact Real Arithmetic

Theorem 1. *Throughout the computation we have* $\|a_n\| \leq 2^{n/2} \min\{\epsilon^{-1}, \lambda(x)\}$. *Moreover,* $\|a_n\| \leq 2^{n/2+1} \lambda(x')$ *holds upon termination.*

For the first time we prove in Theorem 1 that SCFA outputs a relation a_n for the nearby point $x' \neq x$ which has, up to the factor $2^{n/2+1}$, minimal length.
Proof. We let $\overline{b}_1, \ldots, \overline{b}_n, \overline{a}_1, \ldots, \overline{a}_n$ denote the dual bases before and $b_1, \ldots, b_n$, $a_1, \ldots, a_n$ after an arbitrary swap $b_{n-1} \leftrightarrow b_n$ of SCFA. Let $\overline{\mu}_{i,j}$ be the Gram–Schmidt coefficients and $\widehat{\overline{b}}_{i,x}$ be the orthogonal vectors of $x, \overline{b}_1, \ldots, \overline{b}_n$. We have $\widehat{b}_{n-1,x} = \overline{\mu}_{n,n-1} \widehat{\overline{b}}_{n-1,x}$ with $|\overline{\mu}_{n,n-1}| \leq \frac{1}{2}$.
From the characterization of a_{n-1} as $< a_{n-1}, b_i > = \delta_{n-1,i}$, which holds throughout the algorithm, we infer that

$$a_{n-1} = \frac{\widehat{b}_{n-1,x}}{\|\widehat{b}_{n-1,x}\|^2} - \frac{< b_n, \widehat{b}_{n-1,x} >}{\|\widehat{b}_{n-1,x}\|^2} a_n .$$

Applying this equation to the vectors $\overline{b}_1, \ldots, \overline{b}_n$, $\overline{a}_1, \ldots, \overline{a}_n$ and $|\overline{\mu}_{n,n-1}| \leq \frac{1}{2}$ implies the recursion formula

$$\|a_n\| = \|\overline{a}_{n-1}\| = \|\widehat{\overline{b}}_{n-1,x}\|^{-1} + |\overline{\mu}_{n,n-1}| \|\overline{a}_n\|$$
$$\leq \|\widehat{\overline{b}}_{n-1,x}\|^{-1} + \tfrac{1}{2} \|\overline{a}_n\| .$$

From the L^3-reducedness of $\pi_{1,x}(\overline{b}_1), \ldots, \pi_{1,x}(\overline{b}_{n-1})$ and inequality (1) we see that

$$2^{(n-1)/2} \|\widehat{\overline{b}}_{n-1,x}\| \geq \max_{1 \leq i \leq n} 2^{i/2} \|\widehat{\overline{b}}_{i,x}\| \geq 2^{1/2} \lambda(x)^{-1} .$$

Using $\|\widehat{\overline{b}}_{n-1,x}\|^{-1} \leq 2^{n/2-1} \lambda(x)$ and $\|\widehat{\overline{b}}_{n-1,x}\|^{-1} \leq 2^{n/2-1} \epsilon^{-1}$, which follows from correctness property 2, we can rewrite the recursion formula to

$$\|a_n\| \leq 2^{n/2-1} \min\{\epsilon^{-1}, \lambda(x)\} + \tfrac{1}{2} \|\overline{a}_n\| .$$

This inequality holds for every exchange $b_{n-1} \leftrightarrow b_n$ of SCFA. Suppose that there are exactly t such exchanges and using that initially $\|a_n\| = 1$ we obtain the first claim:

$$\|a_n\| \leq 2^{n/2-1} \min\{\epsilon^{-1}, \lambda(x)\} \sum_{j=0}^{t-1} 2^{-j} + 2^{-t} \leq 2^{n/2} \min\{\epsilon^{-1}, \lambda(x)\} . \quad (2)$$

Since the inequality $\|a_n\| \leq 2^{n/2} \min\{\epsilon^{-1}, \lambda(x)\}$ holds after any swap $b_{n-1} \leftrightarrow b_n$ it must always hold because a_n does not change between two swaps.

It remains to prove the second claim in the case $x' \neq x$. For this we use

[RS95] **Lemma 5(3).** *The terminal basis $b_1, \ldots, b_n$ of SCFA satisfies with $x' = x - \pi_n(x)$ the inequalities*

$$\|\widehat{b}_{i,x} - \widehat{b}_{i,x'}\| \leq \|\widehat{b}_{i,x}\| \quad for \quad i = 1, \ldots, n-1. \quad (3)$$

From $\lambda(x') \geq 1/\max_{1 \leq i \leq n} \|\widehat{b}_{i,x'}\|$, $\|a_n\| = \|\widehat{b}_{n,x'}\|^{-1}$ and $\|\widehat{b}_{n-1,x'}\| = 0$ we see that

$$\frac{\|a_n\|}{\lambda(x')} \leq \max_{1 \leq i \leq n} \frac{\|\widehat{b}_{i,x'}\|}{\|\widehat{b}_{n,x'}\|} = \max_{1 \leq i \leq n-1} \{1, \|\widehat{b}_{i,x'}\| \|a_n\|\} .$$

From $\|\widehat{b}_{i,x}\| \leq \epsilon$, which in case $x' \neq x$ holds upon termination of SCFA, and from the inequality $\|\widehat{b}_{i,x'}\| \leq 2 \|\widehat{b}_{i,x}\|$, which follows from (3), we infer

$$\frac{\|a_n\|}{\lambda(x')} \leq \max\{1, 2\epsilon \|a_n\|\} \overset{(2)}{\leq} \max\{1, 2\epsilon \, 2^{n/2} \epsilon^{-1}\} = 2 \cdot 2^{n/2} ,$$

which finishes the proof. □

Theorem 2. *The dual bases $b_1, \ldots, b_n$ and $a_1, \ldots, a_n$, occuring after the L^3-reduction step of SCFA, satisfy for $i = 1, \ldots, n-1$*

1. $\|a_i\| \leq 1.5^{n-i} (\max_{i \leq j < n} \|\widehat{b}_{j,x}\|^{-1} + 2^{n/2} \min\{\epsilon^{-1}, \lambda(x)\})$,

2. $\|b_i\| \leq 2^{n/2} \min\{\epsilon^{-1}, \lambda(x)\} \sum_{j=1}^{i} \prod_{\substack{k=1 \\ k \neq j}}^{n-1} \|\widehat{b}_{k,x}\|^{-1} + \sum_{j=1}^{i} \|\widehat{b}_{j,x}\|$.

Proof. 1. Since SCFA did not terminate previously we know that $\widehat{b}_{n,x} = 0$, and thus $\widehat{b}_{j,x} \neq 0$ holds for $j = 1, \ldots, n-1$. Let $\mu_{i,j}$ be the Gram–Schmidt coefficients of $x, b_0, \ldots, b_n$ and define the $\nu_{i,j}$ by $(\nu_{i,j})_{1 \leq i,j \leq n} := (\mu_{i,j})_{1 \leq i,j \leq n}^{-1}$. We observe that

$$a_i = \sum_{j=i}^{n-1} \nu_{j,i} \frac{\widehat{b}_{j,x}}{\|\widehat{b}_{j,x}\|^2} + \nu_{n,i} a_n \quad (4)$$

holds for $i = 1, \ldots, n$. In fact this formula implies

$$< a_i, b_k > = < \sum_{j=i}^{n-1} \nu_{j,i} \frac{\widehat{b}_{j,x}}{\|\widehat{b}_{j,x}\|^2} + \nu_{n,i} a_n, \sum_{j=0}^{k} \mu_{k,j} \widehat{b}_{j,x} >$$

$$= \sum_{j=1}^{n-1} \nu_{j,i} \mu_{k,j} + \nu_{n,i} < a_n, b_k > = \delta_{i,k} - \nu_{n,i} \mu_{k,n} - \nu_{n,i} < a_n, b_k > = \delta_{i,k}.$$

The L^3-reduction step terminates with $|\mu_{i,j}| \leq \frac{1}{2}$ for $1 \leq j < i \leq n$. Hence

$$|\nu_{i,j}| \leq 1.5^{i-j} \text{ for } 1 \leq j \leq i \leq n.$$

Now, equation (4) yields for $i = 1, \ldots, n$

$$\|a_i\|^2 \leq 1.5^{2(n-i)} \max_{i \leq j < n} \|\widehat{b}_{j,x}\|^{-2} + 1.5^{2(n-i)} \|a_n\|^2.$$

Using the inequality $\|a_n\| \leq 2^{n/2} \min\{\epsilon^{-1}, \lambda(x)\}$ of Theorem 1 this proves the first claim:

$$\|a_i\| \leq 1.5^{n-i} (\max_{i \leq j < n} \|\widehat{b}_{j,x}\|^{-1} + 2^{n/2} \min\{\epsilon^{-1}, \lambda(x)\}).$$

2. We rewrite the equations (4) as

$$[a_1, \ldots, a_n] = \left[\frac{\widehat{b}_{1,x}}{\|\widehat{b}_{1,x}\|^2}, \ldots, \frac{\widehat{b}_{n-1,x}}{\|\widehat{b}_{n-1,x}\|^2}, a_n \right] (\nu_{i,j})_{1 \leq i,j \leq n}.$$

Since the vectors $\widehat{b}_{1,x}, \ldots, \widehat{b}_{n-1,x}$ are pairwise orthogonal there is an orthogonal matrix U, i.e. $U^{-1} = U^{\mathsf{T}}$, such that

$$\left[\frac{\widehat{b}_{1,x}}{\|\widehat{b}_{1,x}\|^2}, \ldots, \frac{\widehat{b}_{n-1,x}}{\|\widehat{b}_{n-1,x}\|^2}, a_n \right] =$$

$$U \begin{bmatrix} 1 & & 0 & a'_{n,1} \\ & \ddots & & \vdots \\ 0 & & 1 & a'_{n,n-1} \\ 0 & \ldots & 0 & a'_{n,n} \end{bmatrix} \begin{bmatrix} \|\widehat{b}_{1,x}\|^{-1} & & 0 & \vdots \\ & \ddots & & 0 \\ 0 & & \|\widehat{b}_{n-1,x}\|^{-1} & \vdots \\ & \ldots & 0 & \ldots & 1 \end{bmatrix},$$

with $(a'_{n,1}, \ldots, a'_{n,n})^{\mathsf{T}} := U a_n$ and $a'_{n,n} = \|\widehat{b}_{1,x}\| \cdot \ldots \cdot \|\widehat{b}_{n-1,x}\|$. From the previous equations and $[b_1, \ldots, b_n]^{\mathsf{T}} = [a_1, \ldots, a_n]^{-1}$ we see that

$$[b_1, \ldots, b_n] =$$

$$U \begin{bmatrix} 1 & & 0 & \vdots \\ & \ddots & & 0 \\ 0 & & 1 & \vdots \\ \overline{a}_{n,1} & \ldots & \overline{a}_{n,n-1} & \overline{a}_{n,n} \end{bmatrix} \begin{bmatrix} \|\widehat{b}_{1,x}\| & & 0 & \vdots \\ & \ddots & & 0 \\ 0 & & \|\widehat{b}_{n-1,x}\| & \vdots \\ & \ldots & 0 & \ldots & 1 \end{bmatrix} (\mu_{i,j})_{1 \leq i,j \leq n}^{\mathsf{T}},$$

where $\bar{a}_{n,n} := a'^{-1}_{n,n}$ and $\bar{a}_{n,i} := -a'_{n,i}/a'_{n,n}$ for $i < n$. Since U is orthogonal $\|b_i\|$ is the length of the i–th column vector of the cofactor of U in this matrix product. From $\bar{a}_{n,n} = \|\widehat{b}_{1,x}\|^{-1} \cdot \ldots \cdot \|\widehat{b}_{n-1,x}\|^{-1}$ and since the matrix $(\mu_{i,j})^{\mathsf{T}}_{1 \le i,j \le n}$ is upper triangular with $|\mu_{i,j}| \le 1$ we have for $i = 1, \ldots, n-1$

$$\|b_i\|^2 \le \|a_n\|^2 \sum_{j=1}^{i} \prod_{\substack{k=1 \\ k \ne j}}^{n-1} \|\widehat{b}_{k,x}\|^{-2} + \sum_{j=1}^{i} \|\widehat{b}_{j,x}\|^2 . \tag{5}$$

Now the claimed upper bound of $\|b_i\|$ follows from the inequality $\|a_n\| \le 2^{n/2} \min\{ \lambda(x),\ \epsilon^{-1} \}$ of Theorem 1. □

Theorem 3. *For real input $x = (x_1, \ldots, x_{n-1}, 1)$ every vector $(p_1, \ldots, p_{n-1}, q) := b_1$ occuring after the L^3–reduction step of SCFA satisfies for $i = 1, \ldots, n-1 : |x_i - p_i/q| \le 2^{\frac{n+2}{4}} \sqrt{1 + x_i^2} / |q|^{1 + \frac{1}{n-1}}$.*

By a theorem of Dirichlet [Di1842] the upper bound $\max_{1 \le i \le n} |x_i - p_i/q| \le 1/|q|^{1 + \frac{1}{n-1}}$ is best possible for diophantine approximations to x in that the term $1/(n-1)$ can in general not be increased. SCFA looses at most the factor $2^{(n+2)/4} \|x\|$ compared to this best general bound. This loss is due to the L^3–reduction. The factor $2^{n/4}$ can be reduced to $(1 + \varepsilon)^{n/4}$ with an arbitrarily small $\varepsilon > 0$ by using block reduced bases [Sc94].

Proof. For $n = 2$ the sequence of rationals p_1/q occuring before a swap $b_{n-1} \leftrightarrow b_n$ corresponds to the continued fraction expansion of x_1. Here we have the stronger inequality $|x_1 - p_1/q| \le 1/|q|^2$. Now consider $n \ge 3$. From Theorem 2, the L^3–reducedness of $\pi_{1,x}(b_1), \ldots, \pi_{1,x}(b_{n-1})$ and $\|\widehat{b}_{1,x}\| \le 1$ we see that

$$\|b_1\| \le 2^{n/2} \epsilon^{-1} \prod_{j=2}^{n-1} \|\widehat{b}_{j,x}\|^{-1} + \|\widehat{b}_{1,x}\| \le 2^{n/2} \epsilon^{-1} \|\widehat{b}_{1,x}\|^{-n+2} 2^{(n-1)(n-2)/4} + 1 .$$

A look into the proofs of Theorem 1 and 2 shows that all inequalities, in particular inequality (2), hold with ϵ^{-1} replaced by $\|\widehat{b}_{1,x}\|^{-1}$, and thus:

$$\|b_1\| \le \|\widehat{b}_{1,x}\|^{-n+1} 2^{n/2} \cdot 2^{(n-1)(n-2)/4} + 1 \overset{n \ge 3}{\le} \|\widehat{b}_{1,x}\|^{-n+1} 2^{(n-1)(n+2)/4} ,$$

$$\|\widehat{b}_{1,x}\| \le \|b_1\|^{-\frac{1}{n-1}} 2^{(n+2)/4} \le |q|^{-\frac{1}{n-1}} 2^{(n+2)/4} .$$

It can easily be seen that every vector $b_1 = (p_1, \ldots, p_{n-1}, q)$ satisfies for $i = 1, \ldots, n-1$

$$|x_i q - x_n p_i| \le \|\widehat{b}_{1,x}\| \sqrt{1 + x_i^2} ,$$

see e.g. equations (18), (19) of [Ju92]. Now the claim follows from the previous upper bound on $\|\widehat{b}_{1,x}\|$ and $x_n = 1$. □

Theorem 4. *If* SCFA *outputs* $x' \neq x$ *then* a_n *is, up to a factor* $2^{n/2+1}$, *a shortest almost relation for* x *in the following sense. If* $m \in \mathbf{Z}^n - 0$ *satisfies* $| < x, m/\|m\| > | < | < x, a_n/\|a_n\| > |/2$ *then* $\|m\| \geq \|a_n\| \, 2^{-n/2-1}$.

Proof. Put $\overline{x} := x - < x, m/\|m\| > m/\|m\|$ and note that $x' := x - < x, a_n/\|a_n\| > a_n/\|a_n\|$. Hence $\|x - \overline{x}\| < \|x - x'\|/2$. From [RS95], Theorem 6 we have $\lambda(\overline{x}) \geq \epsilon^{-1}/2$ whereas $\|a_n\| \leq 2^{n/2} \epsilon^{-1}$ holds by Theorem 1. Hence $\|m\| \geq \lambda(\overline{x}) \geq \|a_n\| \, 2^{-n/2-1}$. $\qquad\square$

Theorem 5. *For rational input* $x = (q_1, \ldots, q_n)/q_n$ *with integers* $q_1, \ldots, q_n$, SCFA *performs at most* $\lfloor \log_2 |q_n| \rfloor$ *swaps* $b_{n-1} \leftrightarrow b_n$.

Proof. The vectors $x, b_1, \ldots, b_{n-1}$ before a swap $b_{n-1} \leftrightarrow b_n$ are linearly independent, they generate a parallelepiped that has the non–zero volume:

$$\|x\| \prod_{j=1}^{n-1} \|\widehat{b}_{j,x}\| = \prod_{j=1}^{n-1} \|\widehat{b}_j\| \, \|\pi_n(x)\| = \|\widehat{b}_n\|^{-1} \|\pi_n(x)\| = \|a_n\| \, \|\pi_n(x)\|.$$

Hence $\prod_{j=1}^{n-1} \|\widehat{b}_{j,x}\| = | < x, a_n > | \, \|x\|^{-1}$. This equation holds for the dual bases $\overline{a}_1, \ldots, \overline{a}_n, \overline{b}_1, \ldots, \overline{b}_n$ before and the dual bases $a_1, \ldots, a_n, b_1, \ldots, b_n$ after the swap. This yields

$$\frac{\|\widehat{b}_{n-1,x}\|}{\|\widehat{\overline{b}}_{n-1,x}\|} = \frac{\prod_{j=1}^{n-1} \|\widehat{b}_{j,x}\|}{\prod_{j=1}^{n-1} \|\widehat{\overline{b}}_{j,x}\|} = \frac{| < x, a_n > |}{| < x, \overline{a}_n > |}.$$

Using $\|\widehat{b}_{n-1,x}\| = |\overline{\mu}_{n,n-1}| \, \|\widehat{\overline{b}}_{n-1,x}\| \leq \frac{1}{2} \|\widehat{\overline{b}}_{n-1,x}\|$ we see that

$$| < x, a_n > | \leq \tfrac{1}{2} | < x, \overline{a}_n > |.$$

Let t be the number of swaps $b_{n-1} \leftrightarrow b_n$. Since initially $< x, a_n >= 1$ we have upon termination of SCFA that $q_n^{-1} \leq | < x, a_n > | \leq 2^{-t}$. This proves the desired bound $t \leq \lfloor \log_2 |q_n| \rfloor$ on the number of swaps $b_{n-1} \leftrightarrow b_n$. $\qquad\square$

Running Time. We refer to the models of computation introduced in section 2. Arithmetic operations are $+, -, \cdot, /, \lceil \; \rfloor$ (the nearest integer function) and $<$ (comparison). In the floating point model we also use $\sqrt{\;}$ (square root).

Exact Real Arithmetic. For real input $x \in \mathbf{R}^n$ SCFA performs $O(n^4 (n + |\log \epsilon|))$ arithmetic operations on real numbers and $O(n^2 (n + |\log \epsilon|))$ many swaps $b_{k-1} \leftrightarrow b_k$ with $2 \leq k \leq n$. This follows from the analysis of the HJLS–algorithm [HJLS89]. The algorithm either uses Gram–Schmidt orthogonalization via the $\mu_{i,j}$ and $\|\widehat{b}_{j,x}\|^2$ or Givens Rotation with square roots via the $\tau_{i,j}$.

Exact Integer Arithmetic. For rational $x = (q_1, \ldots, q_n)/q_n \in \mathbf{Q}^n$ SCFA performs at most $O(n^4 (n + |\log \epsilon|))$ arithmetic operations on integers of bit length $O(n + \max_{1 \leq i \leq n} |\log q_i| + |\log \epsilon|)$. Arithmetic steps use the coordinates of the vectors b_i, a_i and the numerators and denominators of the rational numbers $\mu_{i,j}$, $\|\widehat{b}_{j,x}\|^2$. The algorithm uses Gram–Schmidt orthogonalization. The claimed upper bound on the bit length of all integers follows by adjusting the analysis of the L^3–algorithm in [LLL82] to our algorithm.

5 Numerical Stability of the Floating Point Algorithm

The algorithm is given a rational input $x = (q_1, \ldots, q_n)/q_n \in \mathbb{Q}^n$. Orthonormalization of $x, b_1, \ldots, b_n$ is done by Givens Rotation via the floating point numbers $\tau_{i,j} := \mu_{i,j} \|\widehat{b}_{j,x}\|$. We study floating point errors of the $\tau_{i,j}$ and the correctness of swaps $b_{k-1} \leftrightarrow b_k$.

Every arithmetic operation $+$, $-$, $\cdot$, $/$, $\sqrt{\ }$, $\lceil\ \rfloor$ generates a floating point error when its result is rounded to the nearest floating point number. Let t' denote the floating point value of a real number t with (floating point) *error* $t - t'$ and *relative error* $(1 - t'/t)$. Let r denote the number of precision bits of the floating point arithmetic and 2^{-r} the maximal relative error. We use IEEE 754 double precision format with $r = 53$.

Givens Rotation. The $n \times (n+1)$–matrix $B := (b_{i,j}) = [x, b_1, \ldots, b_n]$ has a unique decomposition $B = U \cdot L^\top$ where L is a lower triangular matrix and U is an orthogonal $n \times n$–matrix. Hence $L = (\tau_{i,j})_{\substack{0 \le i \le n \\ 1 \le j \le n}}$ and

$$U = \left[\frac{x}{\|x\|}, \frac{\widehat{b}_{1,x}}{\|\widehat{b}_{1,x}\|}, \ldots, \frac{\widehat{b}_{n-1,x}}{\|\widehat{b}_{n-1,x}\|} \right].$$

Since U is orthogonal we have $L^\top = U^\top B$. $U^\top = U^{-1}$ is product of *elementary rotations* (ER) $G_{i,j}$ with $1 \le j < i \le n$. If $\overline{B} = (\overline{b}_{i,j}) = [x, \overline{b}_1, \ldots, \overline{b}_n]$ denotes the product of B with all previous ER then $\overline{B} \mapsto G_{i,j} \overline{B}$ puts $\overline{b}_{i,j}$ to zero by transforming the column vectors $\overline{b}_i$, $\overline{b}_j$.

Floating Point Errors. Let $|B| := \max_{0 \le i \le n} \|b_i\|$. Note that $|B| = |\overline{B}|$ holds since the $G_{i,j}$ are orthogonal. Multiplying B with $G_{i,j}$ yields an error $\|\overline{b}_j - \overline{b}_j'\|$, $\|\overline{b}_i - \overline{b}_i'\| \le 7 n \, 2^{-r} |B|$, see pp. 131–139 in [Wi65]. This also holds for the multiplication by several $G_{i,j}$ with pairwise disjoint sets $\{i, j\}$. Following Gentleman [Ge75] we can distribute the $(n-2)(n-1)/2$ many ER $G_{i,j}$ for U into $2n - 3$ stages, each containing pairwise disjoint $G_{i,j}$. This way the error of the $\tau_{i,j}$ produced by Givens Rotation is at most $7(2n - 3) 2^{-r} |B|$, see [Ge75], [H95].

We call a swap $b_{k-1} \leftrightarrow b_k$ *good* if it decreases $\tau_{k-1,k-1}$, i.e. if the swap condition $\tau_{k-1,k-1} > (\tau_{k,k}^2 + \tau_{k,k-1}^2)^{1/2}$ holds before a swap.

Theorem 6. *If $\frac{3}{4} \tau_{k-1,k-1}'^2 > \tau_{k,k}'^2 + \tau_{k,k-1}'^2$ holds for the rounded τ-values of a Givens Rotation for the actual basis B and $195 \, n \, 2^{-r} |B| \le \tau_{k-1,k-1}$ the swap $b_{k-1} \leftrightarrow b_k$ is good.*

Proof. Above, we have shown that the error of the matrix $[\bar{b}_1, \ldots, \bar{b}_n] := (\tau_{i,j})^\top$ satisfies $\|\bar{b}_{k-1} - \bar{b}'_{k-1}\|$, $\|\bar{b}_k - \bar{b}'_k\| \leq 7(2n-3)2^{-r}|B|$ and thus

$$\tau_{k-1,k-1} - (\tau_{k,k}^2 + \tau_{k,k-1}^2)^{1/2} \geq$$

$$(1-\sqrt{\tfrac{3}{4}})\,\tau_{k-1,k-1} + \sqrt{\tfrac{3}{4}}\,\tau'_{k-1,k-1} - (\tau'^2_{k,k-1} + \tau'^2_{k,k})^{1/2} - (\sqrt{\tfrac{3}{4}}+1)\,7(2n-3)2^{-r}|B|.$$

By assumption $\sqrt{\tfrac{3}{4}}\,\tau'_{k-1,k-1} - \sqrt{\tau'^2_{k,k-1} + \tau'^2_{k,k}} + 2^{-r+1} > 0$ holds where 2^{-r+1} bounds the errors in evaluating the swap condition. These inequalities imply $\tau_{k-1,k-1} > (\tau_{k,k}^2 + \tau_{k,k-1}^2)^{1/2}$ since $14(\sqrt{\tfrac{3}{4}}+1)/(1-\sqrt{\tfrac{3}{4}}) < 195$. □

By the argument of Theorem 6 the inequalities $\tfrac{1}{2}\tau_{k-1,k-1}^2 > \tau_{k,k}^2 + \tau_{k,k-1}^2$, $165\,n\,2^{-r}|B| \leq \tau_{k-1,k-1}$ imply that the swap condition $\tfrac{3}{4}\tau'^2_{k-1,k-1} \geq \tau'^2_{k,k-1} + \tau'^2_{k,k}$ holds for the rounded τ-values. (Here we use that $14(\sqrt{\tfrac{3}{4}}+1)/(\sqrt{\tfrac{3}{4}} - \sqrt{\tfrac{1}{2}}) < 165$). Therefore the L^3-reduction step generates a reasonably good approximation of an L^3-reduced basis.

Good Swaps for SCFA. From inequality (5) and since the basis B before a swap is reduced in size we have $|B| \leq \sqrt{n}\,2^{n/2}\epsilon^{-1}(\sum_{i=1}^n q_i^2)^{1/2}$. Hence by Theorem 6 a swap $b_{k-1} \leftrightarrow b_k$ of SCFA is good if $\tau_{k-1,k-1} > \epsilon$ and

$$195\,n^2\,2^{n/2}\,\epsilon^{-2}\,2^{-r}\,\max_i |q_i| \leq 1.$$

E.g. for $r = 53$, $\epsilon^{-1} = 16$, $|q_i| \leq 2^{20}$ this inequality holds up to $n = 18$. Since $\tau_{k-1,k-1} > \epsilon$ holds for all but a few exceptional swaps we see that SCFA is stable for $\epsilon^{-1} = 16$, $\max_i |q_i| \leq 2^{20}$ up to dimension 18. In fact our experiments show that the stability of SCFA is even much better.

Experimental Results. We distinguish the two cases of termination:

- an exchange step $b_{n-1} \leftrightarrow b_n$ results in $\tau_{n,n} \neq 0$ and in an integer relation a_n for x
- $\max_{i=1,\ldots,n} \|\hat{b}_{i,x}\| < \epsilon$ yields a nearby point $x' \neq x$.

The last column of our table reports the number of occurences of the second case. In the first case we check that the vector a_n satisfies $< a_n, x > = 0$.

Each line of the table shows the results for 10 input vectors $x := (q_1, \ldots, q_n)/q_n \in \mathbb{Q}^n$ where the $q_i \in_R [0, 2^Q - 1]$ are random integers. With $Q = 45$ we almost exhaust the 53 precision bits of double precision floating point. We see that the average length of the output vector a_n increases at most linearly with ϵ^{-1}. For $n = 10$ we have $16 < \lambda(x) \leq 58.45$ for all 10 inputs.

n	ϵ^{-1}	Q	av. len	av. dist	av. time	$\sharp\ x'$
10	4	45	7.53	1.987e+7	0.18	10
10	8	45	8.71	20.76	0.28	10
10	16	45	21.46	4.7	0.47	10
10	24	45	20.40	0.54	0.37	9
10	32	45	30.23	0.0	0.37	3
10	64	45	29.95	0.0	0.47	0
20	2	45	4.17	2.256e+4	3.07	10
20	3	45	6.27	1.124e+3	4.26	10
20	4	45	8.68	0.04	5.41	4
20	6	45	7.92	0.0	5.34	0
modified SCFA						
20	16	45	7.92	0.0	5.41	0
40	16	45	7.87	0.0	75.01	0
60	16	45	10.83	0.0	328.14	0
80	16	45	7.45	0.0	958.77	0
100	16	45	6.32	0.0	2268.43	0

n, ϵ : parameters of SCFA

Q : the above parameter

av. len : average length of the output vector a_n per 10 inputs

av. dist : average distance from x to x' per 10 inputs

av. time :average time in seconds per 10 inputs on a HP 715/50 with 62 MIPS

$\sharp\ x'$: number of nearby points $x' \neq x$ for 10 inputs

In dimensions $n \geq 20$ and $Q = 45$ SCFA runs out of numerical precision. To improve the stability we perform before each size–reduction step and before each swap $b_{k-1} \leftrightarrow b_k$ a fresh Givens Rotation for the actual vectors $x, b_1, \ldots, b_n$, so that Theorem 6 applies. With this modification, SCFA becomes numerically stable up to dimension 100 and $Q = 45$.

Reasonable running times demonstrate not only the efficiency but also the stability of SCFA, they show that good swaps $b_{k-1} \leftrightarrow b_k$ prevail. Then the output a_n is most likely correct since a faulty output a_n means a bad swap $b_{n-1} \leftrightarrow b_n$. The fact that SCFA always finds integer relations a_n as expected shows its stability.

References

[Berg80] G. BERGMAN: Notes on Ferguson and Forcade's Generalized Euclidean Algorithm. TR, Department of Mathematics, University of California, Berkeley, CA, 1980.

[Bern71] L. BERNSTEIN: The Jacobi–Perron Algorithm, Lecture Notes in Mathematics 207, Berlin–Heidelberg–New York (1971), pp. 1–161.

[BBP96] D. BAILEY, P. BORWEIN, S. PLOUFFE: On the Rapid Computation of Various Polylogarithmic Constants, Technical Report, Simon Frazer University, Burnaby, B. C., Canada (1996).

[Di1842] G.L. DIRICHLET: Verallgemeinerung eines Satzes aus der Lehre von den Kettenbrüchen nebst einigen Anwendungen auf die Theorie der Zahlen, Bericht über die zur Bekanntmachung geigneten Verhandlungen der Königlich Preussischen Akademie der Wissenschaften zu Berlin (1842), pp. 93–95.

[FB92] H.R.P. FERGUSON and D.H. BAILEY: A Polynomial Time, Numerically Stable Integer Relation Algorithm. RNR Technical Report RNR-91-032, NASA Ames Research Center, Moffett Field, CA (1992).

[FF79] H. FERGUSON and R. FORCADE: Generalization of the Euclidean Algorithm for Real Numbers to all Dimensions Higher than Two, Bull. Amer. Math. Soc., (New Series) 1 (1979), pp. 912–914.

[Ge75] W.M. GENTLEMAN: Error Analysis of QR Decomposition by Givens Transformations. Linear Algebra and its Applications, Vol. 10, pp. 189–197, 1975.

[GL89] G.H. GOLUB and C.F. VAN LOAN: Matrix Computations. The Johns Hopkins University Press, London (1989).

[H95] C. HECKLER: Automatische Parallelisierung und parallele Gitterbasenreduktion. Ph.D. Thesis, University of Saarbrücken, 1995.

[HT93] C. HECKLER and L. THIELE: A Parallel Lattice Basis Reduction for Mesh-connected Processor Arrays and Parallel Complexity. Proceedings of the 5th Symposium on Parallel and Distributed Processing, Dallas (1993).

[HJLS89] J. HASTAD, B. JUST, J.C. LAGARIAS and C.P. SCHNORR: Polynomial Time Algorithms for Finding Integer Relations among Real Numbers. SIAM J. Comput., Vol. 18, No. 5 (1989), pp. 859–881.

[Ja1868] C.G.J. JACOBI: Allgemeine Theorie der Kettenbruchähnlichen Algorithmen, J. Reine Angew. Math. 69 (1868), pp. 29–64.

[Jo93] A. JOUX: A Fast Parallel Lattice Basis Reduction Algorithm. Proceedings of the 2nd Gauss Symposium, Munich (1993).

[Ju92] B. JUST: Generalizing the Continued Fraction Algorithm to Arbitrary Dimensions. SIAM J. Comput., Vol. 21, No. 5 (1992), pp. 909–926.

[LLL82] A.K. LENSTRA, H.W. LENSTRA,JR. and L. LOVÁSZ: Factoring Polynomials with Rational Coefficients. Math. Ann. 21 (1982), pp. 515–534.

[Pe1907] O. PERRON: Grundlagen für eine Theorie des Jacobischen Kettenbruchalgorithmus. Math. Ann. 64 (1907), pp. 1–76.

[RS95] C. RÖSSNER and C.P. SCHNORR: Computation of Highly Regular Nearby Points. Proceedings of the 3rd Israel Symposium on Theory of Computing and Systems, Tel Aviv (1995).

[Sc94] C.P. SCHNORR: Block Reduced Lattice and Successive Minima. Combinatorics, Probablity and Computing 3 (1994), pp. 507–522.

[Sz70] G. SZEKERES: Multidimensional Continued Fractions. Ann. Univ. Sci. Budapest, Eőtvős Sect. Math. 13 (1970), pp. 113–140.

[Wi65] J.H. WILKINSON: The Algebraic Eigenvalue Problem. Oxford University Press (1965).

Algorithms and Extended Formulations for One and Two Facility Network Design

Sunil Chopra

J.L. Kellogg Graduate School of Management, Northwestern University

Itzhak Gilboa

J.L. Kellogg Graduate School of Management, Northwestern University

S. Trilochan Sastry

Indian Institute of Management, Ahmedabad

Abstract

We consider the problem of sending flow from a source to a destination, where there are flow costs on each arc and fixed costs toward the purchase of capacity. Capacity can be purchased in batches of C units on each arc. We show the problem to be NP-hard in general. If d is the quantity to be shipped from the source to the destination, we give an algorithm that solves the problem in time polynomial in the size of the graph but exponential in $\lfloor \frac{d}{C} \rfloor$. Thus for bounded values of $\lfloor \frac{d}{C} \rfloor$ the problem can be solved in polynomial time. This is useful since a simple heuristic gives a very good approximation of the optimal solution for large values of $\lfloor \frac{d}{C} \rfloor$. We also show a similar result to hold for the case when there are no flow costs but capacity can be purchased either in batches of 1 unit or C units. The results characterizing optimal solutions are used to obtain extended formulations in each of the two cases. The LP-relaxations of the extended formulations are shown to be stronger than the natural formulations considered by earlier authors, even with a family of strong valid inequalities added.

1 Introduction

In this paper we consider the one-facility, one-commodity (OFOC) network design problem which can be stated as follows. Consider a directed graph $G = (V, A)$ with a source s and a destination t. Capacity on each arc can be purchased in integer multiples of C units with each batch of C units costing $w_a \geq 0$ on arc a. There is a flow cost of $p_a \geq 0$ per unit of flow on arc a. The total cost of the flow is the flow cost plus the cost of purchasing capacity. The objective is to design a minimum cost network to send d units of flow from s to t.

A more general form of the problem with several sources and sinks arises in the telecommunications and transportation industry. OFOC arises as a subproblem in these instances. OFOC has been studied by Magnanti and Mirchandani(1993) for the special case where the flow cost p_a is zero on each arc. They show that the problem reduces to the shortest path problem and can thus be solved in polynomial time. They also give an inequality description for which they show that all objective functions with $p_a = 0$ for all arcs a, have at least one optimal solution that is integral. A more general case with multiple commodities has been considered by Magnanti, Mirchandani, and Vachani(1995) and Chopra at al. (1995). Another related problem has been considered by Leung, Magnanti, and Vachani(1989), and Pochet and Wolsey(1991), where they study the capacitated lot sizing problem. They provide families of facet defining inequalities for the associated polyhedron.

In this paper we show the problem OFOC to be NP-hard in general when flow costs are present. This is in contrast to the case where all flow costs are zero, which is polynomially solvable (see Magnanti and Mirchandani(1993)). We provide an algorithm to solve OFOC in polynomial time for bounded values of $\lfloor d/C \rfloor$. This is valuable since a simple approximation heuristic is asymptotically optimal.

We also consider the two-facility one-commodity (TFOC) network design problem (see Magnanti and Mirchandani(1993) and Magnanti et al. (1995)). The problem is similar to OFOC except that we assume that capacity can be purchased either in batches of size 1 at a cost of $w_a^1 \geq 0$ or size C at a cost of $w_a^2 \geq 0$. Magnanti and Mirchandani (1993) consider the problem for the case where all flow costs p_a are 0. However the status of the problem in terms of complexity was unresolved. In this paper we show TFOC to be NP-hard for the case where all flow costs are 0. For the case when flow costs are 0, we provide an algorithm to solve TFOC in polynomial time for bounded values of $\lfloor d/C \rfloor$.

We use the results characterizing optimal solutions to OFOC and TFOC to obtain extended formulations in each case. We show that the LP-relaxations of the extended formulations are stronger than the natural formulations considered by earlier authors, even with a family of strong valid inequalities added. We also characterize objective functions for which the LP-relaxations of the extended formulations give integer optima.

In Section 2, we show that OFOC is NP-hard in general. In Section 3, we give an algorithm that allows us to solve OFOC in polynomial time as long as $\lfloor d/C \rfloor$ is bounded. In section 4, we show TFOC to be NP-hard even when all flow costs are 0. An algorithm similar to that given for OFOC allows us to solve TFOC in polynomial time for bounded $\lfloor d/C \rfloor$ if flow costs are 0. Section 5 contains the extended formulations and we show them to be stronger than the natural formulations even with additional facet defining inequalities included.

We assume basic familiarity with graphs and network flows (see for instance Bondy and Murty(1976)). An arc a, directed from u to v will be referred to as (u, v). A vector indexed by the arc set will have variables referred to as x_{uv} or x_a depending upon the context. Given a node set $X \subseteq V$, define $\delta^+(X)$ to be the set of arcs directed from X to $V \setminus X$ and $\delta^-(X)$ to be the set of arcs directed from $V \setminus X$ to X. Given $\bar{A} \subseteq A$, and a vector y indexed by A, define $y(\bar{A}) = \sum_{a \in \bar{A}} y_a$.

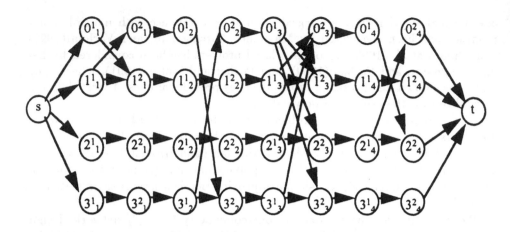

Fig. 1.

2 OFOC is NP-hard

We prove that OFOC is NP-hard by transforming Minimum Cover (see Garey and Johnson(1979)) into an instance of OFOC.

Proposition 2.1 *The problem OFOC is NP-hard.*

Proof: In an instance of Minimum Cover, we are given a collection $F = \{S_j, j = 1, ..., m\}$ of subsets of a finite set $S = \{1, ..., n\}$, and a positive integer $k \leq m$. The question is whether F contains a cover for S of size k or less, i.e., a subset $F' \subseteq F$ with $|F'| \leq k$ such that every element of S belongs to at least one member of F'.

Given the above instance of Minimum Cover, we construct the directed graph $G_F = (V_F, A_F)$, where

$$V_F = \{s, t\} \cup \{j_i^1, j_i^2, i = 1, ..., n; j = 0, 1, ..., m\},$$

$$E_F = \{(s, j_1^1), j = 0, 1, ..., m\} \cup \{(j_i^1, j_i^2), (j_i^2, j_{i+1}^1), j = 1, ..., m, i = 1, ..., n\}$$

$$\cup (0_i^2, 0_{i+1}^1), i = 2, ..., n\} \cup \{(0_i^1, j_i^2), (j_i^1, 0_i^2), \text{if } i \in S_j, \text{ for } i = 1, ..., n\}.$$

In the above description we have assumed that $j_{n+1}^1 = t$ for all j. The graph G_F contains a directed path $P_j, j = 1, ..., n$ from s to t using the nodes j_i^1 and j_i^2 for each subset S_j in F. The arcs $(0_i^1, j_i^2)$ and $(j_i^1, 0_i^2)$ are present if and only if $i \in S_j$.

For the case where $n = 4$, $k = 2$, $m = 3$, $S_1 = \{1, 3\}$, $S_2 = \{3, 4\}$, $S_3 = \{2, 3\}$, the graph G_F is as shown in Figure 1.

On the graph G_F, consider the problem OFOC where $kC + \epsilon$ units of flow is to be sent from s to t, where ϵ is close to 0. Arcs a along the paths P_j have

$w_a = M$ and $p_a = 0$. The arcs of the form $(s, 0_1^1)$ and $(0_i^2, 0_{i+1}^1), i = 2, ..., n$ have $w_a = 0$ and $p_a = 2M$. Here M is a large positive integer ($M = (2n + 1)k$ will suffice). All other arcs have $w_a = 0$ and $p_a = 0$.

Note that to send kC units of flow from s to t we must use k of the paths $P_j, j = 1, ..., n$ (multiple uses of a path are counted as multiple paths), with each path carrying C units, since these are the only directed paths available from s to t. In this case a total cost of $(2n + 1)kM$ is incurred to send the kC units from s to t and we cannot send this portion any cheaper. This leaves ϵ units to be sent from s to t. Note that each arc in a path $P_j, j = 1, ..., n$ that has been used to send C units from s to t can now be used in the reverse direction to send the remaining ϵ units without incurring a cost w_a. A path from s to t can be formed to ship the ϵ units from s to t (where all the arcs from any path $P_j, j = 1, ..., n$ used are in the reverse direction) if and only if for each $i \in \{1, ..., n\}$ there exists a path P_j, with $i \in S_j$, that has been used to send C units from s to t. Each of the arcs $(0_i^2, 0_{i+1}^1)$ is used in this path. If there is no solution to Minimum Cover, the cost incurred to send $k + \epsilon$ units from s to t is at least $(2n + 1)kM + M$, since at least one of the arcs in the paths P_j must be used in the forward direction to carry the ϵ units. On the other hand, if there exists a solution to Minimum Cover, the $k + \epsilon$ units can be sent at a cost of $(2n + 1)kM + 2(n + 1)M\epsilon < (2n + 1)kM + M$, for ϵ sufficiently small. Thus OFOC on G_F has an optimal solution of value $(2n + 1)kM + 2(n + 1)M\epsilon$ if and only if there exists a solution to Minimum Cover. The result thus follows. □

Note that for the example in Figure 1, there is no cover using two or fewer subsets. To send $2 + \epsilon$ units from s to t in the graph G_F in Figure 1, we have to use at least two of the paths P_j (multiple uses of a path being counted as multiple paths) to send 2 units and at least one of the arcs in the paths P_j in the forward direction to send the remaining ϵ units. The total cost incurred in this case is at least $19M$. On the other hand, if we set $S_3 = \{2, 4\}$, there is a cover using two subsets. In the corresponding graph G_F, it is possible to send $2 + \epsilon$ units from s to t at a cost of $18M + 10M\epsilon$.

3 OFOC for bounded $\lfloor \frac{d}{C} \rfloor$

In this section we show that if $d = kC + r, 1 \leq r \leq C - 1$, and k is bounded from above by some constant, then OFOC can be solved in polynomial time.

3.1 Structure of optimal solutions for OFOC

We identify certain structural properties of optimal solutions to OFOC. Consider an optimal solution vector (y^*, f^*) where y_a^* is the capacity installed on arc a and f_a^* is the flow through arc a. Note that since $w_a \geq 0$ and $p_a \geq 0$, given the flow vector f^* in any optimal solution, the optimal capacity installed can be assumed to be given by $y_a^* = \lceil f_a^*/C \rceil$. Given a solution (y^*, f^*), define an arc a with $f_a^* < Cy_a^*$ to be a *free arc*. We now characterize optimal solutions of OFOC with the minimum number of free arcs.

Proposition 3.1 *Let* (y^*, f^*) *be an optimal solution to OFOC with the minimum number of free arcs. All the free arcs defined by* (y^*, f^*) *lie on a path (ignoring direction) from s to t. Free arcs directed forward along this path have a flow from* $\{lC+r\}_{l=0}^k$, *and those directed backward have a flow from* $\{lC-r\}_{l=1}^k$, *in the optimal solution.*

From this point on we restrict attention to optimal solutions to OFOC with the minimum number of free arcs. Further, we can assume that there does not exist another optimal solution $(y', f') \neq (y^*, f^*)$, such that $y'_a \leq y_a^*$ and $f'_a \leq f_a^*$ for all arcs a. Such optimal solutions will be referred to as *extreme optimal solutions*. Given an optimal solution (y^*, f^*) to OFOC, let $G^* = (V^*, A^*)$ be the graph induced by the arcs with $f_a^* > 0$. Since all costs are non-negative, the graph G^* can be assumed to be acyclic. We next prove that the optimal flow f^* can be decomposed into the sum of flows along $2k + 1$ paths $\{P_i\}_{i=1}^{2k+1}$ from s to t, where the first k paths each carry a flow of $C - r$ and the last $k + 1$ paths each carry a flow of r.

Proposition 3.2 *Let* (y^*, f^*) *be an extreme optimal solution to OFOC. There exist* $2k + 1$ *paths* $\{P_i\}_{i=1}^{2k+1}$, *from s to t, such that*

$$f_a^* = (C - r) \sum_{i=1}^k P_i^a + r \sum_{i=k+1}^{2k+1} P_i^a, \tag{1}$$

where $P_i^a = 1$ *if* $a \in P_i$, 0 *otherwise.*

As an example consider the graph in Figure 2. Assume that $d = 17, C = 10$. Assume that $w_{s1} = w_{2t} = 10$, $w_{s2} = w_{1t} = 0$, $p_{s1} = p_{2t} = 1$, $p_{s2} = p_{1t} = 2$, $w_{12} = p_{12} = 0$.

Consider the solution (y^*, f^*) where $f_{s1}^* = f_{2t}^* = 10$, $f_{s2}^* = f_{1t}^* = 7, f_{12}^* = 3$. The flow f^* can be decomposed into a flow of 3 units along the path $\{(s, 1), (1, 2), (2, t)\}$ and a flow of 7 units along the paths $\{(s, 1), (1, t)\}$ and $\{(s, 2), (2, t)\}$.

3.2 Polynomial algorithm to solve OFOC for bounded $\lfloor d/C \rfloor$

We use the results from Section 3.1 to devise an algorithm to solve OFOC. The complexity of the algorithm is polynomial for bounded k, where $d = kC + r$. The algorithm is based on the decomposition of the flow in the optimal solution into a flow along $2k + 1$ paths. As shown in Proposition 3.2, the paths $\{P_i\}_{i=1}^k$, have a flow of $C - r$, while the paths $\{P_i\}_{i=k+1}^{2k+1}$ have a flow of r.

Given the graph $G = (V, A)$, construct an auxiliary graph $H = (N, E)$ to mimic flow along the $2k + 1$ paths. Each node v in N corresponds to a $(2k + 1)$-tuple $(v(1), v(2), ..., v(2k + 1))$, where $v(i) \in V$ for $i = 1, ..., 2k + 1$. The graph H thus contains $|V|^{2k+1}$ nodes. Let s_H be the node in H corresponding to the $(2k + 1)$-tuple $(s, s, ..., s)$, and t_H be the node in H corresponding to the $(2k + 1)$-tuple $(t, t, ..., t)$. In the graph H, the arc directed from node u to node

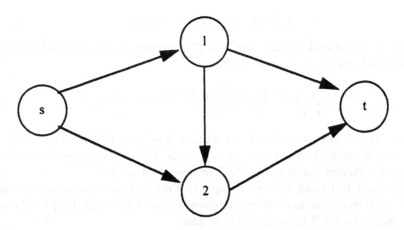

Fig. 2.

v is included if and only if for each $i \in \{1, ..., 2k + 1\}$, either $u(i) = v(i)$, or $(u(i), v(i))$ is an arc in A. Thus a path $P_H = \{(s_H, u_1), (u_1, u_2), ..., (u_r, t_H)\}$ in H defines $2k + 1$ paths $\{P_i\}_{i=1}^{2k+1}$ in G, for

$$P_i = \{(s, u_1(i)), (u_1(i), u_2(i)), ..., (u_r(i), t)\}.$$

Note that our definition allows for $u_j(i) = u_{j+1}(i)$ for some i and j.

Consider the graph in Figure 2. In this case we have $k = 1$. The graph H thus contains 4^3 nodes, each corresponding to a 3-tuple $(u(1), u(2), u(3))$ for $u(i) \in \{1, 2, 3, 4\}, i = 1, 2, 3$. In the graph H there is an arc directed from the node u to node v where u corresponds to the 3-tuple (s,1,2) and v corresponds to the 3-tuple (1,1,t), since $(u(1), v(1)) = (s, 1)$ and $(u(3), v(3)) = (2, t)$ are arcs in G while $u(2) = v(2) = 1$. The path in H corresponding to the 3-tuples (s,s,s), (1,1,s),(2,1,2), (t,t,t), defines three paths in G where

$$P_1 = \{(s, 1), (1, 2), (2, t)\}, \; P_2 = \{(s, 1), (1, t)\}, \; P_3 = \{(s, 2), (2, t)\}.$$

For each pair of arcs $e = (u, v) \in E$ and $a \in A$, define

$$n_1^e(a) = |i \in \{1, ..., k\} : a = (u(i), v(i))|,$$

$$n_2^e(a) = |i \in \{k + 1, ..., 2k + 1\} : a = (u(i), v(i))|.$$

Once again consider the graph G from Figure 2 and the corresponding graph H. Let $e = (u, v) \in E$, where u corresponds to the 3-tuple $(1, 1, s)$ and v corresponds to the 3-tuple $(2, 1, 2)$. For $a = (1, 2) \in A$, we have $n_1^e(a) = 1$ and $n_2^e(a) = 0$, since $(u(i), v(i)) = (1, 2)$ only for $i = 1$.

The arc e in E corresponds to a flow of $C - r$ from $u(i)$ to $v(i)$, $i = 1, ..., k$, and a flow of r from $u(i)$ to $v(i)$, $i = k + 1, ..., 2k + 1$, along the arcs $(u(i), v(i))$, $i = 1, ..., 2k + 1$. If $u(i) = v(i)$, no flow is assumed to have taken place. Define

$$f_a^e = (C - r)n_1^e(a) + rn_2^e(a), \tag{2}$$

for each arc $a \in A$ and $e \in E$. The flow f_a^e corresponds to the total flow along arc a in A defined by arc e in H. Define

$$b_e^p = \sum_{a \in A} p_a f_a^e; b_e^w = \sum_{a \in A} w_a \lceil f_a^e/C \rceil; b_e = b_e^p + b_e^w.$$

b_e^p represents the cost of sending $C - r$ units of flow from $u(i)$ to $v(i)$, $i = 1, ..., k$, and r units from $u(i)$ to $v(i)$, $i = k + 1, ..., 2k + 1$. b_e^w represents the cost of purchasing sufficient capacity for the flow described above.

Once again returning to the example in Figure 2, and considering e to be the arc in H from the node corresponding to the 3-tuple $(1, 1, s)$ to the node corresponding to the 3-tuple $(2, 1, 2)$, we have

$$f_{s1}^e = 0, f_{s2}^e = 7, f_{12}^e = 3, f_{1t}^e = 0, f_{2t}^e = 0.$$

This implies that

$$b_e^p = 2 \times 7 + 0 \times 3 = 14; b_e^w = 0 \times 1 + 0 \times 1 = 0.$$

Consider the shortest path in H from s_H to t_H, using arc costs b_e. We prove that such a shortest path defines the optimal solution to OFOC.

Theorem 3.1 *Given the problem OFOC, let the auxiliary graph H be defined as above. Arc weights b_e are as defined above for $e \in E$. Let P_H^* represent the shortest path in H from s_H to t_H. Let $\{P_i^*\}_{i=1}^{2k+1}$ be the $2k + 1$ paths in G corresponding to P_H^*. Define the flow vector f^*, where*

$$f_a^* = (C - r) \sum_{i=1}^{k} P_i^*(a) + r \sum_{k+1}^{2k+1} P_i^*(a),$$

where $P_i^*(a) = 1$ if $a \in P_i^*$, 0 otherwise. Let $y_a^* = \lceil f_a^*/C \rceil$. The vector (y^*, f^*) is an optimal solution to OFOC.

Once again consider the graph G in Figure 2. In the corresponding graph H, the shortest path from (s, s, s) to (t, t, t) is given by the path corresponding to the node sequence $(s, s, s), (1, 1, s), (2, 1, 2), (t, t, t)$. The length of this path in H is given by $20 + 14 + 34 = 68$. This path corresponds to three paths in G with

$$P_1 = \{(s, 1), (1, 2), (2, t)\}; P_2 = \{(s, 1), (1, t)\}; P_3 = \{(s, 2), (2, t)\}.$$

A flow of $C - r = 3$ is sent along P_1 and a flow of $r = 7$ is sent along each of P_2 and P_3. The total cost of this flow is also 68.

4 The two-facility one-commodity problem

In this section we consider the two-facilty one commodity problem discussed earlier. For the case where flow costs p_a are not all zero, TFOC is clearly NP-hard since it contains OFOC as a special case if we set $w_a^1 = w_a^2$, i.e., it costs the same to buy 1 unit or C units of capacity. We can thus restrict attention to the special case where all flow costs are 0. Note that in this case we may as well assume that d is integer, since if d is fractional the total cost of sending d or $\lceil d \rceil$ units of flow is the same.

TFOC without flow costs is very similar in structure to OFOC with flow costs. In fact all the results from Sections 2 and 3 can be extended to TFOC, with minor modifications. Rather than repeat all the proofs, we simply discuss the minor modifications that can be used to obtain all the results.

Proposition 4.1 *The problem TFOC is NP-hard for the case when all flow costs are 0.*

To obtain an algorithm for TFOC we need structural properties similar to those obtained in Section 3.1. The key result to obtain is one similar to Proposition 3.1, since the rest would then follow as in Section 3. We need to modify some of the definitions in Section 3.1 to obtain such a result. A solution to TFOC is a vector (y_1^*, y_2^*, f^*), where $y_1^*(a)$ corresponds to the number of units of facility 1 purchased (of capacity 1 each) for arc a, $y_2^*(a)$ corresponds to the number of units of facility 2 purchased (of capacity C each) for arc a, and $f^*(a)$ corresponds to the flow on arc a. Given a solution (y_1^*, y_2^*, f^*), define an arc a to be a free arc if $f^*(a) < \lceil f^*(a)/C \rceil C$. Note that this definition is consistent with the definition of free arcs in Section 3.1.

Proposition 4.2 *Let (y_1^*, y_2^*, f^*) be an optimal solution to TFOC with the minimum number of free arcs. All the free arcs defined by (y_1^*, y_2^*, f^*) lie on a path (ignoring direction) from s to t. Free arcs directed forward along this path have a flow from $\{lC+r\}_{l=0}^k$, and those directed backwards have a flow from $\{lC-r\}_{l=1}^k$.*

We can thus prove an equivalent of Proposition 3.2. Define extreme optimal solutions as in Section 3.1.

Proposition 4.3 *Let (y_1^*, y_2^*, f^*) be an extreme optimal solution to TFOC. There exist $2k+1$ paths $\{P_i\}_{i=1}^{2k+1}$, from s to t, such that*

$$f_a^* = (C-r) \sum_{i=1}^{k} P_i^a + r \sum_{i=k+1}^{2k+1} P_i^a,$$

where $P_i^a = 1$ if $a \in P_i$, 0 otherwise.

Given the graph $G = (V, A)$, we can construct an auxiliary graph $H = (N, E)$ exactly as described in Section 3.2. Define f_a^e as in (3). Define b_e^a to be the minimum cost of installing sufficient capacity (using both types of facilities) on arc a to support a flow of f_a^e. Define

$$b_e = \sum_{a \in A} b_e^a.$$

Using a proof identical to that of Theorem 3.1, we can thus prove that the shortest path in H from s_H to t_H, using arc cost b_e, defines the optimal solution to TFOC.

Theorem 4.1 *Given the problem TFOC, let the auxiliary graph H be defined as above. Arc weights b_e are as defined above for $e \in E$. Let P_H^* represent the shortest path in H from s_H to t_H. Let $\{P_i^*\}_{i=1}^{2k+1}$ be the $2k+1$ paths in G corresponding to P_H^*. Define the flow vector f^*, where*

$$f_a^* = (C - r) \sum_{i=1}^{k} P_i^*(a) + r \sum_{k+1}^{2k+1} P_i^*(a),$$

where $P_i^*(a) = 1$ if $a \in P_i^*$, 0 otherwise. Let $y_1^*(a), y_2^*(a)$ be the minimum cost capacity to support a flow of f_a^* on arc a. The vector (y_1^*, y_2^*, f^*) is an optimal solution to OFOC.

From Theorem 4.2 it thus follows that for a fixed $\lfloor d/C \rfloor$, TFOC can be solved in polynomial time if all flow costs are 0.

5 Extended Formulations for OFOC and TFOC

In this section, we use the characterization of extreme optimal solutions from Section 3 to obtain extended formulations for OFOC and TFOC. In each case we show that the LP-relaxation of the extended formulation gives a better lower bound for the integer optimum, compared to the natural formulation, with a family of "cut set" inequalities added. This is valuable because the LP-relaxation of the extended formulation can be solved in polynomial time, while the separation problem for the "cut set" inequalities is hard. In each case, we also characterize objective function coefficients for which the LP-relaxation of the extended formulation gives integer optima.

5.1 Extended Formulation for OFOC

We first consider a natural formulation for OFOC. For each arc (i, j), let f_{ij} be the flow and y_{ij} the batches of capacity installed (each batch provides C units of capacity). OFOC can be formulated using the natural formulation **NFO** (see also Magnanti and Mirchandani(1993)) below:

$$\text{Min} \sum_{(i,j) \in A} w_{ij} y_{ij} + \sum_{(i,j) \in A} p_{ij} f_{ij}$$

s.t.

$$\sum_j f_{ji} - \sum_j f_{ij} = \begin{cases} -kC - r & \text{for } i = s \\ kC + r & \text{for } i = t \\ 0 & \text{otherwise} \end{cases} \tag{3}$$

$$Cy_{ij} - f_{ij} \geq 0 \tag{4}$$

$$y, f \geq 0; y \text{ integer}.$$

Define the polytopes

$$LPO_1 = \{(y, f) \geq 0 | (y, f) \text{ satisfies } (3), (4)\},$$

$$IPO = \{(y, f) \in LPO_1, y \text{ integer}\}.$$

We describe a set of strong valid inequalities for IPO that are similar to, and extend inequalities described by Magnanti and Mirchandani(1993). Given $X \subset V$, define $\delta^+(X)$ ($\delta^-(X)$) to be the arcs in the cut directed out of (into) X. Given $X \subset V, s \in X, t \in V \setminus X$, partition the arcs in $\delta^+(X)$ into the sets B_1 and B_2. For any arc set $D \subset \delta^-(X)$, define the *cut set* inequality

$$\sum_{a \in B_1} f_a + r \sum_{a \in B_2} y_a + (C - r) \sum_{a \in D} y_a - \sum_{a \in D} f_a \geq r\lceil d/C \rceil. \tag{5}$$

For any arc set $S \subset A$, and any vector $x \in R^A$, define $x_S \equiv \sum_{a \in S} x_a$. Given a vector (y, f) satisfying inequalities (5), and a set $S \subseteq A$, define $k_S = \lfloor (f_S - 1)/C \rfloor$ and $r_S = f_S - Ck_S$. We also assume hereafter that $k = \lfloor (d - 1)/C \rfloor$ where $d = kC + r$. Notice that therefore $r > 0$. We now prove that the cut set inequalities are valid for IPO.

Theorem 5.1 *The* cut set *inequalities (5) are valid for IPO.*

Define the polytope

$$LPO_2 = \{(y, f) \in LPO_1 | (y, f) \text{ satisfies } (5)\}.$$

In general, optimizing over LPO_2 is hard because the separation problem for the cut set inequalities is hard.

We now define an extended formulation for OFOC, based on the characterization of extreme optimal solutions in Proposition 3.1. By Proposition 3.1, all free arcs in an extreme optimal solution lie on a single path from s to t. Free arcs directed forward along this path have a flow from $\{lC + r\}_{l=0}^k$, free arcs directed backward on this free path have a flow of $\{lC - r\}_{l=1}^k$. All other arcs have flow that is a multiple of C.

Define variables h_{ij} which takes the value l if the flow on arc (i, j) equals lC, $lC + r$, or $(l + 1)C - r$. In other words, h_{ij} takes the value l if the flow on arc (i, j) is at least lC but less than $(l + 1)C$. The variable $e_{ij}(g_{ij})$ takes the value 1 if the flow on arc (i, j) is $lC + r$ ($lC - r$). The variable y_{ij} is as defined for the naural formulation. OFOC can now be formulated using the extended formulation **EFO** shown below:

$$\textbf{Min} \sum_{(ij)\in A} w_{ij} y_{ij} + \sum_{(ij)\in A} p_{ij}(Ch_{ij} + re_{ij} + (C-r)g_{ij})$$

s.t.

$$\sum_{j}(e_{ji} - g_{ji} - e_{ij} + g_{ij}) = \begin{cases} -1 & \text{for } i = s \\ 1 & \text{for } i = t \\ 0 & \text{otherwise} \end{cases} \tag{6}$$

$$\sum_{j}(h_{ji} + g_{ji} - h_{ij} - g_{ij}) = \begin{cases} -k & \text{for } i = s \\ k & \text{for } i = t \\ 0 & \text{otherwise} \end{cases} \tag{7}$$

$$y_{ij} - e_{ij} - h_{ij} - g_{ij} \geq 0 \tag{8}$$

$$e_{ij}, h_{ij}, g_{ij}, y_{ij} \geq 0, \text{ integer}$$

Define the polytopes,

$$EPO = \{(y, e, g, h) \geq 0 | (y, e, g, h) \text{ satisfies } (6), (7), (8)\},$$

$$EIPO = \{(y, e, g, h) \in EPO | e, g \in \{0, 1\}, h \text{ integer}\}.$$

Let ϕ denote the linear transformation defined by $f_a = re_a + (C - r)g_a + Ch_a$. The next result shows that $IPO = \phi(EIPO)$. It is stated here without proof, since the proof is fairly straightforward.

Theorem 5.2 *Any vector $(y, f) \in IPO$ if and only if there exists a vector $(y, e, g, h) \in EIPO$ where $f_a = re_a + (C - r)g_a + Ch_a$.*

The next result shows that $\phi(EPO) \subseteq LPO_2$, i.e., the linear transformation of the polytope EPO (from the LP-relaxation of the extended formulation) is contained in the polytope LPO_2 (from the LP-relaxation of the natural formulation and all cut set inequalities).

Lemma 5.1 *Given any vector $(y, e, g, h) \in EPO$, the vector $(y, f) \in LPO_2$, where $f_a = re_a + (C - r)g_a + Ch_a$.*

Therefore, EPO provides a LP-relaxation for OFOC that is at least as strong as the one provided by LPO_2 in terms of the lower bound.

Theorem 5.3 *Comparing EPO and LPO_2 we have*

$$min\{\sum_{a\in A} w_a y_a + \sum_{a\in A} p_a(Ch_a + re_a + (C-r)g_a) | (y, e, g, h) \in EPO\} \geq$$

$$min\{\sum_{a\in A} w_a y_a + \sum_{a\in A} p_a f_a | (y, x) \in LPO_2\}.$$

The LP-relaxation of the extended formulation is thus at least as strong as the LP-relaxation of the natural formulation, even after the addition of all cut set inequalities (5).

We now establish conditions on the cost function coefficients, that guarantee integer solutions for OFOC. Let $a(i,j)$ $(b(i,j))$ denote the minimum distance from node i to node j if arc costs are set to $w_{ij} + Cp_{ij}$ $(w_{ij} + rp_{ij})$ and let $P^{\alpha}(i,j)$ $(P^{\beta}(i,j))$ be the shortest such path. The next result gives a condition under which optimizing over EPO results in a integer optimal solution.

Theorem 5.4 *If for every arc* (i,j),
$$a(s,j) - a(s,i) + b(s,i) - b(s,j) \leq w_{ij} + (C-r)p_{ij},$$
then the the optimal solution over EPO (the linear programming relaxation of EFO) is integer.

As a corollary to this result we obtain

Corollary 5.1 *If* $w_{ij}/p_{ij} = q$ *for all arcs* (i,j), *then the optimal solution over EPO is integer.*

5.2 Extended Formulation for TFOC

We first consider a natural formulation for TFOC as in Magnanti and Mirchandani(1993). For each arc (i,j), let f_{ij} be the flow, y^1_{ij} the batches of facility 1 (each batch has 1 unit), and y^2_{ij} the batches of facility 2 (each batch has C units) installed. TFOC can be formulated using the natural formulation NFT shown below:

$$\text{Min} \sum_{(i,j)\in A} w^1_{ij}y^1_{ij} + \sum_{(i,j)\in A} w^2_{ij}y^2_{ij} + \sum_{(i,j)\in A} p_{ij}f_{ij}$$

s.t.

$$\sum_j f_{ji} - \sum_j f_{ij} = \begin{cases} -kC - r & \text{for } i = s \\ kC + r & \text{for } i = t \\ 0 & \text{otherwise .} \end{cases} \tag{9}$$

$$Cy^2_{ij} + y^1_{ij} - f_{ij} \geq 0 \tag{10}$$

$$y^1, y^2, f \geq 0; y^1, y^2 \text{ integer.}$$

Define the polytopes

$$LPT_1 = \{(y^1, y^2, f) \geq 0 | (y^1, y^2, f) \text{ satisfies } (9), (10)\},$$

$$IPT = \{(y_1, y_2, f) \in LPT_1, y^1, y^2 \text{ integer}\}.$$

We describe *cut set* inequalities similar to inequality (5). Given $X \subset V, s \in X, t \in V \setminus X$, let B denote the set of arcs in the cut $\delta^+(X)$. For any arc set $D \subseteq \delta^-(X)$, define the cut set inequality

$$\sum_{a\in B}(y^1_a + ry^2_a) + \sum_{a\in D}(y^1_a + (C-r)y^2_a) - \sum_{a\in D} f_a \geq r\lceil d/C\rceil. \tag{11}$$

The validity of the cut set inequalities (11) can be proved as for the cut set inequalities (5).

Theorem 5.5 *The cut set inequalities (11) are valid for IPT.*

Define the polytope

$$LPT_2 = \{(y^1, y^2, f) \in LPT_1 | (y^1, y^2, f) \text{ satisfies (11)}\}.$$

For the extended formulation, define variables h_{ij}^1 (h_{ij}^2) which take the value l if the flow on arc (i, j) equals lC, $lC + r$, or $(l + 1)C - r$ and the capacity is provided by type 1 (type 2) facilities. The variable $e_{ij}^1 (e_{ij}^2)$ takes the value 1 if the flow on arc (i, j) is $lC + r$ and the capacity for the last r units is provided by type 1 (type 2) facilities. Similarly $g_{ij}^1 (g_{ij}^2)$ takes the value 1 if the flow on arc (i, j) is $lC - r$ and the capacity for the last $C - r$ units of flow is provided by type 1 (type 2) facilities. TFOC can be formulated using the extended formulation **EFT** shown below:

$$\text{Min} \sum_{(ij) \in A} \sum_{u=1}^{2} w_{ij}^u y_{ij}^u$$

s.t.

$$\sum_{j} \sum_{u=1}^{2} (e_{ji}^u - g_{ji}^u - e_{ij}^u + g_{ij}^u) = \begin{cases} -1 & \text{for i = s} \\ 1 & \text{for i = t} \\ 0 & \text{otherwise} \end{cases} \qquad (12)$$

$$\sum_{j} \sum_{u=1}^{2} (h_{ji}^u + g_{ji}^u - h_{ij}^u - g_{ij}^u) = \begin{cases} -k & \text{for i = s} \\ k & \text{for i = t} \\ 0 & \text{otherwise} \end{cases} \qquad (13)$$

$$y_{ij}^1 - re_{ij}^1 - Ch_{ij}^1 - (C - r)g_{ij}^1 \geq 0 \qquad (14)$$

$$y_{ij}^2 - e_{ij}^2 - h_{ij}^2 - g_{ij}^2 \geq 0 \qquad (15)$$

$$e_{ij}^u, h_{ij}^u, g_{ij}^u, y_{ij}^u \geq 0, \text{ integer.}$$

Define the polytopes

$$EPT = \{(y^u, e^u, g^u, h^u; u = 1, 2) \geq 0 | (y^u, e^u, g^u, h^u; u = 1, 2) \text{ satisfies (12)–(15)}\},$$

$$EIPT = \{(y^u, e^u, g^u, h^u; u = 1, 2) \in EPT | e^u, g^u \in \{0, 1\}, h \text{ integer}\}.$$

Let θ denote the linear transformation defined by $f_a = \sum_{u=1}^{2} (Ch_a^u + re_a^u + (C - r)g_a^u)$. The next result (stated without proof) shows that $IPT = \theta(EIPT)$.

Theorem 5.6 *Any vector $(y, f) \in IPT$ if and only if there exists a vector $(y, e, g, h) \in EIPT$ where $f_a = re_a + (C - r)g_a + Ch_a$.*

For any subset of arcs $S \subset A$, and any quantity x_a^u, define $x_S^u \equiv \sum_{a \in S} x_a^u$ for $u = 1, 2$. The next result shows that $\theta(EPT) \subseteq LPT_2$.

Lemma 5.2 *Given any vector $(y^u, e^u, g^u, h^u; u = 1, 2) \in EPT$, the vector $(y^1, y^2, f) \in LPT_2$, where $f_a = \sum_u (re_a^u + (C - r)g_a^u + Ch_a^u)$.*

Therefore, EPT provides a LP-relaxation for TFOC that is at least as strong as the one provided by LPT_2 in terms of the lower bound.

Theorem 5.7 *Comparing EPT and LPT_2 we have*

$$min\{\sum_{a \in A}\sum_{u=1}^{2} w_a^u y_a^u | (y^u, e^y, g^u, h^u; u = 1, 2) \in EPT\} \geq$$

$$min\{\sum_{a \in A}\sum_{u=1}^{2} w_a^u y_a^u | (y^1, y^2, f) \in LPT_2\}.$$

If $w_{ij}^2 \geq C w_{ij}^1$, then we need not consider facility 2 on arc (i, j). Therefore, we assume that $w_{ij}^2 < C w_{ij}^1$. Hence, the minimum cost of sending lC units on any arc (i, j) is always lw_{ij}^2. Let $a(i, j)$ $(b(i, j))$ be the shortest distance from i to j if we set arc costs to w_{kl}^2 (min $\{rw_{kl}^1, w_{kl}^2\}$), and let $w_{ij}(r) = $ min $\{rw_{kl}^1, w_{kl}^2\}$. We now give a sufficient condition under which optimizing over EPT results in an integer solution.

Theorem 5.8 *If for every arc (i, j),*
$a(s, j) - a(s, i) + b(s, i) - b(s, j) \leq$ *min* $\{(C - r)w_{ij}^1, w_{ij}^2\}$
then the optimal solution over EPT is integer.

As a corollary to this result we obtain

Corollary 5.2 *If $w_{ij}^2 / w_{ij}^1 = q$ for all arcs (i, j), then the optimal solution over EPT is integer, with no reverse arcs on the free path.*

Bibliography
1. J. A. Bondy and U. S. R. Murty, *Graph Theory with Applications*, North Holland, Amsterdam (1976).
2. S. Chopra, D. Bienstock, O. Günlük, C.Y. Tsai,"Minimum cost capacity installation for multicommodity network flows," Research Report, Northwestern University, January 1995.
3. M.R. Garey and D.S. Johnson, *Computers and Intractability: A guide to the Theory of NP-Completeness*, W.H. Freeman and Company, New York (1979).
4. J.M.Y. Leung, T.L. Magnanti and R. Vachani, "Facets and algorithms for capacitated lot sizing," *Mathematical Programming*, 45, 331-359.
5. T.L. Magnanti and P. Mirchandani, "Shortest paths, single origin-destination network design and associated polyhedra," *Networks*, Vol. 23, No. 2 (1993) 103-121.
6. T.L. Magnanti, P. Mirchandani, and R. Vachani, "Modeling and solving the two facility capacitated network loading problem," *Operations Research*, Vol. 43, No. 1 (1995) 142-157.
7. Y. Pochet and L.A. Wolsey, "Lot sizing with constant batches: Formulation and valid inequalities," *Mathematics of Operations Research*, 18 (1993) 767-785.

Integer Multicommodity Flow Problems [*]

Cynthia Barnhart[1], Christopher A. Hane[2], Pamela H. Vance[3]

[1] Massachusetts Institute of Technology, Center for Transportation Studies,
Cambridge, MA 02139, USA
[2] CAPS Logistics, Atlanta, Georgia 30334, USA
[3] Auburn University, Industrial and Systems Engineering, Auburn, AL 36849, USA

Abstract. We present a column generation model and solution approach for large integer multicommodity flow problems. We solve the model using branch-and-bound, with bounds provided by linear programs at each node of the branch-and-bound tree. Since the model contains one variable for each origin-destination path, for every commodity, the linear programming relaxation is solved using column generation, i.e., implicit pricing of nonbasic variables to generate new columns or to prove LP optimality. Methods for speeding up the solution of the linear program are presented. Also, we devise new branching rules that allow columns to be generated efficiently at each node of the branch-and-bound tree. Computational results are presented for a set of test problems arising from a transportation application.

1 Introduction

Linear multicommodity flow problems (MCF) are linear programs (LP's) that can be characterized by a set of commodities and an underlying network. The objective is to flow the commodities through the network at minimum cost without exceeding arc capacities. A comprehensive survey of linear multicommodity flow models and solution procedures was presented in Ahuja et al. (1993).

In this paper, we consider the integer multicommodity flow (IMCF) problem, a constrained version of the linear multicommodity flow problem in which flow of a commodity may use only one path from origin to destination. IMCF problems are prevalent in a number of application contexts, including transportation, communication and production. Often in transportation, service considerations dictate that shipments be routed along single paths. Similarly, messages may not be split among paths in a communication network.

We focus our attention on the large-scale problems arising in practice and present a column generation model and a branch-and-bound solution approach involving column generation. In column generation, sets of columns are left out of the LP because there are too many columns to handle efficiently and most of them will have their associated variable equal to zero in an optimal solution anyway. Then to check the optimality of an LP solution, a subproblem, called

[*] This research has been supported by the following grants and contracts: NSF DDM-9058074, NSF DMI-9502502

the pricing problem, which is a separation problem for the dual LP, is solved to try to identify columns to enter the basis. If such columns are found, the LP is reoptimized.

The ability to solve large MCF LP's allows us to consider the solution of large IMCF problems. We design, implement and test a new branch-and-price solution approach. Branch-and-price, which is a generalization of branch-and-bound with LP relaxations, allows column generation to be applied throughout the branch-and-bound tree. (A survey of branch-and-price can be found in Barnhart, et al. (1995).) Branching occurs when no columns price out to enter the basis and the LP solution does not satisfy the integrality conditions.

Contributions

In our view, the contributions of this paper include:

1. Presentation of a modeling and solution framework for large-scale integer multicommodity flow problems. Large IMCF problems are embedded within routing and scheduling applications in transportation, communication and production. These IMCF applications tend to very large-scale and therefore, are best (and maybe only) solved using specialized decomposition approaches.
2. Presentation of methods to speed up the solution of MCF LP's using column generation approaches. Without the ability to solve these LP's quickly, solving IMCF would be impossible.
3. Presentation of an advanced, state-of-the-art solution approach for large-scale IMCF programs that allows column generation within the branch-and-bound solution process. Detailed descriptions are provided of how columns can be generated efficiently within the tree.
4. We demonstrate that when column generation is performed only at the root node and a standard branch-and-bound approach is used, it is not possible to identify feasible IP solutions for large problems. However, our branch-and-price approach finds good feasible solutions in under 1 hour of CPU time on a workstation class computer.

Outline

The remainder of the paper is organized as follows. In Section 1, we present two formulations for the IMCF problem. In Section 2, we describe the solution approach, including speed-up techniques, for the LP relaxation of IMCF. In Section 3, we detail the approach for obtaining IMCF solutions using a branch-and-price solution approach. Branching rules are introduced and details of how to generate columns satisfying the branching decisions are provided. Branch selection, node selection, and branch-and-bound termination are also discussed. Computational results evaluating the strength of of our branching scheme and our branch-and-price procedure are presented in Section 4.

2   IMCF Problem Formulation

We consider two different formulations for the IMCF problem: the *node-arc* or *conventional* formulation and the *path* or *column generation* formulation. We use the conventional formulation to design our branching strategies for the IMCF problem and derive our cutting planes. The column generation formulation is used to solve the IMCF LP relaxation.

The Integer Multicommodity Flow formulation, denoted $IMCF$, is defined over the network G comprised of node set N and arc set A. IMCF contains binary decision variables x, where x_{ij}^k equals 1 if the entire quantity (denoted q^k) of commodity k is assigned to arc ij, and equals 0 otherwise. The cost of assigning commodity k in its entirety to arc ij equals q^k times the unit flow cost for arc ij, denoted c_{ij}^k. Arc ij has capacity d_{ij}, for all $ij \in A$. Node i has supply of commodity k, denoted b_i^k, equal to 1 if i is the origin node for k, equal to -1 if i is the destination node for k, and equal to 0 otherwise.

The conventional or node-arc $IMCF$ formulation is:

$$min \sum_{k \in K} \sum_{ij \in A} c_{ij}^k q^k x_{ij}^k \tag{1}$$

$$\sum_{k \in K} q^k x_{ij}^k \leq d_{ij}, \ \forall ij \in A \tag{2}$$

$$\sum_{ij \in A} x_{ij}^k - \sum_{ji \in A} x_{ji}^k = b_i^k, \ \forall i \in N, \forall k \in K \tag{3}$$

$$x_{ij}^k \in \{0, 1\}, \ \forall ij \in A, \forall k \in K \tag{4}$$

Note that without restricting generality of the problem, we model the arc flow variables x as binary variables. To do this, we scale the demand for each commodity to 1 and accordingly adjust the coefficients in the objective function (1) and in constraints (2).

To contrast, the path-based or column generation $IMCF$ formulation has fewer constraints, and far more variables. Again, the underlying network G is comprised of node set N and arc set A, with q^k representing the quantity of commodity k. $P(k)$ represents the set of all origin-destination paths in G for k, for all $k \in K$. In the column generation model, the binary decision variables are denoted y_p^k, where y_p^k equals 1 if all q^k units of commodity k are assigned to path $p \in P(k)$, and equals 0 otherwise. The cost of assigning commodity k in its entirety to path p equals q^k times the unit flow cost for path p, denoted c_p^k. As before, arc ij has capacity d_{ij}, for all $ij \in A$. Finally, δ_{ij}^p is equal to 1 if arc ij is contained in path $p \in P(k)$, for all $k \in K$; and is equal to 0 otherwise.

The path or column generation $IMCF$ formulation is then:

$$min \sum_{k \in K} \sum_{p \in P(k)} c_p^k q^k y_p^k \tag{5}$$

$$\sum_{k \in K} \sum_{p \in P(k)} q^k y_p^k \delta_{ij}^p \leq d_{ij}, \ \forall ij \in A \tag{6}$$

$$\sum_{p \in P(k)} y_p^k = 1, \ \forall k \in K \tag{7}$$

$$y_p^k \in \{0, 1\}, \ \forall p \in P(k), \forall k \in K \tag{8}$$

3 LP Solution

For large-scale transportation, communication and production applications, the LP relaxation of the conventional IMCF formulation contains a large number of constraints (equal to the number of arcs plus the product of the number of nodes and commodities), and a large number of variables (equal to the product of the number of arcs and commodities.) The column generation LP relaxation, however, contains a moderate number of constraints (one for each commodity and one for each arc) and a huge number of variables (one for each path for each commodity.) Without decomposition, these LP relaxations may require excessive memory and/or runtimes to solve.

Multicommodity flow problems have the Dantzig-Wolfe master program structure, and so, can be solved using specialized solution procedures such as the generalized upper bounding procedure of Dantzig and Van Slyke (1967), or the partitioning procedures of Rosen (1964) or Barnhart et al. (1995). All of these procedures exploit the block-diagonal problem structure and perform all steps of the simplex method on a reduced working basis of dimension m.

Instead, we choose to use the column generation solution approach for two major reasons:

1. Column generation solution approaches for large-scale LP's have been widely and successfully applied. Discussion of the procedure first appeared in Dantzig and Wolfe (1960), with Appelgren (1969) providing one of the first applications of column generation methods (to ship scheduling).

2. Although implementing branch-and-price algorithms (or branch-and-cut algorithms) is still a nontrivial activity, the availability of flexible linear and integer programming systems has made it a less formidable task than it would have been five years ago. Modern simplex codes, such as CPLEX (CPLEX Optimization, 1990) and OSL (IBM Corporation, 1990) not only permit column generation while solving an LP but also allow the embedding of column generation LP solving into a general branch-and-bound structure for solving MIPs. The use of MINTO (Nemhauser, Savelsbergh, and Sigismondi 1994) may reduce the implementation efforts even further. MINTO (Mixed INTeger Optimizer) is a general purpose mixed integer optimizer that can be customized through the incorporation of application functions.

The general idea of column generation is that optimal solutions to large LP's can be obtained without explicitly including all columns (i.e., variables) in the constraint matrix (called the *Master Problem* or *MP*). In fact, only a very small subset of all columns will be in an optimal solution *and* all other (non-basic) columns can be ignored. In a minimization problem, this implies that all columns with positive reduced cost can be ignored. The multicommodity flow column generation strategy, then, is:

Step 0: RMP Construction. Include a subset of columns in a *restricted* MP, called the *Restricted Master Problem, or RMP*;

Step 1: RMP Solution. Solve the RMP;

Step 2: Pricing Problem Solution. Use the dual variables obtained in solving the RMP to solve the pricing problem. The pricing problem either identifies one or more columns with negative reduced cost (i.e., columns that *price out*) or determines that no such column exists.

Step 3: Optimality Test. If one or more columns price out, add the columns (or a subset of them) to the RMP and return to Step 1; otherwise stop, the MP is solved.

For any RMP in Step 1, let $-\pi_{ij}$ represent the non-negative dual variables associated with constraints (6) and σ^k represent the unrestricted dual variables associated with constraints (7). Since c_p^k can be represented as $\sum_{ij \in A} c_{ij}^k \delta_{ij}^p$, the reduced cost of column p for commodity k, denoted $\bar{c}_p^k$, is:

$$\bar{c}_p^k = \sum_{ij \in A} q^k(c_{ij}^k + \pi_{ij})\delta_{ij}^p - \sigma^k, \ \forall p \in P(k), \forall k \in K; \tag{9}$$

For each RMP solution generated in Step 1, the pricing problem in Step 2 can be solved efficiently. Columns that price out can be identified by solving one shortest path problem for each commodity $k \in K$ over a network with arc costs equal to $c_{ij}^k + \pi_{ij}$, for each $ij \in A$. Denote the cost of the *shortest* path $p*$ for any commodity k as c_{p*}^k. Then, if for all $k \in K$,

$$c_{p*}^k q^k - \sigma^k \geq 0, \tag{10}$$

the MP is solved. Otherwise, the MP is not solved and, for each $k \in K$ with

$$c_{p*}^k q^k - \sigma^k < 0, \tag{11}$$

path $p* \in P(k)$ is added to the RMP in Step 3.

3.1 LP Computational Experience

We implemented the column generation algorithm above and evaluated its performance in solving a number of large randomly generated test instances. The problems in the set all had 301 nodes, 497 arcs, and 1320 commodities. The arc costs and capacities were randomly generated. Random problem instances are used for the LP comparisons because the real problem instances we had access to were too small to effectively illustrate the differences between the LP algorithms.

We solve the pricing problems using the Floyd-Warshall shortest path algorithm. With each call to this algorithm, we find the shortest paths from one origin to all nodes in the network. And so, with one execution of the shortest path procedure, we find shortest paths for all commodities originating at the same node. In Step 3, then, we add as many as one column (with negative reduced cost) for each commodity.

Our computational results, including the total time to solve the LP relaxation, the number of iterations (i.e., repetitions of Steps 1-3 above), and the number of columns generated, are reported for these test problems in Table 2.

Table 1. Computational Results: At Most One Path Added per Commodity

problem	iterations	columns generated	total CPU time (secs)
1	3747	9125	240
2	3572	9414	246
3	3772	10119	268
4	3663	10101	289
5	10128	10624	325
6	8509	27041	1289
7	9625	29339	1332
8	7135	22407	842
9	9500	30132	1369
10	7498	23571	833

LP Solution Speed-up. We then considered a modification of the algorithm in which several negative reduced cost paths are added to the LP simultaneously for each commodity. We build upon the *keypath* concept of the Dantzig-Van Slyke, Rosen, and Barnhart procedures above.

To illustrate, consider a basic feasible solution to the multi-commodity flow LP relaxation. Select one basic path, denoted p_k^*, for each commodity k to serve as a *keypath*. Let s^{p_k} denote the symmetric difference between the key path and path p_k for commodity k. The symmetric difference of two paths that share a common origin and destination can be represented as:

+1 for each arc in p_k and not in p_k^*;
-1 for each arc in p_k^* and not in p_k; and
0 for each arc in both or neither p_k and p_k^*.

Then, for any path p_k for any k, we can think of s^{p_k} as the set of the cycles formed between the key path and that path for k. (Additional details are provided in Barnhart et al. (1995).) Let $(c_1, c_2, ..., c_n)$ represent the set of disjoint cycles in s^{p_k}. Observe that the symmetric difference of c_1 and p_k^* is some path, denoted p_1, for commodity k. And in fact, n different paths for k are constructed by taking the symmetric difference of c_j, $j = 1, 2, ..., n$ and p_k^*. We refer to these paths that differ from the keypath by a single cycle as a *simple paths*.

We use the simple paths concept in Step 3 when columns are added to the RMP. Rather than adding only one shortest negative-reduced cost path for a commodity k, we add *all* simple paths for k that can be constructed by considering the symmetric difference of its keypath and the shortest path. The rationale is that since every multicommodity flow solution can be represented as the sum of flow on simple paths, only simple paths need to be added to the constraint matrix. Further, including only simple paths allows *all* multicommodity flow solutions to be represented with far fewer columns.

Table 2. Computational Results: All Simple Paths Added for Each Commodity

problem	iterations	columns added	total CPU time (secs)
1	2455	8855	162
2	2690	10519	199
3	2694	10617	224
4	2511	10496	218
5	2706	11179	234
6	4391	25183	662
7	4208	23880	607
8	3237	17587	398
9	4191	20472	501
10	3633	21926	420

The computational results using *simple path* column generation are shown in Table 3. Compared to the results using a traditional column generation strategy, we were able to reduce significantly (by an average of 40%) the total solution time, primarily because the number of iterations of the column generation algorithm were reduced by a factor of 45%. It is also interesting to note that we do not increase the total number of columns generated. In some cases this quantity actually decreases. These results seem to suggest that the sets of arcs on which flow is assigned in an optimal solution are identified early in the column generation process. The difficulty with the traditional column generation process is that each arc in the generated column must have the same amount of flow assigned to it, unless other columns containing these arcs already exist in the RMP. By adding simple paths, we provide more flexibility by allowing varying amounts of flow to be added to *subsets* of arcs.

4 IP Solution

We use a branch-and-bound approach to solve the IMCF problems, with bounds provided by solving a LP relaxation, called the *subproblem*, at each node of the branch-and-bound tree. Since the multicommodity flow LP's are solved using column generation, our branch-and-bound procedure must allow columns to be generated at each node of the tree. This approach is referred to as *branch-and-price*. For general expositions of branch-and-price methodology see Barnhart et al. (1995), Vanderbeck and Wolsey (1994), and Desrosiers et al. (1994). In branch-and-price, branching occurs when no columns price out to enter the basis and the LP solution does not satisfy the integrality conditions. Branch-and-price, which is a generalization of branch-and-bound with LP relaxations, allows column generation to be applied throughout the branch-and-bound tree.

The key to developing a branch-and-price procedure is identifying a branching rule that eliminates the current fractional solution without compromising the tractability of the pricing problem. Barnhart et al. (1995) develop branching rules for a number of different master problem structures. They also survey specialized algorithms which have appeared in the literature for a broad range of applications.

Parker and Ryan (1994) present a branch-and-price algorithm for the bandwidth packing problem which is closely related to IMCF. The bandwidth packing problem is a version of IMCF where the objective is to choose which of a set of commodities to send in order to maximize revenue. They use a path-based formulation. Their branching scheme selects a fractional path and creates a number of new subproblems equal to the length of the path plus one. On one branch, the path is fixed into the solution and on each other branch one of the arcs on the path is forbidden. They report the solution of problems with as many as 93 commodities on networks with up to 192 nodes and 212 arcs.

4.1 Branching

Applying a standard branch-and-bound procedure to the restricted master problem with its existing columns will not guarantee an optimal (or feasible) solution. After the branching decision modifies RMP, it may be the case that there exists a column for MP that prices out favorably, but is not present in RMP. Therefore, to find an optimal solution we must maintain the ability to solve the pricing problem after branching. The importance of generating columns after the initial LP has been solved is demonstrated for airline crew scheduling applications in Vance et al. (1994). They were unable to find feasible IP solutions using just the columns generated to solve the initial LP relaxation. They developed a branch-and-price approach for crew scheduling problems in which they generaed additional columns whenever the LP bound at a node exceeded a preset IP target objective value.

The difficulty in incorporating column generation with branch-and-bound is that conventional integer programming branching on variables may not be effective because fixing variables can destroy the structure of the pricing problem. To illustrate, consider branching based on variable dichotomy in which one branch forces commodity k to be assigned to path p, i.e., $y_p^k = 1$, and the other branch does not allow commodity k to use path p, i.e., $y_p^k = 0$. The first branch is easy to enforce since no additional paths need to be generated once k is assigned to path p. The latter branch, however, can not be enforced if the pricing problem is solved as a shortest path problem. There is no guarantee that the solution to the shortest path problem is not path p. In fact, it is likely that the shortest path for k is indeed path p. As a result, to enforce a branching decision, the pricing problem solution must be achieved using a *next shortest path procedure*. In general, for a subproblem, involving a set of a branching decisions, the pricing problem solution must be achieved using a kth *shortest path procedure*.

For the multicommodity flow application, our objective is to ensure that the pricing problem for the LP with the branching decisions included can be solved efficiently with a shortest path procedure. That is, our objective is to design a branching rule that does not destroy the structure of the pricing problem. In general, this can be achieved by basing our branching rules on variables in the original formulation, and not on variables in the column generation formulation (Barnhart, et al. (1995), Desrosiers, et al. (1994)). This means that our branching rules should be based on the arc flow variables x_{ij}^k.

Consider then, branching based on variable dichotomy in the original variables. On one branch, we would force flow of commodity k to use arc ij, i.e., $x_{ij}^k = 1$ and on the other branch, we wouldn't allow commodity k to use arc ij, i.e., $x_{ij}^k = 0$. This time, the second branch is easy to enforce in the pricing problem by setting the cost of arc ij for k to a very large value. Enforcing the first branching decision, however, destroys the structure of the pricing problem. While a shortest path containing an arc ij can be found by solving two shortest paths problems, one from node j and one from the origin node of k to node i, it is not possible to find efficiently the shortest path containing a *set* of arcs, as required in subproblems at depths of two or more in the tree.

We propose a new branching strategy that:

1. Is based on the arc flow variables in the original problem formulation; and
2. Is compatible with the pricing problem solution procedure, that is, can be enforced without destroying the structure of the pricing problem.

We derive our branching rule by observing that if commodity k is assigned to more than one path, say for the purposes of this discussion, to two paths, then the two paths differ by at least one arc and further, that the two paths have at least two nodes in common (i.e., the origin and destination nodes are contained in both paths.) We define the first node at which the two paths split as the *divergence node*. Given any two distinct paths $p1$ and $p2$ for k, we can find their divergence node by tracing each path, beginning with the origin node of k, one arc at a time until two different arcs, called $a1$ and $a2$ are identified for each path. The from node of these arcs is the divergence node, denoted d. We denote the set of arcs originating at d as $A(d)$ and let $A(d, a1)$ and $A(d, a2)$ represent some partition of $A(d)$ such that the subset $A(d, a1)$ contains $a1$ and the subset $A(d, a2)$ contains $a2$. We branch creating two subproblems. For the first we require

$$\sum_{a \in A(d, a1)} y_a^k = 0$$

and for the second we require

$$\sum_{a \in A(d, a2)} y_a^k = 0.$$

On the first branch, we do not allow k to use any of the arcs in $A(d, a1)$ and similarly, on the second branch, we do not allow k to use any of the arcs in $A(d, a2)$. Note that these decisions do *not* require that k use *any* of the arcs in $A(d)$, that is, a path for k *not* containing node d is feasible for both of the subproblems.

The resulting division of the problem is valid since:

1. If the LP solution is fractional, we can always find a violated branch, and this branch will eliminate that fractional solution;
2. There are a finite number of branches because there are a finite number of arcs.

A major benefit of our branching rule is that it more evenly divides the problem since we branch on forbidding a *set* of arcs, rather than a *single* arc. Forbidding a set of arcs may achieve faster convergence than forbidding a single arc since the exclusion of a single arc may not have much impact. Note that forbidding a single arc is a special case of our strategy where $|A(d, a1)| = 1$ or $|A(d, a2)| = 1$.

4.2   Subproblem Solution

At each node of the branch-and-bound tree, a restricted multicommodity flow LP, called a subproblem, must be solved. Since the subproblem solution must satisfy the set of branching decisions made along its predecessor path in the tree, it is necessary to restrict the column generation algorithm so that variables violating these rules are not generated in solving the pricing problem. The challenge is to ensure this without increasing the complexity of the pricing problem solution algorithm. The achieve this, observe that *every* branch forbids the assignment of flow of some commodity to one or more arcs. That means, at *any* node in the tree, it is possible to satisfy the branching decisions by restricting flow of possibly several commodities, where the flow restriction for a single commodity is to forbid use of a (possibly large) set of arcs. By setting the commodity's cost on each forbidden arc to a very high value, the pricing problem can still be solved using a shortest path algorithm. As long as a feasible solution exists for that commodity, the shortest path generated will not violate the branching decisions. Then, all of the paths generated for a subproblem will satisfy all of the imposed restrictions.

Branch Selection. Given a fractional LP solution, we select the next branch as follows:

1. Among the commodities whose flow is split, identify the commodity k with the greatest flow, denoted q^k.
2. Identify the two paths p and p' with the greatest fractions y_p^k and $y_{p'}^k$ of the flow of commodity k. Without loss of generality, let path p be shorter than p'.
3. Locate the divergence node d on path p for commodity k. Let arcs $a1$ and $a2$ be incident to d and in paths p and p', respectively.
4. By dividing the set of arcs incident to node d, construct set $A(d, a1)$ containing arc $a1$ and set $A(d, a2)$ with arc $a2$. Let the size of the two sets be roughly equal.
5. Create two new nodes, one where the arcs in $A(d, a1)$ are forbidden for commodity k and one where the arcs in $A(d, a2)$ are forbidden for commodity k.

Node Selection. A depth-first search of the tree is used throughout the algorithm. We choose to search the side of the tree where the shorter path p is still allowed (i.e., we choose the side where the arcs in $A(d, a2)$ are forbidden). In many integer programming algorithms, the nodes are selected in the order of the best LP bound once a feasible solution has been found. We did not choose to switch to best bound in this case because the LP bounds for the different nodes were very close in value. We chose to stay with depth first since additional feasible solutions were more likely to be found deep in the tree.

Branch-and-Price Termination. The branch-and-price solution procedure is terminated when either a provably optimal integer solution is found or the run time exceeds one hour on a workstation class computer.

4.3   Computational Results

We ran several computational trials on a set of ten test problems arising from commercial transportation applications. The characteristics of the test problems are given in Table 3. In problems 3 – 10 the underlying network and set of commodities were identical but the demands for the commodities were scaled differently to create different problem instances.

First to measure the efficacy of our branching rule, we compared a branch-and-bound algorithm using our branching rule to a standard branch-and-bound procedure. Specifically, the standard algorithm was the default branching strategy used by MINTO. In both algorithms, columns were generated to solve the LP at the root node only and branch-and-bound was applied to the resulting IP. The results for branch-and-bound with our branching rule are given in Table 4. The table displays the number of branch-and-bound nodes searched, the LP-IP gap, and the CPU time in seconds on an IBM RS6000/590 using MINTO 2.1 and

Table 3. Computational Results: Problem Characteristics

problem	nodes	arcs	commodities
1	50	97	15
2	91	203	18
3 – 10	50	130	585

CPLEX 3.0. Standard branch-and-bound was unable to find a feasible solution to problems 3 – 10 in the one hour allotted. It required 77 nodes and 0.96 seconds to prove the optimality of the solution for problem 1 - our branching required 12 nodes and 0.43 seconds, and it needed 56,728 nodes and 102.8 seconds to show that problem 2 had no feasible solution - our algorithm required 4264 nodes and 10.2 seconds.

Table 4. Computational Results: Branch-and-Bound Algorithm

problem	nodes	best IP gap	total CPU time (secs)
1	12	0.4 %	0.43
2	12307	infeas.	91.93
3	139869	0.13 %	3600
4	138979	0.5 %	3600
5	126955	1.5 %	3600
6	128489	2.7 %	3600
7	121374	1.5%	3600
8	102360	1.7%	3600
9	96483	4.8 %	3600
10	94742	11.5 %	3600

Table 5 gives computational results for our branch-and-price algorithm. Again the number of nodes searched, LP-IP gap, and CPU time on an IBM RS6000/550 are given. For problem 1, we were able to prove the optimality of the integer solution. It turns out that the optimal solution is the same one identified by the branch-and-bound algorithm, i.e., there was an optimal solution using only columns generated at the root node. Of course this is not true in general. Problem 2 illustrates this point. The branch-and-bound algorithm proved that there was no feasible solution among the columns generated at the root node, but using branch-and-price we were able to find a feasible IP solution within 18% of the LP bound. However, we were unable to prove optimality of this solution within the time limit. For problems 3 – 9 we were able to find good feasible

IP solutions. However, it is interesting to note that for these examples, branch-and-bound was able to find better feasible solutions in the time allowed. This is partly due to the computational demands of the branch-and-price algorithm which requires a great deal of computational effort at each node and is therefore able to search many fewer nodes than the branch-and-bound approach.

Table 5. Computational Results: Branch-and-Price Algorithm

problem	nodes	best IP gap	total CPU time (secs)
1	12	optimal	0.71
2	9575	18.0%	3600
3	10396	0.12%	3600
4	7755	0.6%	3600
5	8525	3.1%	3600
6	7283	2.7%	3600
7	5970	7.6%	3600
8	5202	6.3%	3600
9	4988	36.9 %	3600
10	3076	–	3600

While we were able to find good integer solutions to many of the problems with either our customized branch-and-bound or branch-and-price, proving optimality was difficult. We observed that when we disallowed one commodity from a subset of arcs, the values of the LP solutions for the two subproblems showed little or no change in objective function value. We also found that while the split commodities were changing, the same arcs were showing up repeatedly in the branching decisions. An explanation for this can be seen by examining two subpaths s_1 and s_2, both beginning with some node o and ending with some node d. It is possible that both s_1 and s_2 are contained in origin-destination paths for more than one, and maybe several, commodities. Denote this set of commodities as K'. Assume without loss of generality that s_1 has cost not greater than that of s_2. If in an LP solution, one or more arcs in s_1 is saturated, then it is possible that some of the commodities in K' are assigned to s_2. And in this scenario, it is likely that one of the commodities, called it $k*$, will be assigned to both s_1 and s_2. When the branching decision forces $k*$ off subpath s_1 (s_2), the result is a solution with the same total amount of flow assigned to subpaths s_1 and s_2, with the only difference being that some other commodity $k' \in (K' \setminus \{k*\})$, has its flow split between the two subpaths. As long as arc cost is not differentiated by commodity, the costs of the solutions before and after branching will be the same. This ineffectiveness of the branching strategy results from what is referred to as *problem symmetry*.

The next generation of algorithms for IMCF will have to deal with this problem symmetry if they are to prove optimality. We have demonstrated two algorithms, branch-and-bound with our branching strategy and branch-and-price, which are capable of finding good solutions to large instances of IMCF. In addition branch-and-price is capable of proving optimality for some small instances. One possible avenue to combat this symmetry is to use valid inequalities to strengthen the LP relaxation at each node in the tree. A subsequent paper will consider the use of cutting planes in column generation algorithms for the IMCF.

References

1. R.K. Ahuja, T.L. Magnanti, and J.B. Orlin (1993). *Network Flows: Theory, Algorithms, and Applications.* Prentice Hall, Englewood Cliffs, NJ.
2. L.H. Appelgren (1969). A column generation algorithm for a ship scheduling problem. *Transportation Science 3*, 53-68.
3. C. Barnhart, C.A. Hane, E.L. Johnson, and G. Sigismondi (1991). *An Alternative Formulation and Solution Strategy for Multi-Commodity Network Flow Problems.* Report COC-9102, Georgia Institute of Technology, Atlanta, Georgia.
4. C. Barnhart, E.L. Johnson, G.L. Nemhauser, M.W.P. Savelsbergh, and P.H. Vance (1995). *Branch-and-Price: Column Generation for Solving Huge Integer Programs.* Report COC-9502, Georgia Institute of Technology, Atlanta, Georgia.
5. CPLEX Optimization, Inc. (1990). *Using the CPLEX^{TM} Linear Optimizer.*
6. G.B. Dantzig and R.M. Van Slyke (1967). Generalized Upper Bounding Techniques. *Journal Computer System Sci. 1*, 213-226.
7. G.B. Dantzig and P. Wolfe (1960) Decomposition Principle for Linear Programs. *Operations Research* 8, 108-111.
8. J. Desrósiers, Y. Dumas, M.M. Solomon, and F. Soumis (1994). Time constrained routing and scheduling. M.E. BALL, T.L MAGNANTI, C. MONMA, AND G.L. NEMHAUSER (eds.). *Handbooks in Operations Research and Management Science, Volume on Networks*, to appear.
9. IBM Corporation (1990). *Optimization Subroutine Library, Guide and Reference.*
10. E.L. Johnson (1989). Modeling and Strong Linear Programs for Mixed Integer Programming. S.W. WALLACE (ed.) *Algorithms and Model Formulations in Mathematical Programming.* NATO ASI Series 51, 1-41.
11. G.L. Nemhauser and L.A. Wolsey (1988). *Integer and Combinatorial Optimization.* Wiley, Chichester.
12. G.L. Nemhauser, M.W.P. Savelsbergh, and G.C. Sigismondi (1994). MINTO, a Mixed INTeger Optimizer. *Operations Research Letters* 15, 47-58.
13. M. Parker and J. Ryan (1994). A column generation algorithm for bandwidth packing. *Telecommunications Systems*, to appear.
14. J.B. Rosen (1964) Primal Partition Programming for Block Diagonal Matrices. *Numerische Mathematik 6*, 250-260.
15. D.M. Ryan and B.A.Foster (1981). An integer programming approach to scheduling. A. WREN (ed.) *Computer Scheduling of Public Transport Urban Passenger Vehicle and Crew Scheduling*, North-Holland, Amsterdam, 269-280.
16. P.H. Vance, C. Barnhart, E.L. Johnson, G.L. Nemhauser, D. Mahidara, A. Krishna, and R. Rebello (1994). *Exceptions in Crew Planning.* ORSA/TIMS Detroit, Michigan.
17. F. Vanderbeck and L.A. Wolsey (1994). *An Exact Algorithm for IP Column Generation.* CORE Discussion Papaer, Universite Catholique de Louvain, Belgium.

A Heuristic Algorithm
for the Set Covering Problem

Alberto Caprara[1], Matteo Fischetti[2] and Paolo Toth[1]

[1] DEIS, University of Bologna,
Viale Risorgimento 2, 40136, Bologna, Italy
e-mail: {acaprara,ptoth}@deis.unibo.it
[2] DMI, University of Udine, Italy
e-mail: fisch@dei.unipd.it

Abstract. We present a Lagrangian-based heuristic for the well-known *Set Covering Problem* (SCP). The main characteristics of the algorithm we propose are (1) a dynamic pricing scheme for the variables, akin to that used for solving large-scale LP's, to be coupled with subgradient optimization and greedy algorithms, and (2) the systematic use of column fixing to obtain improved solutions. Moreover, we propose a number of improvements on the standard way of defining the step-size and the ascent direction within the subgradient optimization procedure, and the scores within the greedy algorithms. Finally, an effective refining procedure is proposed. Extensive computational results show the effectiveness of the approach.

1 Introduction

The *Set Covering Problem* (SCP) is a main model for several important applications, including *Crew Scheduling* (CS), where a given set of *trips* has to be covered by a minimum-cost set of *pairings*, a pairing being a sequence of trips that can be performed by a single crew. A widely-used approach to CS works as follows. First, a very large number of pairings are generated. Then an SCP is solved, having as row set the trips to be covered, and as column set the pairings generated. In railway applications, very large scale SCP instances typically arise, involving thousands of rows and millions of columns. The classical methods proposed for SCP meet with considerable difficulties in tackling these instances, concerning both the computing time and the quality of the solutions found. On the other hand, obtaining high-quality solutions can result in considerable savings. For this reason, in 1994 the Italian railway company, *Ferrovie dello Stato SpA*, jointly with the Italian Operational Research Society, *AIRO*, decided to organize a competition called FASTER (Ferrovie Airo Set covering TendER) among the departments of Italian universities, possibly including foreign researchers. Some well-known researchers from all over the world took part in the competition. The code described in this paper won the first prize, and obtained the best solution values on all FASTER instances. After the competition, we tested our code on a wide set of SCP instances from the literature, with remarkably good results: in 92 out of the 94 instances in our test bed we found, within short computing time, the optimal (or the best known) solution. More-over, among the 27 instances for which the optimum is not known, in 6 cases our solution is better than any other solution found by previous techniques.

The main characteristics of the algorithm we propose are (1) a dynamic pricing scheme for the variables, akin to that used for solving large-scale LP's, to be coupled with subgradient optimization and greedy algorithms, and (2) the systematic use of column fixing to obtain improved solutions. Moreover, we propose a number of improvements on the standard way of defining the step-size and the ascent direction within the subgradient optimization procedure, and the scores within the greedy algorithms. Finally, an effective refining procedure is proposed.

2 General Framework

SCP can be formally defined as follows. Let $A = (a_{ij})$ be a 0-1 $m \times n$ matrix, and $c = (c_j)$ be an n-dimensional integer vector. Let $M := \{1, \ldots, m\}$ and $N := \{1, \ldots, n\}$. The value c_j $(j \in N)$ represents the cost of column j, and we assume without loss of generality $c_j > 0$ for $j \in N$. A mathematical model for SCP is then

$$v(\text{SCP}) := \min \left\{ cx : Ax \geq 1, x \in \{0, 1\}^N \right\} \tag{1}$$

For notational convenience, for each row $i \in M$ let $J_i := \{j \in N : a_{ij} = 1\}$, and for each column $j \in N$ let $I_j := \{i \in M : a_{ij} = 1\}$.

It is well known that SCP is NP-hard in the strong sense. The most effec-tive heuristic approaches to SCP are those based on *Lagrangian relaxation* with *subgradient optimization*, following the seminal work by Balas and Ho [2], and then the improvements by Beasley [3], Fisher and Kedia [8], Balas and Carrera [1], and Ceria, Nobili and Sassano [6]. Lorena and Lopes [10] propose an analo-gous approach based on *surrogate relaxation*. Wedelin [11] proposes an algorithm based on Lagrangian relaxation with coordinate search. Recently, Beasley and Chu [5] proposed an effective genetic algorithm.

Our heuristic scheme is based on dual information associated with the widely-used Lagrangian relaxation of model (1). For every vector $u \in R_+^m$ of Lagrangian multipliers associated with the SCP constraints, the Lagrangian subproblem reads:

$$L(u) := \min \left\{ c(u)^T x + \sum_{i \in M} u_i : x \in \{0, 1\}^N \right\}, \tag{2}$$

where $c_j(u) := c_j - \sum_{i \in I_j} u_i$ is the *Lagrangian cost* associated with column $j \in N$. Clearly, an optimal solution to (2) is given by $x_j(u) = 1$ if $c_j(u) < 0$, $x_j(u) = 0$ if $c_j(u) > 0$, and $x_j(u) \in \{0, 1\}$ when $c_j(u) = 0$. The Lagrangian dual problem associated with (2) consists of finding a Lagrangian multiplier vector $u^* \in R_+^m$ which maximizes the lower bound $L(u)$, and is typically approached through subgradient optimization.

For near-optimal Lagrangian multipliers u_i, the Lagrangian cost $c_j(u)$ gives reliable information on the overall utility of selecting column j. Based on this property, we use Lagrangian (rather than original) costs to compute, for each $j \in N$, a *score* σ_j ranking the columns according to their likelihood to be selected in an optimal solution. These scores are given on input to a simple heuristic procedure, that finds in a greedy way a hopefully good SCP solution. Computational experience shows that almost equivalent near-optimal Lagrangian multipliers can produce SCP solutions of substantially different quality. In addition, no strict correlation exists between the lower bound value $L(u)$ and the quality of the SCP solution found. Therefore it is worthwhile applying the heuristic procedure for several near-optimal Lagrangian multiplier vectors.

Our approach consists of three main phases. The first one is referred to as the *subgradient phase*. It is aimed at quickly finding a near-optimal Lagrangian multiplier vector, by means of an aggressive policy. The second one is the *heuristic phase*, in which a sequence of near-optimal Lagrangian vectors is determined and, for each vector, the associated scores are given on input to the heuristic procedure to possibly update the incumbent best SCP solution. In the third phase, called *column fixing*, we select a subset of columns having an estimated high probability of being in an optimal solution, and fix to 1 the corresponding variables. In this way we obtain an SCP instance with a reduced number of columns and rows, on which the three-phase procedure is iterated. According to our experience, column fixing is of fundamental importance to obtain high quality SCP solutions. The overall 3-phase heuristic is outlined next; more details are given in the next sections.

procedure 3-PHASE;
begin
> **repeat**
1.   SUBGRADIENT PHASE:
   find a near-optimal multiplier vector u^*;
2.   HEURISTIC PHASE:
   starting from u^*, generate a sequence of near-optimal multiplier vectors, and for each of them compute a heuristic solution to SCP (the best solution x^* being updated each time a better solution is found);
3.   COLUMN FIXING:
   select a set of "good" columns and fix to 1 the corresponding variables
> **until** a termination condition holds
end.

After each application of procedure 3-PHASE, a *refining procedure* is used, which in some cases produces improved solutions.

3 Pricing

When very large instances are tackled, the computing time spent by the overall algorithm becomes very large. We overcome this difficulty by defining a *core problem* containing a suitable set of columns, chosen among those having the smallest Lagrangian costs. The definition of the core problem is often very critical, since an optimal solution typically contains some columns that, although individually worse than others, must be selected in order to produce an overall good solution. Hence we decided not to "freeze" the core problem. Instead, we use a *variable pricing* scheme to update the core problem dynamically. The approach is in the spirit of the well-known pricing technique for solving large-scale LP's. To our knowledge, however, this approach was never used in combination with Lagrangian relaxation for SCP.

Our core problem is updated by using the dual information associated with the current Lagrangian multiplier vector u^k. To be specific, at the very beginning of the overall algorithm we define a "tentative" core by taking the first 5 columns covering each row. Afterwards, we work on the current core for, say, T consecutive subgradient iterations, after which we re-define the core problem as follows. We compute the Lagrangian cost $c_j(u^k)$, $j \in N$, associated with the current u^k, and define the column set of the new core as $C := C_1 \cup C_2$, where $C_1 := \{j \in N : c_j(u^k) < 0.1\}$, and C_2 contains the 5 smallest Lagrangian cost columns covering each row; if $|C_1| > 5m$, we keep in C_1 only the $5m$ smallest Lagrangian cost columns.

According to our experience, for large scale instances the use of pricing cuts the overall computing time by more than one order of magnitude.

4 Subgradient Phase

As already mentioned, this phase is intended to quickly produce a near-optimal Lagrangian multiplier vector. We use the well-known Held-Karp multiplier updating formula, with a new strategy to control the step-size along the subgradient direction. According to our computational experience, the new strategy leads to a faster convergence to near optimal multipliers, compared with the classical approach; see Figure 1.

In many cases, a large number of columns happen to have a Lagrangian cost $c_j(u)$ very close to zero. In particular, this occurs for large scale instances with costs c_j belonging to a small range, after a few subgradient iterations. In this situation, the Lagrangian problem has a huge number of almost optimal solutions, each obtained by choosing a different subset of the almost zero Lagrangian cost columns. It is known that the steepest ascent direction is given by the minimum-norm convex combination of the subgradients associated with these solutions. However, the exact determination of this combination is very time consuming, as it requires the solution of a quadratic problem. On the other hand, a random choice of the subgradient direction may produce very slow convergence due to zig-zagging phenomena. We overcome this drawback by heuristically selecting a small-norm subgradient direction. Computational results show that this choice leads to a faster convergence of the subgradient procedure; see Figure 2.

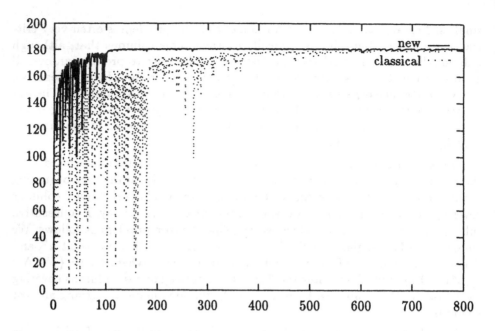

Fig. 1. Comparison between the classical and new updating strategies for the step-size parameter λ on instance RAIL516.

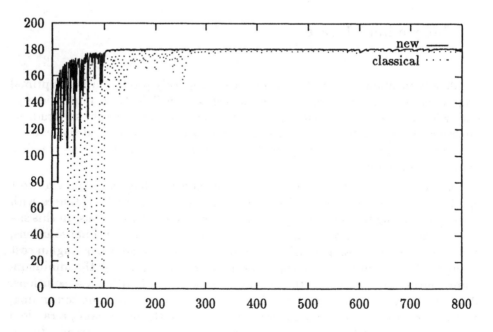

Fig. 2. Comparison between the classical and new definitions of the subgradient directions on instance RAIL516.

5 Heuristic Phase

Let u^* be the best Lagrangian vector found during the subgradient phase. Starting with u^* we generate a sequence of Lagrangian vectors u^k in an attempt to "explore" a neighborhood of near-optimal multipliers. To this end, we update the multipliers as in the subgradient phase, but we do not reduce the subgradient norm, so as to allow for a change in a larger number of components of u^k.

For each u^k, we apply a greedy heuristic procedure, in which at each iteration the column $j \in N$ having the minimum value of a score σ_j is chosen, and the corresponding rows are removed from the problem. The key step of the procedure is the definition of the score. Several rules have been proposed in the literature which define σ_j as a function of c_j and $\mu_j := |I_j \cap M^*|$ (e.g., $\sigma_j := c_j$, or $\sigma_j := c_j/\mu_j$), where M^* is the set of currently uncovered rows. According to our computational experience, these rules produce good results when c_j is replaced by $\gamma_j := c_j - \sum_{i \in I_j \cap M^*} u_i^k$, since this term takes into account the dual information associated with the still uncovered rows M^*. The use of γ_j instead of c_j in a greedy-type heuristic was first proposed by Fisher and Kedia [8]. We have also considered the following new rules: (a) $\sigma_j := \gamma_j/\mu_j$ if $\gamma_j > 0$, $\sigma_j := \gamma_j \mu_j$ if $\gamma_j < 0$; (b) $\sigma_j := (\gamma_j - \min\{\gamma_h : h \in N \setminus S\})/\mu_j$; (c) $\sigma_j := \gamma_j - \alpha\mu_j$ (where $\alpha > 0$ is a given parameter).

An extensive computational analysis showed that rule (a) outperforms all other rules, when our heuristic scheme is applied, hence we decided to only use this rule within our final algorithm.

6 Column Fixing

The heuristic solution available at the end of the heuristic phase can often be improved by fixing in the solution a convenient set of columns, and re-applying the whole procedure to the resulting subproblem. To our knowledge, only the heuristic of Ceria, Nobili and Sassano [6] makes use of variable fixing.

Clearly, the choice of the columns to be fixed is of crucial importance. After extensive computational testing, we decided to implement the following simple criteria. Let u^* be the best multiplier vector found during the subgradient and heuristic phases, and define $Q := \{j \in N : c_j(u^*) < -0.001\}$. We first fix each column $j \in Q$ for which there is a row i covered only by j among the columns in Q, i.e., $J_i \cap Q = \{j\}$. Then, we apply the heuristic procedure described in Section 5, and fix the first $\max\{\lfloor m/200 \rfloor, 1\}$ columns therein chosen.

Table 1 reports computational results for the 4 large scale instances from the FASTER test bed, described in more detail in Section 8.1. In the table, "Name" is the name of the instance, "Size" its size (in the format $m \times n$), and "Best" is the best solution value known. For each instance, we first ran procedure 3-PHASE yielding the solution value reported in column "Col-Fix", and then ran, for the same amount of time (column "Time"), the same procedure with no column fixing, yielding the solution value reported in column "No Col-Fix".

Name	Size	Best	Time	Col-Fix	No Col-Fix
RAIL582	$582 \times 55,515$	211	575.6*	211	213
RAIL507	$507 \times 63,009$	174	634.8*	176	180
RAIL2586	$2,586 \times 920,683$	947	486.1°	952	989
RAIL4872	$4,872 \times 968,672$	1534	854.4°	1550	1606

Table 1. Results of procedure 3-PHASE with and without column fixing – * Time in PC 486/33 CPU seconds – ° Time in HP 735/125 CPU seconds.

7 Refining Procedure

We have defined a simple scheme for refining a given SCP solution. The scheme assigns a score to the chosen columns, fixes to 1 the variables associated with the best scored columns, and re-optimizes the resulting subproblem.

Let x^* define the best SCP solution available, and let u^* be the Lagrangian vector corresponding to the best lower bound computed for the overall problem (that with no fixed column). For each column j with $x_j^* = 1$ we compute an estimate δ_j of the contribution of j to the overall gap, namely

$$\delta_j := \max\{c_j(u^*), 0\} + \sum_{i \in I_j} u_i^* \frac{\sum_{l \in J_i} x_l^* - 1}{\sum_{l \in J_i} x_l^*}. \tag{3}$$

The second term in (3) is obtained by uniformly splitting, for each row i, the gap contribution $u_i^*(\sum_{l \in J_i} x_l^* - 1)$ among all the columns $l \in J_i$ with $x_l^* = 1$. Notice that this score cannot be computed "on line" in a greedy-type heuristic. We define the ordered column set $\{j \in N : x_j^* = 1\} = \{j_1, \ldots, j_p\}$, where $\delta_{j_1} \leq \delta_{j_2} \leq \ldots \leq \delta_{j_p}$, and find the first $j^* \in \{j_1, \ldots, j_p\}$ such that $\frac{|\bigcup_{j=j_1}^{j^*} I_j|}{m} \geq \pi$, where the parameter $\pi \in [0, 1]$ controls the percentage number of rows removed after fixing. We then fix columns $j_1, \ldots, j^*$, and re-optimize the resulting subproblem through procedure 3-PHASE. The overall procedure, named CFT, is iterated until a given time-limit is exceeded.

Value	1550	1549	1547	1546	1544	1543	1542	1541	1540	1538	1537	1536	1535	1534
Time	733.7	866.1	866.3	866.6	868.6	881.6	882.4	937.8	1024.0	1150.0	1150.9	1187.8	4326.0	4556.1

Table 2. Incumbent solution value updatings for instance RAIL4872 – Times in HP 735/125 CPU seconds – The first entry refers to the first application of procedure 3-PHASE.

As an illustration of the effectiveness of the refining procedure, in Table 2 we report the updatings of the incumbent solution value for instance RAIL4872.

8 Computational Results

Algorithm *CFT* was implemented in ANSI FORTRAN 77. Unlike most existing algorithms for SCP, we do not perform any pre-processing on the initial data in order to remove dominated columns and rows. This is motivated by different reasons. First of all, pre-processing is very time consuming for large scale instances. Furthermore, thanks to our pricing procedure the running time of algorithm *CFT* does not change substantially if dominated rows and columns are also present. Finally, the score we use in the greedy heuristic and in the column-fixing phase prevents the choice of dominated columns.

8.1 Instances from the FASTER Competition

					Bologna			Roma	Best Industry
Name	Size	Dens	Range	LB	Sol	Time		Sol	Sol
RAIL582	$582 \times 55,515$ (μ)	1.2%	1-2	210	211	570.0*		211	211
RAIL507	$507 \times 63,009$ (μ)	1.2%	1-2	173	175$^\times$	817.0*		174	174
RAIL2586	$2,586 \times 920,683$ (m)	0.4%	1-2	937	948$^+$	1183.2°		951	952
RAIL4872	$4,872 \times 968,672$ (l)	0.2%	1-2	1509	1534	4556.1°		1534	1538

Table 3. Results on FASTER sample instances – * Time in PC 486/33 CPU seconds – ° Time in HP 735/125 CPU seconds – $^\times$ 174 with an ad hoc tuning – $^+$ 947 with an ad hoc tuning.

Algorithm *CFT* was first tested on the sample instances distributed by Ferrovie dello Stato SpA within the FASTER competition, and Table 3 reports the corresponding results. In the table, "Size" is the size of each instance in the format $m \times n$, "Dens" is the percentage number of entries equal to 1 in the SCP constraint matrix, and "Range" is the cost range in the format min-max. Column "LB" reports the LP lower bound, while "Bologna" reports the solution value found by algorithm *CFT* and the corresponding computing time. Finally, column "Roma" reports the solution value found by the algorithm of Ceria Nobili and Sassano [6], and column "Best Industry" the best solution value found by the SCP packages available on the market, according to [6].

The results for the competition are reported in Table 4, where columns "Bo" and "Rm" report the solution value found by algorithm *CFT* and by the algorithm of Ceria, Nobili and Sassano [6], respectively, while columns "An", "Bo2", "Fi" and "Tn" report the solutions values obtained by the other participants.

Our algorithm provided the best solution for all the instances. It is worth mentioning that for the larger instances, after a few hundred CPU seconds our solution was better than all the other final solutions, with the exception of those obtained by "Rm". Moreover, when 1/6 of the time limit had elapsed, we had already obtained solution values, namely 182 for RAIL516, 691 for RAIL2536, and 1069 for RAIL4284, that were better than all the other final solution values.

Name	Size	Dens	Range	LB	An	Bo	Bo2	Fi	Rm	Tn
RAIL516 *	$516 \times 47,311$ (μ)	1.3%	1-2	182	188	182	$\times$	216	$\times$	$\times$
RAIL2536 $^\circ$	$2,536 \times 1,081,841$ (m)	0.4%	1-2	685	745	691	765	922	692	709
RAIL4284 $^\circ$	$4,284 \times 1,092,610$ (l)	0.2%	1-2	1051	1175	1065	1156	$\times$	1070	1117

Table 4. Results on FASTER competition instances – * Time limit: 3,000 seconds on a PC 486/33 – $^\circ$ Time limit: 10,000 seconds on a HP 735/125.

8.2 Instances from the Literature

In order to analyze the effectiveness and robustness of our algorithm, we considered the SCP instances from the literature. Tables 5 and 6 give the results for the instances from Beasley's OR Library (see Beasley [4]). Since 1990, all the proposed algorithms for SCP have been mainly tested on these instances. We compare algorithm *CFT* with the genetic algorithm *BeCh* by Beasley and Chu [5], yielding the best published solution values on all these instances. Comparison is also made with faster heuristics, namely algorithms *Be* by Beasley [3], *BaCa* by Balas and Carrera [1], *CNS* by Ceria, Nobili and Sassano [6], and *LL* by Lorena and Lopes [10].

Table 5 reports the results for instances in Classes 4, 5, 6, A, B, C, and D, for which the optimal solution value is known (column "Opt"), while Table 6 reports the results for larger instances (Classes E, F, G, and H) for which the optimal solution value is not known, and the rounded-up LP lower bound value is reported in column "LB".

We also tested our algorithm on the SCP instances arising from crew scheduling in some airline companies, mentioned in Wedelin [11]. Comparison is made with the algorithm *Wed* proposed by Wedelin.

Finally, table 8 gives results for the real-world instances mentioned in Balas and Carrera [1]. arising from crew scheduling applications, either in airline (class 'AA') or bus (class 'BUS') companies. We also report the results of Balas and Carrera's heuristic.

Times in the tables are in DECstation 5000/240 CPU seconds, obtained by converting the times of other machines according to the performances reported in Dongarra [7].

The extensive computational results reported in the tables show the effectiveness of our approach. In 92 out of the 94 instances in our test bed we found, within short computing time, the optimal (or the best known) solution. Moreover, among the 27 instances for which the optimum is not known, in 6 cases our solution is better than any other solution found by previous techniques. We do not know of any other commercial/academic code with comparable performances.

References

1. E. Balas and M.C. Carrera, "A Dynamic Subgradient-Based Branch and Bound Procedure for Set Covering", Management Sciences Research Report No. MSSR 568(R), GSIA, Carnegie-Mellon University, October 1991, revised May 1995.
2. E. Balas and A. Ho, "Set Covering Algorithms Using Cutting Planes, Heuristics and Subgradient Optimization: A Computational Study", *Mathematical Programming Study* 12 (1980) 37–60.
3. J.E. Beasley, "A Lagrangian Heuristic for Set Covering Problems", *Naval Research Logistics* 37 (1990) 151–164.
4. J.E. Beasley, "OR-Library: Distributing Test Problems by Electronic Mail", *Journal of the Operational Research Society* 41 (1990) 1069–1072.
5. J.E. Beasley and P.C. Chu, "A Genetic Algorithm for the Set Covering Problem", Working Paper, The Management School, Imperial College, London, July 1994.
6. S. Ceria, P. Nobili, and A. Sassano, "A Lagrangian-Based Heuristic for Large-Scale Set Covering Problems", Working Paper, University of Roma La Sapienza, June 1995.
7. J.J. Dongarra, "Performance of Various Computers Using Standard Linear Equations Software", Technical Report No. CS-89-85, Computer Science Department, University of Tennessee, November 1993.
8. M.L. Fisher and P. Kedia, "Optimal Solutions of Set Covering/Partitioning Problems Using Dual Heuristics", *Management Science* 36 (1990) 674–688.
9. M. Held and R.M. Karp, "The Traveling Salesman Problem and Minimum Spanning Trees: Part II", *Mathematical Programming* 1 (1971) 6–25.
10. L.A.N. Lorena and F.B. Lopes, "A Surrogate Heuristic for Set Covering Problems", *European Journal of Operational Research* 79 (1994) 138–150.
11. D. Wedelin, "An Algorithm for Large Scale 0-1 Integer Programming with Application to Airline Crew Scheduling", *Annals of Operational Research* (to appear) (1995).

Name	Size	Dens	Range	Opt	Be		LL		BaCa		BeCh		CFT	
					Sol	Time	Sol	Time	Sol	Time*	Sol	Time	Sol	Time
4.1	200 × 1,000	2%	1-100	429	429	11.0	429	0.5	429	0.8	429	294.8	429	2.3
4.2	200 × 1,000	2%	1-100	512	512	11.1	512	0.4	512	1.2	512	9.0	512	1.1
4.3	200 × 1,000	2%	1-100	516	516	6.8	516	0.3	516	1.0	516	16.4	516	2.1
4.4	200 × 1,000	2%	1-100	494	495	12.2	495	1.2	494	1.7	494	142.0	494	9.8
4.5	200 × 1,000	2%	1-100	512	512	7.0	512	0.3	512	0.5	512	44.1	512	2.1
4.6	200 × 1,000	2%	1-100	560	561	15.8	560	1.4	560	7.5	560	16.1	560	19.3
4.7	200 × 1,000	2%	1-100	430	430	9.2	430	0.5	430	1.0	430	138.6	430	2.7
4.8	200 × 1,000	2%	1-100	492	493	11.5	493	1.4	492	6.4	492	818.7	492	22.2
4.9	200 × 1,000	2%	1-100	641	641	20.6	641	1.5	641	9.2	641	136.1	641	1.8
4.10	200 × 1,000	2%	1-100	514	514	11.9	514	0.3	514	1.1	514	13.5	514	1.8
5.1	200 × 2,000	2%	1-100	253	255	17.4	253	1.5	254	11.1	253	42.1	253	3.3
5.2	200 × 2,000	2%	1-100	302	304	20.9	302	1.8	307	20.1	302	1332.6	302	2.3
5.3	200 × 2,000	2%	1-100	226	226	10.1	226	0.3	226	1.0	228	11.0	226	2.1
5.4	200 × 2,000	2%	1-100	242	242	11.5	242	1.5	243	11.4	242	10.1	242	1.9
5.5	200 × 2,000	2%	1-100	211	211	7.2	211	0.2	211	2.1	211	14.9	211	1.2
5.6	200 × 2,000	2%	1-100	213	213	11.3	213	0.4	213	1.3	213	29.9	213	0.9
5.7	200 × 2,000	2%	1-100	293	294	18.1	293	1.5	293	7.5	293	194.9	293	15.0
5.8	200 × 2,000	2%	1-100	288	288	20.7	288	1.6	288	4.3	288	3733.3	288	1.6
5.9	200 × 2,000	2%	1-100	279	279	15.7	279	0.8	279	1.5	279	13.5	279	2.6
5.10	200 × 2,000	2%	1-100	265	265	9.8	265	0.2	265	1.1	265	19.2	265	1.3
6.1	200 × 1,000	5%	1-100	138	141	16.8	138	1.7	140	9.2	138	46.1	138	22.6
6.2	200 × 1,000	5%	1-100	146	146	14.5	149	1.9	147	9.2	146	210.5	146	17.8
6.3	200 × 1,000	5%	1-100	145	145	15.0	145	1.6	145	11.5	145	11.8	145	2.3
6.4	200 × 1,000	5%	1-100	131	131	10.3	131	1.2	131	8.3	131	4.8	131	1.8
6.5	200 × 1,000	5%	1-100	161	162	13.3	161	2.1	163	10.4	161	12.1	161	2.2
A.1	300 × 3,000	2%	1-100	253	255	36.0	254	2.7	258	39.0	253	222.4	253	82.0
A.2	300 × 3,000	2%	1-100	252	256	44.2	255	2.9	254	40.9	252	327.9	252	116.2
A.3	300 × 3,000	2%	1-100	232	234	28.1	234	2.6	237	28.6	232	127.0	232	249.9
A.4	300 × 3,000	2%	1-100	234	235	33.5	234	2.4	235	36.3	234	45.5	234	4.7
A.5	300 × 3,000	2%	1-100	236	237	19.0	238	2.2	236	26.2	236	23.7	236	80.0
B.1	300 × 3,000	5%	1-100	69	70	28.4	70	3.0	69	29.0	69	20.0	69	4.0
B.2	300 × 3,000	5%	1-100	76	77	40.8	76	4.0	76	29.0	76	11.6	76	6.1
B.3	300 × 3,000	5%	1-100	80	80	25.4	81	4.4	81	35.1	80	709.7	80	18.0
B.4	300 × 3,000	5%	1-100	79	80	37.0	81	4.3	79	29.0	79	29.9	79	6.3
B.5	300 × 3,000	5%	1-100	72	72	26.0	72	4.1	72	32.6	72	5.3	72	3.3
C.1	400 × 4,000	2%	1-100	227	230	42.4	227	4.0	230	116.2	227	187.9	227	74.0
C.2	400 × 4,000	2%	1-100	219	223	66.0	222	4.1	220	56.1	219	40.7	219	64.2
C.3	400 × 4,000	2%	1-100	243	251	75.1	251	4.9	248	61.7	243	541.3	243	70.2
C.4	400 × 4,000	2%	1-100	219	224	63.4	224	5.4	224	68.1	219	144.6	219	61.6
C.5	400 × 4,000	2%	1-100	215	217	39.9	216	4.1	217	64.6	215	80.6	215	60.3
D.1	400 × 4,000	5%	1-100	60	61	40.9	60	4.8	61	36.6	60	13.8	60	23.1
D.2	400 × 4,000	5%	1-100	66	68	52.7	68	3.5	67	46.6	66	198.6	66	22.0
D.3	400 × 4,000	5%	1-100	72	75	55.8	75	5.8	74	47.2	72	785.3	72	22.6
D.4	400 × 4,000	5%	1-100	62	64	36.5	63	4.8	63	39.8	62	73.5	62	8.3
D.5	400 × 4,000	5%	1-100	61	62	36.7	62	4.5	61	36.2	61	79.8	61	10.3

Table 5. Results on the test instances from Beasley's OR-Library – Times are given in DECstation 5000/240 CPU seconds – * Overall time for the execution of the heuristic algorithm.

Name	Size	Dens	Range	LB	Be Sol	Be Time	BaCa Sol	BaCa Time*	CNS Sol	CNS Time°	BeCh Sol	BeCh Time	CFT Sol	CFT Time
E.1	500 × 5,000	10%	1-100	22	29	72.6	29	55.3	×	×	29	38.2	29	26.0
E.2	500 × 5,000	10%	1-100	23	32	92.7	32	76.0	×	×	30	14647.7	30	408.0
E.3	500 × 5,000	10%	1-100	21	28	92.7	28	80.9	×	×	27	28360.2	27	94.2
E.4	500 × 5,000	10%	1-100	22	30	100.3	29	77.5	×	×	28	539.9	28	26.3
E.5	500 × 5,000	10%	1-100	22	28	80.8	28	61.6	×	×	28	35.0	28	36.6
F.1	500 × 5,000	20%	1-100	9	15	43.9	14	67.5	×	×	14	76.4	14	33.2
F.2	500 × 5,000	20%	1-100	10	16	102.6	15	88.6	×	×	15	78.1	15	31.2
F.3	500 × 5,000	20%	1-100	10	15	124.7	15	76.5	×	×	14	266.8	14	248.5
F.4	500 × 5,000	20%	1-100	9	15	118.2	15	74.8	×	×	14	209.7	14	31.0
F.5	500 × 5,000	20%	1-100	8	14	129.3	14	62.2	×	×	13	13192.6	13	201.1
G.1	1,000 × 10,000	2%	1-100	160	184	287.8	183	325.6	176	4905.5	176	30200.0	176	147.0
G.2	1,000 × 10,000	2%	1-100	142	163	204.9	161	370.1	155	4905.5	155	360.5	154	783.4
G.3	1,000 × 10,000	2%	1-100	148	174	318.2	175	378.6	167	4905.5	166	7841.6	166	978.0
G.4	1,000 × 10,000	2%	1-100	149	176	292.0	176	332.2	170	4905.5	168	25304.7	168	378.5
G.5	1,000 × 10,000	2%	1-100	148	175	277.5	172	262.6	169	4905.5	168	549.3	168	237.2
H.1	1,000 × 10,000	5%	1-100	48	68	317.7	68	488.4	64	4905.5	64	1682.1	63	1451.1
H.2	1,000 × 10,000	5%	1-100	49	66	293.9	67	380.7	64	4905.5	64	530.3	63	887.0
H.3	1,000 × 10,000	5%	1-100	45	65	325.1	63	443.1	60	4905.5	59	1803.5	59	1560.3
H.4	1,000 × 10,000	5%	1-100	44	63	333.5	62	354.7	59	4905.5	58	27241.8	58	237.6
H.5	1,000 × 10,000	5%	1-100	43	60	303.0	58	321.3	55	4905.5	55	449.6	55	155.4

Table 6. Results on the test instances from Beasley's OR-Library – Times are given in DECstation 5000/240 CPU seconds – * Overall time for the execution of the heuristic algorithm – ° Time limit.

Name	Size	Dens	Range	Opt	Wed Sol	Wed Time°	CFT Sol	CFT Time
B727scratch	29 × 157	8.2%	1,600-11,850	94,400	94,400	4.7	94,400	0.3
ALITALIA	118 × 1,165	3.1%	2,200-2,110,900	27,258,300	27,258,300	37.2	27,258,300	6.2
A320	199 × 6,931	2.3%	1,600-2,111,450	12,620,100	12,620,100	216.9	12,620,100	79.5
A320coc	235 × 18,753	1.9%	1,900-1,812,000	14,495,500	14,495,500	1023.7	14,495,600	577.8
SASjump	742 × 10,370	0.6%	4,720-55,849	7,338,844	7,340,777	806.8	7,339,537	396.3
SASD9imp2	1,366 × 25,032	0.3%	3,860-35,200	5,261,088*	5,262,190	1579.7	5,263,640	2082.1

Table 7. Results on the airline instances from Wedelin – Times are given in DECstation 5000/240 CPU seconds – * LP lower bound – ° Overall time for the execution of the heuristic algorithm.

					BaCa		CFT	
Name	Size	Dens	Range	Opt	Sol	Time*	Sol	Time
AA03	106 × 8,661	4.05%	91-3619	33,155	33,157	96.4	33,155	61.0
AA04	106 × 8,002	4.05%	91-3619	34,573	34,573	39.2	34,573	3.6
AA05	105 × 7,435	4.05%	91-3619	31,623	31,623	53.9	31,623	3.1
AA06	105 × 6,951	4.11%	91-3619	37,464	37,464	44.4	37,464	5.2
AA11	271 × 4,413	2.53%	35-2966	35,384	35,478	72.3	35,384	193.7
AA12	272 × 4,208	2.52%	35-2966	30,809	30,815	48.0	30,809	53.8
AA13	265 × 4,025	2.60%	35-2966	33,211	33,211	19.6	33,211	8.3
AA14	266 × 3,868	2.50%	35-2966	33,219	33,222	86.2	33,219	30.3
AA15	267 × 3,701	2.58%	35-2966	34,409	34,510	39.9	34,409	18.8
AA16	265 × 3,558	2.63%	35-2966	32,752	32,858	54.5	32,752	33.6
AA17	264 × 3,425	2.61%	35-2966	31,612	31,717	47.0	31,612	10.9
AA18	271 × 3,314	2.55%	35-2966	36,782	36,866	66.2	36,782	13.5
AA19	263 × 3,202	2.63%	35-2966	32,317	32,317	27.6	32,317	5.9
AA20	269 × 3,095	2.58%	35-2966	34,912	35,160	34.4	34,912	13.6
BUS1	454 × 2,241	1.89%	120-877	27,947	27,947	62.8	27,947	5.0
BUS2	681 × 9,524	0.51%	120-576	67,760	67,868	356.0	67,760	19.2

Table 8. Results on the real-world instances from Balas and Carrera – Times are given in DECstation 5000/240 CPU seconds – * Overall time for the execution of the heuristic algorithm.

An ϵ-Relaxation Method for Generalized Separable Convex Cost Network Flow Problems

Paul Tseng[1] and Dimitri P. Bertsekas[2]

[1] Department of Mathematics, University of Washington, Seattle, WA 98195, USA
[2] Department of Electrical Engineering and Computer Science, Massachusetts
Institute of Technology, Rm. 35-210, Cambridge, MA 02139, USA

Abstract. We propose an extension of the ϵ-relaxation method to generalized network flow problems with separable convex cost. The method
maintains ϵ-complementary slackness satisfied at all iterations and ad-
justs the arc flows and the node prices so to satisfy flow conservation
upon termination. Each iteration of the method involves either a price
change at a node or a flow change at an arc or a flow change around a
simple cycle. Complexity bounds for the method are derived. For one implementation employing ϵ-scaling, the bound is polynomial in the number
of nodes N, the number of arcs A, a certain constant Γ depending on the
arc gains, and $\ln(\epsilon^0/\bar{\epsilon})$, where ϵ^0 and $\bar{\epsilon}$ denote, respectively, the initial
and the final ϵ.

1 Introduction

Consider a directed graph with node set $\mathcal{N} = \{1, \ldots, N\}$ and arc set $\mathcal{A} \subset \mathcal{N} \times \mathcal{N}$,
where N denotes the number of nodes and A denotes the number of arcs. (The
implicit assumption that there exists at most one arc in each direction between
any pair of nodes is made for notational convenience and can be dispensed with.)
We are given, for each node $i \in \mathcal{N}$, a scalar s_i (the *supply* of i) and, for each
arc $(i, j) \in \mathcal{A}$, a positive scalar γ_{ij} (the *gain* of (i, j)) and a convex, lower
semicontinuous function $f_{ij} : \Re \to \Re \cup \{\infty\}$ satisfying $\lim_{\zeta \to -\infty} f_{ij}^-(\zeta) = -\infty$ and
$\lim_{\zeta \to \infty} f_{ij}^+(\zeta) = \infty$, where $f_{ij}^-(\zeta)$ and $f_{ij}^+(\zeta)$ denote, respectively, the left and
right directional derivative of f_{ij} at ζ [Roc84, p. 329]. The generalized separable
convex cost network flow problem is

$$\text{minimize} \quad \sum_{(i,j)\in\mathcal{A}} f_{ij}(x_{ij}) \tag{P}$$

$$\text{subject to} \quad \sum_{\{j|(i,j)\in\mathcal{A}\}} x_{ij} - \sum_{\{j|(j,i)\in\mathcal{A}\}} \gamma_{ji} x_{ji} = s_i, \quad \forall\, i \in \mathcal{N}, \tag{1}$$

where the real variable x_{ij} is referred to as the *flow* of the arc (i, j). We refer to
the vector $x = \{x_{ij} \mid (i, j) \in \mathcal{A}\}$ as the flow vector. An important special case is
the linear cost case, where

$$f_{ij}(x_{ij}) := \begin{cases} a_{ij} x_{ij} & \text{if } b_{ij} \le x_{ij} \le c_{ij} \\ \infty & \text{otherwise} \end{cases}, \tag{2}$$

for given scalars a_{ij}, b_{ij}, c_{ij}. Another important special case is the ordinary network case, where $\gamma_{ij} = 1$ for all $(i, j) \in \mathcal{A}$. This problem and its linear cost case have been much studied (see [AMO93, Chap. 15], [Jew62], [Mur92, Chap. 8] and references therein) and has applications in many areas including financial planning, logistics, hydroelectric power system control (see [AMDZ87], [AMOR95], [GHDS78] and references therein). A flow vector x such that $f_{ij}(x_{ij}) < \infty$ for all $(i, j) \in \mathcal{A}$ and satisfying the conservation of flow constraint (1) is called *feasible*. We will assume that there exists at least one feasible flow vector x satisfying

$$f_{ij}^-(x_{ij}) < \infty \quad \text{and} \quad f_{ij}^+(x_{ij}) > -\infty, \qquad \forall\, (i, j) \in \mathcal{A}. \tag{3}$$

In the linear cost case (2), the condition (3) reduces to the well known capacity condition: $b_{ij} \leq x_{ij} \leq c_{ij}$ for all $(i, j) \in \mathcal{A}$.

For the problem (P), there are available a number of solution methods: primal/dual simplex method, primal-dual method, out-of-kilter method [Roc80] and relaxation method [TsB90]. For the linear cost case, many variants and improvements of the preceding methods have been proposed (see [AMO93, Chap. 15], [Jew62], [Mur92, Chap. 8], [BeT88] and references therein). For the ordinary network case, there have recently been proposed other solution methods, based on minimum mean cycle canceling [KaM93] and on ϵ-relaxation [BPT95], [DMZ95], [Pol95]. For the linear cost ordinary network case, further specialization and improvements of the preceding methods, as well as many other methods, have been proposed (see the recent books [AMO93], [Ber91], [Mur92], [Roc84] and references therein).

Here, we propose an extension of the ϵ-relaxation method studied in [BPT95], [DMZ95], [Pol95] from the ordinary network case to the problem (P), and we present associated complexity bounds for the method. Our interest in extending the ϵ-relaxation method to (P) stems from the good performance of this method on linear/quadratic cost ordinary network flow problems and its suitability for implementation on parallel computers [BCEZ95]. However, the extension is highly nontrivial due to the presence of nonunity arc gains. In particular, flow augmentation along cycles of non-unity gain need to be considered and new techniques need to be developed to deal with the presence of directed cycles in the admissible graph. In fact, even for the linear cost case our algorithm and the associated complexity bounds are new to our knowledge. Previous complexity bounds for the linear cost case (other than those obtained by specializing general linear programming complexity bounds) concerned with either the case of all nodes being supply nodes (i.e., $s_i \geq 0$ for all $i \in \mathcal{N}$) or being demand nodes (i.e., $s_i \leq 0$ for all $i \in \mathcal{N}$) [AdC91], [CoM94] or the case of generalized circulation (i.e., maximizing the flow on a particular arc) [GPT91], [Rad95].

Below we describe the basic idea of ϵ-relaxation for (P), which is steeped in the classical duality framework for (P) and its dual, as developed by von Neumann, Gale, Kuhn, Tucker for the linear cost case and by Rockafellar for the general convex cost case (see [Roc84]). For each $i \in \mathcal{N}$, we associate a Lagrange multiplier p_i (the *price* of node i) to the ith conservation of flow constraint (1). We refer to the vector $p = \{p_i \mid i \in \mathcal{N}\}$ as the *price vector*. The *dual* problem is

$$\text{minimize} \quad q(p) \qquad \qquad \text{(D)}$$
$$\text{subject to} \quad \text{no constraint on } p,$$

where the dual functional q is given by

$$q(p) := \sum_{(i,j)\in\mathcal{A}} q_{ij}(p_i - \gamma_{ij}p_j) - \sum_{i\in\mathcal{N}} s_i p_i,$$

and q_{ij} is derived from f_{ij} by the conjugacy relation

$$q_{ij}(t_{ij}) := \sup_{x_{ij}\in\Re} \{x_{ij}t_{ij} - f_{ij}(x_{ij})\}.$$

We say that a flow vector x and a price vector p satisfy ϵ-*complementary slackness*, or ϵ-CS for short (see [BHT87], [TsB90]), if and only if

$$f_{ij}^-(x_{ij}) < \infty, \quad \text{and} \quad f_{ij}^-(x_{ij}) - \epsilon \le p_i - \gamma_{ij}p_j \le f_{ij}^+(x_{ij}) + \epsilon, \qquad \forall\, (i,j) \in A. \tag{ϵ-CS}$$

It is known [Roc84, p. 360] that, under our assumptions, both the problem (P) and the dual problem (D) have optimal solutions and their optimal costs are the negatives of each other. Moreover, a feasible flow vector x and a price vector p are primal and dual optimal, respectively, if and only if they satisfy 0-CS. For a flow vector x, we define the *surplus* of node i to be the difference between the supply s_i and the net outflow from i:

$$g_i := s_i + \sum_{\{j|(j,i)\in\mathcal{A}\}} \gamma_{ji}x_{ji} - \sum_{\{j|(i,j)\in\mathcal{A}\}} x_{ij}, \tag{4}$$

The idea of ϵ-relaxation is to iteratively adjust the price of the nodes one-at-a-time, with the flow of the arcs adjusted accordingly so as to maintain ϵ-CS while decreasing the surplus of the nodes towards zero. Termination occurs when the surpluses of all the nodes are zero.

In what follows, for any path P, we denote by P^+ the set of forward arcs of P and by P^- the set of backward arcs of P. We let

$$\gamma_P := \left(\prod_{(i,j)\in P^+} \gamma_{ij}\right) \bigg/ \left(\prod_{(i,j)\in P^-} \gamma_{ij}\right), \tag{5a}$$

$$\Gamma_P := \sum_{i\in P} \gamma_{P_i}, \tag{5b}$$

where, for each node $i \in P$, P_i denotes the portion of the path P from the starting node of P to i. We say that a path P is *directed* if $P^- = \emptyset$. A cycle is a path whose starting node equals the ending node. A path is said to be *simple* if it contains no repeated nodes except (in the case where the path is a cycle) for the starting and the ending nodes.

2 The ϵ-Relaxation Method for Generalized Network

Fix any Δ satisfying

$$0 < \Delta < \min\left\{1, \min_{(i,j)\in\mathcal{A}} 1/\gamma_{ij}\right\}.$$

For a flow-price vector pair (x, p) satisfying ϵ-CS, we define for each node $i \in \mathcal{N}$ its *push list* as the set of arcs

$$\{(i,j) \mid -\epsilon \le f_{ij}^+(x_{ij}) - p_i + \gamma_{ij}p_j < -(1-\Delta)\epsilon\}$$
$$\bigcup \{(j,i) \mid (1 - \gamma_{ji}\Delta)\epsilon < f_{ji}^-(x_{ji}) - p_j + \gamma_{ji}p_i \le \epsilon\}. \tag{6}$$

An arc (i,j) [or (j,i)] in the push list of i is said to be *unblocked* if there exists a $\delta > 0$ such that

$$p_i - \gamma_{ij}p_j \ge f_{ij}^+(x_{ij} + \delta),$$

[or $p_j - \gamma_{ji}p_i \le f_{ji}^-(x_{ji} - \delta)$, respectively]. For an unblocked arc, the supremum of δ for which the above relation holds is called the *flow margin* of the arc. It can be verified that the arcs in the push list of a node are unblocked.

For a given pair (x, p) satisfying ϵ-CS, consider an arc set $\mathcal{A}^*$ that contains all push list arcs oriented in the direction of flow change. In particular, for each arc (i,j) in the push list of a node i, we introduce an arc (i,j) in $\mathcal{A}^*$; for each arc (j,i) in the push list of node i, we introduce an arc (i,j) in $\mathcal{A}^*$. The set of nodes $\mathcal{N}$ and the set $\mathcal{A}^*$ define the *admissible graph* $G^* = (\mathcal{N}, \mathcal{A}^*)$. An arc can be in the push list of at most one node, so the admissible graph is well defined.

The ϵ-relaxation method starts with a flow-price vector pair (x, p) satisfying ϵ-CS. The method comprises two phases. In the first phase, only "up" iterations are performed until no node with positive surplus remains. In the second phase, only "down" iterations are performed until no node with negative surplus remains. In a typical up iteration, we select a node i with positive surplus and we perform one of the following three operations:

(a) A *price rise* on node i, which increases the price p_i by the maximum amount that maintains ϵ-CS, while leaving all arc flows unchanged.

(b) A *flow push* (also called a δ-*flow push*) along an arc (i,j) [or along an arc (j,i)], which increases x_{ij} [or decreases x_{ji}] by an amount δ, while leaving all node prices unchanged.

(c) A *flow push* (also called a δ-*flow push*) along a cycle C containing i, which increases x_{kl} [respectively, decreases x_{kl}] by an amount $\gamma_{C_k}\delta$ for all $(k,l) \in C^+$ [respectively, for all $(k,l) \in C^-$], while leaving all node prices unchanged. [Here, C_k denotes the portion of C from i to k, and γ_{C_k} is given by (5a).]

[The effect of operation (c) is to change the surplus at node i by the amount $\delta/(1 - \gamma_C)$ (respectively, 0) if $\gamma_C \ne 1$ (respectively, if $\gamma_C = 1$), while leaving the surplus at all other nodes unchanged.] A δ-flow push along an arc or a cycle in G^* is said to be *saturating* if it changes the flow margin of one of the arcs from positive to zero. An up iteration is as follows: (A down iteration involving price decrease on a node i with negative surplus is defined analogously.)

An "Up" Iteration

Step 1 Select a node i with positive surplus g_i (see (4)). If no such node exists, terminate the first phase of the method; otherwise go to Step 2.

Step 2 If the push list of i is empty, perform a price rise on i and go to the next up iteration. Otherwise, choose any arc in the push list of i and let j be the node at the other end of the arc (so (i,j) is an arc of the admissible graph G^*). In the case where (i,j) belongs to some directed cycle C of G^*, we consider the last up iteration at which j was selected. If at all subsequent up iterations only non-saturating flow pushes were performed, go to Step 4; otherwise, go to either Step 3 or Step 4. In the case where (i,j) does not belong to any directed cycle of G^*, go to Step 3.

Step 3 Perform a δ-flow push towards the opposite node j, where

$$\delta = \begin{cases} \min\{g_i, \text{ flow margin of } (i,j)\} & \text{if } (i,j) \in \mathcal{A} \\ \min\{g_i/\gamma_{ji}, \text{ flow margin of } (j,i)\} & \text{if } (j,i) \in \mathcal{A}. \end{cases}$$

Go to the next iteration.

Step 4 Perform a δ-flow push along C, where

$$\delta = \begin{cases} \min_{(k,l)\in C}\{\text{flow margin of } (k,l)/\gamma_{C_k}\} & \text{if } \gamma_C \geq 1 \\ \min\{g_i/(1-\gamma_C), \min_{(k,l)\in C}\{\text{flow margin of } (k,l)/\gamma_{C_k}\}\} & \text{if } \gamma_C < 1. \end{cases}$$

Go to the next iteration.

In general, finding the cycle C in Step 2 is expensive. However, if we select the node in each up iteration in a special manner, such a cycle can be found fairly easily. More precisely, consider the following implementation of the up iterations:

(a) Select any node i_0 with positive surplus. If no such node exists, terminate the first phase of the method. Else let $k := 0$ and go to (b).

(b) If the push list of $i = i_k$ is empty, perform a price rise on node i_k and go to (a). Otherwise, choose any arc in the push list of i_k and let j be the node at the other end of the arc. If $j = i_l$ for some $l < k$, go to (c). Else perform a δ-flow push along this arc as in Step 3, and if this saturates the arc or if the surplus of j remains nonpositive, go to (a); else let $i_{k+1} := j$, increment k by 1 and go to (b).

(c) A directed cycle in the admissible graph G^* is $C : i_l, i_{l+1}, ..., i_k, i_l$. Perform a δ-flow push along C as in Step 4, and go to (a).

There is also an important modification of the above implementation in which we refrain from performing a δ-flow push until we encounter a node j with negative surplus, at which time we push flow from i_0 along the path $i_0, i_1, ..., i_k, j$ towards j. With this modification, it is possible to mix up and down iterations without affecting the termination or the complexity of the method, and the resulting improvement in the practical performance of the method can be significant.

Finally, we note that, in contrast to the ordinary network case, we need to consider not only flow pushes along arcs, but also flow pushes along cycles. The intuition for this is that a cycle C with $\gamma_C < 1$ is "flow absorbing" when flow is pushed along C and thus C acts like a node with negative surplus; similarly, a cycle C with $\gamma_C > 1$ is "flow generating" when flow is pushed along C and thus C acts like a node with positive surplus.

3   Termination of the ϵ-Relaxation Method

To prove the termination of the ϵ-relaxation method of Section 2, it suffices to prove the termination of its first phase. The first result bounds from below the price rise increments.

Proposition 1. *Each price rise increment in the ϵ-relaxation method is at least $\Delta\epsilon$.*

By using Proposition 1, the next result bounds the total number of price rises. This extends an analogous result for the ordinary network case [BPT95], [Pol95].

Proposition 2. *Assume that for some scalar $K \geq 0$, the initial price vector p^0 for the ϵ-relaxation method satisfies $K\epsilon$-CS together with some feasible flow vector x^0. Then, the ϵ-relaxation method performs at most $\lceil (K+1)\Gamma/\Delta \rceil$ price rises per node, where*

$$\Gamma := \max_{H \text{ a simple path}} \left\{ \gamma_H \left(\max_{\substack{C \text{ a simple cycle} \\ \text{with } \gamma_C < 1}} \frac{\Gamma_C}{1 - \gamma_C} \right) + \Gamma_H \right\},$$

and $\gamma_H, \gamma_C, \Gamma_C, \Gamma_H$ are given by (5).

The result of the preceding proposition shows that the bound on the number of price changes is independent of the cost functions, but depends only on the arc gains and the scalar K^0 given by

$$K^0 := \inf\{K \geq 0 \mid (x^0, p^0) \text{ satisfies } K\epsilon\text{-CS for some feasible flow vector } x^0 \},$$

which is the minimum multiplicity of ϵ by which 0-CS is violated by the starting price together with some feasible flow vector. Note that K^0 is well defined for any p^0 because, for all K sufficiently large, $K\epsilon$-CS is satisfied by p^0 and the feasible flow vector x satisfying Eq. (3).

The next result bounds the number of flow pushes.

Proposition 3. *The number of flow pushes along arcs (respectively, cycles) between two successive price rises (not necessarily at the same node) performed by the ϵ-relaxation method is at most $2N^2A$ (respectively, $2NA$).*

It readily follows from Propositions 2 and 3 that the first phase of the ϵ-relaxation method terminates after at most $O(NK\Gamma/\Delta)$ price rises and at most $O(N^3AK\Gamma/\Delta)$ flow pushes along arcs and at most $O(N^2AK\Gamma/\Delta)$ flow pushes along cycles. An analogous result holds for the second phase. In Section 4, a specific implementation of the method with improved complexity bound will be presented. Upon termination, we have that the flow vector and price vector satisfy ϵ-CS and that the flow vector is feasible since the surplus of all nodes will be zero. The following result from [TsB90, Propositions 7 and 8] shows that this flow vector and price vector are primal optimal and dual optimal within a factor that is essentially proportional to ϵ.

Proposition 4. *For each $\epsilon > 0$, let $x(\epsilon)$ and $p(\epsilon)$ denote any flow and price vector pair satisfying ϵ-CS with $x(\epsilon)$ feasible and let $\xi(\epsilon)$ denote any flow vector satisfying 0-CS with $p(\epsilon)$ ($\xi(\epsilon)$ need not be feasible). Then*

$$0 \le f(x(\epsilon)) + q(p(\epsilon)) \le \epsilon \sum_{(i,j)\in\mathcal{A}} |x_{ij}(\epsilon) - \xi_{ij}(\epsilon)|. \tag{7}$$

Furthermore, $f(x(\epsilon)) + q(p(\epsilon)) \to 0$ as $\epsilon \to 0$.

Proposition 4 does not give us an estimate of how small ϵ has to be in order to achieve a certain degree of optimality. However, in the common case where finiteness of the arc cost functions f_{ij} imply lower and upper bounds on the arc flows:

$$-\infty < b_{ij} := \inf_\xi \{\xi \mid f_{ij}(\xi) < \infty\} \le \sup_\xi \{\xi \mid f_{ij}(\xi) < \infty\} =: c_{ij} < \infty,$$

as in the case of (2), we have that the righthand side of (7) is bounded above by $\epsilon \sum_{(i,j)\in\mathcal{A}} |c_{ij} - b_{ij}|$ (also see [BeT88, §3] for the linear cost case).

4   Sweep Implementation of the ϵ-Relaxation Method

We say that a strongly connected component (abbreviated as scc) of the admissible graph G^* is a *predecessor* of another scc of G^* if there is a directed path in G^* from a node in the first scc to a node in the second scc. (So flow cannot be pushed from a scc to any of its predecessor scc.) The *sweep implementation* of the ϵ-relaxation method, introduced in [Ber86] for the linear cost network flow problem, selects a node for up iteration as follows (an analogous rule holds for selecting a node for a down iteration): Let G^* denote the current admissible graph. Choose any positive scc of G^* (a scc is "positive" if it contains at least one node with positive surplus) whose predecessor scc are all non-positive. In Step 1 of up iteration, select i to be any node in the chosen scc with positive surplus. Also, in Step 2, always go to Step 4 when (i,j) belongs to some directed cycle C of G^*.

For the sweep implementation, we can improve on Proposition 3 as shown below. The intuition for this improvement is that an up iteration on a node i

having a positive predecessor scc may be wasteful since its surplus may be set to zero through a flow push and become positive again by a flow push at a node in the predecessor scc. The sweep implementation avoids performing such an up iteration.

Proposition 5. *For the sweep implementation of the ϵ-relaxation method, the number of non-saturating flow pushes between two successive price rises (not necessarily at the same node) is at most $N + N\tilde{A}$, where $\tilde{A}$ denotes the maximum number of arcs contained in a scc of the admissible graph.*

By using Propositions 2 and 5, we obtain the following improved complexity bound for the sweep implementation of the ϵ-relaxation method.

Proposition 6. *Assume that for some $K \geq 1$ the initial price vector p^0 for the sweep implementation of the ϵ-relaxation method satisfies $K\epsilon$-CS together with some feasible flow vector x^0. Then, the method requires $O(K\Gamma N)$ price changes and $O(K\Gamma N^2(1 + \tilde{A}))$ flow pushes up to termination.*

In the ordinary network case (so $\Gamma = O(N)$), it was shown in [BPT95, Proposition 4] that if in addition we choose $\Delta \leq 1/2$, then the admissible graph is always acyclic, in which case $\tilde{A} = 0$ (each scc always comprises an individual node and hence contains no arc at all) and the complexity bound of Proposition 6 become $O(KN^2)$ price changes and $O(KN^3)$ flow pushes. We can further improve the complexity bound by using ϵ-scaling: initially set $\epsilon = \epsilon^0$ for some ϵ^0, run the ϵ-relaxation method till it terminates with some (x^0, p^0), then decrease ϵ by a certain fixed factor and rerun the ϵ-relaxation method with p^0 as the starting price vector, and so on, till ϵ reaches some target value $\bar{\epsilon}$. This yields an improved complexity bound in which K is replaced by $\ln(\epsilon^0/\bar{\epsilon})$.

Work in Progress. We are presently studying ways to improve the complexity of the ϵ-relaxation method by scaling the arc gains. Computational testing of the method is also in progress.

References

[AdC91]   Adler, I., and Cosares, S., A strongly polynomial algorithm for a special class of linear programs, Operations Research **39** (1991) 955–960

[AMDZ87] Ahlfeld, D.P., Mulvey, J.M., Dembo, R.S., and Zenios, S.A., Nonlinear programming on generalized networks, ACM Transactions on Mathematical Software **13** (1987) 350–367

[AMO93]   Ahuja, R.K., Magnanti, T.L., Orlin, J.B., Network Flows, Prentice-Hall, Englewood Cliffs, 1993

[AMOR95] Ahuja, R.K., Magnanti, T.L., Orlin, J.B., and Reddy, M.R., Applications of network optimization. In: Ball, M.O., Magnanti, T.L., Monma, C.L., and Nemhauser, G.L. (eds.) Network models, North-Holland, Amsterdam, 1995, pp.1–83

[Ber86]   Bertsekas, D.P., Distributed relaxation methods for linear network flow problems, Proceedings of 25th IEEE Conference on Decision and Control (1986) 2101–2106

[Ber91]   Bertsekas, D.P., Linear Network Optimization: Algorithms and Codes, M.I.T. Press, Cambridge, 1991

[BECZ95]   Bertsekas, D.P., Castañon, D., Eckstein, J., and Zenios, S.A., Parallel computing in network optimization. In: Ball, M.O., Magnanti, T.L., Monma, C.L., and Nemhauser, G.L. (eds.) Network models, North-Holland, Amsterdam, 1995, pp.331–399

[BHT87]   Bertsekas, D.P., Hosein, P.A., and Tseng, P., Relaxation methods for network flow problems with convex arc costs," SIAM Journal on Control and Optimization 25 (1987) 1219–1243

[BPT95]   Bertsekas, D.P., Polymenakos, L.C., and Tseng, P., An ϵ-relaxation method for separable convex cost network flow problems, Laboratory for Information and Decision Systems Report, Massachusetts Institute of Technology, Cambridge, 1995

[BeT88]   Bertsekas, D.P., and Tseng, P., Relaxation methods for minimum cost ordinary and generalized network flow problems, Operations Research 36 (1988) 93–114

[CoN94]   Cohen, E., and Megiddo, N., New algorithms for generalized network flows, Mathematical Programming 64 (1994) 325–336

[DMZ95]   De Leone, R., Meyer, R.R., and Zakarian, A., An ϵ-relaxation algorithm for convex network flow problems, Computer Sciences Department Technical Report, University of Wisconsin, Madison, 1995

[GHKS78]   Glover, F., Hultz, J., Klingman, D., and Stutz, J., Generalized networks: a fundamental computer-based planning tool, Management Science 24 (1978) 1209–1220

[GPT91]   Goldberg, A.V., Plotkin, S.A., and Tardos, É., Combinatorial algorithms for the generalized circulation problem, Mathematics of Operations Research 16 (1991) 351–381

[Jew62]   Jewell, W.S., Optimal flow through networks with gains, Operations Research 10 (1962) 476–499

[KaM93]   Karzanov, A.V., and McCormick, S.T., Polynomial methods for separable convex optimization in unimodular linear spaces with applications to circulations and co-circulations in network, Faculty of Commerce Report, University of British Columbia, Vancouver, 1993; SIAM Journal on Computing (to appear)

[Mur92]   Murty, K.G., Network Programming, Prentice-Hall, Englewood Cliffs, 1992

[Pol95]   Polymenakos, L.C. ϵ-relaxation and auction algorithms for the convex cost network flow problem, Electrical Engineering and Computer Science Department Ph.D. Thesis, Massachusetts Institute of Technology, Cambridge, 1995

[Rad95]   Radzik, T., Faster algorithms for the generalized network flow problem, Department of Computer Science Report, King's College London, London, 1995

[Roc84]   Rockafellar, R.T., Network Flows and Monotropic Programming, Wiley-Interscience, New York, 1984

[TsB90]   Tseng, P., and Bertsekas, D.P., Relaxation methods for monotropic programs, Mathematical Programming 46 (1990) 127–151

Finding Real-Valued Single-Source Shortest Paths in $o(n^3)$ Expected Time

Stavros G. Kolliopoulos and Clifford Stein *

Dartmouth College, Department of Computer Science, Hanover, NH 03755-3510.
E-mail: {stavros, cliff}@cs.dartmouth.edu

Abstract. Given an n-vertex directed network G with real costs on the edges and a designated source vertex s, we give a new algorithm to compute shortest paths from s. Our algorithm is a simple deterministic one with $O(n^2 \log n)$ expected running time over a large class of input distributions.

The shortest path problem is an old and fundamental problem with a host of applications. Our algorithm is the first strongly-polynomial algorithm in over 35 years to improve upon some aspect of the running time of the celebrated Bellman-Ford algorithm for arbitrary networks, with any type of cost assignments.

1 Introduction

Given a directed network $G = (V, E, c)$ where c is a real-valued function mapping edges to costs, the single-source shortest path problem (abbreviated SSSP) consists of finding, for each vertex $v \in V$, the simple path of minimum total cost from a designated source vertex s. This is an old and fundamental problem in network optimization with a plethora of applications in operations research, see for example [1]. It also arises as a subproblem in other optimization problems such as network flows.

The classic Bellman-Ford algorithm solves the SSSP problem in an n-vertex m-edge network in $O(nm)$ time [3, 10]. This simple algorithm has been widely used and studied for over 35 years, however, in all that time, no progress has been made in improving the worst case time bound for arbitrary real-valued shortest path problems.

We cannot improve on the Bellman-Ford algorithm in the worst case for all networks. However, in this paper, we will show that on any network with real edge costs chosen from a large class of input distributions, we can provide a deterministic algorithm for SSSP that runs in $O(n^2 \log n)$ expected time. Our algorithm is faster than the Bellman-Ford algorithm in the case when $m = \omega(n \log n)$, and, under the similarity assumption [1], faster than the recent scaling algorithm of Goldberg [15], when $m = \omega(n^{1.5} \log n)$. To our knowledge, this is the first strongly polynomial algorithm that solves SSSP in $o(n^3)$ time for networks

---

* Research partly supported by NSF Award CCR-9308701, a Walter Burke Research Initiation Award and a Dartmouth College Research Initiation Award.

with arbitrary numbers of edges, arbitrary topology, and real-valued costs. In the same time bound our algorithm will detect the existence of a negative cost cycle.

Our model of random edge costs is Bloniarz's *endpoint-independent* model [5]. This is the most general model that is studied in the shortest path literature (see for example [22, 5, 20, 23]) and it can roughly be defined as follows: the distribution according to which the cost of directed edge (u, v) is chosen does not depend on the head v. This model includes the common case of all edge costs drawn from the uniform distribution, and more generally any network in which the only restriction is that all costs of edges emanating from a particular vertex v are chosen from the same distribution. Our method uses ideas from Bellman-Ford and from an algorithm of Moffat and Takaoka [20] originally intended for nonnegative cost assignments, and turns out to be quite simple drawing on the simplicity of these two algorithms.

We also show that in a restricted model of computation, the Bellman-Ford algorithm is the best possible. We consider an *oblivious* model of computation, in which the decisions of which edges to relax are made in advance, before seeing the input. We show that in this model, any algorithm whose basic operation is edge relaxation has to perform $\Omega(nm)$ edge relaxations in the worst case in order to correctly compute shortest paths. The Bellman-Ford algorithm fits into this model.

Previous and Related Work. For arbitrary real costs the existence of negative cost cycles, i.e. paths with negative cost where every vertex has degree 2, makes the SSSP problem NP-hard [14]. In the absence of negative cost cycles the fastest strongly polynomial SSSP algorithm, as mentioned above, is attributed to Bellman and Ford [3, 10] and can be implemented to run in $O(nm)$ time, worst case. This is $O(n^3)$ for dense graphs. Until recently all alternative implementations of Bellman-Ford first solved an assignment problem to find vertex potentials which allows reweighting of edges so that all edge costs are nonnegative. Then Dijkstra's algorithm [9] is applied to the reweighted network. The bottleneck in this approach is the solution of the assignment problem. The first and fastest strongly polynomial time algorithm for the assignment problem is Kuhn's Hungarian algorithm [19]. Implemented with Fibonacci heaps [11], this algorithm runs in $O(nm + n^2 \log n)$ time. Gabow and Tarjan [13] have a scaling algorithm for the assignment problem that runs in $O(\sqrt{n}m \log(nN))$ time, where N is the absolute value of the most negative edge cost. Recently Goldberg [15] has given a scaling algorithm which finds shortest paths without solving an assignment problem first; this algorithm has running time $O(\sqrt{n}m \log N)$. All these algorithms detect the existence of a negative cost cycle. We also note that if the costs are nonnegative, faster algorithms are possible, as Dijkstra's algorithm [9] implemented with Fibonacci heaps [11] runs in $O(n \log n + m)$ time.

We are not aware of any work on the average case complexity of the SSSP problem for real-valued edge costs. However the all pairs shortest path problem with nonnegative edge costs is well studied and the relevant literature spans two decades. In the all pairs shortest path problem, (abbreviated APSP) we are interested in computing the shortest paths between all $n(n-1)$ pairs of vertices. All previous work on the analysis of algorithms for networks with random non-negative edge costs has been for the APSP problem. In most of these papers a SSSP routine is run n times, using each vertex as a source in turn. Because these algorithms require an $O(n^2 \log n)$ time preprocessing sorting phase, the running times are only reported for the APSP problem, but these algorithms do give ideas for the SSSP problem. A first APSP algorithm with expected running time $O(n^2 \log^2 n)$ on networks with independently and identically distributed edge costs was presented in a classical paper by Spira [22] (see [6] for minor corrections). This result was later refined [4] to take into account non-unique edge costs and improved in [23] where an $O(n^2 \log n \log \log n)$ expected time algorithm was given. Bloniarz [5] achieved an expected running time of $O(n^2 \log n \log^* n)$ and relaxed Spira's initial assumption that edge costs are drawn independently from any single but arbitrary distribution. He introduced the endpoint-independent randomness model which is the most general in the shortest path literature. Hassin and Zemel [16] considered the case where the edge costs are uniformly distributed and gave an $O(n^2 \log n)$ expected time algorithm. Frieze and Grimmet [12] gave an $O(n^2 \log n)$ expected time algorithm for the case where edge costs are identically and independently distributed with distribution function F, where $F(0) = 0$ and F is differentiable at 0. The fastest algorithm so far under the endpoint-independent model is due to Moffat and Takaoka [20] and runs in $O(n^2 \log n)$ expected time. Recently some research has been done on randomized algorithms that use ideas from matrix multiplication [2, 21] but, for arbitrary cost assignments, only pseudopolynomial algorithms exist. The extent to which randomization can be used for faster algorithms, as was the case with e.g. minimum cut [18] and minimum spanning tree [17], is an open question.

The outline of the paper is as follows. In Section 2, we start with a high level description of the new algorithm. Subsequently we present the randomness model, and give an implementation with fast average case. In Section 3, we present the lower bound for oblivious algorithms. In Section 4, we conclude with open questions.

2 The Algorithm

In this section we give an algorithm for SSSP with average case running time $O(n^2 \log n)$ on a broad class of networks with random edge costs. We will give the algorithm in two parts. In Section 2.1 we give a modified version of the Bellman-Ford [3, 10] algorithm that reduces solving a shortest path problem to a sequence of n shortest path problems in a simpler network. Then in Section 2.2 we show how to solve this simpler shortest path problem in $O(n \log n)$ time, on average.

We use n to denote $|V|$ and m to denote $|E|$ for the network $G = (V, E, c)$ of interest. The *length* of path $p = (v_1, v_2, \ldots, v_k)$ is $k-1$ while the cost is defined as $c(p) = \sum_{i=1}^{k-1} c(v_i, v_{i+1})$. The shortest path between a pair of vertices u, v is a minimum cost simple path joining the two vertices. This minimum cost is called the *distance*, denoted $d(u,v)$. Our algorithms will concentrate on finding the distance from the source s to each vertex, and we will use $d(v)$ to denote $d(s,v)$. Our method can easily be modified to find the actual paths without asymptotic overhead.

2.1 Skeleton of the new algorithm

We begin by reviewing and modifying the Bellman-Ford algorithm. On a network where negative costs are present Bellman-Ford performs $n-1$ passes. Let $d^i(v)$ denote the distance estimate from s to v that has been found during the first i passes. During pass i, all edges (u,v) are relaxed in an arbitrary order. More precisely, pass i is implemented by first assigning $d^i(v) = d^{i-1}(v)$ for all v and then, for all edges (u,v), if

$$d^i(u) + c(u,v) < d^i(v), \tag{1}$$

then $d^i(v)$ is updated to $d^i(u) + c(u,v)$. This step is called *relaxing* edge (u,v). An extra n-th pass will reveal the existence of a negative cost cycle since any possible vertex cost decrease will be due to a non-simple path. Adding a simple preprocessing phase to Bellman-Ford, in which we sort the edge lists, we get the algorithm FAST_SSSP, which appears in Figure 2.1.

```
procedure FAST_SSSP(Network G, source s)
        for all v ∈ V do
                sort the list of edges out of v;
                d⁰(v):= 0;
        for i= 1 to n-1 do
                dⁱ =PASS(G,dⁱ⁻¹);
        for all edges (u,v) ∈ E do
                if dⁿ⁻¹(u) + c(u,v) < dⁿ⁻¹(v) then
                        report a negative cost cycle and break;
```

Fig. 1. Algorithm FAST_SSSP

It is easy to show by induction that after the i-th pass all shortest paths of length *at most i* have certainly been discovered, and possibly others. More precisely, let $\hat{d}^i(v)$ be the cost of the shortest path from s to v of length at most i. Then it for all i and v, Bellman-Ford maintains the invariant that at the end of pass i, $d^i(v) \leq \hat{d}^i(v)$.

In our statement of FAST_SSSP in Figure 2.1, we treat as a black box procedure PASS, which updates the distance labels. The standard Bellman-Ford algorithm implements PASS by relaxing (applying equation (1) to) all m edges yielding a total running time of $O(nm)$.

Implementing PASS. Our improvements will come from implementing carefully procedure PASS to run in $O(n \log n)$ expected time. We begin by observing that the correctness of Bellman-Ford is not harmed if we only compute the $\hat{d}^i(v)$'s during iteration i; after $n - 1$ iterations, we will then have computed $\hat{d}^{n-1}(v)$, which is exactly what we want.

This allows us to make the following modifications:

Modification 1. *We can relax all edges in "parallel".* Typically, a sequential implementation of Bellman-Ford does not do this, as it only slows the algorithm down. However, we can accomplish this by replacing condition (1) with the following condition:

$$d^{i-1}(u) + c(u, v) < d^i(v). \tag{2}$$

If we apply this rule instead, it will clearly be the case that after pass i, $d^i(v) = \min\{\min_{u \in V}\{d^{i-1}(u) + c(u, v)\}, d^{i-1}(v)\}$. However, since we are only allowing the length of a path to grow by at most one edge in each pass, we can show inductively that we are actually computing $\hat{d}^i(v)$ in iteration i.

Modification 2. *During pass i only vertices already at length exactly $i - 1$ from the source need to have their outgoing edges relaxed.* Consider a vertex v for which $d^{i-1}(v) = d^{i-2}(v)$, i.e. a vertex that is not at length exactly $i - 1$ from the source. Then this vertex had its outgoing edges relaxed at some prior phase $j \leq i - 1$, and hence for all neighbors w,

$$d^i(w) \leq d^j(w) \leq d^j(v) + c(v, w) = d^{i-1}(v) + c(v, w).$$

Thus, for all outgoing edges (v, w), $d^i(w) \leq d^{i-1}(v) + c(v, w)$, and applying (2) to v's outgoing edges will not cause any vertex labels to be updated.

In light of the two observations above, we observe that during pass i, relaxing all edges in the set I_v of incoming edges to vertex v may be too much. Of all the edges in I_v only one is crucial, namely the one that will give the minimum value for $d^i(v)$. Of course, we don't know which edge this is *a priori*, but we will capture a set of edges that will contain it. To do this, we recast the implementation of pass i as a modified SSSP problem in an auxiliary network. Let a *single-source, two-level network* be a directed network $G_\epsilon = (V_\epsilon, E_\epsilon, c_\epsilon)$ where $V_\epsilon = s \cup V_1 \cup V_2$ and edges connect only the source to vertices in V_1 and the vertices in V_1 to vertices in V_2. The cost function c_ϵ ranges over the reals. We show how an auxiliary single source, two-level network can be used in the implementation of PASS.

We define now a particular G_ϵ in terms of the input to PASS, namely $G = (V, E, c)$ and $d^{i-1}(.)$. Let V' be the set of vertices v in G whose current shortest path estimate derives from a path of length exactly $i - 1$, i.e. $d^{i-1}(v) < d^{i-2}(v)$. Then V_1 contains a copy of every vertex in V'. V_2 contains a copy of every vertex $v \in V - s$. We introduce two types of edges in E_ϵ. First, for each pair $(x, y) \in V_1 \times V_2$, where x is a copy of u, y is a copy of v and $(u, v) \in E$, we add an edge (x, y) with $c_\epsilon(x, y) = c(u, v)$. Second, for all $x \in V_1$, x a copy of v, we add an edge (s, x) with $c_\epsilon(s, x) = d^{i-1}(v)$.

We use these definitions in the upcoming lemma.

Lemma 1. *Let $T = T(n, m)$ be the time to solve a SSSP problem in a single-source, two-level network with $\Theta(n)$ vertices and $O(m)$ edges. Then, routine PASS(G, d^{i-1}) can be implemented to compute $d^i(.)$ in $O(T + n)$ time. Hence, FAST_SSSP is correct and has time complexity $O(n^2 + nT + m \log n)$.*

Proof. The implementation of pass i reduces to solving SSSP in the auxiliary network G_ϵ followed by a finishing step. Let SSSP_AUX(G_{aux}, S) be any routine that solves SSSP on a network G_{aux} in time T where S is the set of vertices in G_{aux} with known distances. We implement PASS as follows:

procedure PASS$(G,\ d^{i-1})$
    d_ϵ = SSSP_AUX$(G_\epsilon,\ s \cup V_1)$;
    $d^i(v) = \min\{d_\epsilon(v), d^{i-1}(v)\}$;

By the construction of G_ϵ, we know that the value of $d_\epsilon(v)$ for $v \in V_1$ is equal to $d^{i-1}(v)$. Thus computing distances $d_\epsilon(v)$ in G_ϵ for the vertices in V_2 is equivalent to computing

$$\min_u \{d^{i-1}(u) + c(u, v)\},\ \forall v$$

in G. Note that this is the same as calculating the left hand side of (2) for all edges in G. Thus in the last step of PASS, when $d^i(v)$ is calculated as shown, it is equivalent to the process described above where d^i is first initialized to d^{i-1} and then the test in (2) is applied to every edge. Inductively, if the given $d^{i-1}(.)$ is correct, $d^i(.)$ as output by PASS captures the shortest paths of length at most i in G and thus routine PASS is correct.

For the running time, we notice that G_ϵ does not need to be explicitly constructed. All that is required is to identify the set of vertices whose current shortest path length is exactly $i - 1$, and hence setting up the modified SSSP computation can easily be done in $O(n)$ time. The last step requires an additional $O(n)$ time, so the total time required for PASS is $O(T + n)$ and the lemma follows. □

2.2 An implementation with fast average case

Our goal now is to implement the SSSP_AUX routine to run in $o(n^2)$ time on the particular network G_ϵ. We are unaware of a method achieving such a worst case bound but we show how to do it in $O(n \log n)$ expected time on networks with random edge costs. In this section we define the class of probability measures for which our analysis holds and then present an algorithm by Moffat and Takaoka [20] and show that it can be used to efficiently find shortest paths in G_ϵ.

We define first the randomness model used for the analysis. The definition follows [5] except that we allow negative costs as well. Let $\mathcal{G}_n$ be the set of all n vertex directed networks and suppose P is a probability measure on $\mathcal{G}_n$. We may identify $\mathcal{G}_n$ with the set of all $n \times n$ matrices with entries in $(-\infty, +\infty)$. P is uniquely characterized by its distribution function $F_P : \mathcal{G}_n \to [0, 1]$ defined by

$$F_P(G) = P\{G' \in \mathcal{G}_n | c_{G'}(i, j) \le c_G(i, j) \text{ for } 1 \le i, j \le n\}.$$

We say that P is *endpoint independent* if, for $1 \leq i, j, k \leq n$ and $G \in \mathcal{G}_n$, we have that

$$F_P(G) = F_P(G'),$$

where G' is obtained from G by interchanging the costs of edges (i, j) and (i, k). Intuitively exchanging endpoints of edges emanating from any fixed vertex does not affect the probability.

Several natural edge cost assignments are endpoint independent, including the one used by Spira [22] where edge costs are independently, identically distributed random variables. Another probability measure that meets the endpoint independence criterion is when each source vertex i has a probability measure P_i associated with it and the entries $c_G(i, j)$ of the matrix are drawn independently according to distribution P_i. The reader is referred to [5] for further examples.

We proceed with a high level description of the Moffat-Takaoka method. Moffat and Takaoka give a SSSP algorithm with expected running time $O(n \log n)$ under the above input model for a nonnegative cost assignment, assuming all of the edge lists are sorted. Their algorithm is similar to Dijkstra's [9] and exploits the fact that vertices extracted in increasing cost order from a priority queue cannot have their cost decreased later on; once a vertex v is removed from the priority queue, its distance estimate is the cost of the actual shortest path from the source to v. Of course this is not true, in general, when edges with negative costs are present.

We now review the algorithm of Moffat and Takaoka. For concreteness, we will refer to the algorithm as MT and assume that is takes, as input, a network G, and a set S of vertices of G whose shortest path distances have already been computed. Recall that it relies on all edges having nonnegative costs. The set S of *labeled* vertices is maintained throughout. These are the vertices for which the shortest paths are known. For every element v in S a *candidate* edge (v, t) is maintained, known to be the least cost unexamined edge out of v. Every candidate edge gives rise to a candidate path, there are at most $|S|$ of them at any one time. The candidate paths are maintained in a priority queue. At every step we seek to expand S by picking the least cost candidate path p. If p with final edge (v, t) leads to an unlabeled vertex t, it is deemed to be *useful*, and t is added to S. Otherwise, the path is ignored, but in both cases the candidate path (v, t) is replaced by a different path ending in (v, t') with $c(v, t') \geq c(v, t)$. There are two extremes in the policy for picking this next candidate (v, t'): either select simply the next largest edge out of v or scan the sorted adjacency list until a useful edge (v, t') is found. Based on the policy selection we obtain Spira's [22] and Dantzig's [8] methods respectively. The Moffat-Takaoka routine, MT, uses Dantzig's algorithm up to a critical point with respect to the size of S and then switches to Spira's. Using standard binary heaps we can obtain the following lemma.

Lemma 2 [20]. *Let G be a network of n vertices with nonnegative edge costs drawn from an endpoint-independent distribution and S a set of vertices of G whose shortest path distances have been computed. Then $\mathrm{MT}(G, S)$ solves SSSP on G in $O(n \log n)$ expected time given that the edge lists are presorted by cost.*

Proof. See Theorem 1 in [20]. □

We note that, in general, the nonnegativity condition is necessary for correctness as it is in Dijkstra's algorithm. However, we will now show that for a a single-source, two-level network with real costs, such as G_ϵ, MT computes correctly shortest paths.

Lemma 3. *Let $G_\epsilon = (s \cup V_1 \cup V_2, E_\epsilon, c_\epsilon)$ be the single-source, two-level network defined above. Then $\mathrm{MT}(G_\epsilon, s \cup V_1)$ solves SSSP on G_ϵ in $O(n \log n)$ expected time given that the edge lists of vertices in V_1 are presorted by cost, and the edge costs are drawn from an endpoint-independent distribution.*

Proof. Given a set S of already labeled vertices MT chooses to label next out of the vertices adjacent to S, the one corresponding to the least cost candidate path. In the case of G_ϵ the only paths to vertices in V_2 are the ones passing through V_1 and MT will process each of them in sorted order until all the vertices in V_2 are labeled. Thus it will compute correctly shortest paths. For the running time, it suffices to notice that the analysis in Theorem 1 of [20] relies on the following fact. In an endpoint-independent distribution when $|S| = j$ each candidate leads to each of the $n - j$ unlabeled vertices with equal probability. This fact is not harmed by the negativity of a candidate edge. Thus by Lemma 2 MT on G_ϵ has $O(n \log n)$ expected time.

An alternative self-contained proof is the following. A sufficiently large constant C may be added to all edge costs in G_ϵ to make them nonnegative. This reweighting increases the costs of all paths to $v \in V_2$ by $2C$, and MT can be used to compute the $d_\epsilon(v)$'s correctly in $O(n \log n)$ expected time. Therefore MT also takes $O(n \log n)$ expected time on the original, non-reweighted network. We note that in the context of our FAST_SSSP algorithm the reweighting of all the auxiliary networks can be done in $O(n^2)$ worst case total time; we do not include it though in the algorithm to avoid unnecessary complication. □

By Lemmata 1, 3 we obtain the following result.

Theorem 4. *Let G be a network of n vertices with real edge costs drawn from an endpoint-independent distribution. Algorithm FAST_SSSP solves SSSP on G in $O(n^2 \log n)$ expected time if no negative cost cycle exists. Otherwise it reports the existence of a negative cost cycle in the same time bound.*

The worst case complexity of the Moffat-Takaoka subroutine is $O(n^2)$ therefore the worst case running time of our algorithm is $O(n^3)$.

3 A Lower Bound

For over thirty years Bellman-Ford has been the fastest algorithm for the general SSSP problem with an $O(nm)$ worst case running time. It is thus natural to wonder whether one can show a lower bound on the running time of SSSP in networks with real-valued edge costs. We do not know how to do this, but we can show a lower bound in a restricted model of computation that captures Bellman-Ford-like algorithms.

Observe that the Bellman-Ford algorithm disregards any information about the network except the number of edges, and performs a fixed sequence of edge relaxations until no vertex label can be decreased. Motivated by this we define an *oblivious* algorithms for SSSP to be one that, given as input a network G with m edges numbered $1, \ldots, m$, decides on a sequence $S_m = e_{i_1} e_{i_2} \ldots e_{i_k}$ of k edge relaxations, $1 \leq e_{i_j} \leq m$ on the basis of m alone. An oblivious algorithm performs only the chosen relaxations. Obviously Bellman-Ford falls into the oblivious class, as do generalizations that do not restrict the algorithm to operate in phases. In the following theorem we prove that the Bellman-Ford algorithm performs an asymptotically optimal number of relaxations within this model. We call an edge *optimal* if it belongs to a shortest path.

Theorem 5. *Let A be any oblivious SSSP algorithm. There exists a network G with n vertices and m edges on which A must perform $\Omega(nm)$ edge relaxations worst case in order to correctly output the shortest paths.*

Proof. We restrict our attention to networks G, with $m > n$, in which the maximum length of a shortest path is exactly $n - 1$, i.e. the shortest path tree is a single path. Let k be the length of the relaxation sequence $S_m = e_{i_1} e_{i_2} \ldots e_{i_k}$ that A performs on input G. A *subsequence* σ of S_m is any sequence of the form $\sigma = e_{i_{j_1}} e_{i_{j_2}} \ldots e_{i_{j_l}}$ with $1 \leq j_1 < j_2 < \ldots j_l \leq k$. The correctness of any relaxation-based algorithm comes by guaranteeing that the edges in the shortest path tree are relaxed in an order that is a topological sort of the edges of the tree. In our case, in which the shortest path tree is a path, this means that the edges of the path must be relaxed in the linear order specified by the path. Before seeing the input, it is possible that any sequence s_n of $n - 1$ out of the m edges can be the chain of optimal edges, and there are $\frac{m!}{(m-(n-1))!}$ such chains. A will correctly compute the shortest path distances only if s_n is a subsequence of S_m. The relaxation sequence of length k has exactly $\binom{k}{n-1}$ subsequences of length $n - 1$. In order for the algorithm A to be correct on all networks, it must be that

$$\binom{k}{n-1} \geq \frac{m!}{(m-(n-1))!} \tag{3}$$

For simplicity of presentation we replace $n - 1$ by r in the following. We want to compute k_0, the smallest possible value of k, for which (3) holds. Increasing the left hand side and decreasing the right hand side can only make this lower bound smaller. Let $K(\rho(k))$ denote the minimum k such that a relation $\rho(k)$ holds. We assume, wlog, that $0 \leq n \leq k$. We can upper bound the left-hand side of (3) by using the identity [7]

$$\frac{k^k}{r^r(k-r)^{k-r}} \geq \binom{k}{r}. \tag{4}$$

We can lower bound the right-hand side of (3) by

$$\frac{m!}{(m-r)!} = m(m-1)\ldots(m-r+1) \geq (m-r)^r. \tag{5}$$

By (4) and (5), we get

$$k_0 = K\left(\binom{k}{r} \geq \frac{m!}{(m-r)!}\right) \geq K\left(\frac{k^k}{r^r(k-r)^{k-r}} \geq (m-r)^r\right) = k_1$$

But $k^{k-r}/(k-r)^{k-r} \leq e^r$ thus we have that

$$k_1 \geq K\left(\frac{k^r e^r}{r^r} \geq (m-r)^r\right)$$

$$= K\left(k^r \geq \left(\frac{r(m-r)}{e}\right)^r\right)$$

$$= K\left(k \geq \frac{r(m-r)}{e}\right).$$

Substituting $n-1$ for r gives $k_0 = \Omega(nm)$. $\qquad\qquad\qquad\qquad$ □

4 Conclusion and Open Problems

In this paper we provided the first algorithm in over 30 years that achieves $o(n^3)$ running time for SSSP with arbitrary real costs, although not worst case. The complexity gap between $O(nm)$ time for for dense graphs and $O(n^2 \log n)$ time is considerable. This raises the natural question of whether an $o(nm)$ worst case time algorithm for SSSP exists. If the answer is no, can one reason about a lower bound in a non-oblivious computational model.

Acknowledgements. The first author wishes to thank Donald Johnson for starting him working on shortest paths and for all the valuable discussions. The second author thanks Tom Sundquist for useful discussions.

References

1. R. K. Ahuja, T. L. Magnanti and J. B. Orlin. *Network Flows*, Prentice Hall, Englewood Cliffs, NJ, 1993.
2. N. Alon, Z. Galil and O. Margalit. *On the exponent of the all pairs shortest path problem*, Proc. 32nd IEEE Symp. Found. Comp. Sci., 1991, 569-575.
3. R. Bellman. *On a routing problem*, Quarterly of Appl. Math., **16** (1958), 87-90.
4. P. A. Bloniarz, A. Meyer and M. Fischer. *Some observations on Spira's shortest path algorithm*, Tech. Rept. 79-6, Comp. Sci. Dept., State University of New York at Albany, Albany, NY, 1979.
5. P. A. Bloniarz. *A shortest path algorithm with expected time $O(n^2 \log n \log^* n)$*, SIAM J. Comput., **12** (1983), 588-600.
6. J. S. Carson and A. M. Law. *A note on Spira's algorithm for the all-pairs shortest path problem*, SIAM J. Comput., **6** (1977), 696-699.
7. Thomas H. Cormen, Charles E. Leiserson, and Ronald L. Rivest. *Introduction to Algorithms*. MIT Press, Cambridge, MA, 1990.

8. G. B. Dantzig. *On the shortest route through a network,* Management Sci., **6** (1960), 187-190.

9. E. W. Dijkstra. *A note on two problems in connection with graphs,* Numerische Mathematics, **1** (1959), 269-271.

10. L. R. Ford Jr. and D. R. Fulkerson. *Flows in Networks,* Princeton University Press, Princeton, NJ, 1962.

11. M. L. Fredman and R. E. Tarjan.*Fibonacci heaps and their uses in improved network optimization problems,* J. ACM, **34** (1987), 596-615.

12. A. M. Frieze and G. R. Grimmet. *The shortest-path problem for graphs with random arc-lengths,* Discrete Applied Mathematics, **10** (1985), 57-77.

13. H. N. Gabow and R. E. Tarjan.*Faster scaling algorithms for network problems,* SIAM Journal on Computing, **18** (1989), 1013-1036.

14. M. S. Garey and D. S. Johnson. *Computers and Intractability: A Guide to the Theory of NP-completeness,* W. H. Freeman, San Francisco, CA, 1979.

15. A. V. Goldberg. *Scaling algorithms for the shortest path problem,* SIAM J. Comput., **24** (1995), 494-504. Preliminary version in Proc. 4th Annual ACM-SIAM Symp. on Discrete Algorithms, 1993, 222-231.

16. R. Hassin and E. Zemel. *On the shortest paths in graphs with random weights.* Mathematics of Operations Research, **10** (1985), 557-564.

17. D. R. Karger, P. N. Klein and R. E. Tarjan. *A randomized linear-time algorithm to find minimum spanning trees,* J. of ACM, **42** (1995), 321-328.

18. D. R. Karger and C. Stein. *An $\tilde{O}(n^2)$ algorithm for minimum cuts,* Proc. 25th ACM Symp. on Theory of Computing, 1993, 757-765. To appear in J. ACM.

19. H. W. Kuhn. *The Hungarian method for the assignment problem,* Naval Research Logistics Quarterly, **2** (1955), 83-97.

20. A. Moffat and T. Takaoka. *An all pairs shortest path algorithm with expected time $O(n^2 \log n)$,* SIAM J. Comput., **16** (1987), 1023-1031.

21. R. Seidel. *On the all-pairs-shortest-path problem,* Proc. 24th ACM Symp. on Theory of Computing, 1992, 745-749.

22. P. M. Spira. *A new algorithm for finding all shortest paths in a graph of positive arcs in average time $O(n^2 \log^2 n)$,* SIAM J. Comput., **2** (1973), 28-32.

23. T. Takaoka and A. M. Moffat. *An $O(n^2 \log n \log \log n)$ expected time algorithm for the all shortest distance problem,* in Mathematical Foundations of Computer Science, Lecture Notes in Computer Science, Vol. **88**, P. Dembinski, ed., Springer Verlag, Berlin, 1980, 643-655.

A Network-Flow Technique for Finding Low-Weight Bounded-Degree Spanning Trees

Sándor P. Fekete[1]*, Samir Khuller[2] **, Monika Klemmstein[1]***,
Balaji Raghavachari[3] † and Neal Young[4] ‡

[1] Center for Parallel Computing, Universität zu Köln, D-50923 Köln, Germany.
[2] Computer Science Department and Institute for Advanced Computer Studies,
University of Maryland, College Park, MD 20742.
[3] Department of Computer Science, The University of Texas at Dallas, Box 830688,
Richardson, TX 75083-0688.
[4] Department of Computer Science, Dartmouth College, Hanover NH 03755-3510.

Abstract. Given a graph with edge weights satisfying the triangle inequality, and a degree bound for each vertex, the problem of computing a low weight spanning tree such that the degree of each vertex is at most its specified bound is considered. In particular, modifying a given spanning tree T using *adoptions* to meet the degree constraints is considered. A novel network-flow based algorithm for finding a good sequence of adoptions is introduced. The method yields a better performance guarantee than any previously obtained. Equally importantly, it takes this approach to the limit in the following sense: if any performance guarantee that is solely a function of the topology and edge weights of a given tree holds for *any* algorithm at all, then it also holds for our algorithm. The performance guarantee is the following. If the degree constraint $d(v)$ for each v is at least 2, the algorithm is guaranteed to find a tree whose weight is at most the weight of the given tree times $2 - \min \left\{ \frac{d(v)-2}{\deg_T(v)-2} : \deg_T(v) > 2 \right\}$, where $\deg_T(v)$ is the initial degree of v. Examples are provided in which no lighter tree meeting the degree constraint exists. Linear-time algorithms are provided with the same worst-case performance guarantee.

Choosing T to be a minimum spanning tree yields approximation algorithms for the general problem on geometric graphs with distances induced by various L_p norms. Finally, examples of Euclidean graphs are provided in which the ratio of the lengths of an optimal Traveling Salesman path and a minimum spanning tree can be arbitrarily close to 2.

* E-Mail: sandor@zpr.uni-koeln.de.
** Research supported by NSF Research Initiation Award CCR-9307462 and an NSF CAREER Award CCR-9501355. E-mail: samir@cs.umd.edu.
*** E-Mail: mklemmst@zpr.uni-koeln.de.
† Research supported by NSF Research Initiation Award CCR-9409625. E-mail : rbk@utdallas.edu.
‡ Part of this research was done while at School of ORIE, Cornell University, Ithaca NY 14853 and supported by Éva Tardos' NSF PYI grant DDM-9157199. E-mail : ney@cs.dartmouth.edu.

1 Introduction

Given a complete graph with edge weights satisfying the triangle inequality and a degree bound for each vertex, we consider the problem of computing a low-weight spanning tree in which the degree of each vertex is at most its given bound. In general, it is NP-hard to find such a tree. There are various practical motivations: the problem arises in the context of VLSI layout and network design [8, 12, 20] (such as in the Bellcore software *FIBER OPTIONS*, used for designing survivable optimal fiber networks). The special case of only one vertex with a degree-constraint has been examined [5, 6, 9]; a polynomial time algorithm for the case of a fixed number of nodes with a constrained degree was given by Brezovec et al. [2]. Computational results for some heuristics for the general problem are presented in [14, 19, 21]. Papadimitriou and Vazirani [15] raised the problem of finding the complexity of computing a minimum-weight degree-4 spanning tree of points in the plane. Some geometric aspects are considered in [10, 13, 17].

In this paper, we consider modifying a given spanning tree T, to meet the degree constraints. We introduce a novel network-flow based algorithm that does this optimally in the following sense: if for some algorithm a worst-case performance guarantee can be proved that is solely a function of the topology and edge weights of T, then that performance guarantee also holds for our algorithm. We prove this by showing that our algorithm finds the *optimal* solution for graphs in which the weight of each edge (u, v) equals the cost of the $u \rightsquigarrow v$ path in T.

We also show the following more concrete performance guarantee: If the degree constraint $d(v)$ for each v is at least 2, our algorithm finds a tree whose weight is at most the weight of the T times

$$2 - \min\left\{\frac{d(v) - 2}{\deg_T(v) - 2} : \deg_T(v) > 2\right\},$$

where $\deg_T(v)$ is the initial degree of v. For instance, the degree of each vertex v can be reduced by nearly half, to $1 + \lceil \deg_T(v)/2 \rceil$, without increasing the weight of the tree by more than 50%. For comparison, note that a factor of 2 is straightforward with standard shortcutting techniques. We also describe *linear-time* algorithms that achieve this ratio.

This performance guarantee is optimal in the sense that for any $D \geq d \geq 2$, if T is a complete rooted $(D-1)$-ary tree with unit edge weights and the edge weights in G are those induced by paths in T, then the weight of any spanning tree with maximum degree d is at least the weight of T times $2 - \frac{d-2}{D-2} - o(1)$.

The restriction $d(v) \geq 2$ is necessary to obtain constant performance bounds. Consider the case when T is a simple path of unit weight edges, with the remaining edge weights again induced by T. Any spanning tree in which all but one vertex has degree one is heavier than T by a factor of $\Omega(n)$, the number of vertices in T.

For many metric spaces, graphs induced by points in the space have minimum spanning trees of bounded maximum degree. In such cases our algorithms can be used to find spanning trees of even smaller degree with weight bounded by a factor strictly smaller than 2 times the weight of a minimum spanning tree (MST). For example, in the L_1 metric, a degree-4 MST can be found [17], so that we can find a degree-3 tree with weight at most 1.5 times the weight of an MST. We discuss similar results for the L_1, L_2, and L_∞ norms. For some of these norms, this improves the best current performance guarantees.

Finally, we disprove the following conjecture of [11]: "In Euclidean graphs, perhaps a Traveling Salesman path of weight at most $(2-\varepsilon)$ times the minimum spanning-tree weight always exists..."

Our algorithms modify the given tree by performing a sequence of *adoptions*. Our polynomial-time algorithm performs an optimal sequence of adoptions. Adoptions have been previously used to obtain bounded-degree trees in weighted graphs [10, 16, 18]. The main contributions of this paper are a careful analysis of the power of adoptions and a network-flow technique for selecting an optimal sequence of adoptions. The method yields a stronger performance guarantee and may yield better results in practice. The analysis of adoptions shows that different techniques will be necessary if better bounds are to be obtained.

In the full version of their paper, Ravi et al. [16, Thm. 1.9] (if slightly generalized and improved[5]) gave an algorithm with a performance guarantee of

$$2 - \min\left\{\frac{d(v) - 2}{\deg_T(v) - 1} : v \in V, \deg_T(v) > 2\right\}$$

provided each $d(v) \geq 3$. The performance guarantee of our algorithm is better.

In Euclidean graphs (induced by points in $\mathbb{R}^d$), minimum spanning trees are known to have bounded degree. For such graphs, Khuller, Raghavachari and Young [10] gave a linear-time algorithm to find a degree-3 spanning tree of weight at most 5/3 times the weight of a minimum spanning tree. For points in the plane, the performance guarantee of their algorithm improves to 1.5; if the tree is allowed to have degree four, the ratio improves further to 1.25.

In unweighted graphs, Fürer and Raghavachari [4] gave a polynomial-time algorithm to find a spanning tree of maximum degree exceeding the minimum possible by at most one. In arbitrary weighted graphs, Fischer [3] showed that a minimum spanning tree with maximum degree $O(\delta^* + \log n)$ can be computed in polynomial time, where δ^* is the minimum maximum degree of any minimum spanning tree. He also provided an algorithm that finds a minimum spanning tree with degree $k(\delta^* + 1)$ where k is the number of distinct edge weights.

2 Adoption

Fix the graph $G = (V, V \times V)$ and the edge weights $w : V \times V \to \mathbb{R}$. The algorithm starts with a given tree T and modifies it by performing a sequence of *adoptions*. The adoption operation (illustrated in Figure 1) is as follows:

---

[5] To obtain the improved bound one has to change the proof slightly by upperbounding $c(v_1 v_2) - c(v v_2)$ by $c(v v_1)$ and not $c(v v_2)$ as is done in [16].

ADOPT(u, v)

Precondition: Vertex v has degree at least two in the current tree.

1 Choose a neighbor x of v in the current tree other than the neighbor on the current $u \rightsquigarrow v$ path.
2 Modify the current tree by replacing edge (v, x) by (u, x).

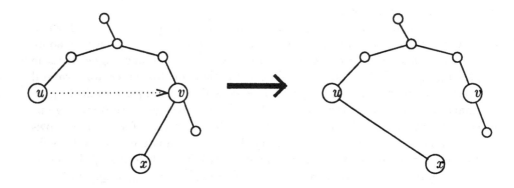

Fig. 1. Vertex u *adopts* a neighbor of v

ADOPT(u, v) decreases the degree of v by one, at the expense of increasing the degree of u by one and increasing the weight of the tree by $w(x, u) - w(x, v) \leq w(u, v)$.

2.1 The Adoption Network

Definition 1. The *deficit* of vertex v with respect to T is $\deg_T(v) - \mathrm{d}(v)$.

Starting with a given tree, consider a sequence of adoptions
ADOPT(u_1, v_1), ADOPT(u_2, v_2), ...

- The sequence is *legal* if the precondition for each adoption is met.
- A sequence is *feasible* if, for each vertex, the change in its degree, i.e., its new degree minus its old degree, is at least its deficit.
- The *cost* of the sequence is $\sum_i w(u_i, v_i)$.

The legal, feasible adoption sequences are precisely those that yield a tree meeting the degree constraints. The cost of a sequence is an upper bound on the resulting increase in the weight of the tree. Our goal is to find a feasible legal sequence of minimum cost. For brevity, we call such a sequence a minimum-cost sequence.

The problem reduces to a minimum-cost flow problem [1] in a flow network that we call the *adoption network* for T. The adoption network is defined as follows. Starting with G, replace each edge (u, v) by two directed edges (u, v)

and (v, u), each with cost $w(u, v)$ and infinite capacity. Assign each vertex a demand equal to its deficit.

A *flow* is an assignment of a non-negative real value (called the flow on the edge) to each edge of the network. For each vertex v, the *surplus* at v is the net flow assigned to incoming edges minus the net flow assigned to outgoing edges. A flow is *legal* if the surplus at each vertex is at most one less than its degree. A flow is *feasible* if the surplus at each vertex is at least its demand. The *cost* of the flow is the sum, over all edges, of the cost of the edge times the flow on the edge.

Since the demands are integers, there exists an integer-valued minimum-cost feasible flow [1]. Assuming that each degree constraint is at least 1, their exists such a flow that is also legal. For brevity, we call such a flow a minimum-cost flow.

Lemma 2. *The following statements are true:*

1. *The adoption sequences correspond to integer-valued flows. The correspondence preserves legality, feasibility, and cost.*
2. *The integer-valued flows correspond to adoption sequences. The correspondence preserves legality and feasibility; it does not increase cost.*

Proof. Given a sequence of adoptions, the corresponding flow f assigns a flow to each edge (u, v) equal to the number of times u adopts a neighbor of v. It can be verified that this correspondence preserves legality, feasibility, and cost.

Conversely, given an integer-valued flow f, modify it if necessary (by canceling flow around cycles) so that the set of edges with positive flow is acyclic. This does not increase the cost. Next, order the vertices so that, for each directed edge (u, v) with positive flow, u precedes v in the order. Consider the vertices in reverse order. For each vertex u, for each edge (u, v) with positive flow, have u adopt $f(u, v)$ neighbors of v. It can be verified that this sequence of adoptions preserves legality and feasibility, and does not increase cost.

3 Polynomial-Time Algorithm

An acyclic, integer, minimum-cost flow can be found in polynomial time [1]. The corresponding legal, feasible adoption sequence can be performed in polynomial time as described in the proof of the second part of Lemma 2. This gives a polynomial-time algorithm.

3.1 Optimality in Tree-Induced Metrics

The following lemma shows that this algorithm is optimal among algorithms that examine only the weights of edges of the given tree.

Lemma 3. *Given a weighted graph $G = (V, E)$ and a spanning tree T such that the weight of each edge in G equals the weight of the corresponding path in T, a minimum-cost sequence of adoptions yields an optimal tree.*

Proof. Fix an optimal tree. Note that the degree of v in the optimal tree is at most $d(v)$; assume without loss of generality that it is exactly $d(v)$. For each subset S of vertices, let $\deg_T(S)$ and $d(S)$ denote the sum of the degrees of vertices in S in T and in the optimal tree, respectively. Define a flow on the edges of T as follows: for each edge (u, v) in T, let $f(u, v) = d(S_u) - \deg_T(S_u)$, where S_u is the set of vertices that are reachable from u using edges in T other than (u, v), provided $f(u, v)$ as defined is non-negative. Inductively it can be shown that for each vertex v, the net flow into it is $\deg_T(v) - d(v)$, so that the adoption sequence determined by the flow f achieves a tree with the same degrees as the optimal tree.

We will show that the cost of the flow, and therefore the cost of the adoption sequence, is at most the difference in the weights of the two trees. This implies that the tree obtained by the adoption sequence is also an optimal tree.

To bound the cost of the flow, we claim that the flow is "necessary" in the following sense: for each edge (u, v) in T, at least $f(u, v) + 1$ edges in the optimal tree have one endpoint in S_u and the other in $V - S_u$. To prove this, let c be the number of edges in the optimal tree crossing the cut $(S_u, V - S_u)$. Note that $\deg_T(S_u) = 2(|S_u| - 1) + 1$. Since the optimal tree is acyclic, the number of edges in the optimal tree with both endpoints in S_u is at most $|S_u| - 1$. Thus $d(S_u) \leq 2(|S_u| - 1) + c = \deg_T(S_u) - 1 + c$. Rewriting gives $c \geq d(S_u) - \deg_T(S_u) + 1 = f(u, v) + 1$. This proves the claim.

To bound the cost of the flow, for each edge (u, v), charge $w(u, v)$ units to each edge in the optimal tree crossing the cut $(S_u, V - S_u)$. By the claim, at least the cost of the flow, plus the cost of T, is charged. However, since the cost of each edge in the optimal tree equals the weight of the corresponding path in T, each edge in the optimal tree is charged at most its weight. Thus, the net charge is bounded by the cost of the optimal tree.

Note that given the exact degrees of the desired tree (for instance, if the degree constraints sum to $2(|V| - 1)$), the optimal flow in Lemma 3 can be computed in linear time.

3.2 Worst-Case Performance Guarantee

The next theorem establishes a worst-case performance guarantee for the algorithm in general graphs satisfying the triangle inequality.

Theorem 4. *Given a graph $G = (V, E)$ with edge weights satisfying the triangle inequality, a spanning tree T, and, for each vertex v, a degree constraint $d(v) \geq 2$, the algorithm produces a tree whose weight is at most the weight of T times*

$$2 - \min\left\{\frac{d(v) - 2}{\deg_T(v) - 2} : v \in V, \deg_T(v) > 2\right\}.$$

Proof. The increase in the cost of the tree is at most the cost of the best sequence. By Lemma 2, this is bounded by the cost of the minimum-cost flow. We exhibit a fractional feasible, legal flow whose cost is appropriately bounded. The minimum-cost flow is guaranteed to be at least as good.

Root the tree T at an arbitrary vertex r. Push a uniform amount of flow along each edge towards the root as follows. Let $p(v)$ be the parent of each non-root vertex v. For a constant c to be determined later, define

$$f(u,v) = \begin{cases} c & \text{if } v = p(u) \\ 0 & \text{otherwise.} \end{cases}$$

The cost of the flow is c times the weight of T. Let v be any vertex. The surplus at v is at least $c(\deg_T(v) - 2)$. We choose c just large enough so that the flow is feasible.

There are three cases. If $\deg_T(v) = 1$, the deficit at v will be satisfied provided $c \leq 1$ and $d(v) \geq 2$. If $\deg_T(v) = 2$, the deficit at v will be satisfied provided $d(v) \geq 2$. For $\deg_T(v) > 2$, the deficit will be satisfied provided

$$c \geq \frac{\deg_T(v) - d(v)}{\deg_T(v) - 2} = 1 - \frac{d(v) - 2}{\deg_T(v) - 2}.$$

Thus, taking

$$c = 1 - \min\left\{\frac{d(v) - 2}{\deg_T(v) - 2} : v \in V, \deg_T(v) > 2\right\}$$

gives the result.

4 Optimality of Performance Guarantee

In this section, we show that the worst-case performance guarantee established in Theorem 4 is the best obtainable.

Lemma 5. *Consider an n-vertex weighted graph G with a spanning tree T such that the weight of each edge in T is 1 and the weight of each remaining edge is the weight of the corresponding path in T. If T corresponds to a complete rooted $(D-1)$-ary tree, then the weight of any spanning tree with maximum degree d is at least the weight of T times*

$$2 - \frac{d - 2}{D - 2} - o(1),$$

where $o(1)$ tends to 0 as n grows.

Proof. Fix any spanning tree T' of maximum degree d. Let S_i denote the vertices at distance i from the root in T. Consider the flow on the edges of T corresponding to T', as defined in the proof of Lemma 3. The proof shows that at least $|S_i|(d - D) - 1$ units of flow cross the cut $(V - S_i, S_i)$. Thus the net cost of the flow is at least $\sum_{i=0}^{k-1} |S_i|(d - D) - 1$. The cost of T is $\sum_{i=0}^{k-1} |S_{i+1}| - |S_i|$. Since

$|S_{i+1}| = |S_i|(D-1)+1$, so $|S_{i+1}| - |S_i| = |S_i|(D-2)+1$, the ratio of the cost of the flow to the cost of T is at least

$$\frac{\sum_{i=0}^{k-1} |S_i|(d-D)-1}{\sum_{i=0}^{k-1} |S_i|(D-2)+1}.$$

Simplifying shows that the ratio is at least $(d-D)/(D-2) - o(1)$. Since the ratio of the cost of T' to the cost of T is 1 more than this, the result follows.

Next we observe that the $d(v) \geq 2$ constraint is necessary to obtain any constant performance guarantee:

Lemma 6. *Consider an n-vertex weighted graph G with a spanning tree T such that the weight of each edge in T is 1 and the weight of each remaining edge is the weight of the corresponding path in T. If T corresponds to a path of length n with endpoint r, then the weight of any spanning tree in which each vertex other than r has degree 1 is at least the weight of T times $n/2$.*

The proof is straightforward.

5 Linear-Time Algorithms

Note that to obtain the worst-case performance guarantee a minimum-cost flow is not required. It suffices to find a feasible integer flow of cost bounded by the cost of the fractional flow f defined in the proof of Theorem 4. We describe two methods to find such a flow, and to implement the corresponding sequence of adoptions, in linear time.

Algorithm 1: Let f be the fractional flow defined in Theorem 4. Modify f by repeatedly performing the following short-cutting step: choose a maximal path in the set of edges with positive flow; replace the (c units of) flow on the path by (c units of) flow on the single new edge (u,v), where the path goes from u to v. Let $q(u)$ be the child of v on the path. Stop when all paths have been replaced by new edges. This phase requires linear time, because each step requires time proportional to the number of edges short-cut.

In the resulting flow, the only edges with positive flow are edges from leaves of the (rooted) tree T to interior vertices. Round the flow to an integer flow as follows. Consider each vertex v with positive deficit, say D. Using a linear-time selection algorithm, among the edges (u,v) sending flow to v, find the D smallest-weighted edges. Assign one unit of flow to each of these D edges. The resulting flow is integer-valued, feasible, legal, and has cost bounded by the cost of f. This phase requires linear time.

Assume that each vertex maintains a doubly linked list of its children. Given a pointer to any vertex, we can obtain its sibling in constant time. As adoptions are done, this list is maintained dynamically. Perform the adoptions corresponding to the flow in any order: for each edge (u,v) with a unit of flow, have u adopt the right sibling of $q(u)$ (in the original tree T). The tree remains connected because $d(v) \geq 2$, so at least one child of v is not adopted.

Algorithm 2: Consider the following restricted adoption network. Root the tree T as in the proof of Theorem 4. Direct each edge (u, v) of the tree towards the root. (Non-tree edges are not used.) Assign each edge a capacity of 1 and a cost equal to its weight. Assign each vertex a demand equal to its deficit.

We show below that an integer-valued minimum-cost flow in this network can be found in linear time. Because the fractional flow defined in the proof of Theorem 4 is a feasible legal flow in this network, the minimum-cost flow that we find is at least as good.

Find the flow via dynamic programming. For each vertex v, consider the sub-network corresponding to the subtree rooted at v. Let $C_j(v)$ denote the minimum cost of a flow in this subnetwork such that the surplus at v exceeds its demand D by j, for $j = 0, 1$. For each child u of v, Let $\delta(u)$ denote $w(u, v) + C_1(u) - C_0(u)$ — the additional cost incurred for v to obtain a unit of flow along edge (u, v). Let U_j denote the $D + j$ children with smallest $\delta(u)$, for $j = 0, 1$. Then, for $j = 0, 1$,

$$C_j(v) = \sum_{u \in U_j} \delta(u) + \sum_u C_0(u).$$

Using this equation, compute the C_j's bottom-up in linear time. The cost of the minimum-cost flow in the restricted network is given by $C_0(r)$, where r is the root. The flow itself is easily recovered in linear time.

To finish, shortcut the flow as in the first phase of the previous algorithm and perform the adoptions as in the last phase of that algorithm.

6 Geometric Problems

Our general result has several implications for cases of particular distance functions where it is possible to give a priori bounds on the maximum degree of an MST. For the case of L_2 distances in the plane, there always is an MST of maximum degree 5 [13]; for the case of L_1 or L_∞ distances there always exists a MST of maximum degree 4 [13, 17]. Without using any specific structure of the involved distance functions, we note as a corollary:

Corollary 7. *Let $T_{\min}$ be an MST and T_k be a tree whose maximal degree is at most k. For L_1 or L_∞ distances in $\mathbb{R}^2$, we get a degree-3 tree T_3 with*

$$- \quad w(T_3) < \tfrac{3}{2} w(T_{\min}).$$

For the case of Euclidean distances in the plane, we get bounded degree trees that satisfy

$$- \quad w(T_3) < \tfrac{5}{3} w(T_{\min})$$
$$- \quad w(T_4) < \tfrac{4}{3} w(T_{\min}).$$

The latter two bounds are worse than those shown by Khuller, Raghavachari and Young [10] using the geometry of point arrangements. (It was shown that $\tfrac{3}{2}$ and $\tfrac{5}{4}$ are upper bounds.) We conjecture that the following are the optimal ratios:

Conjecture 8. *For the case of Euclidean distances in the plane, we conjecture that there exist bounded degree trees that satisfy*

$$- \frac{w(T_3)}{w(T_{\min})} \leq \frac{\sqrt{2}+3}{4} \approx 1.103\ldots$$

$$- \frac{w(T_4)}{w(T_{\min})} \leq \frac{2\sin(\frac{\pi}{10})+4}{5} \approx 1.035\ldots$$

For L_1 and L_∞ distances in $\mathbb{R}^2$, we conjecture

$$- \frac{w(T_3)}{w(T_{\min})} \leq \frac{5}{4}$$

The best known lower bounds (yielding the ratios of Conjecture 8) are shown in Figure 2. (Note that the example for L_∞ metric is obtained by rotating the arrangement in (c) by 45 degrees.)

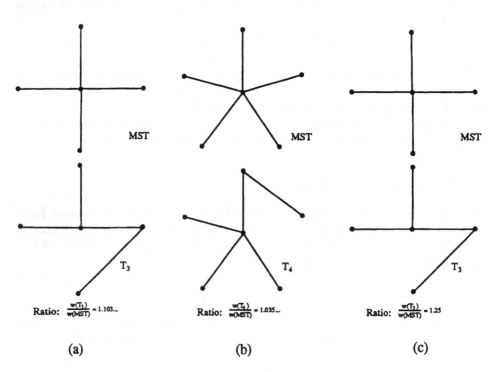

Ratio: $\frac{w(T_3)}{w(MST)} = 1.103\ldots$ Ratio: $\frac{w(T_4)}{w(MST)} = 1.035\ldots$ Ratio: $\frac{w(T_3)}{w(MST)} = 1.25$

(a)   (b)   (c)

Fig. 2. The worst known examples for: (a) $\frac{w(T_3)}{w(T_{\min})}$, L_2 distances

(b) $\frac{w(T_4)}{w(T_{\min})}$, L_2 distances

(c) $\frac{w(T_3)}{w(T_{\min})}$, L_1 distances.

6.1 Geometric Hamiltonian Paths

We conclude this paper by settling a question raised in [11], in the negative:

"In Euclidean graphs, perhaps a Traveling Salesman path of weight at most $(2 - \varepsilon)$ times the minimum spanning-tree weight always exists and can be found in polynomial time."

Theorem 9. *For an arrangement of points in the plane with Euclidean distances, the ratio $\frac{w(T_2)}{w(T_{\min})}$ can be arbitrarily close to 2.*

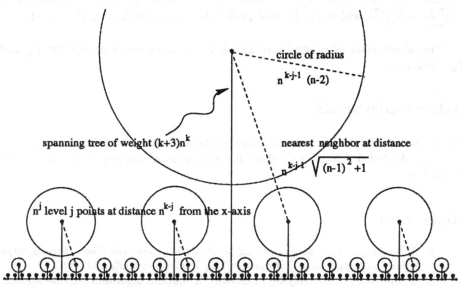

circle of radius

n^{k-j-1} (n-2)

spanning tree of weight $(k+3)n^k$

nearest neighbor at distance

$n^{k-j-1} \sqrt{(n-1)^2 + 1}$

n^j level j points at distance n^{k-j} from the x-axis

basis of length $2\,n^k$, containing base points

Fig. 3. A class of examples showing $\frac{w(T_2)}{w(T_{\min})} \to 2$

Proof. Let n and k be sufficiently large. Construct a point set as follows (see Figure 3):

Take base points at $(0,0)$ and $(2n^k, 0)$.

For $j = 0, \ldots, k$, add points as follows:

For $i = 1, \ldots, n^j$

Add *level j* point at $((2i - 1)n^{k-j}, n^{k-j})$;

add *base* point at $((2i - 1)n^{k-j}, 0)$.

The points at level j, i.e., at height n^{k-j} have nearest neighbors at distance at least $n^{k-j-1}(n - 1)$. To prove the lower bound, we draw a circle centered at each point at level $j < k$. For the points at level j, the radius of the circle is $(n^{k-j-1}(n - 2))$. The circles corresponding to two points do not intersect. Since each point has degree two in a Hamilton cycle, twice the sum of the radii of the circles gives us a lower bound on the length of the Hamilton cycle. This can be computed as follows (observe that we can always pick $n \geq 2k$).

$$2 \sum_{j=0}^{k-1} n^j (n^{k-j-1}(n-2)) = 2kn^{k-1}(n-2) \geq 2(k-1)n^k.$$

Since no edge can have length more than $2n^k$, we conclude that no Hamilton path can have a weight smaller than $2(k-1)n^k - 2n^k = 2(k-2)n^k$.

It can be verified that there is a tree of weight $(k+3)n^k$ that spans the points. Hence this is an upper bound on the weight of $T_{\min}$. It follows that $\frac{w(T_2)}{w(T_{\min})} > \frac{2(k-2)}{k+3}$, which can be arbitrarily close to 2, concluding the proof.

The above class of examples establishes the same lower bound for L_1 and L_∞ distances.

Acknowledgements

We thank Joe Mitchell for establishing the transatlantic connection between the authors. We thank Chandra Chekuri for asking us about degree 3 trees in the L_1 metric.

References

1. R. K. Ahuja, T. L. Magnanti and J. B. Orlin. *Network flows (theory, algorithms and applications)*. Prentice Hall, Englewood Cliffs, NJ, 1993.
2. C. Brezovec, G. Cornuéjols, F. Glover. A matroid algorithm and its application to the efficient solution of two optimization problems in graphs. *Math. Programming* **42** (1988), pp. 471–487.
3. T. Fischer. *Optimizing the degree of minimum weight spanning trees*. Tech. Rep. 93-1338, Dept. of Computer Science, Cornell University, April 1993.
4. M. Fürer and B. Raghavachari. Approximating the minimum-degree Steiner tree to within one of optimal. *J. Algorithms* **17** (1994), pp. 409–423.
5. H. N. Gabow. A good algorithm for smallest spanning trees with a degree constraint. *Networks* **8** (1978), pp. 201–208.
6. H. N. Gabow and R. E. Tarjan. Efficient algorithms for a family of matroid intersection problems. *J. Algorithms* **5** (1984), pp. 80–131.
7. M. R. Garey and D. S. Johnson. *Computers and intractability: a guide to the theory of NP-completeness*. Freeman, San Francisco, CA, 1979.
8. B. Gavish. Topological design of centralized computer networks — formulations and algorithms. *Networks* **12** (1982), pp. 355–377.
9. F. Glover, D. Klingman. Finding minimum spanning trees with a fixed number of links at a node. In: B. Roy (ed.), *Combinatorial Programming: Methods and Applications*. D. Reidel Publishing Company, Dordrecht-Holland, 1975. pp. 191–201.
10. S. Khuller, B. Raghavachari and N. Young. Low degree spanning trees of small weight. *Proc. of 26th Annual ACM Symp. on the Theory of Computing*, pp. 412–421, May 1994. To appear in *SIAM J. Comput.*

11. S. Khuller, B. Raghavachari, N. Young, Balancing minimum spanning trees and shortest-path trees. *Algorithmica* **14** (1995), pp. 305–321.
12. C. L. Monma, D. Shallcross. Methods for designing communication networks with certain two-connected survivability constraints. *Oper. Res.* **37** (1989), pp. 531–541.
13. C. L. Monma, S. Suri, Transitions in geometric minimum spanning trees. *Discrete & Computational Geometry* **8** (1992), pp. 265–293.
14. S. C. Narula and C. A. Ho. Degree-constrained minimum spanning tree. *Comput. Ops. Res.* **7** (1980), pp. 239–249.
15. C. H. Papadimitriou, U. V. Vazirani, On two geometric problems related to the traveling salesman problem. *J. Algorithms* **5** (1984), pp. 231–246.
16. R. Ravi, M. V. Marathe, S. S. Ravi, D. J. Rosenkrantz and H. B. Hunt III. Many birds with one stone: multi-objective approximation algorithms. Manuscript. A preliminary version appeared in *Proc. 25th Annual ACM Symp. on the Theory of Computing*, pp. 438–447, May 1993.
17. G. Robins and J. S. Salowe. Low-degree minimum spanning trees. *Discrete and Computational Geometry* **14** (1995), pp. 151–166.
18. J. S. Salowe. Euclidean spanner graphs with degree four. *Discrete Appl. Math.* **54** (1994), pp. 55–66.
19. M. Savelsbergh and A. Volgenant. Edge exchanges in the degree-constrained minimum spanning tree problem. *Comput. Ops. Res.* **12** (1985), pp. 341–348.
20. M. Stoer. *Design of survivable networks*. Lecture Notes on Mathematics # 1531. Springer, Heidelberg, 1992.
21. A. Volgenant. A Lagrangean approach to the degree-constrained minimum spanning tree problem. *Europ. J. Ops. Res.* **39** (1989), pp. 325–331.

Approximating k-Set Cover and Complementary Graph Coloring

Magnús M. Halldórsson*

Science Institute, University of Iceland, IS-107 Reykjavik, Iceland.

Abstract

We consider instances of the Set Cover problem where each set is of small size. For collections of sets of size at most three, we obtain improved performance ratios of $1.4 + \epsilon$, for any constant $\epsilon > 0$. Similar improvements hold also for collections of larger sets. A corollary of this result is an improved performance ratio of 4/3 for the problem of minimizing the unused colors in a graph coloring.

1 Introduction

A *set system*, or a *hypergraph*, consists of a finite base set and a collection of subsets of the base set. A *cover* of a set system is a sub-collection whose elementwise union equals the base set. The SET COVER problem is given an arbitrary set system, find a minimum cardinality cover. We consider instances where each set in the system is of size at most k. The k-SET COVER (k-SC) problem is then to find a cover of a a system of k-sets.

The set cover problem is of fundamental importance in combinatorial optimization, with innumerable applications in operations research and other fields. The bounded set cover problem is an important special case; for instance, Goldschmidt, Hochbaum and Yu [5] cite an application in manufacturing.

Even when the maximum set size k is three, the k-SET COVER problem is known to be NP-hard, and approximating it within $1 + \delta$ is NP-hard, for some fixed $\delta > 0$ [3]. On the other hand, when maximum set size is two, the problem becomes a graph problem known as EDGE COVER, which can be solved optimally via a straightforward transformation to MAXIMUM MATCHING. It is therefore interesting to study how well we can solve the boundary case, 3-SET COVER.

We are interested in efficient heuristics that always find solutions close to optimal. Efficiency dictates that the algorithm run at least in polynomial time, and effectiveness is measured by the *performance ratio* of the algorithm, which is the maximum ratio of the size of the obtained cover to that of the optimal cover. An algorithm is said to be *ρ-approximate* if its performance ratio is at most ρ.

---

*mmh@rhi.hi.is. Work done in part at Japan Advanced Institute of Science and Technology - Hokuriku, Japan.

The approach considered in this paper is local search. Given a particular solution, we search for small subsets that can we swapped in and out of the solution and thereby in some sense improving it. Once no further improvement operations can be found, the solution will be moderately large in comparison with some abstract optimal solution. We shall be looking for several different types of improvements.

The premier form of a local improvement is *t-change*, where we have t sets in the solution that can be covered by $t - 1$ sets outside the solution.

A second form of an improvement is an *augmenting path*. Here we have a sequence of 2-sets alternately from outside and inside the solution that connect two 1-sets from the solution. This improvement operation may involve arbitrarily many sets and thus may not be "local" in the usual sense, but what matters most is that it can be found, or its non-existence confirmed, in reasonable amount of time.

Both of the above improvement types decrease the cardinality of the solution. We also use two types of improvements that do not affect the solution size but have the effect of increasing the number of 3-sets in the solution.

Our main result is that local search of the form described approximates 3-SET COVER within a factor of $7/5 + \epsilon$, in time polynomial in $\log \epsilon^{-1}$. It extends to a performance ratio of $\mathcal{H}_k - 1/3$ for k-set covers. When local search is used with t-change improvements only, we get a ratio of $3/2 + \epsilon$, which follows from a result of Hurkens and Schrijver [8]. Previous published ratios for this problem are 11/6 [9, 10], 10/6 [5], and 11/7 [6], which are also based on local improvements of greedy initial solutions.

One application of approximations of covers of small sets is in the coloring problem. A non-standard measure of a coloring is the number of colors saved over the trivial allocation of a different color for each vertex. Using the set cover algorithm, we obtain a ratio of 4/3, improving previous ratios of 2 [4], 1.5 [7] and 1.4 [6].

2 The problem and the notation

The k-SET COVER problem is defined by: given a collection C of sets of at most k elements each drawn from a finite domain S, find a subcollection $C' \subseteq C$ of minimum cardinality such that $\bigcup C' \overset{def}{=} \cup_{X \in C'} X$ equals S.

In order to simplify much of the presentation, we shall be working with a slightly different problem, which we name *minimum exact cover of a monotone collection*. We require that the solution be a partition of the domain S, i.e. that the sets be disjoint in addition to forming a cover. The input collection, however, is now assumed to be monotone, in that all subsets of a set in the collection are also in the collection.

A monotone collection can be produced from the original input by adding all subsets. If explicitly represented this may increase the size of the input by a factor of at most $2^k - 1$, but it may also be possible to represent this implicitly. Given that maximum set size k is for us a small constant, this causes at most a constant factor overhead. A solution in the monotone system can be easily mapped back to a solution in the original system with the same cardinality.

Intersection hypergraph Our analysis proceeds by comparing the solution found $\mathcal{A}$ to an arbitrary cover $\mathcal{B}$. The interaction between the two solutions is often well illustrated by the *intersection hypergraph* $G(\mathcal{B}, \mathcal{A})$ of the two set collections. The sets in $\mathcal{B}$ form the vertices of the hypergraph, and the edges are given by $\{B_j \in \mathcal{B} \mid A_i \cap B_j \neq \emptyset\}$, for each A_i in $\mathcal{A}$. Since both $\mathcal{A}$ and $\mathcal{B}$ are covers with 3-sets, G has the properties that each vertex is incident on at most three edges and each edge is incident on at most three vertices. The ratio between the solution then equals the density or average degree of the graph.

Notation The following notation is used throughout.

$\mathcal{A}$ refers to the cover output by the algorithm in question, $\mathcal{B}$ is any other cover. We partition $\mathcal{A}$ into $\mathcal{A}_i$, $i = 1, 2, 3$, of the sets in $\mathcal{A}$ of size i. Partition $\mathcal{B}$ into $\mathcal{B}_1$ and $\mathcal{B}_0$, where $\mathcal{B}_1$ consists of the sets in $\mathcal{B}$ with non-empty intersections with sets in $\mathcal{A}_1$. That is, $\mathcal{B}_1 = \{B \in \mathcal{B} \mid B \cap \bigcup \mathcal{A}_1 \neq \emptyset\}$, and $\mathcal{B}_0$ contains the remaining sets $\mathcal{B} - \mathcal{B}_1$.

We define $\mathcal{A}_0$ to be the restriction of $\mathcal{A}$ to $\bigcup \mathcal{B}_0$, or $\mathcal{A}_0 = \{A_i \cap \bigcup \mathcal{B}_0 \mid A_i \in \mathcal{A}\} - \emptyset$. We shall motivate this definition when we get to the analysis.

The cardinalities of the above and other collections are denoted by the respective lower case letter, e.g. a, b. The number of elements of the base set, or the vertices of the hypergraph, is denoted by n (but also by s), and the number of sets in the input, or the edges of the hypergraph, by m.

We casually use "singleton", "doubleton", and "triplet" to refer to sets of size one, two and three, respectively, and also call them 2-sets, 3-sets etc.

An example set system Figure 1 contains an example set system. The elements of the five singleton sets in $\mathcal{A}_1$ are all marked with "o" and the region is marked off with a bold line for clarity. The doubleton sets are marked by x and y, while the triplets from $\mathcal{A}_3$ are marked by A, B, C and D. In addition to these 11 sets, the columns form the sets of the optimal solution $\mathcal{B}$. The first five columns are $\mathcal{B}_1$, as those sets contain the singleton sets of $\mathcal{A}_1$. $\mathcal{A}_0$ contains six single element sets, of the elements in the last two columns.

3 Local Search Algorithms

This paper focuses on a local improvement strategy for producing small 3-set covers. The strategy is to start with an arbitrary initial cover and iteratively search for a solution of smaller cardinality that differs only in few elements.

Definition 1 *A* t-change improvement *of a cover* $\mathcal{A}$ *is formed by sets* D_1, D_2, *...,* D_t *in* A *and sets* $E_1, E_2, \ldots E_{t-1}$ *in* C *(normally outside of* A*) such that the symmetric difference*

$$\mathcal{A}' = \mathcal{A} \oplus \{D_1, \ldots D_t, E_1, \ldots E_{t-1}\} = (\mathcal{A} - \{D_1, \ldots, D_t\}) \cup \{E_1, \ldots E_{t-1}\}$$

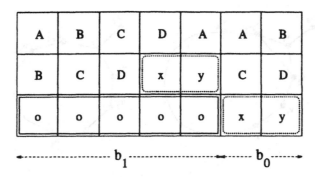

Figure 1: An example of a set system and its covers.

is also a cover.

Given any solution, finding an improvement or verifying local optimality can be done in polynomial time for any fixed value of t.

Definition 2 *An* augmenting path *with respect to a set cover solution $\mathcal{A}$ is a set of edges (v_0), (v_0, v_1), $(v_1, v_2) \ldots, (v_{2k}, v_{2k+1})$, (v_{2k+1}) in C where (v_0), (v_{2k+1}) and (v_{2i-1}, v_{2i}) are in $\mathcal{A}$, for $1 \leq i \leq k$.*

Both t-change and augmenting path improvements decrease the size of the cover, primarily by reducing the number of singleton sets. The following types do not reduce the count, but have other beneficial effects. Namely, they give preference to 3-sets, and also increase the singletons at the cost of doubletons. This will then, intuitively, make further reductions more likely.

Definition 3 *A* lasso *with respect to a set cover solution $\mathcal{A}$ is a set of edges forming a path (v_0, v_1), (v_1, v_2), $\ldots (v_{2k}, v_{2k+1})$, $k \geq 1$, a triplet (v_1, v_2, v_{2k+1}) and a singleton v_0 such that (v_{2i}, v_{2i+1}) is in $\mathcal{A}$, for $0 \leq i \leq k$.*

Definition 4 *A* fat singleton *with respect to a set cover solution $\mathcal{A}$ is a set of edges (v_0), (v_1, u_1), (v_2, u_2), (v_0, v_1, v_2), (u_1), (u_2) where the first three are contained in $\mathcal{A}$.*

Figure 2 pictures an arbitrary lasso. On the left is the set representation, with dotted lines connecting elements in the same set outside of $\mathcal{A}$. On the right is the hypergraph of $\mathcal{A}$ onto the cover given by the dotted edges.

If T is a lasso or a fat singleton with respect to $\mathcal{A}$, then the symmetric difference $\mathcal{A} \oplus T$ satisfies:

1. The cardinality of $\mathcal{A} \oplus T$ equals the cardinality of $\mathcal{A}$, and

2. The number of 3-sets in $\mathcal{A} \oplus T$ is one greater than the number of 3-sets in $\mathcal{A}$.

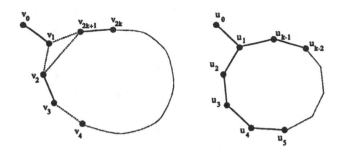

Figure 2: A lasso, as a set system, and its intersection hypergraph.

Finding the improvements Fat singletons are easily found by inspection, by marking the elements of S according to membership in $\mathcal{A}_i$, $i = 1, 2, 3$, and inspecting each 3-set in $\mathcal{C}$. We now indicate how the other types of improvements can be discovered efficiently.

Form a directed bipartite graph $G = (U, V, E)$ where U and V are formed by the elements of $\bigcup(\mathcal{A}_1 \cup \mathcal{A}_2)$. For an element x, let x_U (x_V) denote the corresponding vertex in U (V), respectively. The edges are given by

$$E = \{\langle x_U, y_V \rangle \mid (x, y) \in \mathcal{A}_2\} \cup \{\langle y_V, z_U \rangle \mid (y, z) \in \mathcal{C} - \mathcal{A}_2\}$$

(Notice that $\mathcal{A}_2$ contains unsorted sets, giving rise to pairs of edges in E.) Edges in one direction correspond to doubletons in the solution $\mathcal{A}$, while edges in the other direction correspond to 2-sets outside of $\mathcal{A}$. We have then a correspondence between (directed) paths in G and alternating paths in the set graph.

There is a lasso with respect to $\mathcal{A}$ iff for some set (x, y, z) in $\mathcal{C} - \mathcal{A}$ and contained in $\bigcup \mathcal{A}_2$, there is a path from x_U to y_V in G. (Hint: View x, y, z as v_2, v_{2k+1}, v_1 of Figure 2.)

There is an augmenting path iff for some x, y in $\bigcup \mathcal{A}_1$, there is a path from x_V to y_U in G.

The existence of a lasso improvement can be discovered in time proportional to finding the transitive closure of G, or $O(|E(G)||V(G)|)$. An augmenting path can be found in time $O(|E(G)|)$, via a single breadth-first search.

Algorithm Our main algorithm, LI_t, short for "local improvement", is indexed by a parameter t indicating the depth of the t-change local search.

> $\mathsf{LI}_t(S, \mathcal{C})$:
>   Start with a Greedy initial solution
>   Find optima under augmenting paths
>   Find optima under t-change
>   Find optima under lassos and fat singletons
>   Repeat search as necessary
> end

Lassos and fat singletons preserve the solution size and increase the number of 3-sets. Augmenting paths and t-change decrease the solution size while increasing the number of 3-sets by at most $\lfloor (t-4)/2 \rfloor$. Hence, fewer than $tn/2$ iterations are ever performed.

4 Analysis

For the analysis of the algorithm, we derive a sequence of equations on the relative size of the various partitions of the optimal and heuristic solutions. Linear combinations of these inequalities combine to give the desired performance ratio.

Basic properties of set collections Let us consider some basic bounds on these set collections.

1. The sum of the cardinalities of partition classes equals the cardinality of the whole collection.

$$a_1 + a_2 + a_3 = a, \qquad b_1 + b_0 = b.$$

2. The total number of elements in a cover is s.

$$a_1 + 2a_2 + 3a_3 = s. \tag{1}$$

3. Each set trivially contains at most three elements. Thus,

$$s \leq 3b, \tag{2}$$

which is tight only when $\mathcal{B}$ is an exact cover.

Properties of small improvements Systems without small improvements obey additional bounds.

1. Suppose a set in $\mathcal{B}_1$ intersects more than one set in $\mathcal{A}_1$. Then, the three sets form a 2-improvement. Since each set in $\mathcal{B}_1$ does by definition intersect a set in $\mathcal{A}_1$, the following holds for 2-optimal covers:

$$a_1 = b_1 \tag{3}$$

2. Suppose a set in $\mathcal{A}_2$ or $\mathcal{A}_3$ contains all its elements within $\mathcal{B}_1$ (and thus none within $\mathcal{B}_0$). Then, there exists a 3- or 4-improvement. Thus, each such set in a 4-optimal solution contains a non-empty portion in $\bigcup \mathcal{B}_0$, forming a set in $\mathcal{A}_0$:

$$a_2 + a_3 = a_0. \tag{4}$$

Relating larger improvements The set system $\mathcal{A}_0$ was defined to be the assemblage of the parts of the sets in $\mathcal{A}$ that were contained in $\bigcup \mathcal{B}_0$. In other words, those parts not contained in $\bigcup \mathcal{B}_1$. The motivation for that definition is the following observation. Suppose that a t-change local search picks up a set $B \in \mathcal{B}_1$. It has then already covered a set in $\mathcal{A}$ (namely, the singleton set from $\mathcal{A}_1$ intersecting B) and thereby "paid for itself". Further, it additionally covers the other elements in the set, which are then taken care of. Thus, suppose we have obtained an improvement in $\mathcal{A}_0$, i.e. covered the $\mathcal{B}_0$-portion of several sets in $\mathcal{A}$, we can extend it to a covering of the whole of these sets by including the appropriate sets from $\mathcal{B}_1$ while maintaining the balance between the sets to be added and the sets to be deleted.

This is formalized in the following lemma, which shows that an improvement in the derived collection $(\mathcal{A}_0, \mathcal{B}_0)$ implies the existence of an improvement, albeit larger one, in the original collection $(\mathcal{A}, \mathcal{B})$.

Lemma 4.1 *If there is an r-change improvement in $(\mathcal{A}_0, \mathcal{B}_0)$, then there is a $(2r+2)$-change improvement in $(\mathcal{A}, \mathcal{B})$.*

Proof. Consider an improvement in $(\mathcal{A}_0, \mathcal{B}_0)$ of minimal size. That is, we have sets $D_1, \ldots, D_h$ in $\mathcal{A}_0$ which intersect exactly the sets $E_1, \ldots, E_{h-1}$ in $\mathcal{B}_0$, where $h \leq r$. Let d_q, $q = 1, 2, 3$ be the number of sets D_j of size q.

Consider the intersection hypergraph with vertices E_j, $j = 1, \ldots h - 1$, and hyperedges $\{E_j \mid D_i \cap E_j \neq \emptyset\}$, for $i = 1, \ldots, h$. By minimality, this graph is connected. Each hyperedge of size q can reduce the number of components by at most $q - 1$, while together they reduce the number of components from $h - 1$ of the empty graph to 1 in the full connected graph. Thus,

$$d_2 + 2d_3 \geq h - 2.$$

Since $h = d_1 + d_2 + d_3$, the total number of elements in the D_j's is at least $2h - 2$:

$$\sum_{i=1}^{\bullet} |D_i| = d_1 + 2d_2 + 3d_3 \geq 2h - 2. \tag{5}$$

Now let us map this improvement from $\mathcal{A}_0$ to $\mathcal{A}$. Let D'_i be the[2] set in $\mathcal{A}$ that is a superset of D_i, for $i = 1, \ldots h$, and E'_j be the superset of E_j in $\mathcal{B}$, for $j = 1, \ldots h - 1$. There are at most $3h$ elements contained in these sets combined. By (5), at least $2h - 2$ of those elements are covered by $E'_1, \ldots, E'_{h-1}$. Each of the remaining $h' \leq h + 2$ elements is outside $\mathcal{A}_0$ and $\mathcal{B}_0$ and thus, by definition, contained in a unique set in $\mathcal{B}_1$. This set in $\mathcal{B}_1$ also intersects a (single) set in $\mathcal{A}_1$. Our improvement is a union of these sets: $D'_1, \ldots D'_h$ and the h' sets in $\mathcal{A}_1$ on one side; $E'_1, \ldots, E'_{h-1}$ and the h' sets in $\mathcal{B}_1$ on the other side. ∎

---

[2]This set is unique, under our assumption that a cover contain only disjoint sets and the input collection be monotone.

Bounds for t-change improvements t-change improvements turn out to be surprisingly powerful, even when applied to an arbitrary starting solution. The following is a simple application of a result of Hurkens and Schrijver [8]. We can obtain the same $3/2 + \epsilon$ ratio directly from Lemma 4.1 but with a worse dependence on ϵ.

Theorem 4.2 Let $\mathcal{A} = \{A_1, A_2, \ldots A_a\}$ be a 3-set cover with no t-change improvement, and let $\mathcal{B} = \{B_1, B_2, \ldots B_b\}$ be any other 3-set cover. Then,

$$\frac{a}{b} \leq \alpha_t \overset{def}{=} \frac{3}{2} + \begin{cases} \frac{1}{2^{p+1}/3-2}, & \text{if } t = 2p+1; \\ \frac{1}{2^p-2}, & \text{if } t = 2p+2. \end{cases} \tag{6}$$

The first few values of α_t are 3, 2, 9/5, 5/3, 21/13 and 11/7.

Proof. Consider the intersection hypergraph $(V, \mathcal{E})$:

$$V = \{B_1, \ldots, B_b\} \quad \text{and} \quad E_i = \{B_j | B_j \cap A_i \neq 0\} \text{ for } i = 1, \ldots, a.$$

We have, by t-change local optimality, that:

For any $h \leq t$, any h of the sets in $\mathcal{E}$ cover at least h elements of V.

By the König-Hall Theorem, this can be restated as:

Any collection of at most t sets in $\mathcal{E}$ has a system of distinct representatives.

Thus, $(V, \mathcal{E})$ satisfies the t-SDR property of [8], and the theorem thus follows from Theorem 1 of [8].   ∎

Hurkens and Schrijver [8] also gave a construction that yields a matching lower bound on the performance of this local improvement algorithm. The same bounds were also obtained in [6] in the context of similar local improvement algorithms for set packing problems.

We can also obtain stronger bounds for hypergraphs of bounded degree, via the ideas of Berman and Fürer [1], as in the case of the Set Packing problem [6].

Theorem 4.3 *There is a polynomial time algorithm for approximating 3-Set Cover in hypergraphs of degree Δ within a factor of $4/3 + \epsilon$ in time $n^{poly(\Delta, 1/\epsilon)}$.*

Properties of other improvement types

Lemma 4.4 *Suppose $\mathcal{A}$ is a solution that has no lasso, no augmenting path, and no fat singleton. Then,*

$$a_1 + a_2 \leq b \tag{7}$$

Proof. Form the intersection graph H of $\mathcal{A}_1 \cup \mathcal{A}_2$ onto $\mathcal{B}$. This is a multigraph whose vertices are the sets in $\mathcal{B}$. It has an edge for each set in $\mathcal{A}_2$ connecting the corresponding sets in $\mathcal{B}$ it intersects, and a self-loop for each set in $\mathcal{A}_1$. The number of vertices of H is b and the number of edges is $a_1 + a_2$.

Suppose there exists a connected component in H with more edges than vertices. The component must include at least two loops, either as circuits or self-loops. Thus, at least one of the three following cases must occur:

Two self-loops. Then there is an augmenting path with respect to $\mathcal{A}$.

A vertex with a self-loop adjacent to two other vertices Then, there is a fat singleton with respect to $\mathcal{A}$.

A simple cycle with one vertex of degree 3. Let $(u_1, u_2, \ldots u_k = u_1)$ be a cycle with u_1 additionally adjacent to u_0. The edges map to sets in $\mathcal{A}_2$; (u_i, u_{i+1}) is the set (v_{2i}, v_{2i+1}). The vertices map to sets in $\mathcal{B}$; if we look at the subsets spanned by $v_0, \ldots v_{2k}$, we have u_0 corresponding to v_0, u_1 corresponding to (v_0, v_1, v_{2k}), and u_i, $2 \leq i \leq k$, corresponding to (v_{2i-1}, v_{2i}). Thus, we have identified a lasso with respect to $\mathcal{A}$.

We have thus proved the converse of the statement of the lemma. ∎

Main result

Theorem 4.5 *Let $\mathcal{A}$ be an output of LI_{2r+2} and $\mathcal{B}$ be a cover. Then,*

$$a \leq \frac{(\alpha_r - 1)s + (2\alpha_r - 1)b}{3\alpha_r - 2}. \tag{8}$$

Proof. Since $(\mathcal{A}, \mathcal{B})$ contains no $2r + 2$-improvement, $(\mathcal{A}_0, \mathcal{B}_0)$ contains no r-improvement, by Lemma 4.1. Applying Theorem 4.2 to $(\mathcal{A}_0, \mathcal{B}_0)$, we have:

$$a_0 \leq \alpha_r b_0. \tag{9}$$

Combine (3), (4), and (9):

$$a \leq b_1 + \alpha_r b_0, \tag{10}$$

and (3), (7), and (1):

$$3a \leq b_1 + s + b. \tag{11}$$

Now add (10) and $(\alpha_r - 1)$ times (11) to yield the theorem. ∎

Recalling that $s \leq 3b$ and substituting for α_r in the statement of the previous theorem yields our main claim.

Corollary 4.6 *The performance ratio of LI_{2r+2} is at most*

$$\frac{5\alpha_r - 4}{3\alpha_r - 2} = \frac{7}{5} + O(2^{-r/2}).$$

For $r = 1, 2, 3$, the ratios are $11/7$, $3/2$, and $25/17$, respectively.

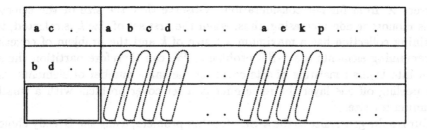

Figure 3: A construction for a ratio 1.4 for any LI_t algorithm

Limitations Figure 1 shows that that the 11/7 bound for LI_4 is tight. Rather than constructing examples for each value of t, we focus on the asymptotic case.

Figure 3 illustrates a construction that contains no lassos, fat singletons, augmenting paths, and no t-change improvements, for any fixed value of t. B_1 is one-fifth of B and contains one element from each set in A_3 as well as the singleton sets. B_0 contains b_0 doubletons, and two elements from each triplet, properly arranged.

The intersection hypergraph of A_0 on B_0 in this example is a 3-regular simple graph with a high girth g; there do exist such graphs with girth $\log n - 1$ [2]. Any improvement to A must contain an improvement of A_0, since the singletons are the only sets fully contained in B_1. Any improvement must contain a cycle in the intersection hypergraph, and thus must be of cardinality at least g.

Complexity The naive bound for LI_t would give complexity of $n^t m^{t-1}$ per iteration; showing $O(n^t)$ is not too difficult. We sketch briefly how LI_{2r+2} can be implemented in time $O(n^r)$ per iteration for $r \geq 2$, and $O(mn)$ for $r = 1$.

Observe from our analysis, particularly Lemma 4.1, that we need only a special type of $2r + 2$-change improvement: r sets in $A_2 \cup A_3$, along with up to $r + 2$ sets in A_1 (and appropriate sets from B). We examine all r different sets in $A_2 \cup A_3$. These contain only $O(r)$ elements, so we can try all possible partitions to see if they form sets in $C - A$, and to match parts of these partition classes with elements in A_1. For the latter, we will have precomputed for each singleton or doubleton, the number of elements in A_1 that combine to form sets in $C - A_1$.

LI_4 can be implemented using Greedy, followed by augmenting paths improvements, followed by restricted 4-change improvements. It can also be done in time $O(m)$ per iteration (or $O(n^2)$, if the instance is dense), for a total complexity of $O(mn)$.

k-Set Cover The best studied heuristic for k-SET COVER is the Greedy algorithm, which has a performance ratio of $\mathcal{H}_k$ [9, 10], or 11/6 for 3-SC. The algorithm operates by iteratively adding a set X to the solution that covers the most number of (yet uncovered) elements and then eliminating those sets that intersect X from further consideration. Observe that the size of the selected sets is monotone non-increasing; thus, when the first set of size k is selected, the remaining collection has a maximum set size of k and the problem of covering the remaining elements is a k-SC problems. We can therefore partition the solution into layers consisting of sets covering the same number of elements, and after peeling off one layer the remainder is a set cover problem with a smaller maximum set size.

Our results generalize to the k-SET COVER problem, using the Greedy choices of sets of size 4 and more, and optimizing the smaller sets with the preceding technique.

Theorem 4.7 *The performance ratio of* $\mathsf{LI_6}$ *for* k-SET COVER *is at most* $\mathcal{H}_k - 1/3$.

Proof. When $k = 3$, the statement is trivial, so assume $k \geq 4$.

Recall that the Greedy algorithm chooses sets of non-increasing size, resulting in a sequence of subproblems i-SC, where i goes from k down to 2. Let N_i, $3 \leq i \leq k$, denote the number of elements to be covered in the i-SC subproblem. Then, $N_i - N_{i-1}$ represents the number of elements covered by sets of size i, and $N_k = s$. Further, let X_i denote the number of sets in the optimal solution (for k-SET COVER) that contain i or more elements. Then,

$$N_i \leq X_1 + X_2 + \ldots + X_i.$$

Notice that the size of the optimal solution for i-SC is at most b, the size of the optimal solution for k-SC, since Greedy leaves no set of size $i + 1$ or more.

The number of sets used by $\mathsf{LI_6}$ on the 3-SC subproblem is bounded by $(N_3 + 3b)/4$, by Theorem 4.5.

The number of sets that the combined algorithm uses is bounded by:

$$
\begin{aligned}
a &\leq \frac{N_k - N_{k-1}}{k} + \frac{N_{k-1} - N_{k-2}}{k-1} + \cdots + \frac{N_4 - N_3}{4} + \frac{N_3}{4} + \frac{3b}{4} \\
&= \frac{N_k}{k} + \sum_{i=4}^{k-1} \frac{N_i}{(i+1)i} + \frac{3b}{4} \\
&\leq \frac{N_k}{k} + \sum_{i=4}^{k-1} \frac{1}{(i+1)i} \sum_{j=1}^{i} X_j + \frac{3b}{4} \\
&= \frac{N_k}{k} + \sum_{j=4}^{k-1} X_j \sum_{i=j}^{k-1} \frac{1}{(i+1)i} + (X_1 + X_2 + X_3) \sum_{i=4}^{k-1} \frac{1}{(i+1)i} + \frac{3b}{4} \\
&= \frac{N_k}{k} + \sum_{j=4}^{k-1} X_j \left(\frac{1}{j} - \frac{1}{k}\right) + (X_1 + X_2 + X_3)\left(\frac{1}{4} - \frac{1}{k}\right) + \frac{3b}{4} \\
&\leq \sum_{j=4}^{k} \frac{X_j}{j} + \frac{X_1 + X_2 + X_3}{4} + \frac{3b}{4}.
\end{aligned}
$$

We have obtained a bound that is stronger than what the lemma claims, which improves on the corresponding bound of [9, 10, 5]. Since $X_i \leq b$,

$$a \leq (\mathcal{H}_k - 1/3)b.$$

∎

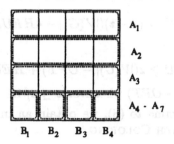

Figure 4: A construction with a ratio of 7/4 on 4-set systems.

This bound is tight for the 4-set system on sixteen elements that appears in Figure 4. The construction generalizes easily to a matching bound for $k > 4$.

5 Applications to Graph Coloring

The *graph coloring* problem is that of assigning a discrete value, or color, to each vertex such that adjacent vertices get different colors. Typically, the measure is the number of colors used $HEU(G)$, for which the objective is to minimize. Demange, Grisoni, and Paschos [4] considered finding a coloring with the objective of maximizing the number of *unused* colors, or $|V(G)| - HEU(G)$. The *complementary performance ratio* of a coloring algorithm is defined to be

$$\max_G \frac{|V(G)| - OPT(G)}{|V(G)| - HEU(G)}.$$

A fairly complicated algorithm with a complementary performance ratio of 2 was given in [4]. Improved results were obtained independently by Hassin and Lahav [7] and ourselves [6] using very similar ideas. They include a trivial algorithm with ratio of 2, and an algorithm with a ratio of 1.5. A connection of this problem to set cover was made explicit in [6], and the improved 3-set cover algorithm obtained there used to improve the coloring ratio to 1.4. We take it one step further to a ratio of 4/3 via the set cover algorithm of previous section.

Lemma 5.1 *Suppose there is an algorithm that finds a 3-set cover with at most* $(1 - z)s + zb$ *sets, where* $0 < z \leq 3/4$. *Then, there is a coloring algorithm with a complementary performance ratio of at most* $1/z$.

Proof. First find a maximal collection of disjoint 4-independent sets in the graph, and color each with a different color. From the remaining graph G', form the set system consisting of all independent sets, which are necessarily of size at most three. Apply the set cover algorithm, and color each of the sets in the cover with a different color.

Let HEU_4 denote the number of 4-sets found, and let $s = |V(G')| = |V(G)| - 4HEU_4$ be the remaining elements. The number of color classes is at most

$$HEU \leq HEU_4 + (1 - z)(|V(G)| - 4HEU_4) + zOPT.$$

Thus,

$$|V(G)| - HEU \geq z(|V(G)| - OPT) + HEU_4(3 - 4z)$$

which is at least $z(|V(G)| - OPT)$ whenever $z \leq 3/4$. ∎

Applying the above lemma to (8), we obtain an improved complementary performance ratio for GRAPH COLORING.

Theorem 5.2 *Using* Ll_6 *on the set system formed by the independent sets of size at most 4 yields a coloring with a complementary performance ratio of* $4/3$.

6 Discussion

The same bounds can be obtained for the Set Multi-Cover problem, where elements are to be covered possibly more than once. Also recall the immediate applications via standard reductions to the Dominating Set problem when maximum degree is bounded by $k - 1$ (or Total Dominating Set of degree k), as well as the Hitting Set problem when each element occurs at most k times.

An important open problem is to determine the approximability of these or similar algorithms on the weighted version of the Set Cover problem.

Acknowledgments

I would like to thank Martin Fürer and Rafi Hassin for helpful comments.

References

[1] P. Berman and M. Fürer. Approximating maximum independent set in bounded degree graphs. In *Proc. Fifth Ann. ACM-SIAM Symp. on Discrete Algorithms*, pages 365–371, Jan. 1994.

[2] B. Bollobás. *Extremal Graph Theory*. Academic Press, 1978.

[3] P. Crescenzi and V. Kann. A compendium of NP optimization problems. Dynamic on-line survey available at nada.kth.se, 1995.

[4] M. Demange, P. Grisoni, and V. T. Paschos. Approximation results for the minimum graph coloring problem. *Inform. Process. Lett.*, 50:19–23, 1994.

[5] O. Goldschmidt, D. S. Hochbaum, and G. Yu. A modified greedy heuristic for the set covering problem with improved worst case bound. *Inform. Process. Lett.*, 48:305–310, 1993.

[6] M. M. Halldórsson. Approximating discrete collections via local improvements. In *Proc. Sixth ACM-SIAM Symp. on Discrete Algorithms*, pages 160–169, Jan. 1995.

[7] R. Hassin and S. Lahav. Maximizing the number of unused colors in the vertex coloring problem. *Inform. Process. Lett.*, 52:87–90, 1994.

[8] C. A. J. Hurkens and A. Schrijver. On the size of systems of sets every t of which have an SDR, with an application to the worst-case ratio of heuristics for packing problems. *SIAM J. Disc. Math.*, 2(1):68–72, Feb. 1989.

[9] D. S. Johnson. Approximation algorithms for combinatorial problems. *J. Comput. Syst. Sci.*, 9:256–278, 1974.

[10] L. Lovász. On the ratio of optimal integral and fractional covers. *Discrete Math.*, 13:383–390, 1975.

On Minimum 3-Cuts and Approximating k-Cuts Using Cut Trees

Sanjiv Kapoor*

Abstract. This paper describes two results on graph partitioning. Our first result is a non-crossing property of minimum 3-cuts. This property generalises the results by Gomory-Hu on min-cuts (2-cuts) in graphs. We also give an algorithm for finding minimum 3-cuts in $O(n^3)$ Max-Flow computations. The second part of the paper describes a Performance Bounding technique based on Cut Trees for solving Partitioning Problems in weighted, undirected graphs. We show how to use this technique to derive approximation algorithms for two problems, the Minimum k-cut problem and the Multi-way cut problem. Our first illustration of the bounding technique is an algorithm for the Minimum k-cut which requires $O(kn(m+n\log n))$ steps and gives an approximation of $2(1-1/k)$. We then generalise the Bounding Technique to achieve the approximation factor $2 - f(j,k)$ where $f(j,k) = j/k - (j-2)/k^2 + O(j/k^3), j \geq 3$. The algorithm presented for the Minimum k-cut problem is polynomial in n and k for fixed j. We also give an approximation algorithm for the planar Multi-way Cut problem.

1 Introduction

In this paper we consider Graph Partitioning Problems. These problems are of practical interest with application to VLSI and minimization of communication costs in parallel computing systems[S].

In the first part of the paper we consider properties related to minimum 3-cuts in a graph G. Suppose we are given a graph $G = (V, E)$, $|V| = n$ and $|E| = m$, with non-negative weights. A minimum 3-cut is a set of edges of minimum total weight, the removal of which separates the graph into 3 non-empty components. The minimum 2-cut problem has been extensively studied [GH, SW]. The problem can be solved via repeated applications of the maximum flow problem. The maximum flow between two vertices s and t gives the minimum cut separating s and t. A fundamental result by Gomory-Hu shows that there are $n - 1$ distinct s-t cuts in the graph. The result is based on a *non-crossing* property of s-t cuts. These cuts can be represented by a tree, called the Gomory-Hu Cut Tree. We show that a somewhat similar non-crossing property also holds for minimum 3-cuts.

---
* Part of this work was done while the author was a visitor at the Max-Planck-Institute für Informatik, Saarbrücken, Germany. The author (e-mail: skapoor@cse.iitd.ernet.in) is with Dept. of Computer Science and Engineering, Indian Institute of Technology, Haus Khas, New Delhi, India.

Unfortunately, this does not immediately result in an efficient algorithm for finding minimum 3-cuts. We are able to reduce the problem to that of finding minimum 3-cuts in a special graph. Previous approaches to solving the minimum 3-cut problem [GH] have resulted in a solution requiring $O(n^4 T_F(n, m))$ steps, where $T_F(n, m)$ is the time required to solve the maximum flow problem on a graph with n vertices and m edges. In the first part of this paper we also modify the strategy in [GH] to give an algorithm for minimum 3-cuts requiring $O(n^3 T_F(n, m))$ steps.

In the second part of the paper we apply techniques based on Cut Trees to compute approximate solutions to the following two problems defined on an undirected graph with non-negative edge weights.

1. The Minimum k-cut problem: Find a set $S \in E$ of minimum weight the removal of which leaves k connected components.
2. Multi-way cut problem: Given $G = (V, E)$ and k specified vertices, called terminals, find a minimum weight set of edges, $E' \subseteq E$, such that the removal of E' disconnects each terminal from all the others.

The above two problems have been extensively studied in the literature. The Minimum k-cut Problem is solvable for fixed k for arbitrary graphs. The running time is $O(n^{k^2/2 - 3k/2 + 4} T_F(n, m))$ [GH], where $T_F(n, m)$ is the time required to find a minimum s-t cut in the graph. For $k = 3$ and unweighted planar graphs a faster scheme $(O(n^2))$ is known [HS], [He]. Several variants also have polynomial time schemes [Ch , Cu]. Our result on 3-Cuts improves the k-Cut algorithm, though marginally.

The Multi-way cut Problem is NP-Hard for all fixed $k \geq 3$ when the graph is arbitrary. For fixed k, the planar Multi-way cut problem is solvable in polynomial time. The time complexity is exponential in k $(O((4k)^k n^{2k-1} \log n))$. These results may be found in [DJPSY]. In [CR] it has been observed that the Multi-k-Way cut problem can be solved in linear time for trees using dynamic programming.

It is thus of interest to compute approximate solutions to these problems. These problems have been studied in the past where for each of the problems a $2(1 - 1/k)$ approximation algorithm is obtained [SV],[DJP].

We first show how to use Cut-Trees trees for constructing bounds to optimal solutions to these cut problems. We then show how simple approximation algorithms arise naturally for these problems. We show the use of Gomory-Hu cut trees [GoHu] to devise simple algorithms. We obtain the same approximation ratio, i.e. $2(1 - 1/k)$. The algorithm for approximating the Minimum k-cut problem requires $O(k)$ computations of the minimum cut in a weighted undirected graph. This approach to the approximation has been presented before in [SV] and they require $O(kn)$ maximum flow computations. The minimum cut problem has been recently solved in $O(n(m + n \log n))$ steps where n and m are the number of vertices and edges , respectively, in the graph. The implementation of the approximation algorithm presented here uses the minimum cut algorithm and is shown to require $O(kn(m + n \log n))$ steps. The main intent, however, is the development of a unified bounding technique for the two problems.

We next extend our bounding technique to give a $2 - f(j,k), j \geq 3$, where $f(j,k) = j/k - (j-2)/k^2 + O(j/k^3)$, approximation algorithms for :

1. The Minimum k-cut problem
2. Planar Multi-way cut problem

The algorithms for these problems rely on polynomial time algorithms for fixed j. The first problem is approximated in time polynomial in k and n when j is fixed. The second problem is approximated by a simple algorithm which is polynomial in n and k for fixed j when $k = O(\log n)$. And also by an algorithm which is polynomial in n and k. We thus show that we can improve on the 2-2/k approximation. In fact a range of approximations are possible as a function of j. The running time however is exponential in j. The question whether a polynomial time approximation scheme exists remains unanswered.

This paper is organised as follows. In Section 2 we present our results on minimum 3-cuts in a graph. In Section 3 we present a Bounding Technique and our first approximation algorithm for the Minimum k-Cut problem and for the Multi-way cut problem. In Section 4 we generalize the bounding technique to give better approximations for the Minimum k-cut problem and the planar Multi-way cut problem.

2   On Minimum 3-Cuts in Graphs

In this section we first show a non-crossing property of 3-cuts in graphs.

Consider a graph $G = (V, E)$ with weights on the edges and a minimum 3 cut, $3\text{-}MC \subseteq E$, in this graph. Let P_1 , P_2 and P_3 be the three parts of the graph. Each part is a subgraph induced by the vertices in the components obtained by removing the edges in $3\text{-}MC$. We let $MC(u,v,w)$ be the minimum weight 3-way cut between the vertices u, v and w. We define a cut to be *contained* in a set of edges if it is a subset of the edge set. Also, we shall refer to the edge set of a subgraph G' by G' itself. We first show that the minimum 3-way cut between any three vertices in any part, say P_1, is contained in the part itself or in $3\text{-}MC$, i.e. no edges outside the part or the set of edges forming the minimum 3-cut, $3\text{-}MC$ are used. This generalizes the Gomory-Hu result which shows a non-crossing property for 2-way cuts. We let $wt(C)$ be the weight of a cut C. And we let $wt(A, B)$, $A, B \subseteq V$ be the sum of the weights of edges with one vertex in A and the other in B.

Lemma 2.1 Non-Crossing Property: *Let $3\text{-}MC$ be the minimum 3-cut in G. Let $\mathcal{P} = \{P_1, P_2, P_3\}$ be the set of three parts of G obtained from $3\text{-}MC$ and let u, v, w be three vertices in $P_j \in \mathcal{P}$. Then the minimum 3-way cut separating u, v, w is contained in $P_j \cup 3\text{-}MC$.*

Proof: We prove the result by contradiction. Let P_1 contain u, v and w. We let C be the minimum 3-way cut separating u, v and w. Suppose C divides G into the parts S_u, S_v and S_w which are contained in a set, S. We consider the cases that arise depending on how the parts separating u, v and w intersect P_1, P_2, P_3.

First consider the case when at least one part in S, say S_v, has a non-empty intersection with both P_2 and P_3, i.e. $S_v \cap P_2$ and $S_v \cap P_3$ are both non-empty. This can be ensured by appropriate renaming of u, v and w. Let S_1 be the subset of P_2 associated with u, i.e. $S_1 = P_2 \cap S_u$, and let S_2 be the subset of P_3 associated with u. Let S_3 and S_4 be the subsets of P_2 and P_3, respectively, that are associated with, say, w. We shall show that transferring the subset S_1, S_2, S_3 and S_4 to S_v will give a 3-cut, C', of smaller cost, thus contradicting the minimality of C. Towards this end we first show that

$$wt(S_1, P_2 - S_1) \geq wt(P_1, S_1).$$

To prove this, consider the 3-cut, MC' which divides the graph into three parts, $S_1 \cup P_1$, $P_2 - S_1$ and P_3. Transferring S_1 from the first part to the second part changes the cost of the 3-cut by Δ where

$$\Delta = wt(P_1, S_1) - wt(P_2 - S_1, S_1)$$

Since $wt(MC') \geq wt(MC)$, $\Delta \leq 0$ and the claim follows.

Similarly, $wt(S_2, P_3 - S_2) \geq wt(P_1, S_2)$, $wt(S_3, P_2 - S_3) \geq wt(P_1, S_3)$ and $wt(S_4, P_3 - S_4) \geq wt(P_1, S_4)$.

Transferring parts S_1, S_2, S_3 and S_4 to v we obtain a change in cost of the cut such that

$$
\begin{aligned}
wt(C') - wt(C) \leq\ & wt(P_1 - P_v, S_1) - wt(S_1, P_2 - S_1) - wt(S_1, P_3 - S_2) \\
&+ wt(P_1 - P_v, S_2) - wt(S_2, P_3 - S_2) - wt(S_2, P_2 - S_1) \\
&+ wt(P_1 - P_v, S_3) - wt(S_3, P_2 - S_3) - wt(S_3, P_3 - S_4) \\
&+ wt(P_1 - P_v, S_4) - wt(S_4, P_3 - S_4) - wt(S_4, P_2 - S_3) \\
\leq\ & wt(P_1, S_1) - wt(S_1, P_2 - S_1) + wt(P_1, S_2) - wt(S_2, P_3 - S_2) \\
&+ wt(P_1, S_3) - wt(S_3, P_2 - S_3) + wt(P_1, S_4) - wt(S_4, P_3 - S_4) \\
\leq\ & 0
\end{aligned}
$$

Next, consider the case when there does not exist a part in S with non-empty intersection with both P_2 and P_3. Then one part, say P_3, is contained within a set in S, say S_w. The other part, P_2, has non-empty intersection with S_u and S_v but not with S_w. Let S_1 be the subset of P_2 in S_u. Then, as before
$$wt(S_1, P_2 - S_1) \geq wt(P_1, S_1).$$

Transfering S_1 to S_v gives another 3-cut C' such that:
$$wt(C') - wt(C) \leq wt(P_1, S_1) - wt(S_1, P_2 - S_1) \leq 0.$$
Thus we get a contradiction to the optimality of C.

The lemma follows. ■

The above Lemma also applies when a 2-way cut is to be found between two vertices in one of the parts. We let $MC(u, v)$ be the minimum cut separating u and v with P_u and P_v the two parts.

Lemma 2.2 *Let $3\text{-}MC$ be the minimum 3-cut in G. Let $\mathcal{P} = \{P_1, P_2, P_3\}$ be the three parts of G obtained from the minimum 3-cut. Let u, v be two vertices in $P_j \in \mathcal{P}$. Then the minimum 2-way cut separating u and v is contained in $P_j \cup 3\text{-}MC$.*

Proof: Assume w.l.o.g. that u and v are in P_1. We prove the lemma by contradiction. Suppose the cut, $MC(u, v)$, also partitions P_2 and P_3.

First, suppose that one part, say P_v, i.e. has a non-zero intersection with both P_2 and P_3. Let S_1 and S_2 be the parts $P_u \cap P_2$ and $P_u \cap P_3$, respectively. S_1 and S_2 may be empty or non-empty. Note that from the minimality of 3-MC, $wt(S_1, P_2 - S_1) \geq wt(P_1, S_1)$, and $wt(S_2, P_3 - S_2) \geq wt(P_1, S_2)$. Thus transferring the two parts S_1 and S_2 to P_v reduces the cost of the cut giving a contradiction.

Next, consider the case when no one part of $MC(u, v)$ intersects both P_2 and P_3, i.e. if P_1 intersects P_v then $P_2 \cap P_v = \Phi$ and $P_2 \cap P_u \neq \Phi$. In this case, the lemma is immediately valid. ∎

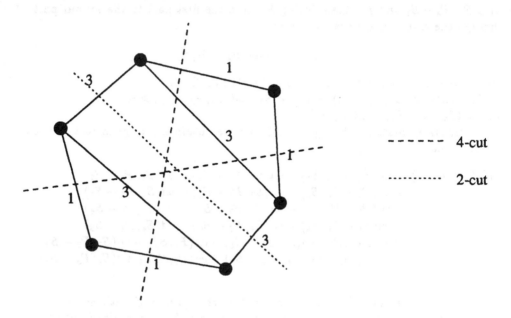

4-cut

2-cut

Figure 1 Crossings in 4-cuts

This second lemma does not extend in a natural way to k-cuts, $k \geq 4$ as is illustrated in the counter example in Figure 1. The example can be extended to hold for any (large) size of the graph. The second lemma also defines a non-crossing property.

We next show another interesting property of 3-cuts. Suppose that 3-MC partitions G into P_1, P_2 and P_3. Furthermore, let C_1, C_2 and C_3 be the cost of the edge sets that separate P_1, P_2 and P_3, respectively, from the rest of the graph. Asssume w.l.o.g. that $C_1 \leq C_2 \leq C_3$.

Lemma 2.3 *Let $MC(u, v)$ be a 2-cut. And let P_u and P_v be two parts into which $MC(u, v)$ partitions G. Then C_1 is either $MC(u, v)$ or is contained in one of $P_u \cup MC(u, v)$ and $P_v \cup MC(u, v)$.*

Proof: We prove this by contradiction. Assume that C_1 is not $MC(u, v)$.

First assume that u and v are both in one part, say P_1. Then Lemma 2.2 shows that $MC(u, v)$ will not partition any other part. Furthermore, $MC(u, v)$ cannot partition C_1 so that a subset of C_1 is in P_u and another subset is in P_v. This is proven by contradiction: Suppose $MC(u, v)$ does partition C_1. Then $MC(u, v)$ includes edges separating P_2 from P_3. Asume w.l.o.g. that $P_3 \in P_v$. Let A be the set of edges from P_u to P_2 and let B be the set of edges from P_2 to P_3. $wt(A) < wt(B)$ since C_1 has minimum cost. Transferring P_2 to P_v reduces the cost of $MC(u, v)$. This contradicts the optimality of $MC(u, v)$.

Next, suppose that u and v are in different parts. W.l.o.g. assume that $u \in P_1$ and $v \in P_2$. First assume that only one of P_u and P_v has a non-empty intersection with P_3. Assume that $P_v \cap P_3 \neq \Phi$. There are two cases. Suppose that $P_2 \cap P_u \neq \Phi$. Note that $P_2 \cap P_v \neq \Phi$. Let S_1 be $P_2 \cap P_u$. Transferring S_1 to part P_v reduces the value of $MC(u, v)$. The proof is similar to the proof of the above two lemmas. This leads to a contradiction. If $P_2 \cap P_u = \Phi$ then the lemma is immediately valid.

Alternatively, suppose that both P_u and P_v have non-null intersection with P_3. Let S_2 be $P_3 \cap P_u$. First suppose that $S_1 = P_2 \cap P_u \neq \Phi$ and $P_2 \cap P_v \neq \Phi$. Transferring both S_1 and S_2 to P_v again reduces the cost of $MC(u, v)$ thus leading to a contradiction. If P_2 does not intersect both P_u and P_v then there are two cases. Consider P_1. If P_1 intersects both P_u and P_v then a similar argument reduces the cost of $MC(u, v)$. Alternatively, neither P_1 nor P_2 intersect both P_u and P_v. Then again transferring $S_2 = P_3 \cap P_u$ to P_v reduces the cost of $MC(u, v)$. ■

We attempt to exploit the above lemma in order to find minimum 3-cuts in G. We let $GHT(G)$ denote the Gomory-Hu cut tree of G. The algorithm proceeds as follows: Let C be a cut in $GHT(G)$ which partitions G into P_a and P_b. If C has not been considered already then

1. Find the minimum 3-cut extending C. This is done by finding the minimum cuts in P_a and P_b which together with C give the 3-cuts, C_a and C_b.
2. Shrink all vertices in P_a into one vertex, V_a. Find a minimum 3-cut in the shrunk graph excluding C_a. Let it be $C1_a$.
3. Shrink all vertices in P_b into one vertex, V_b. Find a minimum 3-cut in the shrunk graph excluding C_b. Let it be $C1_b$.
4. The minimum 3-cut w.r.t. C is the minimum of C_a, C_b, $C1_a$ and $C2_b$.

Let G/P_a be the graph obtained by shrinking vertices in P_a. Note that $GHT(G/P_a)$ is a subtree of $GHT(G)$ and can be obtained by contracting the vertices in P_a into one vertex in $GHT(G)$. The algorithm is applied recursively on $GHT(G/P_a)$. When the algorithm cannot be applied further the resulting graph has a GHT which is a star graph. We call such a graph as a STAR GHT graph. The algorithm reduces the problem of finding a 3-MC in G to that of finding a minimum 3-cut in a graph with a STAR GHT. The additional work required is that of finding $O(n)$ minimum 2-cuts. Thus:

Lemma 2.4 *Finding a minimum 3-cut in G can be reduced to finding a minimum 3-cut in a STAR GHT graph in $O(n)$ minimum 2-cut operations.*

An Improved Algorithm :

We next show a lemma, a modification of Theorem 2 [GH], which allows us to improve the bounds for finding minimum 3-cuts.

Lemma 2.5 *Let p_1 be a vertex in G. Then $\exists 3$ vertices p_2, p_3, p_4 such that $MC((p_1, p_2), (p_3, p_4))$, i.e. the minimum cut between the pair of vertices, (p_1, p_2) and (p_3, p_4) is C_1.*

Proof: Let MC' be the minimum 3-cut with part P_1 of maximum size from amongst all 3-cuts. There are two cases.

Firstly, assume that p_1 is in P_1. Consider all choices of $p_2 \in P_1$. Let $p_3 \in P_2$ and $p_4 \in P_3$ be two other vertices. By a proof similar to that in [GH], it follows that $MC((p_1, p_2), (p_3, p_4))$ does not separate vertices in $P_2 \cup P_3$. Consider the vertex p_2 such that $MC' = MC((p_1, p_2), (p_3, p_4))$ is a maximum size maximal minimum 2-cut separating the pairs $(p_1, q), q \in P_1$, and (p_3, p_4). Suppose this cut is not C_1. Then this cut must be of value $< C_1$. (We use C_1 itself as the value of the cut C_1). Also there is vertex, $p' \in P_1$, which is not in the same part as p_1 in MC'. The cut $MC((p_1, p'), (p_3, p_4))$ crosses MC' and is of value $< C_1$. We can thus obtain a 3-cut, $3 - MC$ of lesser cost.

Secondly, assume that p_1 is in P_2. (A similar analysis follows when $p_2 \in P_3$.) Then p_2 can be chosen from P_3 and two other vertices p_3 and p_4 chosen from P_1 such that $MC((p_1, p_2), (p_3, p_4))$ is C_1. The proof is similar that used above. ■

The following scheme suggests itself for computing the minimum 3-cut.

1. Pick a vertex p_1.
2. For each triple (p_2, p_3, p_4), compute $MC((p_1, p_2), (p_3, p_4))$. Extend each such cut into a 3-cut by finding a minimum 2-cut in each part. The minimum 3-cut thus obtained is $3 - MC$.

The above scheme requires $O(n^3)$ Maximum Flow computations.

3 A Bounding technique for approximating graph partitioning

Given a graph $G = (V, E)$ with each edge, e, being assigned a cost, $c(e)$, consider a k-partition, P, of G which partitions G into k components, $P_1, P_2 \ldots P_k$. For each component Q we let $C(Q) = \Sigma_{e \in E(Q)} c(e)$ where $E(Q)$ is the set of edges with one endpoint in Q and the other endpoint in $V - Q$.

We first define *Partition Cut Trees*. Given a partition P, construct a tree, $T(P)$, with weighted edges such that for each component Q of the partition P we associate a node V_Q in $T(P)$. Each tree edge is a edge, called a *cut edge*, which separates the vertices of the tree into two parts. This induces a partition of the components in P since each vertex in T is a component in G. The cut edge thus represents a 2-cut in the graph G. The weight on the edge is the sum of the weights of the edges in the 2-cut of the graph G. Such a tree is called a *Partition Cut Tree*.

Also for a given partition P, we define a specific partition cut tree, $SCT(P)$, as follows: For each component Q construct a node V_Q. Let P_1 be the component of the partition with the largest value of $C(Q)$. Construct edges $E_i = (V_{P_i}, V(P_1)), 2 \leq i \leq k$, with E_i having a weight equal to $C(P_i)$.

Finally, consider the specific partition tree obtained from the optimal k-partition, OP. Let this tree be called a k-partition cut tree $SCT(OP)$. It has the property that the sum of the weights on the edges is at most twice the value of the optimal cut denoted by $wt(OP)$. We define *weight* of a partition cut tree to be the sum of the weights of all the edges in the cut.

Lemma 3.1 *The k-partition cut tree, $SCT(OP)$, of an optimal partition OP has weight $\leq 2(1 - 1/k)wt(OP)$.*

Proof: Each edge of a cut, except the ones incident onto P_1 contributes twice to the edge weights in $SCT(OP)$. ∎

It follows from the above lemma that constructing a partition with cost less than $wt(SCT(OP))$ suffices to give a good partition, i.e. one which is atmost twice the cost of the optimal.

We next consider two cut problems and show how to construct optimal partition trees for these problems.

3.1 Minimum k-cut problem

In this problem we construct a partition tree using the Gomory-Hu Cut tree. Let OPT_k be the set of edges with minimum weight, the removal of which partitions the graph into k parts. And $SCT(OPT_k)$ the specific k-partition tree. Consider the following algorithm:

ALGO k-parts(G)
    Pick the k-1 best cuts in the Gomory-Hu tree of G.

Removing these cuts from the G-H tree gives k components. Shrink the components into a vertex. Add an edge of the GHT tree between two components if they are connected by one of the cuts. We call the resultant tree, the $KGHT$ tree.

Lemma 3.2 $wt(KGHT) \leq wt(SCT(OPT_k))$

Proof: To prove this, pick a vertex from each component in OPT_k and construct a cut tree, $RGHT$, on the k vertices such that each edge (u, v) represents a minimum cut between u and v in the graph G with the minimum cut between two vertices a and b being given by the minimum cost edge on the path between a and b in the tree, $RGHT$. $wt(RGHT) \leq wt(SCT(OPT_k))$. Moreover each cut in *RGHT* is present in the Gomory-Hu cut tree of G. Since the best $k-1$ cuts are chosen to give $KGHT$, $wt(KGHT) \leq wt(RGHT)$ and the result follows. ∎

From Lemma 3.1 and Lemma 3.2 we get:

Theorem 3.1 *The partition generated by ALGO k-parts is within* $2(1 - 1/k)wt(OPT_k)$.

Algorithm :

We next describe an algorithm which finds cuts with weight less than the best $k-1$-cuts in the undirected weighted graph, $G = (V, E)$.

The algorithm is based on branch and bound. We start with the graph $G = (V, E)$. Firstly, the Min-cut algorithm of [SW] is applied to determine the minimum cut (2-cut). We will refer to 2-cut and cut interchangeably. This cut divides the graph into two parts G_1 and G_2. The minimum cut in these parts is added to a set of cuts, F, called the frontier cuts. At each step, the cut in F with minimum cost is selected to partition the graph further. New cuts are found in the parts generated. The process continues until $k-1$ cuts are found. We formalise this algorithm below:

Definition: $MC(G)$ is the minimum 2-cut in G.

Definition: $Findmin(G)$ is a function that finds the minimum 2-cut in the graph G.

Algorithm K-Best
$MC(G) = Findmin(G);\ F \leftarrow \{MC(G)\};$
Repeat
    $Cmin \leftarrow$ Pick Minimum cut from F;
    Let G_j be the part of which $Cmin$ is the minimum cut;
    Let G_{j1} and G_{j2} be the two parts into
      which G_j is partitioned by $Cmin$;
    $F \leftarrow F \cup Findmin(CG_{j1});$
    $F \leftarrow F \cup Findmin(CG_{j2});$
Until $k-1$ cuts are found.

Let $C_1, C_2 \ldots C_{k-1}$ be the $k-1$ cuts picked from F. Cut C_i is picked in a graph G_i. And let $B_1, B_2 \ldots B_{k-1}$ be the $k-1$ best cuts in a Gomory-Hu tree of G.

Lemma 3.3 *Algorithm K-Best correctly computes* $k-1$ *cuts of total value less than* $\Sigma_{1 \leq i \leq k-1} B_i$.

Proof: omitted here. Also see [SV]. ∎

The procedure to pick the minimum cut in a graph is the algorithm given in [SW]. We thus have the following result:

Theorem 3.2 *Algorithm K-BEST divides the graph into* k *parts with cost no more than* $2(1 - 1/k)wt(OPT_k)$ *in* $O(kn(m + n \log n))$ *steps.*

3.2   Multi-Way Cut Problems.

In this problem, given graph $G = (V, E)$, let OP_m be the optimal Multi-way partition on k specified vertices forming set $S = \{s_1, s_2, \ldots s_k\}$. Consider a tree similar to the GH cut-tree, $SGHT$, on the k specified vertices, i.e. vertices in the set S. For each vertex in S, we construct a vertex in $SGHT$. Each edge in the tree represents a minimum weight cut between the vertices it connects. The tree has the property that the minimum cut between any two vertices s_i and s_j is given by the least weighted edge in the path from s_i to s_j. Moreover, this tree can be constructed using $k - 1$ max-flow computations in a fashion similar to the construction of the Gomory-Hu cut tree for the graph.

Lemma 3.4 $wt(SGHT) \leq wt(SCT(OP_m))$

Proof: Let C_1 be a cut represented by an edge incident onto a vertex v in $SCT(OP_m)$. This cut partitions the component representing v from all the other components. As such the weight of this edge is greater than any one edge incident onto w in $SGHT$, where w is the vertex representing the component containing v in $SGHT$. Assigning a 1-1 correspondence between edges in $SCT(OP_m)$ and $SGHT$ gives the result. ■

Using Lemma 3.1 and Lemma 3.4 we get

Theorem 3.3 *The partition generated by $SGHT$ has weight* $\leq$ $2(1 - 1/k)wt(OP_m)$.

4   Improving the Approximation Performance bound

In this section we describe a scheme with an improved performance bound for the Multi-way cut problem. We first show how to construct partition cut trees with improved cost. We then show how to construct cuts with a corresponding partition tree of cost less than the partition cut tree constructed from the optimal solution.

4.1   A better bounding strategy

The basic idea behind this construction is that of combining partitions. Consider a k partition, P of the graph G. Let the components be $P_1, P_2 \ldots P_k$. Let the weight of edges incident onto P_i be $C(P_i)$. We next define a *j-Partition Cut Tree*. W.l.o.g let P_1 be the component with the largest value of $C(P_i)$ for $1 \leq i \leq k$. Consider sets of components $S_1, S_2 \ldots S_p$, $p = k/(j - 1)$ such that each set contains $j - 1$ components excluding P_1. Note that when k is not a multiple of $j - 1$, S_p is allowed to contain less components. We let $C(S_i)$ be the weight of edges with only one vertex in $C_j, C_j \in S_i$. The other vertex is in a component contained in some other set. And we let $IC(S_i)$ be the weight of edges which connect vertices in different components within S_i. We construct a graph j-$PCT(P) = (V_T, E_T)$ as follows: For P_1 add a vertex V_1 to V_T and a vertex V_{s_i}

for each set S_i. Add edges between V_{si} and V_1 with cost equal to $C(S_i)+IC(S_i)$. We define $wt(j\text{-}PCT(P))$ to be the cost of all edges in $j\text{-}PCT(P)$.

Given an optimal partition OPT_k which solves the Multi-way Cut problem we let $j\text{-}PCT(OPT_k)$ represent the corresponding $j\text{-}Partition\ Cut\ Tree$. In fact we consider a specific partition tree $j\text{-}SPCT(OPT_k)$ which is a partition tree with the following property:

No-Swap Property: A $j\text{-}Partition\ Cut\ Tree$, $j\text{-}PCT$, satisfies the No-Swap property if no two components can be interchanged between sets without increasing $wt(j\text{-}PCT)$.

We start with the following lemma where
$OUTS = \Sigma_{1 \le i \le p} C(S_i)/2$ and $INS = \Sigma_{1 \le i \le p} IC(S_i)$.

Lemma 4.1 $INS \ge OUTS(1 - 1/j')/(p-1)$ for a $j\text{-}SPCT$, $j' = j-1, j \ge 3$, which satisfies the No-Swap Property.

Proof: We first construct a graph, $G_c = (V_c, E_c)$ where each component is represented by a vertex and edges between components exist with weight being the sum of the weights of the edges between vertices in the two components. Furthermore, we let EO be the set of edges between components in different sets of components and we let IO be the set of edges between components in the same set. Also, consider two components C_r and C_s such that $C_r \in S_a$ and $C_s \in S_b$. Let A be the weight of edges between C_r and components in S_b, B be the weight of edges between C_s and components in S_a. Let C be the weight of edges between C_r and other components in S_a and D be the weight of edges between C_s and other components in S_b. Then because of the No-Swap Property

$$C + D \ge A + B - 2wt(C_r, C_s) \tag{1}$$

where $wt(C_r, C_s)$ is the weight of edges between C_r and C_s.

Consider a minimum weight matching in the bipartite subgraph of G_c induced by vertices which represent components in the two sets S_a and S_b with edges chosen from EO, represented by $EO(S_a, S_b)$. The weight of the matching, $MINWM(S_a, S_b)$, satisfies

$$MINWM(S_a, S_b) \le OUTS(S_a, S_b)/j'$$

where $OUTS(S_a, S_b)$ is the weight of edges in $EO(S_a, S_b)$. This follows because in the worst case all edges are equally weighted. Applying inequality (1) for all pairs of vertices in the matching we obtain

$$2INS(S_a, S_b) \ge 2OUTS(S_a, S_b) - 2OUTS(S_a, S_b)/j'$$

where $INS(S_a, S_b) = IC(S_a) + IC(S_b)$. Summing over all pairs of sets we get

$$2(p-1)INS \ge 2OUTS - 2OUTS/j'$$

which gives the desired result. ∎

We next bound the weight of the partition tree constructed. For this purpose let $f(j, k) = j/k - (j-2)/k^2 + O(j/k^3)$

Lemma 4.2 $wt(j\text{-}SPCT(OPT_k)) \leq wt(OPT_k)(2 - f(j,k))$ *for a* $j\text{-}SPCT$, $j \geq 3$, *which satisfies the No-Swap Property.*

Proof: Since $j\text{-}SPCT(OPT_k)$ is constructed after excluding the largest weighted component, $INS + OUTS + C(P_1) = wt(OPT_k)$, where $C(P_1) \geq 2wt(OPT_k)/k$. Also

$$INS \geq OUTS(1 - 1/j')/(p-1)$$

Thus $INS + C(P_1) \geq (wt(OPT_k) + (f_j - 1)C(P_1))/f_j$
where $f_j = (p - 1/j')/(1 - 1/j')$.
Thus
$$wt(j - SPCT(OPT_k)) = INS + 2OUTS + C(P_1) \leq 2wt(OPT_k) - (INS + C(P_1))$$
$$\leq wt(OPT_k)(2 - 1/f_j - 2/k + 2/(kf_j))$$
and thus the result follows. ∎

4.2 Improved Algorithms

In this section we first describe an improved algorithm for the Minimum k-Cut problem. We achieve the bounds specified in the previous section.

Minimum k-Cut Problem :

We let OPT_k be the optimum k-cut. The following algorithm successively partitions the graph using q-cuts, $2 \leq q \leq j$.

Algorithm K-Good-Cuts(G)
> $PARTS \leftarrow G$
>   **Repeat**
>>   Extend-Parts by finding cuts to increase $|PARTS|$ by $j - 1$, i.e.
>> $PARTS \leftarrow Extend - Parts(PARTS)$;
>>   Remove edges in cuts found
>   **Until** $|PARTS| = k$

To extend the number of parts in $PARTS$ we use dynamic programming to find the minimum weight set of cuts which will increase $|PARTS|$ by $j - 1$.

Let $P_1, P_2 \ldots P_l, \cup_{1 \leq i \leq l} P_i = G$ be the parts in $PARTS$. Let $MWC(p, PARTS)$ be the minimum weight q-cuts, $1 \leq q \leq j$ which increases $|PARTS|$ by p creating the partition, $P'_1, P'_2 \ldots P'_{l+p}, \cup_{1 \leq i \leq l+p} P'_i = G$. And let $MC(q, G), q \geq 1$ be the minimum q-cut in G. Note that $MC(1, G) = G$. The following recurrence characterises $MWC(p, PARTS)$.

$$MWC(p, \{P_1, P_2 \ldots P_l\}) = MIN_{1 \leq q \leq j}(MC(q, P_1)$$
$$+ MWC((p - q + 1), PARTS - \{P_1\})) \quad (2)$$
$$MWC(1, PARTS) = 0 \quad (3)$$

Using the above recurrence it is easy to devise a dynamic programming algorithm which requires time polynomial in j and l. The complexity of the scheme is $O(j^2 l + jlT(j, n))$ where $T(j, n)$ is the time required to find a j-cut in a graph with n vertices.

The correctness of the above scheme is next proven. Let $C_1, C_2 \ldots C_q$ be the j-cuts in j-$SPCT(OPT_k)$. And let $M_1, M_2, \ldots M_q$ be the cuts found in the iterations of the algorithm described above with M_i being the union of cuts in the ith iteration. We let $wt(C_i)$ be the weight of the cut C_i. Furthermore, we let $wt(PARTS)$ be the weights of the edges in the cuts which partition G into $PARTS$. The cuts themselves are in $MPARTS$. We show the following property:

Lemma 4.3 *At the end of the ith iteration of Algorithm K-Good-Cuts, $PARTS$ is updated with cuts M_i such that $wt(M_i) + wt(PARTS) \leq \Sigma_{1 \leq j \leq i} wt(C_j)$.*

Proof: The proof is by induction. When $i = 1$ the proof follows immediately from the fact that M_1 is the minimum weight j-cut in G. Let $i = l + 1$. Firstly note that $C_1, C_2 \ldots C_{l+1}$ partitions the graph G into $(l + 1)(j - 1) + 1$ parts. Let $ACUTS(i) = \cup_{1 \leq r \leq i} C_r$, $wt(ACUTS(l))$ be the weight of cuts in $ACUTS(l)$ and $PARTS_l$ be the partition at the lth iteration obtained by cuts from $MPARTS_l$. We let G_i be the graph obtained by removing edges of the cuts $M_1 \ldots M_i$. Also assume inductively that there is a 1-1 correspondence, $\mathcal{F}_i$ between cuts generating the partition $PARTS_i$ and cuts in $ACUTS(i)$ such that $wt(PARTS_i) \leq wt(ACUTS(i))$.

Consider the graph G_l. Since G_l has only $l(j - 1) + 1$ parts there are $j - 1$ additional parts in G_{l+1}. These parts are obtained from q cuts, where each cut is at most a j-cut, which further create $j - 1$ partitions together with $M_1 \ldots M_l$. Consider addition of the j-cut, C_{l+1} to the set of cuts. We construct a new correspondence $\mathcal{F}_{l+1}$ using $\mathcal{F}_l$. If any new cut, added using C_{l+1}, is the same as a cut or a union of cuts amongst $MPARTS_l$, say CP_1, then the 1-1 correspondence changes with a cut in $ACUTS(l)$, say AC_1, being replaced by the new cut to generate the correspondence between $ACUTS(l + 1)$ and $MPARTS_{l+1}$. This leaves a cut in $ACUTS(l)$ without a correspondence and this cut, say AC_1, is chosen to be one of the cuts in a candidate set CD. Note that this cut AC_1 cannot be the same as any other cut in $MPARTS_l$ otherwise it would be in 1-1 correspondence with that cut. Thus the cut generates partitions and is added to CD. Alternatively, the new cut is itself a candidate for CD. Since the algorithm computes a j-cut, M_{l+1}, which create $j - 1$ more partitions, the 1-1 correspondence can be extended to $ACUTS(l + 1)$ and the new set $MPARTS_{l+1}$ by repeated application of arguments above and by corresponding cuts in the set $CD = \{D_1, D_2 \ldots D_{j-1}\}$ with the new cuts introduced by M_{l+1}. The cuts chosen to form M_{l+1} are less in total weight than $\Sigma_{1 \leq r \leq p} D_r$ since the cuts are of minimum weight. Thus

$$wt(PARTS_{l+1}) \leq \Sigma_{1 \leq r \leq p} D_r + wt(PARTS_l) \leq \Sigma_{1 \leq s \leq l+1} C_s. \qquad (4)$$

since $wt(PARTS_l) \leq \Sigma_{1 \leq i \leq l} C_i$. Moreover $\mathcal{F}_{l+1}$ is a valid 1-1 correspondence between cuts in $Parts_{l+1}$ and $ACUTS(l + 1)$. The lemma follows. ∎

Using the fact that $T(j, n)$ is $O(n^{j^2/2 - 3j/2 + 4} T(n, m))$ where $T(n, m)$ is the time required to solve the maximum flow problem, and $l \leq k$ we get the result

Theorem 4.1 *The Minimum k-cut problem can be approximated to within a factor of $2 - f(j, k)$ in time polynomial in n and k.*

Planar Multi-way Cut Problem :

For this problem note that the Multi-j-way Cut Problem is itself NP-Hard for $j \geq 3$. This rules out approximations similar to the previous scheme. However, for planar graphs the Multi-j-way Cut Problem can be solved in polynomial time ($O(j^j n^{2j-1} \log n)$) when j is fixed. We can thus devise schemes to give improved approximation schemes.

Given a planar graph $G = (V, E)$ with weights on the edges and k specified vertices $S = \{s_1, s_2 \ldots s_k\}$, let OP_m be the optimal partition that separates vertices in S. The first scheme is as follows:

1. Compute a minimum Multi-j-way cut for all j subsets of S.
2. Use dynamic programming to compute a j-PCT partition tree with minimum cost. The minimum cost j-PCT tree for a set of vertices, $SSET$, is computed from minimum cost j-PCT trees for subsets of $SSET$.

The above algorithm obviously computes a tree of cost less than a j-$SPCT(OP_m)$ with the **No-Swap Property**. The time complexity is determined by the dynamic programming scheme and requires $O(k^j (2^k + TP(j, n)))$ where $TP(j, n)$ is the time required to solve the planar Multi-j-way cut problem. We thus get the result

Theorem 4.2 *The planar Multi-k-way cut problem can be approximated to within $2 - f(j, k)$ in $O(k^j (2^k + j^j n^{2j-1} \log n))$ operations.*

We can also apply the technique of the previous section to obtain a more complicated algorithm which computes minimum separating Multi-j-way cuts in partitions repeatedly. This will use dynamic programming at every step as in the algorithm for the Minimum k-cut problem.

The algorithm is similar to that in the previous section. A set of parts, $PARTS$, is extended at every step by adding cuts to create $j - 1$ new parts until the set S of vertices is completely separated.

We let $PARTS = \{P_1 \ldots P_l\}$ be the partition of G, which separates l vertices of G. We let $MMC(r, G)$ be the minimum multi-r-way cut in G separating vertices in a set S. Note that $MMC(1, G) = G$. And we let $MMWC(p, PARTS)$ be the minimum weight multi-q-way cuts, $1 \leq q \leq j$, which add p parts to $PARTS$ with each new part containing at least one vertex from S.

To compute $MMC(r, G)$ we consider all r subsets from S and compute the minimum cut separating the vertices in a subset. This requires time polynomial in n for fixed r. The following recurrence characterizes $MMWC(p, PARTS)$:

$$MMWC(p, \{P_1, P_2 \ldots P_l\}) = MIN_{1 \leq q \leq j}(MMC(q, P_1)$$
$$+ MMWC(p - q + 1), PARTS - \{P_1\})) \quad (5)$$
$$MMWC(1, PARTS) = 0 \quad (6)$$

The above recurrence can be computed by a dynamic programming algorithm which requires time polynomial in n and j.

The correctness of the scheme can be proved in a fashion similar to the proof of Lemma 4.3. We thus obtain the result

Theorem 4.3 *The planar Multi-k-way cut problem can be approximated to within* $2 - f(j, k)$ *in time polynomial in* n *and* k.

Note that the first algorithm is polynomial in n when $k = O(\log n)$.

5 Conclusions

In the first part of this paper we showed a non-crossing property for 3-cuts. It would be of interest to consider applications of this property. In the second part we have presented improved polynomial time approximation algorithms for graph partitioning problems. We introduced the *Partition Cut Tree* as a technique for bounding costs of the optimal solution and showed how to obtain cut trees of lesser cost in polynomial time. We hope this technique will find more applications.

The author would like to thank K. Mehlhorn for motivating this research and S. Sen and V. Vazirani for helpful discussions.

References

[Ch]   V. Chvátal, "Tough graphs and Hamiltonian circuits", *Discrete Mathematics*, Vol. 5, 1973, pp. 215–228.

[Cu]   W. H. Cunningham, "Optimal attack and reinforcement of a network", *JACM*, Vol. 32, No. 3, 1985, pp. 549–561.

[CR]   S. Chopra and M.R. Rao, "On the Multi-way cut Polyhedron", NET-WORKS(21), 1991, 51-89.

[DJPSY] E. Dalhaus, D. S. Johnson, C. H. Papadimitriou, P. Seymour and M. Yannakakis, "The complexity of the multiway cuts", Proc. 24th Annual ACM Symposium on the Theory of Computing, 1992, pp. 241–251.

[GoHu] R. Gomory and T. C. Hu, "Multi-terminal network flows", *J. SIAM*, Vol. 9, 1961, pp. 551–570.

[GH]   O. Goldschmidt and D. S. Hochbaum, "Polynomial algorithm for the k-cut problem", Proc. 29th Annual Symp. on the Foundations of Computer Science, 1988, pp. 444–451.

[He]   Xin He, "On the planar 3-cut problem", *J. Algorithms*, 12, 1991, pp. 23–37.

[HS]   D. Hochbaum and D. Shmoys, "An $O(V^2)$ algorithm for the planar 3-cut problem," SIAM J. on Alg. and Discrete Methods, 6:4:707 - 712, 1985.

[SV]   H. Saran and V. Vazirani, "Finding k-cuts within twice the optimal," Proc. 32nd Annual Symp. on Foundation of Computer Science,1991, 743-751.

[SW]   M. Stoer and F. Wagner, "A Simple Min Cut Algorithm", Technical Report, Fachberich Mathematik and Informatik, Freie Universitat, Berlin.

[S]   H.S. Stone, "Multiprocessor scheduling with the aid of Network flow algorithms", IEEE Trans.on Software Engg., SE-3, 1977, 85-93.

Primal-Dual Approximation Algorithms for Feedback Problems in Planar Graphs

Michel X. Goemans[1] and David P. Williamson[2]

[1] Dept. of Mathematics, Room 2-382, M.I.T., Cambridge, MA 02139.
Email: goemans@math.mit.edu.
[2] IBM T.J. Watson Research Center, P.O. Box 218, Yorktown Heights, NY, 10598.
Email: dpw@watson.ibm.com.

Abstract. Given a subset of cycles of a graph, we consider the problem of finding a minimum-weight set of vertices that meets all cycles in the subset. This problem generalizes a number of problems, including the minimum-weight feedback vertex set problem in both directed and undirected graphs, the subset feedback vertex set problem, and the graph bipartization problem, in which one must remove a minimum-weight set of vertices so that the remaining graph is bipartite. We give a $\frac{9}{4}$-approximation algorithm for the general problem in planar graphs, given that the subset of cycles obeys certain properties. This results in $\frac{9}{4}$-approximation algorithms for the aforementioned feedback and bipartization problems in planar graphs. Our algorithms use the primal-dual method for approximation algorithms as given in Goemans and Williamson [14]. We also show that our results have an interesting bearing on a conjecture of Akiyama and Watanabe [2] on the cardinality of feedback vertex sets in planar graphs.

1 The problems

We consider the following general problem: given a graph $G = (V, E)$, non-negative weights w_i on the vertices $i \in V$, and a collection C of cycles of G, find a minimum-cost set of vertices F such that every cycle in C contains some vertex of F. We call this problem the *hitting cycle* problem, since we must hit every cycle in C. The hitting cycle problem generalizes several other problems we will study in this paper. If C is the set of all cycles in G, then the hitting cycle problem is equivalent to the problem of finding a minimum-weight *feedback vertex set* in a graph; that is, the problem of finding a minimum-weight set $F \subseteq V$ such that the graph $G[V - F]$ induced by $V - F$ is acyclic. The feedback vertex set problem will be abbreviated by FVS. If G is a directed graph (digraph), and C the set of all directed cycles in G, then we have the minimum-weight feedback vertex set problem in directed graphs (D-FVS). If we are given a set of *special* vertices and C is all cycles of an undirected graph G that contain some special vertex, then we have the *subset* feedback vertex set problem (S-FVS). Finally, if C contains all odd cycles of G, then we have the *graph bipartization* problem (BIP); that is, the problem of finding a minimum-weight subset F such that $G[V - F]$ is bipartite. All these problems are also special cases of *vertex deletion problems*: that is, find a minimum-weight (or minimum cardinality) set of vertices whose deletion gives a graph satisfying a given property.

We will restrict our attention to the versions of these problems in which the input graph is planar and simple. Yannakakis [24] has given a general NP-hardness proof for almost all vertex deletion problems restricted to planar graphs; his results apply to the planar (directed, undirected or subset) feedback vertex set problem and to the planar graph bipartization problem. In addition, the planar D-FVS is NP-hard even if both the indegree and outdegree of every vertex is no more than 3 [10, p. 192].

We consider *approximation algorithms* for these problems. An α-approximation algorithm for a minimization problem runs in polynomial time and produces a solution of weight no more that α times the weight of an optimal solution. We call α the *performance guarantee* of the algorithm. In this paper, we give a $\frac{9}{4}$-approximation algorithm for a general class of planar hitting cycle problems which includes the planar feedback vertex set problem in undirected or directed graphs, the planar subset feedback vertex set problem in undirected graphs, and the planar graph bipartization problem.

Our algorithms are based on the primal-dual method for approximation algorithms. This method has proven useful over the past few years in designing algorithms for network design problems (see, for example, [13, 11, 18, 23]). The authors have written a survey of this method [14] which gives a generic algorithm and theorem for deriving approximation algorithms for the hitting set problem, of which the hitting cycle problem is a special case. The algorithm and analysis here are an application of the algorithm and theorem given in the survey.

We now briefly review previously known work. For FVS in general undirected graphs, two slightly different 2-approximation algorithms were given recently by Becker and Geiger [6] and Bafna, Berman, and Fujito [4]; see Hochbaum for an overview [15]. These algorithms improve on a log n-approximation algorithm of Bar-Yehuda, Geiger, Naor, and Roth [5], where n is the number of vertices. Bar-Yehuda et al. also gave a 10-approximation algorithm for the case of undirected planar graphs, which we can show to be a 5-approximation algorithm for this case. For all three other problems we consider, the best known approximation algorithms for general graphs have polylogarithmic guarantees; because of space limitations, we simply refer the reader to the relevant papers [8, 9, 12, 17, 19, 21]. In the case of planar graphs, the only additional result we are aware of is an approximation algorithm of Stamm [22] for D-FVS, but its performance guarantee can be linear.

Although our result for the undirected feedback vertex set problem on planar graphs is worse than the known approximation algorithm for general undirected graphs, it still turns out to be interesting. Our result implies that the LP relaxation of the cycle formulation of all four problems is within a factor of 9/4 of the corresponding optimum value for planar graphs. This is known to be false for general graphs (the ratio can be logarithmic in n [21, 9]). We in fact conjecture that the ratio cannot be greater than 3/2. This would have very interesting combinatorial consequences that we discuss in Section 6. For example, it would imply a conjecture of Akiyama and Watanabe [2] and Albertson and Berman [3] that any undirected planar graph on n vertices contains a feedback vertex set of size no more than $n/2$. Our bound of $9/4$ implies the existence of a feedback vertex set of size at most $3n/4$. This follows easily from the 4-color theorem, but we don't know any other proof. A coloring result of Borodin [7] shows that any planar graph has a feedback vertex set of size no more than $3n/5$; however, Jensen and Toth [16, p. 6] call the proof reminiscent of the proof of the 4-color theorem, partly because it involves 450 reducible configurations.

The paper is structured as follows. In Section 2 we begin with some preliminary concepts and definitions. Section 3 reviews the generic primal-dual algorithm and its analysis from Goemans and Williamson [14]. In Section 4, we show how the algorithm leads to a 3-approximation algorithm for a class of hitting cycle problems, and in Section 5 we improve the algorithm and its analysis to give a $\frac{9}{4}$-approximation algorithm. We comment on the integrality gap of the linear programming relaxation and its relation to Akiyama and Watanabe's conjecture in Section 6. The implementation of the algorithms is described in Section 7, and we conclude in Section 8.

2 Preliminaries

When we refer to a cycle C of an undirected graph $G = (V, E)$, we refer to its vertex set, even though this is somewhat ambiguous. If we would like to refer to its edge set, we will write $E(C)$.

Recall the hitting cycle problem defined in the previous section. Let G be an undirected graph, let $w_i \geq 0$ be the weight of vertex i, and let $\mathcal{C}$ be a collection of cycles of G. The hitting cycle problem is that of finding a minimum-weight set F of vertices such that F intersects every member of $\mathcal{C}$. In most cases, when we will refer to a *cycle*, we will implicitly mean a cycle of $\mathcal{C}$, unless stated otherwise.

We will restrict our attention to families $\mathcal{C}$ satisfying the following property. We abuse notation slightly here by referring to cycles C as both sets of edges and of vertices. Paths P are sets of edges; for directed graphs, the set of edges is a path for the underlying undirected graph.

Property A For any two cycles $C_1, C_2 \in \mathcal{C}$, let P_2 be a path in C_2 which intersects C_1 only at the endpoints of P_2. Let P_1 be a path in C_1 between the endpoints of P_2. Then either $P_1 \cup P_2 \in \mathcal{C}$ and $(C_1 - P_1) \cup (C_2 - P_2)$ contains a cycle in $\mathcal{C}$, or $(C_1 - P_1) \cup P_2 \in \mathcal{C}$ and $(C_2 - P_2) \cup P_1$ contains a cycle in $\mathcal{C}$.

We will refer to families satisfying Property A as *uncrossable*. Our approximation algorithms will apply to any uncrossable hitting cycle problem for input graphs restricted to be planar, given that we can compute efficiently certain minimal cycles which we will define in a moment.

We claim that the problems we are interested in correspond to uncrossable families. First notice that the graph $H = E(C_1) \cup E(C_2)$ is Eulerian, i.e. every vertex has even degree, or every vertex has indegree equal to outdegree in the case of D-FVS. Also, when removing a cycle C from H, the resulting graph remains Eulerian (assuming C is directed in the case of D-FVS). It can therefore be decomposed into cycles. This shows that Property A is clearly satisfied for FVS. For D-FVS, Property A is also satisfied. Let a and b be the two endpoints of the path P_2. Then either P_2 is directed from a to b (and $C_2 - P_2$ is directed from b to a) or vice versa. Thus, either $P_1 \cup P_2$ or $(C_1 - P_1) \cup P_2$ defines a directed cycle C, and $H - E(C)$ contains a directed cycle since it is Eulerian. For S-FVS, there must be a special vertex on either P_1 or $C_1 - P_1$ and also on either P_2 or $C_2 - P_2$. Therefore, we can make sure that the Eulerian graph $H - E(C)$ still contains a special vertex, so that one of the two cases of Property A must hold. For BIP, we observe that $P_1 \cup P_2$ and $(C_1 - P_1) \cup P_2$ have different parities, and therefore one of them must be odd. Moreover, $H - E(C)$ is Eulerian and has an odd number of edges if C is odd, and therefore must contain an odd cycle in any cycle decomposition. So, once again, Property A holds.

The FVS, S-FVS and BIP also satisfy an additional property:

Property B For any cycle $C \in \mathcal{C}$ and any path P intersecting C only at the endpoints of P, let C_1, C_2 be the two cycles defined by C and P. Then either C_1 or C_2 (or both) belongs to $\mathcal{C}$.

Observe that this is not the case for D-FVS since there is no guarantee that P is a directed path. Property B will be useful for implementation purposes.

Our approximation algorithms for uncrossable hitting cycle problems will depend on the embedding of the planar graph. Given a plane graph G (i.e. a planar graph with an embedding), any cycle C partitions the plane into two regions, the interior and exterior regions. We will associate to any cycle C the set $f(C)$ of faces in the interior region of C. Observe that the exterior face of the embedding of G never belongs to $f(C)$. We will say that cycle C_1 contains cycle C_2 and write $C_1 \supseteq_f C_2$ or $C_2 \subseteq_f C_1$ if $f(C_1) \supseteq f(C_2)$. Two cycles C_1 and C_2 are said to be crossing if $f(C_1)$ and $f(C_2)$ cross[3], i.e. $f(C_1) \cap f(C_2) \neq \emptyset$, $f(C_1) - f(C_2) \neq \emptyset$ and $f(C_2) - f(C_1) \neq \emptyset$. Similarly, we say that a collection of cycles form a *laminar family* if no two cycles are crossing.

We say that a cycle $C \in \mathcal{C}$ is *face-minimal* if there does not exist a cycle $C' \in \mathcal{C}$, $C' \neq C$, with $f(C') \subseteq_f f(C)$. The collection of face-minimal cycles will play a central role in our approximation algorithms. The following lemma shows that face-minimal cycles form a laminar family.

Lemma 1. *Let $\mathcal{C}$ satisfy Property A and let $C_1, C_2 \in \mathcal{C}$. If C_1 is a face-minimal cycle then C_1 and C_2 do not cross.*

Proof. The proof follows immediately from Property A. If the two cycles were to cross, then by choosing P_2 to be a path in C_2 which lies in the interior of C_1,

---
[3] Observe that the exterior face is never in $f(C_1) \cup f(C_2)$, and thus the notions of crossing and intersecting are equivalent.

the two cycles $P_1 \cup P_2$ and $(C_1 - P_1) \cup P_2$ would both be contained in C_1. This is a contradiction since at least one of them belongs to C and C_1 is face-minimal.

Whenever C satisfies Property B, the face-minimal cycles have a very simple structure given by the next lemma.

Lemma 2. *Let C satisfy Property B. Then the face-minimal cycles are the boundaries of the interior faces which are simple cycles.*

Proof. Suppose C is a face-minimal cycle of C which is not given by the boundary of an interior face. Then there must be a path P in the interior of C that only intersects C at its endpoints. Using property B, one of the two cycles defined by C and P must be in C. But this cycle must be contained in C, which contradicts the face-minimality of C.

In particular, for families satisfying Property B, this lemma shows that the face-minimal cycles are the boundaries of *all* interior faces corresponding to cycles in C if the graph is 2-connected. The lemma is not true for the directed feedback vertex set problem, which does not satisfy Property B.

3 The primal-dual framework

The uncrossable hitting cycle problem is a special case of the general *hitting set problem* in which one needs to find a minimum-weight set hitting every set in a given collection of sets. More precisely, given a ground set of elements E, weights c_e for all $e \in E$, and sets $T_1, \ldots, T_p \subseteq E$, the hitting set problem is that of finding a minimum-weight $A \subseteq E$ such that $A \cap T_i \neq \emptyset$ for $i = 1, \ldots, p$. In a recent survey [14], we have developed a general methodology to derive approximation algorithms for hitting set problems based on the so-called *primal-dual method*. This was motivated by a sequence of papers [1, 13, 18, 23] developing the technique for network design problems. In the survey, we propose a generic primal-dual method for deriving approximation algorithms for hitting set problems, with a generic proof of the performance guarantee. We illustrate in [14] the technique on a variety of problems, and also claim that the method can be applied to many more problems. As we show here, the technique directly applies to any uncrossable hitting cycle problem in planar graphs.

A hitting cycle problem can be formulated by the following integer program (IP):

$$\text{Min} \sum_{i \in V} w_i x_i$$

(IP) subject to:

$$\sum_{i \in C} x_i \geq 1 \qquad \text{cycles } C \in C$$

$$x_i \in \{0, 1\} \qquad i \in V.$$

The primal-dual method simultaneously constructs a feasible solution to this hitting set problem, and a solution feasible for the dual of the linear programming relaxation of (IP). The dual of the LP relaxation is:

$$
\text{Max} \sum_{C \in \mathcal{C}} y_C
$$

(D) subject to:

$$
\sum_{C: i \in C} y_C \leq w_i \qquad i \in V
$$

$$
y_C \geq 0 \qquad C \in \mathcal{C}.
$$

The generic primal-dual method developed in [14] is described in Figure 1. It is specified by the oracle VIOLATION(S) which given a set of vertices S outputs a specific set of cycles in $\mathcal{C}$ which are not hit by S. The algorithm begins with an empty set of vertices S and a dual solution $y = 0$. While S is not a feasible solution to the hitting cycle problem, it increases the dual variables on the cycles returned by VIOLATION(S) until one of the dual packing constraints becomes tight for some vertex $i \in V$. This vertex is added to S and the process continues. When S becomes feasible, the algorithm performs a "clean-up" step. It goes through the vertices in the reverse of the order in which they were added and removes any vertex which is not necessary for S to remain feasible.

In [14], it is proved that the performance guarantee of this algorithm can be obtained by using the following theorem. In this theorem, a *minimal augmentation* F of S means a feasible solution F containing S such that for any $v \in F - S$, $F - v$ is not feasible.

Theorem 3 (Goemans and Williamson [14]). *The primal-dual algorithm described in Figure 1 delivers a solution of cost at most $\gamma \sum_C y_C \leq \gamma z_{OPT}$, where z_{OPT} denotes the weight of an optimum solution, if γ satisfies that for any infeasible set $S \subset V$ and any minimal augmentation F of S*

$$
\sum_{C \in \mathcal{V}(S)} |F \cap C| \leq \gamma |\mathcal{V}(S)|,
$$

where $\mathcal{V}(S)$ denotes the collection of violated sets output by the VIOLATION oracle on input S.

Therefore, we only need to specify what the VIOLATION oracle does, compute the value of γ given by Theorem 3, and prove that the algorithm runs in polynomial time in order to obtain a γ-approximation algorithm. Observe that by considering $G - S$, we can assume without loss of generality that, in Theorem 3, $S = \emptyset$ and F is a minimal feasible solution.

One possibility is that the VIOLATION oracle returns only one cycle. This is essentially the approach used by Bar-Yehuda et al. [5] for FVS in general graphs. They gave a 10-approximation algorithm for this problem in planar graphs by simply finding a "short" cycle in the graph, but their analysis can be improved.

```
1    y ← 0
2    S ← ∅
3    l ← 0
4    While S is not feasible
5       l ← l + 1
6       V ← VIOLATION(S)
7       Increase y_C uniformly for all C ∈ V until ∃v_l ∉ S : ∑_{C:v_l∈C} y_C = w_{v_l}
8       S ← S ∪ {v_l}
9    For j ← l downto 1
10      if S − {v_j} is feasible then S ← S − {v_j}
11   Output S (and y)
```

Fig. 1. Primal-dual algorithm for uncrossable hitting cycle problems.

We give below a brief sketch of their VIOLATION oracle and of the improved analysis. Given the planar graph G, we can first assume that G has no degree 1 vertex since such vertices can be deleted without affecting the cycles of G. We claim that the resulting graph has a cycle with at most 5 vertices of degree 3 or higher; moreover, this cycle can be chosen to be (part of) the boundary of a face. It is then easy to see that γ can be chosen to be 5 in Theorem 3. To prove the claim, observe that, if the graph is 2-connected, the claim is equivalent to the existence of a vertex of degree at most 5 in the dual graph, a well-known fact (since the sum of the degrees is at most $6|V| - 12$). If the graph is not 2-connected, we consider an endblock of the graph (i.e. a block with at most one cutvertex) and use the same argument. The only slight problem is that the resulting cycle may contain the cutvertex and this cutvertex may have degree 2 in the endblock. This however can be dealt with by using the fact that a planar graph has more than one vertex of degree at most 5. The idea of having the VIOLATION oracle return only one cycle does not seem to work for S-FVS, D-FVS or BIP.

4 A 3-approximation algorithm

In this section, we consider the VIOLATION oracle which, on input S, returns the set of face-minimal cycles of $G - S$ (with respect to C). We will refer to this oracle as FACE-MINIMAL. We show that the corresponding value of γ is 3. In the following section, we give a refined oracle for which the corresponding γ is 9/4. These performance guarantees are tight for D-FVS, S-FVS and BIP.

In order to prove that FACE-MINIMAL has a γ value of 3, we need to show the following result (applied to the graph $G - S$).

Theorem 4. *Let G be a planar graph and let M be the collection of face-minimal cycles corresponding to an uncrossable family C. Consider any minimal solution*

F. *Then*

$$\sum_{C \in \mathcal{M}} |F \cap C| \leq 3|\mathcal{M}|.$$

Since F is a minimal solution, we know that for every $v \in F$, $F - v$ is not feasible, implying the existence of a cycle $C_v \in \mathcal{C}$ such that $C_v \cap F = \{v\}$. We call such a cycle C_v a *witness cycle* (for v). A *family* of witness cycles is a collection of witness cycles $C_v \in \mathcal{C}$, one for each $v \in F$.

Lemma 5. *There exists a laminar family of witness cycles* $C_v \in \mathcal{C}$, $v \in F$.

Proof. Consider any family of witness cycles and assume the existence of two witness cycles C_u and C_v that cross for $u, v \in F$. By assumption $F \cap C_u = \{u\}$ and $F \cap C_v = \{v\}$. The assumption implies that u and v have degree 2 in $H = E(C_v) \cup E(C_u)$ and that no other vertices of H are in F. Since the cycles cross there is some path P_u of C_u in the interior of C_v which intersects C_v only at its endpoints. By Property A, C_u and C_v can be replaced by two cycles such that one is in $\mathcal{C}$, call it C', and the other contains a cycle say C'' in $\mathcal{C}$. Say that C_v is replaced by C'; by property A, it will contain strictly fewer faces than C_v. Since F is feasible, both C' and C'' must be hit by F. However, since u and v have degree 2, it must be the case that C' and C'' each have exactly one of u and v and are witness cycles for u and v.

In order to show the existence of a laminar family of witness cycles, we need to prove that the crossing pairs of cycles being replaced can be selected in such a way that the replacing process terminates with a laminar family. For this purpose, we fix an ordering of the vertices in F, say $F = \{1, 2, \ldots, k\}$. We start by repeatedly replacing C_1 with all the other witness cycles that cross it. Notice that this must terminate since C_1 is always replaced by a cycle which encloses fewer faces. Thus at termination C_1 does not cross any of the other witness cycles; we say that we have *uncrossed* C_1. After uncrossing C_i with all the other cycles ($i = 1, \ldots$), we uncross C_{i+1} with all the other cycles. The important observation to make is that as we replace a crossing pair C_i and C_j as explained in the first part of the proof, if C_k does not cross either C_i or C_j, then C_k still does not cross the new witness cycles C' and C'' for i and j. This follows from the fact that $f(C_k)$ must either be contained entirely in one of the faces of $H = E(C_i) \cup E(C_j)$ or must contain all the interior faces of H. Therefore, as we replace C_i with the other witness cycles that cross it, we don't need to consider any C_k for $k < i$. Therefore, this uncrossing process terminates with a laminar family of witness cycles.

A laminar family $\mathcal{F} = \{C_v \in \mathcal{C} : v \in F\}$ of witness cycles can be represented by a tree or more precisely by a forest by considering the partial order imposed by $\subseteq_f$. To simplify the exposition, we can add a root node $\mathbf{r}$ which is connected to all maximal sets in the family, and thus obtain a tree $\mathbf{T}$. Notice that any vertex in $\mathbf{T}$ is either $\mathbf{r}$ or corresponds to a cycle C_v for $v \in F$. Thus for each vertex $v \in F$ we will correspond a vertex $\mathbf{v} \in \mathbf{T}$.

The crucial (and only) properties of $\mathcal{M}$ we will be using are the following:

1. No element of $\mathcal{M}$ crosses any element of $\mathcal{F}$. This follows from Lemma 1.
2. Every element of $\mathcal{F}$ (and therefore the cycles corresponding to the leaves of **T**) contains at least one element of $\mathcal{M}$.

For the analysis, and because of these two properties, we assign every element of $\mathcal{M}$ to some node in the tree **T**: cycle $C \in \mathcal{M}$ is assigned to the vertex of **T** corresponding to the smallest set in $\mathcal{F}$ (inclusion-wise) which contains it. For $\mathbf{v} \in \mathbf{T}$, let $\mathcal{M}_{\mathbf{v}}$ denote the set of cycles of $\mathcal{M}$ assigned to node $\mathbf{v}$ of **T**. Observe that $\mathcal{M}_{\mathbf{r}}$ may be non-empty, and that some $\mathcal{M}_{\mathbf{v}}$ may be empty. However, because of property 2, $\mathcal{M}_{\mathbf{v}}$ is non-empty for every leaf $\mathbf{v}$ of **T**.

In order to prove Theorem 4, we first derive an upper bound on $\sum_{C \in \mathcal{M}_{\mathbf{v}}} |F \cap C|$ for every $\mathbf{v} \in \mathbf{T}$. Fix $\mathbf{v} \in \mathbf{T}$, and let $F_{\mathbf{v}}$ denote the subset of vertices of F corresponding to $\mathbf{v}$ (unless $\mathbf{v} = \mathbf{r}$) and the children (if any) of $\mathbf{v}$ in **T**. Observe that $F \cap C = F_{\mathbf{v}} \cap C$ for any $C \in \mathcal{M}_{\mathbf{v}}$. Thus, $\sum_{C \in \mathcal{M}_{\mathbf{v}}} |F \cap C| = \sum_{C \in \mathcal{M}_{\mathbf{v}}} |F_{\mathbf{v}} \cap C|$. By definition of $F_{\mathbf{v}}$, its cardinality is equal to the degree $deg(\mathbf{v})$ of node $\mathbf{v}$ in **T**. In order to get an upper bound on $\sum_{C \in \mathcal{M}_{\mathbf{v}}} |F_{\mathbf{v}} \cap C|$, we construct a bipartite graph B. B has a vertex for every $u \in F_{\mathbf{v}}$ and for every $C \in \mathcal{M}_{\mathbf{v}}$, and an edge between u and C iff $u \in C$. Therefore, $\sum_{C \in \mathcal{M}_{\mathbf{v}}} |F_{\mathbf{v}} \cap C|$ is precisely the number of edges of B. Observe that B is planar, since a planar embedding of B can be obtained from the embedding of G by placing the vertex corresponding to $C \in \mathcal{M}_{\mathbf{v}}$ in the interior of C. But the number of edges of a simple bipartite planar graph is at most twice the number of vertices minus four, unless the graph consists simply of a single vertex or of two vertices with one edge. Notice that B can only be a single vertex if $\mathbf{v} = \mathbf{r}$. Also, B can be an edge on two vertices; this can occur only if $\mathbf{v}$ is a leaf of **T** or $\mathbf{v} = \mathbf{r}$. We have therefore derived that

$$\sum_{C \in \mathcal{M}_{\mathbf{v}}} |F_{\mathbf{v}} \cap C| \le 2|\mathcal{M}_{\mathbf{v}}| + 2|F_{\mathbf{v}}| - 4 = 2|\mathcal{M}_{\mathbf{v}}| + 2deg(\mathbf{v}) - 4, \qquad (1)$$

unless $\mathbf{v}$ is a leaf of **T** in which case

$$\sum_{C \in \mathcal{M}_{\mathbf{v}}} |F_{\mathbf{v}} \cap C| \le 2|\mathcal{M}_{\mathbf{v}}| + 2deg(\mathbf{v}) - 3,$$

or v corresponds to $\mathbf{r}$ in which case

$$\sum_{C \in \mathcal{M}_{\mathbf{r}}} |F_{\mathbf{v}} \cap C| \le 2|\mathcal{M}_{\mathbf{r}}| + 2deg(\mathbf{r}) - 2.$$

Summing over all $\mathbf{v} \in \mathbf{T}$, we derive that

$$\sum_{C \in \mathcal{M}} |F \cap C| \le 2|\mathcal{M}| + 2 \sum_{\mathbf{v} \in \mathbf{T}} deg(\mathbf{v}) - 4|\mathbf{T}| + l + 2,$$

where l denotes the number of leaves of **T**. Since **T** is a tree, $\sum deg(\mathbf{v})$ is equal to twice the number of nodes of the tree minus two. This implies that

$$\sum_{C \in \mathcal{M}} |F \cap C| \le 2|\mathcal{M}| - 2 + l.$$

Moreover, because of property 2, the number l of leaves is upper bounded by $|\mathcal{M}|$. This therefore shows that

$$\sum_{C \in \mathcal{M}} |T \cap C| \le 3|\mathcal{M}| - 2,$$

proving Theorem 4.

For FVS, the worst instance we are aware of for our primal-dual algorithm with the oracle FACE-MINIMAL achieves a performance ratio of 2. However, for the other problems, namely D-FVS, S-FVS and BIP, the performance guarantee of 3 is tight. Instances achieving this ratio are given in Figure 2; the same figure applies to all three problems. There are k white vertices and they have a weight of 3, and the other (black) vertices have a weight of $1 + \epsilon$. In the case of S-FVS, the special vertices are denoted by (black) squares, while for D-FVS the orientation of the arcs along two of the faces are explicitly given on the figure (the orientation of the other arcs are such that the shaded faces define directed cycles). The face-minimal cycles are the boundaries of the shaded faces, and the algorithm will select all white vertices in the solution for a total weight of $3k$. However, in all three cases, the black squares constitute a feasible solution of weight $(k + 2)(1 + \epsilon)$, giving the desired bound as k gets large and ϵ tends to 0. The analysis of our algorithm in fact indicates that bad examples arise only when there are two cycles in $\mathcal{M}$ with several points in common. The improved VIOLATION oracle we develop in the next section deals precisely with such cases.

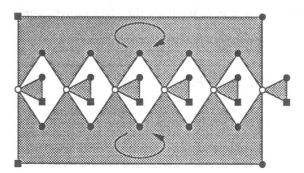

Fig. 2. A bad example for the 3-approximation algorithm applied to BIP, to D-FVS, or to S-FVS.

5   A 9/4-approximation algorithm

We first need some preliminaries. Two (face)-disjoint[4] cycles C_1 and C_2 partition the plane into one or several regions; excluding the interiors of C_1 and C_2, each

---

[4] that is, $f(C_1) \cap f(C_2) = \emptyset$.

remaining region corresponds to a connected component of the dual graph after having removed $f(C_1) \cup f(C_2)$. One of these regions contains the exterior face, and we refer to the others as the *pockets* between C_1 and C_2. The boundary of any pocket is defined by two vertices common to C_1 and C_2, say u and v, and consists of two paths between u and v, one from C_1 and one from C_2. If there exist k non-empty pockets between C_1 and C_2 then C_1 and C_2 must have at least $k + 1$ vertices in common. We say that two disjoint cycles C_1 and C_2 *surround* a cycle C_3 if $f(C_3)$ is contained in one of the pockets between C_1 and C_2.

Our improved algorithm is based on the following oracle which returns a subset $\mathcal{V}$ of the family $\mathcal{M}$ of face-minimal cycles. If $\mathcal{M}$ does not contain two cycles which surround a third one then the oracle returns $\mathcal{M}$. Otherwise, the oracle outputs a non-empty subset $\mathcal{V}$ of $\mathcal{M}$ such that (i) there do not exist two cycles C_1 and C_2 in $\mathcal{V}$ which surround a third cycle of $\mathcal{V}$, and (ii) $\mathcal{V}$ consists of all cycles of $\mathcal{M}$ in one of the pockets between two cycles C_1 and C_2 of $\mathcal{M}$. This is always possible since the oracle can simply recursively select the non-empty set of cycles of one of the pockets between two cycles C_1 and C_2 until the remaining collection satisfies (i).

Theorem 6. *Let G be a planar graph and let $\mathcal{V}$ be as defined in the paragraph above. Consider any minimal feasible solution F. Then*

$$\sum_{C \in \mathcal{V}} |F \cap C| \leq \frac{9}{4} |\mathcal{V}|.$$

The structure of the proof is similar to the one in the previous section, the main difference being the proof of a sharper version of inequality (1). The basic idea is to exploit the fact that the bipartite graph constructed does not have any cycle of length 4. However, the proof is somewhat more complicated and is omitted for space reasons.

The performance guarantee of 9/4 is tight for D-FVS, S-FVS and BIP, but again we are not aware of an instance with a performance worse than 2 for FVS.

6 Worst-case duality gaps

In this section, we discuss the worst-case ratio between the value of the problem considered and the optimum value of the linear programming relaxation of (IP) (or the value of its dual (D)), the worst-case being taken over all non-negatively weighted planar instances. The results of the previous section in fact immediately imply that this worst-case ratio ρ is at most 9/4 for any uncrossable hitting cycle problem.

Before considering the worst-case ratio for hitting cycle problems in more detail, we investigate the vertex cover problem. In the vertex cover problem, one would like to find a minimum-weight set of vertices S such that for every edge at least one of its endpoints is in S. A classical linear programming relaxation of this problem is given below:

$$\text{Min} \sum_{i \in V} w_i x_i$$

(LP)   subject to:

$$x_i + x_j \geq 1 \qquad\qquad (i,j) \in E$$
$$x_i \geq 0 \qquad\qquad i \in V.$$

It is well-known that the ratio between the value of the vertex cover problem and the value of (LP) is upper bounded by 2, and this can be approached arbitrarily by general graphs. However, we show below that the worst-case ratio is exactly 3/2 for planar instances by using the 4-color theorem.

Theorem 7. *For planar graphs, $\rho_{VC} = \frac{3}{2}$.*

Proof. For K_4 with unit weights, the minimum vertex cover has size 3, but the LP value is 2 and this is obtained by setting all x_i's to 0.5. This shows that $\rho_{VC} \geq \frac{3}{2}$.

To prove the other inequality, we use the 4-color theorem and a result about the structure of the extreme points of (LP). It is known that at the extreme points of (LP), $x_i \in \{0, \frac{1}{2}, 1\}$ for all i [20]. Given a four-coloring of the graph and an optimal extreme point of (LP), we find the color class $\mathcal{X}$ which maximizes $\sum_{i \in \mathcal{X}: x_i = 1/2} w_i$. Consider then the integral solution

$$x_i^* = \begin{cases} 1 \text{ if } x_i = 1 \text{ or } x_i = \frac{1}{2}, i \notin \mathcal{X} \\ 0 \text{ if } x_i = 0 \text{ or } x_i = \frac{1}{2}, i \in \mathcal{X} \end{cases}$$

By construction $\sum_i w_i x_i^* \leq \frac{3}{2} \sum_i w_i x_i$. Furthermore, x^* corresponds to a vertex cover since for any edge (i,j) with $x_i = x_j = \frac{1}{2}$, both i and j cannot be in $\mathcal{X}$.

A proof of this result not based on the 4-color theorem would be very nice. Indeed, since the solution $x_i = 0.5$ for all i is always feasible for the linear programming relaxation, the above theorem implies the existence of a vertex cover of size at most $3n/4$ (or an independent set of size at least $n/4$), which follows immediately from the 4-color theorem, but no other proof of this result is known.

The K_4 instance for the vertex cover problem leads to bad instances for many hitting cycle problems. Consider FVS, for example. If we replace in K_4 every edge by a triangle (introducing one new vertex) and if we keep all weights to be equal to 1, then the optimum solution still has value 3, and a feasible solution to the linear programming relaxation of the hitting cycle formulation (IP) can be obtained by setting the original vertices to have $x_i = 0.5$ and the new vertices to have $x_i = 0$. This shows that the worst-case ratio ρ_{FVS} for FVS on planar instances is at least $\frac{3}{2}$. The same construction shows that that $\rho_{BIP} \geq 3/2$ and $\rho_{D-FVS} \geq 3/2$ for BIP and D-FVS both in the planar case.

We conjecture that these bounds are tight.

Conjecture 8. $\rho_{FVS} = \frac{3}{2}, \rho_{D-FVS} = \frac{3}{2}$ and $\rho_{BIP} = \frac{3}{2}$.

If any part of this conjecture was true, this would have some interesting combinatorial implications. Consider first FVS. If $\rho_{FVS} = \frac{3}{2}$, then since the solution with $x_i = \frac{1}{3}$ is feasible for the LP relaxation, this implies the existence of a feedback vertex set of size at most $n/2$, a statement conjectured by Akiyama and Watanabe [2] and Albertson and Berman [3]. For BIP, the conjecture that $\rho_{BIP} = \frac{3}{2}$ would similarly imply the existence of at most $n/2$ vertices whose removal makes the graph bipartite. This follows easily from the 4-color theorem (removing the two smallest color classes), but we are not aware of any proof of this statement not based on the 4-color theorem. We should point out that in the worst case one cannot remove less than half the vertices for either FVS or BIP (consider K_4 or multiple copies of K_4). For D-FVS on simple planar digraphs, the same reasoning would imply the existence of a feedback vertex set of size at most $n/2$, which would follow clearly from Akiyama and Watanabe's or Albertson and Berman's conjecture. It seems possible in fact that $n/3$ vertices are enough for simple digraphs.

7 Implementation

In this section we sketch how our 3-approximation algorithms can be implemented in $O(n^2)$ time, where $n = |V|$. For all problems considered, the FACE-MINIMAL oracle can easily be implemented in linear time as follows. For the three undirected problems (FVS, S-FVS and BIP), we can first decide whether the boundary of any face is a cycle of C in time proportional to the length of this cycle. Over all faces, this gives a linear running time to compute the set $\mathcal{M}$ of face-minimal cycles in C (since the total length of all faces is equal to twice the number of edges, which is at most $3n - 6$). To implement the FACE-MINIMAL oracle in the case of D-FVS, we consider the planar dual G^* of the graph G. It is not difficult to see that the face-minimal cycles correspond to sources and sinks in a DAG formed by collapsing the strongly connected components of G^*. The planar dual, its strongly connected components and the sources and sinks can easily be found in linear time, and as a result we can implement FACE-MINIMAL in linear time also for D-FVS. Notice that the FACE-MINIMAL oracle can also be used to implement the "clean-up" phase (line 10 of Figure 1): a set S is feasible if the oracle does not return any cycle. As we build $\mathcal{M}$ for any of these problems, we can also compute for each vertex v the quantity $r(v) = |\{C \in \mathcal{M} : v \in C\}|$ which represents the rate of growth of the left-hand-side of the dual constraint corresponding to v. This is useful in order to select the next vertex to add to S. Indeed, if we keep track of $a(v) = \sum_{C:v \in C} y_C$ for each vertex v then the next vertex selected by the algorithm is the one minimizing $\epsilon = \min_v (w_v - a(v))/r(v)$. We can then update $a(v)$ by setting $a(v) \leftarrow a(v) + \epsilon \cdot r(v)$. As we add a vertex to S (and remove it from the graph), we can easily update the planar graph in linear time as well. Since both loops of Figure 1 are executed $O(n)$ times, this gives a total running time of $O(n^2)$.

8 Conclusion

The most pressing question left open by this work is whether one can derive an α-approximation algorithm for FVS in planar graphs using the primal-dual technique for $\alpha \leq 2$. Such a result would immediately imply that planar graphs have feedback vertex sets of size at most $\alpha n/3$, which we think would be interesting even for $\alpha = 2$, since alternate proofs invoke the four color theorem or similar results. To prove such a result, one would "simply" need to find some subset of cycles $\mathcal{N}$ such that for any minimal fvs F, $\sum_{C \in \mathcal{N}} |F \cap C| \leq \alpha |\mathcal{N}|$. However, we have not yet been able to find such a subset. Note that in order to prove a bound on the size of a feedback vertex set, the subset would not necessarily have to be polynomial-time computable.

Acknowledgements

We thank Seffi Naor for pointing out reference [22]. The research of the first author was supported in part by IBM, NSF contract 9302476-CCR, a Sloan fellowship, and ARPA Contract N00014-95-1-1246. This research was conducted in part while the first author was visiting IBM.

References

1. A. Agrawal, P. Klein, and R. Ravi. When trees collide: An approximation algorithm for the generalized Steiner problem on networks. *SIAM Journal on Computing*, 24:440–456, 1995.
2. Akiyama and Watanabe. Research problem. *Graphs and Combinatorics*, 3:201–202, 1986.
3. M. Albertson and D. Berman. A conjecture on planar graphs. In J. Bondy and U. Murty, editors, *Graph Theory and Related Topics*. Academic Press, 1979.
4. V. Bafna, P. Berman, and T. Fujito. Constant ratio approximation of the weighted feedback vertex set problem for undirected graphs. In J. Staples, P. Eades, N. Katoh, and A. Moffat, editors, *ISAAC '95 Algorithms and Computation*, volume 1004 of *Lecture Notes in Computer Science*, pages 142–151, 1995.
5. R. Bar-Yehuda, D. Geiger, J. Naor, and R. M. Roth. Approximation algorithms for the vertex feedback set problem with applications to constraint satisfaction and Bayesian inference. In *Proceedings of the 5th Annual ACM-SIAM Symposium on Discrete Algorithms*, pages 344–354, 1994.
6. A. Becker and D. Geiger. Approximation algorithms for the loop cutset problem. In *Proceedings of the 10th Conference on Uncertainty in Artificial Intelligence*, pages 60–68, 1994.
7. O. Borodin. On acyclic colorings of planar graphs. *Discrete Mathematics*, 25:211–236, 1979.
8. G. Even, J. Naor, B. Schieber, and M. Sudan. Approximating minimum feedback sets and multi-cuts in directed graphs. In E. Balas and J. Clausen, editors, *Integer Programming and Combinatorial Optimization*, volume 920 of *Lecture Notes in Computer Science*, pages 14–28. Springer-Verlag, 1995.

9. G. Even, J. Naor, B. Schieber, and L. Zosin. Approximating minimum subset feedback sets in undirected graphs with applications to multicuts. Manuscript, 1995.

10. M. R. Garey and D. S. Johnson. *Computers and Intractability*. W.H. Freeman and Company, New York, 1979.

11. N. Garg, V. Vazirani, and M. Yannakakis. Primal-dual approximation algorithms for integral flow and multicut in trees, with applications to matching and set cover. In *Proceedings of the 20th International Colloquium on Automata, Languages and Programming*, 1993. To appear in *Algorithmica* under the title "Primal-dual approximation algorithms for integral flow and multicut in trees".

12. N. Garg, V. V. Vazirani, and M. Yannakakis. Approximate max-flow min-(multi)cut theorems and their applications. In *Proceedings of the 25th Annual ACM Symposium on Theory of Computing*, pages 698–707, 1993. To appear in *SIAM J. Comp.*

13. M. X. Goemans and D. P. Williamson. A general approximation technique for constrained forest problems. *SIAM Journal on Computing*, 24:296–317, 1995.

14. M. X. Goemans and D. P. Williamson. The primal-dual method for approximation algorithms and its application to network design problems. In D. S. Hochbaum, editor, *Approximation Algorithms for NP-hard Problems*, chapter 4. PWS, Boston, 1996. Forthcoming.

15. D. S. Hochbaum. Good, better, best, and better than best approximation algorithms. In D. S. Hochbaum, editor, *Approximation Algorithms for NP-hard Problems*, chapter 9. PWS, Boston, 1996. Forthcoming.

16. T. R. Jensen and B. Toft. *Graph Coloring Problems*. John Wiley and Sons, New York, 1995.

17. P. Klein, S. Rao, A. Agrawal, and R. Ravi. An approximate max-flow min-cut relation for undirected multicommodity flow, with applications. *Combinatorica*, 15:187–202, 1995.

18. P. Klein and R. Ravi. When cycles collapse: A general approximation technique for constrained two-connectivity problems. In *Proceedings of the Third MPS Conference on Integer Programming and Combinatorial Optimization*, pages 39–55, 1993. Also appears as Brown University Technical Report CS-92-30.

19. T. Leighton and S. Rao. An approximate max-flow min-cut theorem for uniform multicommodity flow problems with applications to approximation algorithms. In *Proceedings of the 29th Annual Symposium on Foundations of Computer Science*, pages 422–431, 1988.

20. G. L. Nemhauser and L. E. Trotter Jr. Vertex packing: Structural properties and algorithms. *Mathematical Programming*, 8:232–248, 1975.

21. P. D. Seymour. Packing directed circuits fractionally. *Combinatorica*, 15:281–288, 1995.

22. H. Stamm. On feedback problems in planar digraphs. In R. Möhring, editor, *Graph-Theoretic Concepts in Computer Science*, number 484 in Lecture Notes in Computer Science, pages 79–89. Springer-Verlag, 1990.

23. D. P. Williamson, M. X. Goemans, M. Mihail, and V. V. Vazirani. A primal-dual approximation algorithm for generalized Steiner network problems. *Combinatorica*, 15:435–454, 1995.

24. M. Yannakakis. Node and edge-deletion NP-complete problems. In *Proceedings of the 10th Annual ACM Symposium on Theory of Computing*, pages 253–264, May 1978.

Cone-LP's and Semidefinite Programs: Geometry and a Simplex-Type Method

Gábor Pataki *

Department of Combinatorics and Optimization
University of Waterloo
Waterloo, Ontario N2L 3G1, Canada

Abstract. We consider optimization problems expressed as a linear program with a cone constraint. Cone-LP's subsume ordinary linear programs, and semidefinite programs. We study the notions of basic solutions, nondegeneracy, and feasible directions, and propose a generalization of the simplex method for a large class including LP's and SDP's. One key feature of our approach is considering feasible directions as a sum of *two* directions. In LP, these correspond to variables leaving and entering the basis, respectively. The resulting algorithm for SDP inherits several important properties of the LP-simplex method. In particular, the linesearch can be done in the current face of the cone, similarly to LP, where the linesearch must determine only the variable leaving the basis.

1 Introduction

Consider the optimization problem

$$
\begin{aligned}
Min \ &cx \\
s.t. \quad &x \in K \\
&Ax = b
\end{aligned}
\qquad (P)
$$

where K is a closed cone in $\mathcal{R}^k$, $A \in \mathcal{R}^{m \times k}$, $b \in \mathcal{R}^m$, $c \in \mathcal{R}^k$. (P) is called a *linear program over a cone*, or a *cone-LP*. It models a large variety of optimization problems; in fact, every convex programming problem can be cast in this form, see ([14], pg. 103). Cone-LP's have been introduced in the fifties as a natural generalization of linear programs; probably the first paper studying duality is due to Duffin [7]. Two interesting special cases are:

- When $K = \mathcal{R}^k_+$ (P) is an ordinary linear program (LP).
- When $k = n(n+1)/2$, $K = \{x \in \mathcal{R}^k :$ the symmetric matrix formed of x is positive semidefinite$\}$, we get a *semidefinite program (SDP)*, a problem, that received a lot of attention recently, see e.g. [14], [2], [10].

---

* Updated full versions of this paper will be available by anonymous ftp at orion.math.uwaterloo.ca in pub/gabor

As shown by Nesterov and Nemirovskii [14], cone-LP's can be solved in polynomial time by interior-point methods, provided K is equipped with an efficiently computable *self-concordant barrier* function. The semidefinite cone does have this property; in recent years, several interior-point methods for SDP have been proposed citeAli95, [11], [22]. It is a natural question, (and one, that could be asked 40 years ago, when Duffin's paper appeared) whether one could generalize the simplex method for solving a reasonably large class of cone-LP's. As the structure of the semidefinite cone has been thoroughly studied, (see e.g. [4]) SDP seems to be a natural candidate. Such a generalization is obviously of interest from a theoretical viewpoint.

Also, semidefinite programming is quickly becoming a tool to solve practical problems, see e.g. [18], [22]. For large-scale problems, the cost of interior-point methods is frequently prohibitive. Since the primal and dual matrix iterates are (by the nature of the algorithm) full rank, one must perform costly, full factorizations at every step. A simplex-type method in which the iterates are low rank, may be able to avoid this drawback. In fact, active set methods for solving eigenvalue-optimization problems (a subclass of SDP) have been used with a reasonable amount of success [6], [15]. However, these methods did not rely on the notions of basic solution, nondegeneracy and extreme rays of the cone of feasible directions, as the simplex method for LP.

Surprisingly, the literature on generalizations of the simplex method for cone-LP's is scant. The only comprehensive work we are aware of is the book of Anderson and Nash [3]; they describe simplex-type methods for several classes of cone-LP's, however, their treatment does not work for finite-dimensional, non-polyhedral cones, such as the semidefinite cone. First, let us clarify, which are the main features of the simplex method, that one wishes to carry over. Given a basic feasible solution, the simplex method

1. Constructs a complementary dual solution.
2. If this solution is feasible to the dual problem, (i.e. the slack is nonnegative) it declares optimality.
3. If not, it finds a negative component, and constructs an improving extreme ray of the cone of feasible directions.
4. After a linesearch in this direction, it arrives at a new basic solution.

Also, we are allowed to distinguish basic solutions to be "nondegenerate", and "degenerate", and at first assume that our basic solutions encountered during the algorithm are nondegenerate, provided nondegeneracy is a generic property (that is, the set of degenerate solutions is of measure zero in an appropriate model). We can then deal with the degenerate case separately (let's say, using a perturbation argument).

Therefore, when looking for an appropriate generalization, one must answer these questions:

1. How to characterize basic solutions?
2. How to define nondegeneracy, and show that it is indeed a generic property?

3. How to characterize directions emanating at a current solution, which are in a sense extreme?

In this paper we show, how to address these issues. In particular, we define nondegeneracy of a solution by giving a common generalization of a definition of Shapiro [21], and Alizadeh et. al. [1] for SDP, and the usual definition for LP. Of the above three questions, the last seems the least natural. If K is not polyhedral, the cone of feasible directions in (P) at a current solution $\bar{x}$ is usually not even closed, much less does it have extreme rays. There is a simple way to overcome this difficulty. We decompose every feasible direction into the sum of two directions. The first corresponds to moving in the current face of K, until we hit its boundary (corresponding to a variable leaving the basis in LP). The second component takes us to a higher dimensional face of K (corresponding to a variable entering the basis in LP).

The rest of the paper is structured as follows. In Section 2 we introduce the necessary notation and review preliminaries. In Section 3 we derive our results on the geometry of cone-LP's: we study basic solutions, complementarity and nondegeneracy, and feasible directions. In Section 4 we present two algorithms: the first is a purification algorithm (using the terminology of [3]) to construct a basic feasible solution. Finally, we present our simplex-type method. Most proofs in the paper are either simplified, or omitted; full proofs appear in the full-length version.

In our study we consider general cone-LP's. All our results hold when specialized to LP and SDP; for LP, they are well-known. We use the properties of the semidefinite cone, only when necessary. The reason for this is twofold: First, a large part of our results hold for more general cone-LP's. We also want to extract the simple geometric properties of the positive orthant that make the simplex method so successful in solving linear programs.

2 Preliminaries

We shall heavily use tools from convex analysis. The standard reference is the book of Rockafellar [17] (a more introductory level text is [5]). In the following we let K to be a closed cone in $\mathcal{R}^k$. We call the set

$$K^* = \{y : yx \geq 0 \text{ for all } x \in K\}$$

the *polar cone* of K. If K is closed, so is K^* and $K^{**} = K$. We assume that K is *facially exposed*, that is, every face of K is the intersection of K with a supporting hyperplane. Then there is a natural correspondence between the faces of K and K^*. For F a face of K, the *conjugate face of F* is

$$F^\triangle = \{y \in K^* : yx = 0 \text{ for all } x \in F\}$$

Applying conjugacy twice gives back the original face, i.e. $F^{\triangle\triangle} = F$. The faces of K and K^* form a lattice, under the operations $\vee$ and $\wedge$ where $F \vee G$ is the smallest face containing both F and G and $F \wedge G$ is the intersection of F and G. The relative interior of a convex set S is denoted by ri S.

The positive orthant in n-space is denoted by $\mathcal{R}^n$. The set of n by n symmetric matrices is denoted by $\mathcal{S}^n$.

The nullspace and rangespace of a matrix A are denoted by $N(A)$ and $R(A)$. The cone of n by n symmetric, positive semidefinite matrices is denoted by $\mathcal{S}^n_+$. For simplicity, the elements of $\mathcal{S}^n_+$ are denoted by small letters; however, for $x \in \mathcal{S}^n_+$ we still write $N(x)$ and $R(x)$ for the null- and rangespaces. These two cones are *self-polar*, i.e. $K^* = K$. For the positive orthant, trivially, faces are in one-to-one correspondence with subsets of indices corresponding to 0 components of the vectors in the face, and the conjugate face is associated with the complement subset.

In the semidefinite cone the faces are in one-to-one correspondence with the *subspaces* of $\mathcal{R}^n$. Precisely, F is a face of $\mathcal{S}^n_+$ if and only if

$$F = F(L) = \{x : x \in \mathcal{S}^n_+ , R(x) \subseteq L\}$$

for some subspace L of $\mathcal{R}^n$ (see e.g. [4]). For $F(L)$ we have

$$\text{ri } F(L) = \{x : x \in \mathcal{S}^n_+ , R(x) = L\}$$
$$\text{lin } F(L) = \{x : x \in \mathcal{S}^n , R(x) \subseteq L\}$$

Let $r = \dim L$. Then the rank of matrices in $F(L)$ is at most r, and the dimension of $F(L)$ is

$$t(r) := r(r + 1)/2$$

($t(r)$ is the r'th "triangular number"). Also, the *conjugate* face corresponds to the subspace *orthogonal* to L, i.e.

$$F^\triangle = F(L^\perp)$$

Note that in $\mathcal{S}^n_+$, the dimensions of faces can take only n distinct values, while the dimension of $\mathcal{S}^n_+$ is $t(n)$. Also, $\dim F + \dim F^\triangle = t(r) + t(n - r) \neq t(n) = \dim K$, except in the trivial cases $r = 0$ and $r = n$. These two phenomena (the missing dimensions and the linear hull of the conjugate faces not spanning the whole space) are present in all nonpolyhedral cones.

In the following we shall also consider the the *dual* of (P) defined as

$$\begin{aligned} &Max \; yb \\ &s.t. \quad z \in K^* \\ &\qquad A^t y + z = c \end{aligned} \qquad\qquad (D)$$

It is easy to see, that taking the dual of (D) again we obtain (P). Weak duality between (P) and (D) is easy to prove; to obtain strong duality, when K is not polyhedral, one needs additional assumptions, see e.g. [2], [23], [19] for the case of SDP. We assume in the rest of the paper, that strong duality holds between (P) and (D).

3 Geometry

3.1 Basic solutions

In this subsection we define basic feasible solutions for cone-LP's, and derive several equivalent characterizations. Consider the dual pair of cone-LP's

$$
\begin{array}{ll}
& Min\ cx \\
(P)\ s.t. & x \in K \\
& Ax = b
\end{array}
\qquad\qquad
\begin{array}{ll}
& Max\ yb \\
(D)\ s.t. & z \in K^* \\
& A^t y + z = c
\end{array}
$$

We denote by *Feas(P)* and *Feas(D)* the feasible sets of (P) and (D), respectively.

Definition 3.1 *We call the extreme points of Feas(P) and Feas(D) primal, and dual basic feasible solutions, (bfs's), resp.*

Theorem 3.2 *Let $\bar{x} \in Feas(P)$, F the smallest face of K containing $\bar{x}$. Then the following statements are equivalent.*

1. *$\bar{x}$ is a basic feasible solution.*
2. *$\mathcal{N}(A) \cap \mathrm{lin}\ F = \{0\}$.*
3. *$\bar{x} = F \cap \{x : Ax = b\}$ and for all F' proper faces of F, $F' \cap \{x : Ax = b\} = \emptyset$.*

Moreover, if $\bar{x}$ is a bfs, then

4. *$\dim F \leq m$.*

$\square$

When specialized to linear programs, 3. states that $\bar{x}$ is a bfs if and only if the submatrix of A corresponding to nonzero components has of $\bar{x}$ has linearly independent columns. Also, 4. characterizes a bfs as a solution with *minimal support*.

Similarly, one can obtain a characterization of dual basic feasible solutions, as (D) is also a cone-LP (the cone is $\mathcal{R}^m \times K^*$ and the constraint-matrix is (A^t, I)). For brevity, we state only the results corresponding to 2. and 4. above.

Theorem 3.3 *Let $(\bar{y}, \bar{z})$ be a bfs of (D), G the smallest face of K^* that contains $\bar{z}$. Then $(\bar{y}, \bar{z})$ is a basic feasible solution if and only if*

1. *$R(A^t) \cap \mathrm{lin}\ G = \{0\}$.*

Moreover, if $(\bar{y}, \bar{z})$ is a bfs, then

2. *$\dim G \leq k - m$.*

$\square$

In the case of LP, in the last statement of Theorems 3.2 and 3.3 we get $\dim F \leq m$ and $\dim G \leq k - m$, resp., i.e. we recover the well-known bound on the number of nonzeros in basic feasible solutions.

In semidefinite programming the bounds on $\dim F$ and $\dim G$ yield a bound on the *rank* of extreme matrices.

Corollary 3.4 *Let* $K = S_+^n$.

1. *Let* $\bar{x}$ *be a basic solution of (P),* rank $\bar{x} = r$. *Then* r *satisfies* $t(r) \leq m$.
2. *Let* $(\bar{y}, \bar{z})$ *be a basic solution of (D),* rank $\bar{z} = s$. *Then* s *satisfies* $t(s) \leq t(n) - m$.

□

Again, notice that in the last statements of Theorems 3.2 and 3.3 we cannot expect equality in general, if K is not polyhedral. The reason is, that not all possible numbers in $\{0, 1, ..., k\}$ appear as the dimension of some face in K.

3.2 Complementarity and nondegeneracy

This section is motivated by the recent papers of Shapiro [21] and Alizadeh et. al. [1], where they define the notion of strict complementarity and nondegeneracy for SDP's and study their properties. We present a simple, common generalization of their definition and the corresponding definitions for LP, and show that most results corresponding to linear programs hold in this more general setting.

Let x and (y, z) be feasible solutions of (P) and (D), respectively. Since $cx = (yA + z)x = yb + zx$, x and (y, z) are both optimal, if and only if $xz = 0$. Therefore, x and z must lie in conjugate faces of K and K^* respectively. We introduce the following

Definition 3.5 The pair x and (y, z) is *strictly complementary*, if

$$(SC) \quad x \in \text{ri } F \text{ and } z \in \text{ri } F^\Delta$$ □

In LP, strict complementarity requires that the sum of the number of nonzeros in the primal and dual slacks be equal to n. In SDP, it requires that the sum of the *ranks* of the primal and dual slack matrices be n. Contrary to the case of LP, in general cone-LP's a strictly complementary solution-pair may not always exist. A counterexample for *SDP* is given in ([1]).

Definition 3.6 *Let* x *be feasible for* (P), F *the smallest face of* K *that contains* x. *We say that*

$$(PND) \quad x \text{ is nondegenerate if } \text{R}(A^t) \cap \text{lin } F^\Delta = \{0\}$$

Notice, that the above definition of nondegeneracy can be formally obtained from the characterization of a bfs in 3. of Theorem 3.2 by replacing $N(A)$ by $R(A^t)$ and F by F^Δ; extremity and nondegeneracy are complementary notions. In linear programs, the above definition requires the submatrix of A corresponding to nonzero components of x to have linearly independent *rows*. Nondegeneracy of a dual feasible solution can be defined in a similar manner. Exactly as in LP, we get

Theorem 3.7 *Let* x *and* (y, z) *be optimal solutions of* (P) *and* (D), *resp.*

1. *If* x *is nondegenerate, then* (y, z) *is basic.*

2. *Suppose that x and (y, z) satisfy (SC). Then x is basic if and only if (y, z) is nondegenerate.*

□

As it is well-known, nondegeneracy is a *generic* property in LP's, i.e. a random vertex of a randomly generated polyhedron is nondegenerate with probability one. As recently shown in [1], a similar property is true for SDP's: a randomly chosen extreme point of a random SDP is nondegenerate with probability one.

If x is a nondegenerate basic solution, then

$$\dim F \leq m$$
$$\dim F^{\Delta} \leq k - m \qquad (3.3)$$

In the case of LP, these two bounds imply $\dim F = m$, i.e., as known, the number of nonzeros in a nondegenerate basic solution must be *exactly* m. If $K = S_+^n$, (3.3) gives upper and lower bounds on the rank of x (observed also in [1]). If rank $x = r$, then (3.3) is equivalent to

$$t(r) \leq m$$
$$t(n - r) \leq t(n) - m \qquad (3.4)$$

These bounds allow a range of possible values of r. For example, $n = 10$, $m = 15$ implies $2 \leq r \leq 5$.

In linear programs with a special structure, an upper bound on the number of the nonzeros frequently yields combinatorial results. Similarly, using the upper bound on the extreme ranks in structured SDP's one obtains interesting corollaries, that we may group together under the name *semidefinite combinatorics*. Several examples (detailed in the full-length paper) are

1. A lower bound on the multiplicity of critical eigenvalues in eigenvalue-optimization, see [16].
2. A lower bound on the number of tight constraints in quadratic programs, where both the constraints and the objective function are convex.
3. The polynomial-time solvability of (possibly nonconvex) quadratic programs with few constraints.
4. An upper bound on the dimension of optimal orthonormal representations in the θ_1 and θ_4 formulations in the Lovász theta-function.

3.3 Feasible directions

Consider the primal problem

$$Min \ cx$$
$$s.t. \quad x \in K \qquad (P)$$
$$Ax = b$$

For F, a face of K define the set

$$D_F = \{(f, g) : f \in \text{lin } F, g \in K, A(f + g) = 0\}$$

The following simple lemma is crucial.

Lemma 3.8 *Let x be a feasible solution to (P), and F the smallest face of K that contains x. Then d is a feasible direction for x if and only if $d = f + g$ for some $(f, g) \in D_F$.*

Proof (If) Suppose $d = f + g$ for some $(f, g) \in D_F$. Since $f \in \operatorname{lin} F$, $x + \alpha f \in F$ for some $\alpha > 0$, and clearly $(x + \alpha f) + \alpha g \in K$.

(Only if) Suppose $x + \alpha d = g \in K$ for some $\alpha > 0$. Then $d = \frac{1}{\alpha} g + \frac{1}{\alpha}(-x)$, and clearly $\frac{1}{\alpha} g \in K$, $\frac{1}{\alpha}(-x) \in \operatorname{lin} F$. □

Remark 3.9 When $K = \mathcal{R}_+^n$, in the decomposition of d f corresponds to moving in the current face of the positive orthant, while g corresponds to a direction that moves into a higher dimensional face. The cone of feasible directions in the usual sense is a projection of D_F. When K is polyhedral, so is D_F, and also its projection. This is no longer true, when K is not polyhedral. In fact, as shown recently by Ramana et. al. [19], when $K = S_+^n$ and F is a face, the set

$$\{f + g : f \in \operatorname{lin} F, g \in K\}$$

is *never closed* ! (Of course, its intersection with the constraints $A(f+g) = 0$ *may* be closed; however, this result shows, that an approach considering directions in the projected cone is hopeless in general.) However, as we shall see, when designing a simplex method, it suffices to consider the (f, g) pairs in D_F, there is no need to project.

Theorem 3.10 *Let x and F be as in Lemma 3.8. Then the following hold.*

1. *If x is a bfs, then D_F has extreme rays.*
2. *Let $(f, g) \in D_F$, and G the smallest face of K that contains g. Let*

$$\alpha^* = \max\{\alpha : x + \alpha(f + g) \in K\}$$

   Then $F \vee G$ contains the feasible segment $[x, x + \alpha^*(f+g)]$ and it is a minimal face of K with this property.
3. *Assume that K satisfies the following property.*

   Property 1 *Let F and G be faces of K. Assume $y \in \operatorname{lin} F \setminus F, g \in G$, $y + g \in K$. Then $F \wedge G \neq \emptyset$.*

   Let (f, g) be an extreme ray of D_F, and define α^* as above. Then

$$\alpha^* = \max\{\alpha : x + \alpha f \in F\}$$

Proof of 1 D_F is clearly closed. To show that it has extreme rays, it suffices to prove that $D_F \cap (-D_F) = \{0\}$. For simplicity, assume $K \cap (-K) = \{0\}$, and let $(f, g) \in D_F \cap (-D_F)$. Then $g \in K \cap (-K)$ hence $g = 0$. Therefore $f \in \operatorname{lin} F \cap \mathcal{N}(A)$, and since x is basic, $f = 0$. □

Remark 3.11 Property 1 looks somewhat artificial, hence it is worth checking its validity, when $K = S_+^n$. Suppose that $F = F(L)$ and $G = F(J)$, where L and J are subspaces of $\mathcal{R}^n$. The condition $y \in \operatorname{lin} F \setminus F$ is satisfied if and only if $R(y) \subseteq L$, but y is not psd. As $y + g$ *is* psd, the rangespace of g must have a nontrivial intersection with $R(y)$.

Remark 3.12 Suppose (f, g) is an extreme ray of D_F. It is natural to ask, whether g must be an extreme ray of K. The answer is *no* in general. Suppose that G is the smallest face of K that contains g, and let us calculate an upper bound on $\dim G$. Similarly to Theorem 3.2 we obtain

$$\dim G \leq m - \dim F + 1 \qquad (3.6)$$

(as D_F is defined by $(k - \dim F + m$ constraints, the number of unconstrained variables is k, and (f, g) must be in a face of dimension 1 of D_F). First, let us consider the case of LP. Here $\dim F$ is equal to the number of nonzero components of x. If $\dim F = m$, i.e. x is nondegenerate, then we can move away from x by increasing the value of only one nonbasic variable (this may eventually decrease the value of a basic variable to zero). If x is degenerate, i.e. $\dim F < m$, then we may have to increase the values of $m - \dim F + 1$ variables from zero to a positive value in order to move away from x. In SDP, as explained earlier, we cannot expect $m = \dim F$, even in the nondegenerate case. For instance, consider an SDP with $m = n$. Such semidefinite programs do arise in practice, e.g. the max-cut relaxation SDP, whose feasible set has been termed the *elliptope* and studied in depth by Laurent and Poljak [13]. Then, if a basic solution x is nondegenerate, and of rank r, then r must satisfy

$$t(r) \leq n, \; t(n - r) \leq t(n) - n \Rightarrow 1 \leq t(r) \leq n$$

Thus $t(r) - n$ can be of order n, hence the best upper bound we can give on the rank of g in an extreme ray (f, g) is of order $\sqrt{2n}$. Therefore, in SDP's we cannot expect to be able to move away from the current solution by increasing the rank by 1, i.e. "bringing a one-dimensional subspace into the basis".

4 Algorithms

The following assumption will remain in force throughout the rest of the paper.

Assumption 4.1 *1. Given $x \in K$ one can find generators for* $\text{lin } F$ *and* $\text{lin } F^\triangle$, *where F is the smallest face of K that contains x.*

2. The separation problem is solvable for the polar cone K^*. Namely, given $z \in \mathcal{R}^k$ *we can either*

(a) Assert $z \in K^*$ OR

(b) Assert $z \notin K^*$, and find $v \in K$, satisfying $vz < 0$.

Assumption 4.1 obviously holds for the positive orthant. For S_+^n , 1. can be done in polynomial time by determining the rangespace of a positive semidefinite matrix. Separation for S_+^n , (recall, that S_+^n is self-polar) can also be done in polynomial time as described e.g. in [9].

Remark 4.2 For simplicity, we shall assume, that all computations are done in exact arithmetic.

4.1 Finding a basic feasible solution

Theorem 4.3 *The following algorithm finds a basic feasible solution of (P) in finitely many iterations.*

Algorithm 1.
Input: A, b, c, $x \in K$ satisfying $Ax = b$.
Output: $\bar{x} \in K$ satisfying $A\bar{x} = b$ and $c\bar{x} \leq cx$.
1. Let F be the smallest face of K that contains x.
2. Find $f \neq 0$ s.t

$$Af = 0$$
$$f \in \text{lin } F \tag{4.7}$$

If no such f exists, set $\bar{x} = x$ and STOP.
3. If $cf > 0$, set $f = -f$. Determine

$$\alpha^* = \max \alpha$$
$$\text{s.t} \quad x + \alpha f \in F \tag{4.8}$$

4. Set $x = x + \alpha^* f$, and go to 1.

Proof The correctness of Algorithm 1. follows from Theorem 3.2. Since the dimension of the smallest face containing the current x strictly decreases in every iteration, finiteness is obvious. □

4.2 A Simplex-type Method

In this section we first state our simplex procedure. We assume, that we are given x, a basic, nondegenerate solution of (P).

First, we need the following lemma.

Lemma 4.4 *Let x be a feasible solution to (P), and F the smallest face of K that contains x. Then d is in the closure of feasible directions for x if and only if $d = f + g$ for some $f \in (\text{lin } F^\triangle)^\perp, g \in K$. Moreover, if g is in the relative interior of K, then d is a feasible direction.*

□

Remark 4.5 Since in a general cone-LP lin F and lin $F^\triangle$ together do not span the whole space, the cone of feasible directions is not closed in general. Adding a $g \in \text{ri } K$ to f amounts to perturbing f towards the interior of K.

Algorithm Simplex
Input: A, b, c, K, $x \in K$ s.t. $Ax = b$.
1. **Construct a complementary dual solution.** Let F be the smallest face of K containing x, $F^\triangle$ is the conjugate of F. Construct subspaces M_P, M_D satisfying
   i. $N(A) \oplus (\text{lin } F + M_P) = \{0\}$
   ii. $\mathcal{R}(A^t) \oplus (\text{lin } F^\triangle + M_D) = \{0\}$

Find (y, z) s.t.

$$z \in \text{lin } F^\triangle + M_D$$
$$A^t y + z = c \tag{4.9}$$

2. **Check optimality.** If $z \in K^*$ STOP; x and (y, z) are optimal. Else, find $g \in K$ s.t. $zg < 0$.
3. **Find improving direction.** Construct (f, h), an extreme ray of D_F s.t. $zh < 0$.
4. **Line-search.** Determine

$$\alpha^* = max \ \alpha$$
$$s.t \quad x + \alpha f \in F \tag{4.10}$$

Replace x by $x + \alpha^*(f + h)$ and goto 1.

The difference, as opposed to the LP-simplex method is, that we cannot obtain an extreme ray of D_F in one step. We can proceed as follows.

First, perturb g found in Step 2. to obtain $g_1 \in \text{ri } K$, $zg_1 < 0$. Solve the system

$$f \in \text{lin } F \oplus M_P,$$
$$A(f + g_1) = 0 \tag{4.11}$$

Notice, that in this system the number of variables is the same as the number of constraints. This would *not* be true, if we replaced lin $F \oplus M_P$ by lin F. By Lemma 4.4 $f + g_1$ is a feasible direction, therefore we can find $x_1 = x + \epsilon(f + g_1) \in K$ for some small $\epsilon > 0$. Then $(x_1 - x, x_1) \in D_F$, although it is in general not an extreme ray. However, we can use an algorithm similar to Algorithm 1. to convert it into one.

The main difference between our approach and the LP-simplex method is, that we may not get an extreme ray of D_F in one step. It can be shown, that in the case of SDP, one does not have to choose g_1 to be in ri K, that is of full rank. The rank of g_1 needed to get $(f, g_1) \in D_F$ depends on the dimension of the subspace M_P. The smaller the dimension of M_P, the smaller the rank of g_1 needs to be. If dim $M_P = 0$, g_1 can be chosen to be equal to g, and no purification is needed to get an extreme ray of D_F.

We conjecture that when the current solution x is basic, and (f, g) is an extreme ray of D_F, then the new solution will also be basic, at least in the case of SDP. So far, we haven't been able to prove this conjecture. What we can state is an upper bound on the rank of the new solution. Clearly the rank of the new solution is at most rank x + rank $h - 1$, and both the rank of x and h can be bounded from above, as x is basic, and (f, h) is an extreme ray of D_F.

5 Conclusion

In this work we presented a generalization of the simplex method for a class of cone-LP's, including semidefinite programs. The main structural results we needed to derive, were

- A characterization of basic solutions.
- Defining nondegeneracy, and deriving some properties of nondegenerate solutions, (building on the results of [21], [1]).
- Characterizing extreme feasible directions in an appropriate higher dimensional space.

These structural results are of some independent interest.
It seems that the success of a simplex-type method for solving SDP's will depend on two issues.

- Its convergence properties.
- An efficient implementation. The advantage of our method, as opposed to an interior-point algorithm is, that our matrices, since they are basic solutions, are low rank. Also, when we move along an extreme ray of D_F the rangespace of the current iterate does not change by much. Therefore, it may be possible to design an efficient update scheme analogous to the update scheme of the revised simplex method for LP.

These issues will be dealt with in the full version of this article.

Acknowledgement I would like to thank László Lovász for suggesting me to work on semidefinite programming and my thesis advisor Egon Balas for his patience and support while I was working on this project. Many thanks are due to Adrian Lewis, Levent Tunçel and Henry Wolkowicz at the University of Waterloo for their support and many helpful discussions on the subject.

References

1. F. Alizadeh, J.-P. Haeberly and M. Overton: Complementarity and Nondegeneracy in Semidefinite Programming, manuscript, 1995
2. F. Alizadeh: Combinatorial Optimization with Semi-Definite Matrices *SIAM Journal of Optimization*, 5 (1), 1995
3. E. Anderson andf P. Nash: Linear Programming in Infinite Dimensional Spaces, John Wiley and Sons, 1987
4. G. P. Barker and D. Carlson: Cones of Diagonally Dominant Matrices, *Pacific Journal of Mathematics*, 57:1, 15-32, 1975
5. A. Brondsted: An Introduction to Convex Polytopes, Springer-Verlag, 1983
6. J. Cullum, W.E. Donath and P. Wolfe: The minimization of certain nondifferentiable sums of eigenvalue problems, *Mathematical Programming Study*, 3:35-55, 1975
7. R. J. Duffin: Infinite Programs, in A. W. Tucker, editor, *Linear Equalities and Related Systems*, 157-170, Princeton University Press, Princeton, NJ, 1956
8. A. Frieze and M. Jerrum: Approximation algorithms for the k-cut, and graph bisection problems using semidefinite programming, *4'th Conference on Integer Programming and Combinatorial Optimization, Copenhagen, 1995*
9. M. Grötschel, L. Lovász and A. Schrijver: Geometric Algorithms and Combinatorial Optimization, Springer-Verlag, 1988

10. M.X.Goemans and D. Williamson: Improved Approximation Algorithms for Maximum Cut and Satisfiability Problems Using Semidefinite Programming, *to appear in Journal of ACM*

11. C. Helmberg, F. Rendl, R. Vanderbei and H. Wolkowicz: An Interior-Point Method for Semi-Definite Programming, *to appear in SIAM Journal of Optimization*

12. D. Karger, R. Motwani and M. Sudan: Approximate graph coloring by semidefinite programming, Technical report, Stanford University

13. M. Laurent and S. Poljak: On a Positive Semidefinite Relaxation of the Cut Polytope, Technical Report, LIENS 93-27

14. Yu. Nesterov and A. Nemirovskii: Interior Point Algorithms in Convex Programming, SIAM, Philadelphia, 1994

15. M. L. Overton: Large-scale Optimization of Eigenvalues, *SIAM J. Optimization*, 2(1):88-120

16. G. Pataki: On the Rank of Extreme Matrices in Semidefinite Programs and the Multiplicity of Optimal Eigenvalues, *Management Science Research Report MSRR-#604, GSIA, Carnegie Mellon University*

17. Rockafellar, R. T.: Convex Analysis, Princeton University Press, Princeton, N.J., 1970

18. F. Rendl and H. Wolkowicz: A Semidefinite Framework for Trust Region Subproblems with Applications to Large Scale Minimization, *Technical Report CORR 94-32, University of Waterloo, 1994*

19. M. Ramana, L. Tunçel and H. Wolkowicz: Strong duality in semidefinite programming, Technical Report CORR 95-12, University of Waterloo, *to appear in SIAM Journal of Optimization*

20. A. Shapiro and M.K.H. Fan: On eigenvalue optimization, *SIAM Journal of Optimization* 3(1995), 552-568

21. A. Shapiro: First and second order analysis of nonlinear semidefinite programs, *Math. Programming, Ser. B to appear*

22. L. Vandenberghe and S. Boyd: Positive definite programming, Technical Report, Electrical Engineering Department, Stanford University, 1994

23. H. Wolkowicz: Some applications of optimization in matrix theory, *Linear Algebra and its Applications*, 40:101-118, 1981

Quadratic Knapsack Relaxations Using Cutting Planes and Semidefinite Programming

C. Helmberg[1], F. Rendl[2], and R. Weismantel[1]

[1] Konrad Zuse Zentrum für Informationstechnik Berlin,
Heilbronnerstraße 10, D-10711 Berlin, Germany
(http://www.zib-berlin.de/KombOpt/)
[2] Technische Universität Graz, Institut für Mathematik,
Steyrergasse 30, A-8010 Graz, Austria

Abstract. We investigate dominance relations between basic semidefinite relaxations and classes of cuts. We show that simple semidefinite relaxations are tighter than corresponding linear relaxations even in case of linear cost functions. Numerical results are presented illustrating the quality of these relaxations.

1 Introduction

The quadratic knapsack problem is the easiest case of constrained 0/1 quadratic programming and is extremely difficult to solve by linear programming alone. Semidefinite programming is well known to provide powerful relaxations for quadratic 0/1 programming [7, 1, 4] and, as we intend to show, it is very useful for quadratic knapsack problems as well. We compare several possibilities for setting up initial relaxations and show that in the special case of linear cost functions some are even better than the canonical linear relaxation. We discuss possible strengthenings of these relaxations by polyhedral cutting plane approaches in theory and in practice. The main practical difficulty with semidefinite approaches is the high computational cost involved. These stem from the factorization of a completely dense symmetric positive definite matrix with dimension equal to the number of constraints. To keep the number of constraints small it is of major importance to understand the interaction and dominance relations between different classes of cuts. We give several theoretical results in this direction. Finally, we present computational results of this approach on practical data.

Let $N = \{1, \ldots, n\}$ be a set of items, $a \in \mathbb{N}^n$ a vector of weights, $b \in \mathbb{N}$ a capacity, and $C \in \mathbb{R}^{n \times n}$ a matrix of costs. The quadratic knapsack problem reads

$$(\text{QK}) \quad \begin{array}{l} \text{Maximize } x^t C x \\ \text{subject to } a^t x \leq b \\ \qquad x \in \{0,1\}^n . \end{array}$$

We can interpret this problem in graph theoretic terms: Given the complete graph on n vertices with node weights a_i and profit c_{ii} for all $i = 1, \ldots, n$. Every edge ij in the complete graph is assigned an objective function coefficient

c_{ij}. Find a set of nodes S with sum of the node weights not greater than the threshold b that maximizes the profit $\sum_{i \in S} c_{ii} + \sum_{i,j \in S, i < j} 2c_{ij}$. As in the case of the linear knapsack problem the quadratic knapsack problem often appears as a subproblem to more complex optimization problems. Typical applications arise in VLSI- and compiler design [3, 6].

Our approach builds up on [4], which concentrates on the quadratic 0/1 programming aspects. Here, we investigate quadratic representations of a linear constraint, as suggested in [7, 1, 4] and discuss various aspects of knapsack specific inequalities.

The paper is structured as follows. Section 2 introduces several semidefinite relaxations obtained by different representations of the knapsack constraint and analyzes their strength. Section 3 surveys both well known and some new polyhedral concepts for generating knapsack specific cuts. In Section 4 we deal with the dominance relation between these cuts. In Section 5 implementational issues are discussed. We also present our numerical results.

2 Semidefinite Relaxation

(QK) is a constrained quadratic 0/1 programming problem. The usual approach for designing relaxations is to linearize the quadratic cost function by switching to "quadratic space". To this end we introduce variables y_{ij} for $i \leq j$ which are used to model the products $x_i x_j$. In the unconstrained case the convex hull of all feasible points in quadratic space is referred to as the boolean quadric polytope. The knapsack constraint cuts off part of this polytope. Although the convex hull of the restricted set of feasible integral points may differ substantially from the boolean quadric polytope it seems natural to start with a strong relaxation for the boolean quadric polytope and add knapsack specific inequalities on top.

Relaxation for the Boolean Quadric Polytope. As a relaxation for the boolean quadric polytope we use the semidefinite framework of [4] which is based on [7] and [1]. We model the dyadic product xx^t by a (symmetric) matrix variable Y. We denote the diagonal of this matrix by y. Using this notation the feasible set of matrices can be restricted to those satisfying $Y - yy^t \succeq 0$, i.e. $Y - yy^t$ must be positive semidefinite. This condition is equivalent to

$$\begin{pmatrix} Y & y \\ y^t & 1 \end{pmatrix} \succeq 0.$$

The diagonal elements y_i are obviously bounded by 0 and 1 and correspond to x_i. Looking at the determinant of a 3×3 principal minor containing the last row we get

$$y_i y_j - \sqrt{y_i y_j (1 + y_i y_j - y_i - y_j)} \leq y_{ij} \leq y_i y_j + \sqrt{y_i y_j (1 + y_i y_j - y_i - y_j)} \quad (1)$$

which yields an absolute lower bound of $-\frac{1}{8}$ for y_{ij}.

Numerous facet defining inequalities are known for the boolean quadric poly-tope [9] and can be added to sharpen the relaxation. Some of the most popular inequalities are (for all possible i, j and k)

$$y_{ij} \geq 0 \tag{2}$$
$$y_{ij} \leq y_{ii} \tag{3}$$
$$y_{ii} + y_{jj} \leq 1 + y_{ij} \tag{4}$$
$$y_{ik} + y_{jk} \leq y_{kk} + y_{ij} \tag{5}$$
$$y_{ij} + y_{ik} + y_{jk} + 1 \geq y_{ii} + y_{jj} + y_{kk} \tag{6}$$

These correspond to the triangle inequalities of the max-cut polytope [2].

Modelling the Knapsack Constraint. The easiest way to model the knap-sack constraint $a^t x \leq b$ on Y is to restrict the diagonal elements of Y, yielding our first semidefinite relaxation,

$$\text{(SQK1)} \quad \text{Maximize } \text{tr}(CY)$$
$$\text{subject to } \text{tr}(\text{Diag}(a)Y) \leq b$$
$$Y - yy^t \succeq 0.$$

Can we do better than (SQK1) by choosing some other representation of the knapsack inequality? Let us first consider a generic approach [7]. $b - a^t x \geq 0$ implies

$$(b - a^t x)(b - a^t x) = b^2 - 2ba^t x + a^t xx^t a \geq 0.$$

So a possible representation for the knapsack inequality could read

$$b^2 - 2ba^t y + a^t Y a \geq 0.$$

However, this inequality is already implied by the semidefinite constraint $Y - yy^t \succeq 0$. On the other hand exploiting the fact that $a^t x \geq -b$ on the feasible set we get a very useful representation in a very similar manner. We square both sides of $a^t x \leq b$ and get

$$a^t xx^t a \leq b^2.$$

Replacing xx^t by Y we call this the square representation of the inequality and use it to form a second relaxation

$$\text{(SQK2)} \quad \text{Maximize } \text{tr}(CY)$$
$$\text{subject to } \text{tr}(aa^t Y) \leq b^2$$
$$Y - yy^t \succeq 0.$$

Lemma 1. *(SQK2) is tighter than (SQK1).*

Proof. With $Z = Y - yy^t$ we get

$$a^t Z a + (a^t y)^2 \leq b^2 \tag{7}$$

which implies $a^t y \leq b$ by the positive semidefiniteness of Z. □

This proof suggests the following corollary.

Corollary 2. *If $a^t y = b$ for some Y satisfying $tr(aa^t Y) \leq b^2$ and $Y - yy^t \succeq 0$, then a is in the null space of $Z = Y - yy^t$.*

Another possibility to construct quadratic representations is to multiply the inequality by either x_i or $(1 - x_i)$ [7, 1]. If, for some fixed i, we sum up the two inequalities

$$\sum_{j=1}^{n} a_j y_{ij} \leq b y_i \tag{8}$$

$$\sum_{j=1}^{n} a_j (y_j - y_{ij}) \leq b(1 - y_i) \tag{9}$$

we get $a^t y \leq b$.

Lemma 3. *The relaxation obtained by replacing $tr(Diag(a)Y) \leq b$ of (SQK1) with a pair of inequalities (8) and (9) for some i is tighter than (SQK1).*

By including all n inequalities of type (8) and one additional inequality of type (9) we get

$$
\begin{aligned}
\text{(SQK3)} \quad & \text{Maximize } tr(CY) \\
& \text{subject to } \sum_{j=1}^{n} a_j y_{ij} \leq b y_i \qquad i = 1 \ldots n \\
& \qquad\qquad \sum_{j=1}^{n} a_j (y_{jj} - y_{1j}) \leq b(1 - y_1) \\
& \qquad\qquad Y - yy^t \succeq 0.
\end{aligned}
$$

Lemma 4. *(SQK3) is tighter than (SQK2).*

Proof. By multiplying inequality i of type (8) with a_i

$$\sum_{j=1}^{n} a_i a_j y_{ij} \leq b a_i y_i$$

and summing up over all n inequalities, we obtain $a^t Y a \leq b a^t y \leq b^2$. The right hand side inequality follows from Lemma 3. □

In practice it is more efficient to start with (SQK2) and to add Inequalities (8) and (9) in case of violation only.

Comparison With a Linear Relaxation. We investigate the special case of a linear cost function $C = Diag(c)$, i.e. $C_{ij} = 0$ for $i \neq j$. The standard linear relaxation reads

$$
\begin{aligned}
\text{(LK)} \quad & \text{Maximize } c^t x \\
& \qquad\quad a^t x \leq b \\
& \qquad\quad 0 \leq x_i \leq 1 \qquad i = 1, \ldots, n.
\end{aligned}
$$

(SQK1) is equivalent to (LK) because for any feasible x vector there is a feasible matrix Y having x as its diagonal. However, this is not true for (SQK2).

Lemma 5. *Let Y^* be an optimal solution of (SQK2) for $C = Diag(c)$. If (LK) has a unique optimal solution x^* which is not integral then $tr(Y^*C) < c^t x^*$.*

Proof. First notice that if x^* is not integral then $a^t x^* = b$. As x^* is unique it has exactly one element x_i with $0 < x_i < 1$. Consider a matrix Y satisfying $y = \text{diag}(Y) = x^*$ and $Y - yy^t \succeq 0$. Because of (1) $y_{ij} = y_i y_j$ for $i \neq j$. Therefore the only non zero element of $Z = Y - yy^t$ is $z_{ii} = x_i - x_i^2$. Obviously $a^t Z a = a_i^2 z_{ii}$ is greater than zero. Thus, by Corollary 2, Y is not feasible for (SQK2). Finally, the fact that $y^* = \text{diag}(Y^*)$ is feasible for (LK) completes the proof. $\square$

Because of this result we can expect that for numerous linear 0/1 programming problems we get better relaxations by simply translating the linear relaxation to the semidefinite representation.

3 Cutting Planes

In this section we introduce several classes of valid inequalities for the polyhedra associated with the linear and the quadratic representation. These classes serve as the basis for an algorithm to tighten bounds obtained from the semidefinite relaxation of a knapsack problem, see Section 5.

The Linear Knapsack Polyhedron. Our starting point is the polyhedron

$$\mathcal{P} := \text{conv}\{x \in \{0,1\}^n : \sum_{i \in N} a_i x_i \leq b\}.$$

A typical example of valid inequalities for $\mathcal{P}$ are cover inequalities. Let S be a subset of N with $\sum_{i \in S} a_i > b$, then the cover inequality with respect to the cover S

$$\sum_{i \in S} x_i \leq |S| - 1$$

is valid for $\mathcal{P}$. The original weights are completely ignored by cover inequalities.

Definition 6 (weight inequalities). Let $T \subseteq N$ with $a(T) < b$ be given and set $r := b - a(T)$. The *weight inequality with respect to T* is defined as

$$\sum_{i \in T} a_i x_i + \sum_{i \in N \setminus T} \max\{0, (a_i - r)\} x_i \leq a(T).$$

The name *weight inequality* expresses that the coefficients of the items in T equal their weights. The symbol $r := b - a(T)$ corresponds to the remaining capacity of the knapsack when $x_t = 1$ for all $t \in T$. The right hand side of the inequality is the weight of the set T. Hence, if for an item $i \in N \setminus T$ $a_i \leq r$ holds, then $x_t = 1$ for all $t \in T$ and $x_i = 1$ is a feasible solution. Therefore, the coefficient of i equals 0 in this case. For an item $i \in N \setminus T$ such that $a_i > r$, the value $a_i - r$ corresponds to the weight by which the knapsack capacity b is exceeded if we set $x_i = 1$ and $x_t = 1$ for all $t \in T$. These arguments can be made precise to yield Proposition 7.

Proposition 7. *[10]* *For $T \subseteq N$, $a(T) < b$ and $r := b - a(T)$, the weight inequality with respect to T is valid for $\mathcal{P}$.*

Example 1. For the knapsack polyhedron

$$SP_7 := \text{conv}\{x \in \{0,1\}^7 : x_1 + x_2 + x_3 + 2x_4 + 2x_5 + 3x_6 + 4x_7 \leq 6,$$

the following weight inequalities are easily seen to be facet-defining:

$$x_1 + x_2 + x_3 + 2x_4 + x_5 + 2x_6 + 3x_7 \leq 5$$

$$x_i + x_j + 2x_k + x_6 + 2x_7 \leq 4 \text{ for all } i,j \in \{1,2,3\},\ i \neq j \text{ and for all } k \in \{4,5\}$$

$$x_1 + x_2 + x_3 + x_7 \leq 3.$$

The idea of weight inequalities can be extended to more general cases. Instead of taking the values of the weights of the items into account, we introduce "relative" weights for all the items and derive an analogon of weight inequalities for these relative weights.

For disjoint subsets T and I such that $a(T \cup I) \leq b$, $a_t \leq a_i$ for all $t \in T$ and $i \in I$ and $a(T) \geq a_i$ for all $i \in I$, we define the relative weight c_u of an item $u \in T \cup I$ as follows:

$$c_u := 1 \text{ if } u \in T; \qquad c_u := \min\{|S| : S \subseteq T,\ a(S) \geq a_u\} \text{ if } u \in I.$$

In words, we first normalize the weights of the items in T to the value 1; thereafter an item $i \in I$ obtains as a new weight the value that counts the number of items in T that one needs in order to cover the original weight a_i. Under these assumptions we define for $z \in N \setminus (T \cup I)$ the extended weight inequality with respect to $T \cup I \cup \{z\}$ as follows:

Definition 8 (extended weight inequalities). For $r := b - a(T) - a(I)$, the extended weight inequality with respect to $T \cup I \cup \{z\}$ is of the form

$$\sum_{i \in T} x_i + \sum_{i \in I} c_i x_i + c_z x_z \leq |T| + \sum_{i \in I} c_i,$$

where $c_z := \min\{|S| + \sum_{j \in J} c_j : S \subseteq T,\ J \subseteq I,\ a(S \cup J) \geq a_z - r\}$.

Extended weight inequalities have been introduced and analyzed in [10]. For the purpose of this paper the following proposition is needed.

Proposition 9. *[10] The extended weight inequality defined for $T \cup I \cup \{z\}$ is valid for $\mathcal{P}$.*

Example 2. We continue analyzing the knapsack polyhedron $\mathcal{SP}_7$. Setting $T :=$ $\{1, 2, 4\}$, $I = \emptyset$ and $z = 6$, the extended weight inequality for T, I and $\{z\}$ is the inequality

$$x_1 + x_2 + x_4 + x_6 \leq 3,$$

that happens to be a minimal cover inequality. This inequality lifts to the facet-defining inequality

$$x_1 + x_2 + x_4 + x_6 + x_7 \leq 3$$

of $\mathcal{SP}_7$. The extended weight inequality with respect to $T = \{1, 4\}$, $I = \{6\}$ and $z = 7$ is the inequality

$$x_1 + x_4 + 2x_6 + 3x_7 \leq 4.$$

After computing lifting coefficients according to the sequence $(7, 2, 5, 3)$, for instance, we obtain the inequality

$$x_1 + x_2 + x_4 + x_5 + 2x_6 + 3x_7 \leq 4.$$

This induces a facet of $\mathcal{SP}_7$.

It was also shown in [10] that for any extended weight inequality lifting coefficients can always be computed in polynomial time. In particular, the exact lifting coefficient of an item coincides either with a certain lower bound or its value equals this lower bound plus 1.

The Quadratic Knapsack Polyhedron. In the following we will study the polyhedron

$$\mathcal{Q} := \operatorname{conv}\{y \in \{0, 1\}^{n(n+1)/2} : \sum_{i \in N} a_i y_{ii} \leq b, \; y_{ij} = y_{ii}y_{jj} \; \forall i < j\}.$$

that we obtain by lifting the original polyhedron to the space of quadratic variables. In this higher dimensional space, there are novel ways to construct relaxations of $\mathcal{Q}$ that, itself, allow for generating valid inequalities for $\mathcal{Q}$.

Lemma 10. *Let $N_1, \ldots, N_k$ be a partition of N. For every $v \in \{1, \ldots, k\}$ we choose a spanning tree (N_v, T_v) in the complete graph $K(N_v)$ on the node set N_v. By deg_i^v we denote the degree of node i in the tree (N_v, T_v). The polyhedron $\operatorname{conv}\{y \in \{0, 1\}^{n(n+1)/2} : \sum_{v=1}^{k}(\sum_{i \in N_v} a_i)[\sum_{ij \in T_v} y_{ij} + \sum_{i \in N_v}(1 - deg_i^v)y_{ii}] \leq b\}$ contains all the feasible points of $\mathcal{Q}$.*

Proof. Let $y \in \mathcal{Q}$, integral, be given and choose a tree (N_v, T_v), $v \in \{1, \ldots, k\}$. We want to show that the corresponding summand $\sum_{ij \in T_v} y_{ij} + \sum_{i \in N_v}(1 - deg_i^v y_{ii})$ is one if $y_{ii} = 1$ for all $i \in N_v$ and not greater than zero otherwise. First suppose $y_{ii} = 1$ for all $i \in N_v$. In this case

$$\sum_{ij \in T_v} 1 + \sum_{i \in N_v} 1 - \sum_{i \in N_v} deg_i^v = |N_v| - 1 + |N_v| - 2|N_v - 1| = 1.$$

Otherwise edges with $y_{ij} = 0$ decompose the tree into several subtrees which again satisfy $y_{ii} = 1$ for all nodes belonging to the same subtree and are maximal in this respect. However, the value deg_i^v of at least one node in such a subtree must exceed the degree within the subtree by at least one. This is due to the fact that each subtree is connected by at least one edge to a vertex $j \in N_v$ with $y_{jj} = 0$. Therefore the contribution of a subtree to the summand is not greater than zero. $\qquad\qquad\qquad\qquad\qquad\qquad\qquad\qquad\qquad\qquad\qquad\qquad\qquad$ $\square$

Lemma 10 allows us to derive valid inequalities for Q via the following scheme: Generate a relaxation Q' of Q as stated in the Lemma. Find valid inequalities, like cover inequalities, weight inequalities or extended weight inequalities for Q'. These inequalities are also valid for Q.

Example 3. Consider the knapsack polyhedron

$$\text{conv}\{x \in \{0,1\}^6 : 5x_1 + 6x_2 + 7x_3 + 8x_4 + 9x_5 + 12x_6 \leq 21\}.$$

Partitioning into sets $\{1,2\}, \{3,4\}, \{5,6\}$ and choosing the edge set of the complete graphs on two nodes for all elements in the partition yields the knapsack polyhedron

$$Q' := \text{conv}\{y \in \{0,1\}^{21} : 11y_{1,2} + 15y_{3,4} + 21y_{5,6} \leq 21\}.$$

A valid inequality for this polyhedron is given by the cover inequality $y_{1,2} + y_{3,4} + y_{5,6} \leq 1$. Partitioning N into the sets $\{1,2,3\}, \{4\}, \{5\}, \{6\}$ and choosing the edges $(1,2), (1,3)$ in the complete graph with vertices $1, 2, 3$ yields another knapsack polyhedron

$$Q'' := \text{conv}\{y \in \{0,1\}^{21} : 18[y_{1,2} + y_{1,3} - y_{1,1}] + 8y_{4,4} + 9y_{5,5} + 12y_{6,6} \leq 21\}.$$

A valid inequality for Q'' is given, for instance, by the constraint $2[y_{1,2} + y_{1,3} - y_{1,1}] + y_{4,4} + y_{5,5} + y_{6,6} \leq 2$.

In the remainder of this paper we sometimes refer to special relaxations of Q. These are obtained by partitioning a subset $S = \{i_1, \ldots, i_s\}$ of N of even cardinality into elements of cardinality two, $S^1, \ldots S^{\frac{s}{2}}$, $S^1 = \{i_1, i_2\}$, $S^2 = \{i_3, i_4\}$, $\ldots$, $S^{\frac{s}{2}} = \{i_{s-1}, i_s\}$, for instance. In other words, we choose a perfect matching M in the complete graph with node set S, or a matching M in the complete graph with node set N. By Lemma 10 the polyhedron

$$\text{conv}\left\{y \in \{0,1\}^{n(n+1)/2} : \sum_{ij \in M} (a_j + a_i) y_{ij} + \sum_{i \in N \setminus S} a_{ii} y_{ii} \leq b\right\}$$

is a relaxation of Q. The knapsack inequality

$$\sum_{ij \in M} (a_j + a_i) y_{ij} + \sum_{i \in N \setminus S} a_{ii} y_{ii} \leq b$$

is called the *matching-knapsack*-constraint associated with the matching M in the complete graph with node set N. We will refer to a cover inequality based on a matching-knapsack-constraint as *matching-cover*-constraint.

We conclude this section by introducing a quadratic representation for linear cover inequalities. Let $S \subset N$ define a valid cover inequality for $\mathcal{P}$ and choose any hamiltonian cycle C_S in the complete graph over the vertex set S. Then

$$\sum_{ij \in C_S} y_{ij} \leq |C_S| - 2$$

is a valid inequality for $\mathcal{Q}$. We refer to this type of inequalities as cycle inequalities [3].

4 Various Aspects of Cutting Planes

In general (SQK2) and (SQK3) will not be tight enough to provide provably optimal solutions but it is possible to improve these semidefinite relaxations by adding further inequalities. We have already mentioned generic cuts from the boolean quadric polytope in Section 2. In this section we will consider knapsack specific inequalities.

Linear Cutting Planes. We start with valid inequalities for $\mathcal{P}$ as defined in Section 3. These are again linear constraints which have to be transformed into some quadratic representation. In principal we have the same possibilities as for modeling the knapsack inequality and the same results apply. Note, that in case of multiplication with x_i it may be worth to postpone the lifting procedure. Multiplication of $a^t x \leq b$ with x_i corresponds to a conditional inequality, which is effective only if $x_i > 0$,

$$x_i \sum_{j \neq i} a_j x_j \leq (b - a_i) x_i.$$

So for an extended weight inequality multiplied with x_i we can lift the remaining coefficients with respect to the reduced knapsack inequality $\sum_{j \neq i} a_j x_j \leq b - a_i$ instead of the original $a^t x \leq b$.

Example 4. For the knapsack polyhedron

$$\mathcal{SP}_4 := \text{conv}\{x \in \{0, 1\}^4 : 4x_1 + 5x_2 + 6x_3 + 7x_4 \leq 16\},$$

lifting the inequality $x_1 + x_2 + x_3 \leq 3$ with respect to the original inequality yields $x_1 + x_2 + x_3 + x_4 \leq 3$. By multiplying with x_3 we get

$$y_{13} + y_{23} - 2y_{33} + y_{34} \leq 0.$$

Lifting $x_1 + x_2 \leq 2$ with respect to $4x_1 + 5x_2 + 7x_3 \leq 10$ yields

$$y_{13} + y_{23} - 2y_{33} + 2y_{34} \leq 0.$$

It is also interesting to investigate the dominance relation between different representations if we include triangle inequalities (2) to (6) in the basic relaxation. Consider the extended weight inequality for $\mathcal{P}$

$$\sum_{i \in T}(1 - x_i) + \sum_{i \in I} c_i(1 - x_i) - c_z x_z \geq 0. \tag{10}$$

Multiplication with x_z yields the quadratic representation

$$\sum_{i \in T}(y_{zz} - y_{iz}) + \sum_{i \in I} c_i(y_{zz} - y_{iz}) - c_z y_{zz} \geq 0. \tag{11}$$

We subtract this inequality from the diagonal representation of (10) (replace x_i with y_{ii}) and get

$$\sum_{i \in T}(1 - y_{ii} - y_{zz} + y_{iz}) + \sum_{i \in I} c_i(1 - y_{ii} - y_{zz} + y_{iz}) \geq 0.$$

If we require the triangle inequalities (4) to hold, the latter expression is clearly nonnegative and (11) dominates the diagonal representation of (10). Intuitively, the triangle inequalities help to spread the influence of an inequality defined on a single row over the whole matrix.

Quadratic Cutting Planes. We now turn towards valid inequalities for the polyhedron $\mathcal{Q}$. One question in terms of computations is to choose a relaxation of the original problem that allows to derive strong valid cuts for the quadratic knapsack problem. If we restrict the discussions to cuts that are cover inequalities, a precise statement can be made for a comparison of the polyhedra $\mathcal{Q}$ and

$$\mathcal{C} := \text{conv}\{y \in \{0, 1\}^{n(n+1)/2} : \sum_{i \in N} a_i^2 y_{ii} + \sum_{i < j,\, i,j \in N} 2a_i a_j y_{ij} \leq b^2\}$$

that we associate with the form $a^t x a^t x \leq b^2$ of the given quadratic knapsack problem.

Lemma 11. *For (SQK1) combined with the triangle inequalities (3) every cover inequality that is valid for $\mathcal{C}$ is dominated by a matching-cover-constraint.*

Proof. Consider a subset S of the set of variables $\bigcup_{i \in N}\{i\}$ and a subset T of the set of variables $\bigcup_{i < j,\, i,j \in N}\{ij\}$ such that $\sum_{i \in S} a_i^2 + \sum_{ij \in T} 2a_i a_j > b^2$. The cover inequality associated with S and T reads

$$\sum_{i \in S} y_{ii} + \sum_{ij \in T} y_{ij} \leq |S| + |T| - 1, \tag{12}$$

and is obviously valid for $\mathcal{C}$. Let I denote the set of indices appearing in S or, as an endpoint, in T. We first show that $\sum_{i \in I} a_i > b$. Assume otherwise, then

$$b^2 \geq (\sum_{i \in I} a_i)^2 \geq \sum_{i \in S} a_i^2 + \sum_{ij \in T} 2a_i a_j,$$

which is a contradiction to our assumptions on S and T. Now consider a maximum cardinality matching $M \subset T$ in the graph (I, T). We denote by $J \subset I$ the indices not covered by an edge of M. Obviously,

$$\sum_{i \in J} y_{ii} + \sum_{ij \in M} y_{ij} \le |J| + |M| - 1$$

is a valid matching-cover-constraint for Q. Except for the variables y_{ii} with $i \in J \setminus S$ all variables of this inequality also appear in (12). For any $i \in J \setminus S$ there is at least one j such that $ij \in T$. Since we require (3) to hold and for all ij $y_{ij} \le 1$ by the semidefiniteness constraint the matching-cover-constraint indeed dominates (12). $\qquad\square$

The next lemma is another indication that matching-knapsack-constraints yield useful relaxations for deriving valid inequalities.

Lemma 12. *Let $S \subset N$ be a cover. The square representation of the cover inequality with respect to S is dominated by the diagonal representation combined with*

(a) matching-cover-inequalities if $|S|$ is even,
(b) cycle inequalities if $|S|$ is odd.

Proof. Let $S \subset N$ be a cover for $\mathcal{P}$. The square representation reads

$$\sum_{i \in S} y_{ii} + \sum_{i,j \in S, i < j} 2y_{ij} \le (|S| - 1)^2. \tag{13}$$

We first assume that $|S|$ is even. For M a perfect matching in the complete graph with vertex set S we obtain the matching-cover constraint

$$\sum_{ij \in M} y_{ij} \le \frac{|S|}{2} - 1.$$

Let $M_1, M_2, \ldots, M_{|S|-1}$ be a partition of the edge set of the complete graph with node set S into perfect matchings. Then

$$\sum_{k=1}^{|S|-1} \sum_{ij \in M_k} y_{ij} \le (|S| - 1)(\frac{|S|}{2} - 1).$$

To cover the y_{ii} terms of (13) we add the diagonal representation of the underlying cover inequality and get

$$\sum_{i \in S} y_{ii} + \sum_{i,j \in S, i < j} 2y_{ij} = 2 \sum_{k=1}^{|S|-1} \sum_{ij \in M_k} y_{ij} + \sum_{i \in S} y_{ii}$$

$$\le 2(|S| - 1)(\frac{|S|}{2} - 1) + |S| - 1 = (|S| - 1)^2.$$

If $|S|$ is odd then we partition the edge set of the complete graph with node set S into $\frac{|S|-1}{2}$ hamiltonian cycles, $C_1 \ldots, C_{\frac{|S|-1}{2}}$. With each cycle C_k we associate the cycle constraint

$$\sum_{ij \in C_k} y_{ij} \leq |S| - 2.$$

Summing up over all cycles we get

$$\sum_{k=1}^{\frac{|S|-1}{2}} \sum_{ij \in C_k} y_{ij} + y_{vv} = \sum_{i,j \in S, i<j} y_{ij} \leq \frac{|S-1|}{2}(|S| - 2).$$

Multiplying by two and adding the diagonal representation yields (13). □

5 Implementation

For solving the semidefinite programs we use the primal-dual path-following interior point algorithm of [5]. To guarantee that there is no duality gap between primal and dual optimal solutions we have to ensure that at least one of both has a feasible point satisfying all inequalities strictly. To this end we add the constraint $y_{ij} = 0$ whenever $a_i + a_j > b$ for some $i \neq j$. The arithmetic mean of all zero, one, and two items solutions is now such a feasible point.

Each iteration of the interior point code requires the factorization of a dense positive definite matrix. The dimension of this matrix is the number of constraints of the semidefinite program. More than 60% of the overall computation time are spent in this routine. It is therefore extremely important to keep the set of constraints as small as possible. Even expensive separation routines will pay off if they help to achieve this goal.

To illustrate the quality of the relaxations we give numerical results in Table 1 for some compiler design problems taken from [6]. For all problems the cost matrix is nonnegative and sparse (e.g. the problem of dimension 61 has just 187 nonzeros), and both, costs and weights, vary over a wide range. We emphasize that we do not exploit the sparsity of the problem at all. For each example we compute solutions for right hand sides $450, 512$, and 600.

The first column of Table 1 gives the dimension of the problem, the second the value of the right hand side of the knapsack constraint, the third gives the best feasible solution we found. All other columns display the relative gap

$$\% = \left(\frac{\text{upper bound}}{\text{feasible solution}} - 1 \right) \times 100$$

of the upper bound — obtained after at most 30 minutes of CPU time[3] using the specific relaxation — with respect to this feasible solution. Whenever the gap between feasible solution and upper bound is closed (relative gap $< 5 \cdot 10^{-6}$) we mark this by ▷ and give the computation time, instead.

---

[3] Computation times refer to a Sun Sparcstation 10.

Table 1. Compiler design problems taken from [6]

dim	rhs	feas. sol.	SQK1 %	SQK2 %	SQK3 %	Δ-ineq. %/mm:ss	lin. cuts %/mm:ss	matching %/mm:ss
30	450	1580	41	17	13	0.23	▷ 00:19	▷ 00:24
	512	1802	39	20	17	5.60	0.35	▷ 26:10
	600	2326	24	12	11	2.64	▷ 1:55	▷ 1:31
45	450	2840	16	8.7	8.4	2.90	▷ 13:54	▷ 13:43
	512	3154	30	12.7	12.7	3.07	1.61	1.58
	600	3840	22	8.2	8.2	▷ 20:16	▷ 3:45	▷ 3:02
47	450	1732	7	5.9	5.8	2.51	▷ 15:02	▷ 31:10
	512	1932	30	12	11	1.30	▷ 16:59	▷ 8:20
	600	2186a	31	18	17	8.89	6.02	4.09
61	450	26996	3.7	2.4	2.4	0.40	▷ 3:17	▷ 3:23
	512	29492	2.9	2.0	2.0	1.34	▷ 28:53	0.02
	600	32552	2.6	1.9	1.9	1.04	0.42	0.33

Columns (SQK1), (SQK2), and (SQK3) refer to the respective relaxations of Section 2. The performance of (SQK1) is rather poor, (SQK2) halves the gap of (SQK1) requiring the same amount of computation time, (SQK3) is just a little bit better than (SQK2) but takes about twice as long to compute.

For column Δ-*ineq.* we start the algorithm with (SQK2) as initial relaxation and compute its optimal solution. Then we improve the relaxation by adding n cutting planes of type (2)–(6), (8), or (9) and iterate. All these cutting planes are applicable for all quadratic 0/1 problems and do not exploit any special properties of the knapsack problem. Yet the bound is already acceptable.

In column *lin. cuts* we also include cutting planes which are quadratic representations of valid inequalities for the linear knapsack polytope. In particular we consider weight inequalities and extended weight inequalities. In our experiments (8) was clearly the most successful quadratic representation. Representation (9) will be of importance if the cost matrix contains negative entries as well. As we can see in Table 1 the addition of these cuts was sufficient to close the gap for most of our test problems within half an hour. In general the final relaxation included just a few knapsack specific cutting planes and lots of triangle inequalities (2)–(6).

Finally, column *matching* gives the results if we separate matching inequalities as well. We use a preliminary version of the maximum weighted matching code of LEDA [8] to find reasonable starting sets and derive extended weight inequalities for these sets. Although this separation procedure is computationally quite expensive total computation time is roughly the same. In view of the fact that we started only recently to experiment with matching inequalities these results are very promising.

6 Conclusions

We presented several basic semidefinite relaxations for the 0/1 quadratic programming problem and compared them with respect to their quality in theory and in practice. The straight forward approach of modelling the constraint on the diagonal (SQK1) yields a rather poor bound. At the same computational cost we can get a much better bound by using the square representation (SQK2). For this relaxation we proved that in case of linear cost functions it is superior to the canonical linear relaxation. Slightly better than (SQK2) is the relaxation formed by using all representations obtained by "multiplying" the knapsack inequality with x_i for all i (SQK3). However, computationally it is more efficient to start with (SQK2) and to compute (SQK3) by successively adding violated inequalities.

Generic 0/1 cutting planes such as the triangle inequalities (2)–(6) significantly improve these relaxations and yield surprisingly good bounds without requiring any special knowledge about the nature of the problem itself. However, special polyhedral knowledge is indispensable to close the gap and speed up the computation.

A good way to model valid inequalities of the linear knapsack polyhedron in quadratic space is to multiply it with some x_i (8). In case of negative elements in the cost matrix it is worth considering representation (9) as well. Combining the semidefinite framework with quadratic representations of weight inequalities and extended weight inequalities yields a reasonable approach to solve rather dense, small to medium sized problems.

In Section 3 we introduced a large new class of valid inequalities for the quadratic knapsack polyhedron which allow for direct derivation of quadratic cutting planes. We presented some theoretical and computational evidence for their importance in cutting plane approaches. However, there is much room for improvement, and it can be expected that the quadratic knapsack polytope has still a lot to offer.

The good quality of the bounds, even without exploiting special properties of the problem at hand, gives rise to the hope that — in spite of the high computational cost involved — semidefinite programming will become a useful tool to model and solve difficult subproblems in integer programming.

References

1. E. BALAS, S. CERIA and G. CORNUEJOLS. A lift-and-project cutting plane algorithm for mixed 0/1 programs, *Mathematical Programming* 58:295–324, 1993.
2. C. DE SIMONE. The cut polytope and the boolean quadric polytope. *Discrete Mathematics*, 79:71–75, 1989.
3. C. E. FERREIRA, A. MARTIN, C. DE SOUZA, R. WEISMANTEL and L. WOLSEY. Formulations and valid inequalities for the node capacitated graph partitioning problem, *CORE discussion paper* No. 9437, Université Catholique de Louvain, 1994. To appear in *Mathematical Programming*.

4. C. HELMBERG, S. POLJAK, F. RENDL, and H. WOLKOWICZ. Combining Semidefinite and Polyhedral Relaxations for Integer Programs. *Lecture Notes in Computer Science*, 920:124–134, 1995, Proceedings of IPCO 4 (E. Balas, J. Clausen eds).

5. C. HELMBERG, F. RENDL, R. J. VANDERBEI, and H. WOLKOWICZ. An interior–point method for semidefinite programming. *SIAM Journal on Optimization*. To appear.

6. E. JOHNSON, A. MEHROTRA and G.L. NEMHAUSER. Min-cut clustering, *Mathematical Programming* 62:133–152, 1993.

7. L. LOVÁSZ and A. SCHRIJVER. Cones of matrices and set functions and 0-1 optimization. *SIAM J. Optimization*, 1(2):166–190, 1991.

8. S. NÄHER and C. UHRIG. The LEDA User Manual Version R 3.3 beta. Max-Planck-Institut für Informatik, Im Stadtwald, Building 46.1, D–66123 Saarbrücken, Germany. (http://www.mpi-sb.mpg.de/LEDA/leda.html)

9. M. W. PADBERG. The boolean quadric polytope, *Mathematical Programming* 45:132–172, 1989.

10. R. WEISMANTEL. On the 0/1 knapsack polytope, *Preprint* SC 94-01, Konrad-Zuse-Zentrum Berlin, 1994.

A Semidefinite Bound for Mixing Rates of Markov Chains

Nabil Kahale*

AT&T Bell Laboratories, Murray Hill, NJ 07974. Email: kahale@research.att.com.

Abstract. We study the method of bounding the spectral gap of a reversible Markov chain by establishing canonical paths between the states. We provide natural examples where improved bounds can be obtained by allowing variable length functions on the edges. We give a simple heuristic for computing good length functions. Further generalization using multicommodity flow yields a bound which is an invariant of the Markov chain, and which can be computed at an arbitrary precision in polynomial time via semidefinite programming. We show that, for any reversible Markov chain on n states, this bound is off by a factor of order at most $\log^2 n$, and that this can be tight.

1 Introduction

Let $(X_m), m \geq 0$, be an irreducible Markov chain on a finite state space V with transition matrix P and stationary distribution π. We assume that P is reversible, that is

$$\pi(x)P(x,y) = \pi(y)P(y,x) = Q(x,y), \text{ for all } x,y \in V.$$

Under these conditions, all the eigenvalues of P are real, and will be denoted by $1 = \lambda_0 \geq \lambda_1 \geq \cdots \geq \lambda_{n-1}$, where $n = |V|$.

It is known that the rate of convergence of a Markov chain can be bounded using the eigenvalues of P (see, e.g., [5]). Informally, the chain is rapidly mixing if $\max(\lambda_1, |\lambda_{n-1}|)$ is small compared to 1. Since we can assume the eigenvalues to be non-negative by replacing P with $\frac{1}{2}(I + P)$, it suffices to upper-bound $\tau = 1/(1 - \lambda_1)$ in order to bound the mixing rate of the (tranformed) Markov chain. Random walks have been used to establish approximation algorithms for hard combinatorial problems (see [10, 15, 12], and references therein). General results on Markov chains can be found in [1].

Let $G = (V, E)$ be the graph on the vertex set V, with $(x, y) \in E$ if and only if $Q(x, y) > 0$ and $x \neq y$. We assume that the elements of E are *directed*

---

* Based on DIMACS Tech Report 95-41, September 95. This work was partly done while the author was at the Massachusetts Institute of Technology, at the Institute for Mathematics and its Applications, at DIMACS and at XEROX Palo Alto Research Center.

edges. Following [10], Diaconis and Stroock [5] proposed a method for bounding τ by establishing a simple path γ_{xy} in G from x to y, for each $(x, y) \in V \times V$, $x \neq y$. Given a length function l that assigns a non-negative real number to each element of E, let

$$|\gamma_{xy}|_l = \sum_{e \in \gamma_{xy}} l(e).$$

Diaconis and Stroock showed the following Poincaré inequality:

$$\tau \leq \max_e \sum_{\gamma_{xy} \ni e} |\gamma_{xy}|_{Q^{-1}} \pi(x)\pi(y), \tag{1}$$

where $|\gamma_{xy}|_{Q^{-1}}$ is the length of the path γ_{xy} according to the length function $l(e) = Q^{-1}(e)$. They also provided several examples where Eq. 1 gives a good bound on τ. Along the same lines, Sinclair [14] showed that

$$\tau \leq \max_e \frac{1}{Q(e)} \sum_{\gamma_{xy} \ni e} |\gamma_{xy}| \pi(x)\pi(y), \tag{2}$$

where $|\gamma_{xy}|$ is the number of edges in γ_{xy}. The two bounds in Eqs. 1 and 2 coincide in the case of the nearest-neighbor random walk on a graph, but are in general incomparable.

As observed in [6, 16], the bounds 1 and 2 can be generalized as follows. Let Γ be the set of paths γ_{xy}.

Proposition 1. *For any positive length function l, $\tau \leq \kappa(\Gamma, l)$, where*

$$\kappa(\Gamma, l) = \max_e \frac{1}{l(e)Q(e)} \sum_{\gamma_{xy} \ni e} |\gamma_{xy}|_l \pi(x)\pi(y).$$

Proposition 1 can be shown by a straightforward generalization of the proofs of the aformentioned bounds. We give a different proof of Proposition 1 in Section 2, by using the theory of non-negative matrices. Under a general condition, this allows us to characterize the length functions l that minimize $\kappa(\Gamma, l)$. Most of the material in Section 2 has been obtained independently in [6, 16].

In Section 3, we give examples where the bound given by a simple length function improves upon the bounds 1 and 2. We provide a heuristic for finding such a length function. In Section 4, we extend our results to the case where Γ is a multicommodity flow. Multicommodity flow has been used in [14] in the case of the unit length function. We show that the best bound μ on τ that can be obtained via this method, over all flows and length functions, can be computed at an arbitrary precision in polynomial time by a reduction to a semidefinite program. The bound μ is an invariant of the Markov chain. It coincides with τ in the case of a birth-death chain. A simple expression for μ is derived for edge-transitive graphs. We also show that the gap between τ and μ is at most of order $\log^2 n$, and that for expander graphs the gap is actually of order $\log^2 n$. We give an application to the exclusion process. Our results can be easily adapted to bound the second smallest eigenvalue of the Laplacian of a graph, essentially by replacing $\pi(x)$ by 1 and $Q(e)$ by n, as discussed in Section 5.

2 Bounding τ via Paths and Length Functions

Let Q be the diagonal matrix indexed by E, with $Q_{e,e} = Q(e)$, and Π the diagonal matrix indexed by $(V \times V) - \{(x,x) : x \in V\}$, with $\Pi_{(x,y),(x,y)} = \pi(x)\pi(y)$. We represent Γ by the matrix, also denote by Γ, whose rows are indexed by $(V \times V) - \{(x,x) : x \in V\}$ and columns indexed by E, with

$$\Gamma_{(x,y),e} = \begin{cases} 1 & \text{if } e \in \gamma_{xy}, \\ 0 & \text{otherwise.} \end{cases}$$

We say that a non-negative matrix is irreducible if its associated graph is strongly connected.

Theorem 2. *All the eigenvalues of the matrix $M = Q^{-1}\Gamma^t \Pi \Gamma$ are real. Let $\mu_0(\Gamma)$ be the largest one. Then*

$$\tau \le \mu_0(\Gamma). \tag{3}$$

Moreover,

$$\mu_0(\Gamma) = \min_l \kappa(\Gamma, l), \tag{4}$$

where l ranges over the set of positive length functions. If M is irreducible then $\mu_0(\Gamma) = \kappa(\Gamma, l)$ if and only if l is a positive eigenvector of M.

Proof. M is similar to the symmetric matrix $\sqrt{Q}^{-1} \Gamma^t \Pi \Gamma \sqrt{Q}^{-1}$, and thus all its eigenvalues are real. Let δ be the linear operator that maps every $\phi \in L^2(V)$ to $\phi' \in L^2(E)$, with $\phi'((x,y)) = \phi(x) - \phi(y)$. For $\phi \in L^2(V)$, we have $(\Gamma\phi)(x,y) = \phi(x) - \phi(y)$. Therefore, by the variational characterization of λ_1,

$$
\begin{aligned}
\tau &= \max_\phi \frac{\sum_{(x,y)\in V \times V} \pi(x)\pi(y)(\phi(x) - \phi(y))^2}{\sum_{(x,y)\in V \times V} Q(x,y)(\phi(x) - \phi(y))^2} \\
&= \max_\phi \frac{\|\sqrt{\Pi}\Gamma\phi'\|^2}{\|\sqrt{Q}\phi'\|^2} \\
&\le \max_{l \in L^2(E) - \{0\}} \frac{\|\sqrt{\Pi}\Gamma l\|^2}{\|\sqrt{Q}l\|^2} \\
&= \max_{l \in L^2(E) - \{0\}} \frac{\|\sqrt{\Pi}\Gamma\sqrt{Q}^{-1} l\|^2}{\|l\|^2}
\end{aligned}
$$

where, in the first two equations, ϕ ranges over the non-constant functions in $L^2(V)$. The last term is equal to the largest eigenvalue of $\sqrt{Q}^{-1}\Gamma^t \Pi \Gamma \sqrt{Q}^{-1}$, which is $\mu_0(\Gamma)$, since M and $\sqrt{Q}^{-1}\Gamma^t \Pi \Gamma \sqrt{Q}^{-1}$ have the same eigenvalues. Eq. 3 follows.

To show Eq. 4, we first note that

$$(Ml)(e) = \frac{1}{Q(e)} \sum_{\gamma_{xy} \ni e} |\gamma_{xy}|_l \pi(x)\pi(y). \tag{5}$$

Thus,

$$\kappa(\Gamma, l) = \max_e \frac{(Ml)(e)}{l(e)}. \tag{6}$$

for all $e \in E$. From the theory of non-negative matrices [13], any non-negative matrix B has a largest eigenvalue (in module) which is non-negative real, and equal to $\min_l \max_e (Bl)(e)/l(e)$, where l ranges over the set of positive vectors. (For the rest of this paper, we will refer to this eigenvalue simply as the largest eigenvalue of B.) Together with Eq. 6, this implies that the largest eigenvalue of M is equal to $\min_l \kappa(\Gamma, l)$, where l ranges over the set of positive length functions. Hence Eq. 4.

Finally, if l is a positive eigenvector of M, it corresponds to $\mu_0(\Gamma)$, by the theory of non-negative matrices. Thus, $(Ml)(e) = \mu_0(\Gamma)l(e)$. By Eq. 6, this implies that $\kappa(\Gamma, l) = \mu_0(\Gamma)$. Conversely, if M is irreducible and $\kappa(\Gamma, l) = \mu_0(\Gamma)$, then $Ml \leq \mu_0(\Gamma)l$, where inequality holds for every coordinate. By the theory of non-negative matrices, since M is irreducible and l positive, it follows that l is an eigenvector of M corresponding to $\mu_0(\Gamma)$.

Remark. Eq. 3 can be generalized as follows: For every i, $1 \leq i \leq n - 1$,

$$\mu_{|E|-n+i} \leq \frac{1}{1 - \lambda_i} \leq \mu_{i-1},$$

where μ_{i-1} is the i-th largest eigenvalue of M. The proof is similar to the case $i = 1$, and uses the usual minimax characterization of eigenvalues

$$\frac{1}{1 - \lambda_i} = \min_H \max_{\phi \in H} \frac{\sum_{(x,y) \in V \times V} \pi(x)\pi(y)(\phi(x) - \phi(y))^2}{\sum_{e \in E} Q(e)\phi'(e)^2},$$

and a similar relation involving μ_i, where H ranges over the set of vector sub-spaces of $L_0^2(V)$ of dimension $n - i$, and $L_0^2(V)$ denotes the set of $\phi \in L^2(V)$ such that $\sum_{x \in V} \pi(x)\phi(x) = 0$. Note that H and $\delta(H)$ have the same dimension since the restriction of δ to $L_0^2(V)$ is injective. The lower bound follows similarly.

3 Good Length Functions

We assume in this section that we are given a set of paths in G between every pair of vertices. Theorem 2 shows that, under the general condition that M is irreducible, a length function that gives the best bound on τ is an eigenvector of M corresponding to its largest eigenvalue. Such an eigenvector is generally not easy to compute, however. We give in this section a method for finding a length function that tends to give a good bound on τ.

Proposition 3. *For any positive length function l, $\kappa(\Gamma, Ml) \leq \kappa(\Gamma, l)$. More-over,*

$$\mu_0(\Gamma) \geq \min_e \frac{(Ml)(e)}{l(e)}. \tag{7}$$

Proof. Let $l' = Ml$. By Eq. 6, $\kappa(\Gamma, l)$ is the smallest number such that $l' \leq \kappa(\Gamma, l)l$. Since M is a non-negative matrix, it follows that $Ml' \leq \kappa(\Gamma, l)Ml = \kappa(\Gamma, l)l'$. Using Eq. 6 again, we conclude that $\kappa(\Gamma, l') \leq \kappa(\Gamma, l)$, as desired. Eq. 7 follows from the fact that, for any non-negative matrix B and any positive vector l, the largest eigenvalue of B is at least $\min_e (Bl)(e)/l(e)$.

Proposition 3 provides a method for computing better and better bounds on $\mu_0(\Gamma)$ by starting with any positive vector l, computing Ml via Eq. 5, and bounding $\mu_0(\Gamma)$ by $\max_e (Ml)(e)/l(e)$. We can obtain a better bound by repeating the same process, starting with the length function Ml. In the limit, the bound that we get is exactly $\mu_0(\Gamma)$ since, for any non-negative matrix B and any positive vector l, $\max_e (B^{t+1}l)(e)/(B^t l)(e)$ converges to the largest eigenvalue of B as t goes to infinity.

In practice, it suffices to calculate Ml within a constant multiplicative factor. Indeed, if l' approximates Ml within a constant multiplicative factor, then $\kappa(\Gamma, Ml)$ and $\kappa(\Gamma, l')$ differ only by a constant multiplicative factor.

Observe that if γ_{xy} and γ_{yx} consist of the same sequence of edges, but in reverse order, for any $(x, y) \in (V \times V) - \{(x, x) : x \in V\}$, then there exists a symmetric length function l that minimizes $\kappa(\Gamma, l)$. This can be seen by replacing a function l_1 that minimizes $\kappa(\Gamma, l_1)$ with $l_1(x, y) + l_1(y, x)$. Note also that if l is symmetric, then so is the improved length function Ml.

3.1 Examples

An important reversible Markov chain is the nearest-neighbor random walk on a graph $G = (V, E)$. For this Markov chain, $P(x, y) = 1/d_x$, for $(x, y) \in E$, where d_x is the degree of x in G. Also, $\pi(x) = d_x/|E|$ and $Q(e) = 1/|E|$, where $|E| = \sum_{x \in V} d_x$ is the number of *directed edges* in G.

d-ary tree. In a *d*-ary tree T of depth h, each node which is not a leave has exactly d descendants. The case $d = 2$ was treated in [5], where it was shown that Eq. 1 gives a bound on τ off by a multiplicative factor of h. Indeed, the number of nodes in T is $n = (d^{h+1} - 1)/(d - 1)$. There is a unique simple path between every pair of points in the tree. For a symmetric length function l, let $\text{diam}_l(T)$ be the maximum l-length of any simple path in the tree. For any edge $e = (u, v)$ such that the depth of u is i and the depth of v is $i-1$, with $1 \leq i \leq h$, we have

$$(Ml)(e) = \frac{1}{Q(e)} \sum_{\gamma_{xy} \ni e} |\gamma_{xy}|_l \pi(x)\pi(y) \tag{8}$$

$$\leq |E| \text{diam}_l(T) \sum_{x : \exists y, \gamma_{xy} \ni e} \pi(x)$$

$$\leq 2\text{diam}_l(T) \frac{d^{h+1-i}}{d - 1}.$$

If l is the unit length function $l(e) = 1$, Eq. 8 shows that $(Ml)(e) \leq 4hd^{h+1-i}/(d-1)$. This upper bound on $(Ml)(e)$ is tight, up to a constant factor. It implies that

$\kappa(\Gamma, l) \leq 4hd^h/(d-1)$. To obtain a better bound on μ_0, we replace l by Ml, which is (up to a constant factor) equivalent to replacing l by l_1, with $l_1(e) = d^{-i}$. By Eq. 8, $(Ml_1)(e) \leq 4d^{h+1-i}/(d-1)^2$, and so $\kappa(\Gamma, l_1) \leq 4d^{h+1}/(d-1)^2$. On the other hand, if ϕ is the vector that takes value 1 on a subtree rooted at one of the d children of the root of T, and value -1 at another subtree, and value 0 elsewhere, the numerator in the variational expression of τ is at least $1/(2d)$, the denominator is $4/|E|$ and so $\tau \geq (d^h - 1)/(4(d-1))$. In conclusion,

$$\frac{d^h - 1}{4(d-1)} \leq \tau \leq \mu_0 \leq \frac{4d^{h+1}}{(d-1)^2}.$$

This determines τ and μ_0 up to a constant factor. The constants in the above inequality can be easily improved, and we have made no attempt to optimize them.

Birth-death chain. In a birth-death chain, G consists of a single line, whose vertices are labeled from 0 to $n-1$. This chain is always reversible. In this case too, the paths are uniquely determined. Moreover, $\tau = \mu_0$. This is because

$$M = \begin{pmatrix} M_1 & 0 \\ 0 & M_1 \end{pmatrix},$$

where the first $n(n-1)/2$ rows are indexed by the pairs $(i, i+1)$, $0 \leq i \leq n-2$, and the last $n(n-1)/2$ rows are indexed by the pairs $(i+1, i)$, $0 \leq i \leq n-2$. In the proof of Theorem 2, we can restrict the summations to $(x, y) \in V \times V$, with $x < y$, and replace all matrices, vectors and operators involved with their restriction to the indices $(x, y) \in V \times V$, $x < y$, and $(i, i+1)$, $0 \leq i \leq n-2$. We conclude that τ is at most the largest eigenvalue of M_1. But, since any vector l indexed by $(i, i+1)$, $0 \leq i \leq n-2$, is of the form $(\phi(i) - \phi(i+1))$, for $\phi \in L^2(V)$, the third equation is in fact an equality, and so τ is equal to the largest eigenvalue of M_1, which is equal to μ_0. (Note that the equality $\tau = \mu_0$ does not hold for all trees; it is shown in Subsection 4.1 that the two quantities differ for stars with more than 2 leaves.)

Consider now the particular case where $P(i-1, i) = \beta = 1 - P(i, i-1)$ for $1 \leq i \leq n-1$, and $P(n-1, n-1) = \beta = 1 - P(0, 0)$, where $0 < \beta < 1/2$ is a constant. The chain is ergodic and reversible, with stationary distribution $\pi(i) = r^i/Z$, where $r = \beta/(1-\beta)$, and $Z = (1 - r^n)/(1-r)$. This example was considered in [14] where it is shown that, when r is fixed, the bound on τ given by Eq. 2 is linear in n, whereas τ is of constant order. Indeed, $Q(i-1, i) = \beta r^{i-1}/Z$, for $1 \leq i \leq n-1$. For the unit length function l that assigns 1 to every edge, we have $(Ml)(i-1, i) = \Theta(i)$, where the constants behind Θ may depend on β. Thus the maximum of $(Ml)(e)/l(e)$ differs from its mimimum by a factor of order n. A similar situation happens for the length function Ml, since $(M^2l)(i-1, i) = \Theta(i^2)$. The reverse situation happens for the length function $l_{DS}(e) = 1/Q(e)$, where $(Ml_{DS})(i-1, i)/l_{DS}(i-1, i) = \Theta(n-i)$. This suggests that an intermediate length function of the form $l_\alpha(i-1, i) = \alpha^{i-1}$, with $1 < \alpha < r^{-1}$, would give a good bound on τ. By bounding $|\gamma_{xy}|_{l_\alpha}$ by $\alpha^y/(\alpha-1)$, a simple calculation shows that

$$\frac{(Ml_\alpha)(e)}{l_\alpha(e)} \le \frac{(1+r)\alpha}{(\alpha-1)(1-r\alpha)}.$$

Thus, any such α gives a constant bound on τ. This bound is minimum for $\alpha = r^{-1/2}$, in which case it yields $\lambda_1 \le 2r^{1/2}/(1+r)$. This is very close to the exact value of λ_1, which is equal to $2r^{1/2}\cos(\pi/n)/(1+r)$. Note that the *Cheeger inequality* (see [14]) gives a constant but weaker bound on τ. Examples similar to this one can be found in [4].

3.2 Bounded-degree Graphs

Further simplification can be acheived for the nearest neighbor walk on a bounded-degree graph. Let d be the maximum degree of G, l a positive length function, Γ a set of simple paths between every pair of distinct points, $\text{cong}(e)$ the number of pairs (x, y) such that $e \in \gamma_{xy}$, and $\text{diam}_l(G, \Gamma)$ the maximum of $|\gamma_{xy}|_l$ over all pairs x and y. By Eq. 5,

$$(Ml)(e) \le \frac{\text{diam}_l(G, \Gamma)d^2}{|E|} \text{cong}(e). \tag{9}$$

In many cases, the right-hand side of Eq. 9 provides a rather accurate estimate on Ml. For example, for d-ary trees, it is off by a factor of order d^2. Thus, by using the length function $cong$ that assigns $\text{cong}(e)$ to e, we expect to get a good bound on $\mu_0(\Gamma)$. By Eq. 9,

$$\kappa(\Gamma, cong) \le \frac{\text{diam}_{cong}(G, \Gamma)d^2}{|E|}.$$

Thus,

Proposition 4. *For the nearest-neighbor walk on a graph with maximum degree d,*

$$\tau \le \frac{\text{diam}_{cong}(G, \Gamma)d^2}{|E|}.$$

3.3 Examples

Proposition 4 often gives a good bound on τ, as shown in the examples below.

d-ary tree. The congestion of an edge at distance i from the root is at most $(n-1)d^{h-i}/(d-1)$. Thus, $\text{diam}_{cong}(T) \le 2(n-1)d^{h+1}/(d-1)^2$. By Proposition 4, it follows that $\tau \le d^{h+3}/(d-1)^2$, which is off from the exact value by a factor of order d^2.

Elongated grid. Consider the subgraph of an $(au+1) \times (bv+1)$–grid induced on the vertices (i, j), with $0 \le i \le au$ and $0 \le j \le bv$, such that i is multiple of a or j is multiple of b. We establish a path from (i, j) to (i', j') by first going to $(i, \lfloor j/b \rfloor b)$, then to $(\lfloor i'/a \rfloor a, \lfloor j/b \rfloor b)$, then to $(\lfloor i'/a \rfloor a, j')$, then to (i', j'). Each time, we use a geodesic (or empty) path. A simple calculation shows that the congestion of a horizontal edge is at most $u(a + b)n$, where n is the number of vertices, and the congestion of a vertical edge is at most $v(a + b)n$. Thus, $\text{diam}_{cong}(G, \Gamma) \leq (au^2 + bv^2)(a + b)n$. Since $|E| \geq 2n$, Proposition 4 shows that $\tau \leq 8(au^2 + bv^2)(a + b)$.

On the other hand, let $\phi \in L_0^2(V)$ be the function that assigns $2i - au$ to (i, j). Then the numerator in the variational expression of τ is at least $a^2u^2/2$, and the denominator is $8auv/|E|$, and so $\tau \geq au|E|/(16v) \geq au^2(a + b)/8$. Similarly, $\tau \geq bv^2(a + b)/8$, and so $\tau \geq (au^2 + bv^2)(a + b)/16$. Thus, τ is of order $(au^2 + bv^2)(a + b)$. Here again, we have made no attempt to optimize the constants in the calculation.

Note that the bound on τ given by Eqs. 1 and 2 (using the same paths) may be of different order than τ. For example, when $u = b = 1$ and $a = v^2$, τ is of order v^4, but the bound given by Eqs. 1 and 2 is of order v^5.

Double-grid. Consider two $a \times a$ grids having one common origin O. We use Proposition 4 to show that $\tau = \Theta(a^2 \log a)$. To see this, we start by showing that there exist geodesic paths that connect every point in the double-grid to O such that the congestion of an edge at distance h from O is $O(a^2/(h+1))$. This can be done by backward induction on the distance from O. Indeed, label the nodes in one grid by (i, j), $0 \leq i, j \leq a - 1$, the origin O being labeled $(0, 0)$. Starting with the node $(a - 1, a - 1)$, recursively route an $i/(i + j)$ fraction of the paths traversing (i, j) to $(i - 1, j)$, and the remaining fraction to $(i, j - 1)$. Rounding is done in favor of the first set of paths. If c_h is the maximum congestion of a node at distance h from O then

$$c_{h-1} \leq \frac{h + 1}{h} c_h + 2,$$

which implies that $c_h \leq 4a^2/(h+1)$, as desired. By connecting every pair of points by first connecting them to O, the congestion of an edge at distance h from O becomes of order $a^4/(h + 1)$. Proposition 4 then implies that $\tau = O(a^2 \log a)$. A lower bound of order $a^2 \log a$ follows using the variational expression of τ and the function that assigns $\log(h + 1)$ to the vertices at distance h from O in one specified grid, and $-\log(h + 1)$ to their symmetric counterparts. Diaconis and Saloff-Coste [2] have obtained independently the same estimate on τ using a different approach.

4 A Semidefinite Relaxation of $\mu_0(\Gamma)$

The results and definitions in the preceding sections can be generalized in an obvious manner by replacing paths with flows of unit value, along [14]. The use of flow will allow us to apply general results from optimization of convex functions, as shown below. Let F be the set of non-negative real matrices Γ whose rows are indexed by $(V \times V) - \{(x, x) : x \in V\}$ and columns indexed by E, and such that each row indexed by (x, y) represents a flow in G from x to y with unit value. In other words, $\Gamma_{(x,y)}$ is a convex combination of the indicator functions of simple paths in G from x to y. As before, let $\mu_0(\Gamma)$ be the largest eigenvalue of $M = Q^{-1}\Gamma^t\Pi\Gamma$. For a non-negative length function l on E, denote by $\text{dist}_l(x, y)$ the length according to l of a shortest path from x to y.

Lemma 5. *If C is a closed convex set of non-negative matrices in $R^{p \times q}$, then*

$$\min_{B \in C} ||B|| = \max_{l \in (\mathcal{R}^+)^q : ||l|| \leq 1} \min_{B \in C} ||Bl||, \tag{10}$$

where $||B|| = \max_{l \in \mathcal{R}^q : ||l|| \leq 1} ||Bl||$.

Proof. The right-hand side of Eq. 10 is upper-bounded by the left-hand side by weak duality. We show that equality holds.

First, assume that any matrix in C is positive, i.e, all entries are strictly positive. Since C is closed convex, the norm function attains its minimum on C at some matrix B_0. Since the matrix $B_0^t B_0$ is positive, it follows from the theory of non-negative matrices [13] that it has a unique (positive) largest eigenvalue. Therefore, the norm function is differentiable at B_0, since $||B||$ is equal to the square root of the largest eigenvalue of $B^t B$. Moreover, its gradient at B_0 is equal to uv^t, where $||u|| = ||v|| = 1$ and $B_0 v = ||B_0|| u$ (see, e.g., [18].) Since the norm function and C are convex, it follows that $\text{tr}((B - B_0)vu^t) \geq 0$ for all $B \in C$. Indeed, if a convex function f (the norm function, in this case) attains its minimum on a convex set C at x_0, and if f is differentiable at x_0, then $(x - x_0) \cdot \nabla_{x_0} f \geq 0$, for any x in C. Here the dot product of two matrices B_1 and B_2 is $\text{tr}(B_1 B_2^t)$. But $\text{tr}(B_0 vu^t) = ||B_0||\text{tr}(uu^t) = ||B_0||$, and $\text{tr}(Bvu^t) \leq ||Bv|| \, ||u|| = ||Bv||$. Thus, $||Bv|| \geq ||B_0||$. Since $||B_0 v|| = ||B_0||$, it follows that $\min_{B \in C} ||Bv|| = ||B_0||$. We conclude that Eq. 10 holds under the assumption that any matrix in C is positive (clearly, l can be assumed to be non-negative.) The general case can be reduced to the previous case by compactness and by considering the sets $C_i = C + (1/i)\mathbf{1}$, where i is a positive integer. Indeed, if l_i is a vector that maximizes the right-hand side of Eq. 10 for the set C_i, and l is the limit of a subsequence of l_i, it is not hard to show that $\min_{B \in C} ||B|| = \min_{B \in C} ||Bl||$.

We note that Eq. 10 may not hold if C is an arbitrary closed convex set of matrices in $R^{p \times q}$. For example, if C is the set of $p \times p$ matrices whose trace is 1, then the left-hand side is equal to $1/p$, but the right-hand side is equal to 0.

Theorem 6. *Let $\mu = \min_{\Gamma \in F} \mu_0(\Gamma)$. Then $\tau \leq \mu$. For any $\epsilon > 0$, μ can be computed within a factor of $1 + \epsilon$ in polynomial time in n, $\log(1/\epsilon)$ and the bitsize of the input (the transition probabilities). Moreover,*

$$\mu = \max_l \frac{\sum_{(x,y) \in V \times V, x \neq y} \pi(x)\pi(y)\text{dist}_l(x, y)^2}{\sum_{e \in E} Q(e)l(e)^2}, \tag{11}$$

where l ranges over the set of non-negative symmetric length functions on E.

Proof. A proof similar to that of Theorem 2 shows that $\tau \leq \mu_0(\Gamma)$, for any $\Gamma \in F$. This is because $(\Gamma(\delta(\phi)))(x,y) = \phi(x) - \phi(y)$. Hence $\tau \leq \mu$. Given Q and Π, the problem of minimizing $\mu_0(\Gamma) = ||\sqrt{\Pi}\Gamma\sqrt{Q}^{-1}||^2$ can be reduced to a semidefinite program since, as noted in e.g. [17], $||B|| \leq t$ if and only if the matrix

$$\begin{pmatrix} tI & B \\ B^t & tI \end{pmatrix}$$

is positive semidefinite. Under general conditions that can be easily checked here, semidefinite programs can be solved at an arbitrary precision in polynomial time [9]. Thus, μ can be computed at an arbitrary precision in polynomial time using, e.g., the ellipsoid method.

On the other hand,

$$\mu = \min_{\Gamma \in F} ||\sqrt{\Pi}\Gamma\sqrt{Q}^{-1}||^2$$

$$= \max_{l:||l||\leq 1} \min_{\Gamma \in F} ||\sqrt{\Pi}\Gamma\sqrt{Q}^{-1}l||^2$$

$$= \max_{l:||\sqrt{Q}l||\leq 1} \min_{\Gamma \in F} ||\sqrt{\Pi}\Gamma l||^2$$

$$= \max_{l:||\sqrt{Q}l||\leq 1} \sum_{(x,y)\in V\times V, x\neq y} \pi(x)\pi(y)\mathrm{dist}_l(x,y)^2$$

$$= \max_{l} \frac{\sum_{(x,y)\in V\times V, x\neq y} \pi(x)\pi(y)\mathrm{dist}_l(x,y)^2}{\sum_{e\in E} Q(e)l(e)^2}.$$

The second equation follows from Lemma 5. The fourth equation follows from the fact that a flow from x to y is a convex combination of simple paths from x to y. By convexity of the norm function, there exists $\Gamma \in F$ that minimizes $||\sqrt{\Pi}\Gamma\sqrt{Q}^{-1}||$ and such that $\Gamma_{(x,y),(u,v)} = \Gamma_{(y,x),(v,u)}$. By the proof of Lemma 5 and the remark following Proposition 3, we can assume that l is symmetric.

Note that, in general, it is easier to compute eigenvalues than to solve semidefinite programs. However, whereas other related quantities associated with a Markov chain such as minimum bisection are NP-hard to compute [8], μ can be computed at an arbitrary precision in polynomial time.

It is easy to show directly that τ is upper-bounded by the right-hand side of Eq. 11. This can be seen by considering $l(x,y) = |\phi(x) - \phi(y)|$, where ϕ is an eigenvector of P corresponding to λ_1. We now show that the best bound on τ using the Poincaré inequalities is off by a factor of order at most $\log^2 n$. We apply a technique used in [11] in the context of multicommodity flow.

Theorem 7 [11]. *There exists a universal constant c such that, for any metric space (X,d) on n vertices, there exists an embedding ψ of X in R^p, for some integer p, such that*

$$d(x,y) \leq ||\psi(x) - \psi(y)|| \leq cd(x,y)\log n,$$

for any elements $x, y \in X$.

Theorem 8. *For any irreducible, reversible Markov chain on n states, $\mu \leq c^2 \tau \log^2 n$.*

Proof. By Theorem 6, it suffices to show that, for any non-negative symmetric length function l,

$$\frac{\sum_{(x,y) \in V \times V, x \neq y} \pi(x)\pi(y)\mathrm{dist}_l(x,y)^2}{\sum_{e \in E} Q(e)l(e)^2} \leq c^2 \tau \log^2 n. \tag{12}$$

By continuity, we can assume without loss of generality that l is positive, so that dist_l defines a metric on V. By Theorem 7, it suffices for our needs to show that, for any embedding ψ of V in R^p,

$$\frac{\sum_{(x,y) \in V \times V, x \neq y} \pi(x)\pi(y)\|\psi(x) - \psi(y)\|^2}{\sum_{(x,y) \in V \times V, x \neq y} Q(x,y)\|\psi(x) - \psi(y)\|^2} \leq \tau. \tag{13}$$

Let $\phi_i(x)$ be the ith coordinate of $\phi(x)$, $1 \leq i \leq p$. To show Eq. 13 it suffices to prove that, for $1 \leq i \leq p$,

$$\frac{\sum_{(x,y) \in V \times V, x \neq y} \pi(x)\pi(y)(\psi_i(x) - \psi_i(y))^2}{\sum_{(x,y) \in V \times V, x \neq y} Q(x,y)(\psi_i(x) - \psi_i(y))^2} \leq \tau.$$

But this follows immediately from the variational definition of τ. This concludes the proof.

A graph is edge-transitive if, for any $(x,y), (x',y') \in E$, there exists an automorphism of the graph that either takes x to x' and y to y', or that takes x to y' and y to x'.

Corollary 9. *For any reversible Markov chain, we have $\mu \geq D$, where D is the expected squared distance between two random points in G chosen independently according to the stationary distribution. Equality $\mu = D$ occurs for the nearest-neighbor walk on an edge-transitive graph.*

Proof. The inequality $\mu \geq D$ follows from Theorem 6 by considering the unit length function l.

Assume now that the graph is edge-transitive. By the definition of μ and Eq. 4, $\mu \leq \kappa(\Gamma, l)$ for any $\Gamma \in F$ and any l. Furthermore, Diaconis and Stroock [5] produce l (unit length function) and $\Gamma \in F$ (average combination of geodesic paths) so that $\kappa(\Gamma, l) = D$. Thus $\mu \leq D$.

4.1 Examples

We calculate μ for the complete bipartite graph $K_{a,b}$ and, up to a constant factor, for expander graphs, and give an application to the exclusion process.

Complete bipartite graph $K_{a,b}$. This graph is edge-transitive, and so $\mu = D = 5/2 - 1/a - 1/b$, whereas $\tau = 1$ for $a + b \geq 3$. Corollary 9 shows that μ

is equal to $\kappa(\Gamma, l)$, where l is the unit length function, and Γ is determined by assigining the same weight to all shortest paths between two given points. In contrast, in the case $a = 2$, deterministic paths will give an $\Omega(b)$ bound on τ, uniformly in all length functions. This example was treated in [14] for $a = 2$ and the unit length function (see also [3]).

Expander Graphs. A d-regular graph G is a (d, λ_1)-expander if all the eigenvalues of the adjacency matrix of G, besides d, are at most $\lambda_1 d$, where $\lambda_1 < 1$ is a fixed constant. It is known [14, Th. 8] that, for any (d, λ_1)-expander on n vertices, one can establish simultaneously a flow of unit value between every pair of distinct points such that the total amount of flow traversing any given edge is $O(n \log n)$, where the constant behind O depends on λ_1. Moreover, every such flow is a convex combination of paths of length $O(\log n)$. Using the unit length function, we conclude that $\mu_0(\Gamma) = O(\log^2 n)$. A lower bound on μ can be obtained using Corollary 9. Indeed, since most of the vertices in the graph are at distance $\Omega(\log_{d-1} n)$ from any given point, $D = \Omega(\log_{d-1}^2 n)$. Thus $\mu = \Theta(\log^2 n)$, where the constants behind Θ depend on d and λ_1, whereas $\tau = \Theta(1)$. This example shows the sharpness of Theorem 8. Note that the Cheeger inequality provides a constant bound on τ in this case.

The exclusion process. Given a connected graph $G = (V, E)$ and $1 \le r < n$, the exclusion process is the Markov chain on all r-sets of V defined as follows. If the current state is at A, pick a random element x of A with probability proportional to its degree, pick a neighbor y of x uniformly at random, and move to $A' = A - \{x\} \cup \{y\}$. Note that A' may be equal to A. It is known (see [3] and references therein) that this Markov chain is reversible. Bounds on the mixing rate of the exclusion process were established in [7, 3]. In particular, it was shown in [3] that when G is regular, for any set Γ of canonical paths in G,

$$\tau(r) \le r\kappa(\Gamma, 1), \tag{14}$$

where $\tau(r)$ is the inverse of the eigenvalue gap for the exclusion process. Here the Markov chain on G is the usual nearest-neighbor walk on G. The proof-technique in [3] shows that in fact, when G is regular, $\tau(r) \le r\mu$. Together with Theorem 8, this implies that $\tau(r) \le c^2 r\tau \log^2 n$. Up to a constant factor, this inequality also holds for nearly-regular graphs. Diaconis and Saloff-Coste [3] have raised the question whether $\tau(r) \le \gamma r\tau$, for some universal constant γ and all graphs. They showed [3] a lower bound $\tau(r) = \Omega(r\tau)$ for nearly-regular graphs.

5 Concluding Remarks

1. Our results can be easily adapted to bound the second smallest eigenvalue of the Laplacian of a graph. The Laplacian is the matrix indexed by the vertices which for $i \ne j$, has a 0 in entry (i, j) if vertices i and j are not connected, and a -1 if they are connected; while on the diagonal, the degrees of the vertices appear in the corresponding order. Essentially, this can be done by replacing $\pi(x)$ by 1 and $Q(e)$ by n in the proofs and results. In particular,

if η_{lap} is the inverse of the second smallest eigenvalue of the Laplacian (the smallest one is 0), then

$$n\eta_{lap} \leq \max_e \frac{1}{l(e)} \sum_{\gamma_{xy} \ni e} |\gamma_{xy}|_l,$$

for any positive length function l, and so

$$n\eta_{lap} \leq \text{diam}_{cong}(G, \Gamma).$$

The best bound on η_{lap} that can be obtained using multicommodity flow and a variable length function is

$$\mu_{lap} = \max_l \frac{\sum_{(x,y) \in V \times V} \text{dist}_l(x,y)^2}{n \sum_{e \in E} l(e)^2},$$

where l ranges over the set of non-negative symmetric length functions. Moreover, $\mu_{lap} \leq O(\log^2 n)\eta_{lap}$.

2. It is easy to show that, for any set Γ of canonical paths and any positive length function l, $\kappa(\Gamma, 1) \leq (\max_{x \neq y} |\gamma_{xy}|)\kappa(\Gamma, l)$. A similar remark holds for any flow matrix Γ. Thus, it is not possible to improve the bound on τ by more than a factor of $\max_{x \neq y} |\gamma_{xy}|$ by using an optimal length function rather than the unit length function. Our birth-death chain and tree examples show that such an improvement can be achieved for some chains.

6 Acknowledgments

The author is grateful to David Aldous, Noga Alon, Stephen Boyd, Persi Diaconis, Jim Fill, Joel Friedman, Stephen Guattery, Gary Miller, Laurent Saloff-Coste, and Alistair Sinclair for helpful discussions.

References

1. D. Aldous. *Reversible Markov Chains and random walks on graphs*. Book in preparation.
2. P. Diaconis and L. Saloff-Coste. Personal Communication, 1995.
3. P. Diaconis and L. Saloff-Coste. Comparison theorems for reversible Markov Chains. *The Annals of Applied Probability*, 3:696–730, 1993.
4. P. Diaconis and L. Saloff-Coste. What do we know about the Metropolis algorithm? In *27th Annual ACM Symposium on Theory of Computing*, pages 112–129. ACM Press, 1995.
5. P. Diaconis and D. Stroock. Geometric bounds for eigenvalues of Markov Chains. *The Annals of Applied Probability*, 1:36–61, 1991.
6. J. Fill. Unpublished manuscript, July 1990.
7. J. Fill. Eigenvalue bounds on convergence to stationarity for nonreversible markov chains with an application to the exclusion process. *The Annals of Applied Probability*, 1:62–87, 1991.

8. M. R. Garey and D. S. Johnson. *Computers and intractability: a guide to the theory of NP-completeness.* Freeman and Company, San Fransisco, 1979.

9. M. Grötschel, L. Lovász, and A. Schrijver. The ellipsoid method and its consequences in combinatorial optimization. *Combinatorica*, 1:169–197, 1981.

10. M. Jerrum and A. Sinclair. Approximating the permanent. *SIAM J. on Comput.*, 18:1149–1178, 1989.

11. N. Linial, E. London, and Y. Rabinovich. The geometry of graphs and some of its algorithmic applications. *Combinatorica*, 15:215–245, 1995.

12. L. Lovász and M. Simonovits. Random walks in a convex body and an improved volume algorithm. *Random Strutures & Algorithms*, 4(4):359–412, 1993.

13. E. Seneta. *Non-negative matrices and Markov Chains.* Springer-Verlag, 1981.

14. A. Sinclair. Improved bounds for mixing rates of Markov Chains and multicommodity flow. *Combinatorics, Probability and Computing*, 1:351–370, 1992.

15. A. Sinclair and M. Jerrum. Approximate counting, uniform generation, and rapidly mixing Markov Chains. *Information and Computation*, 82:93–113, 1989.

16. A. D. Sokal. Optimal Poincaré inequalities for the spectra of Markov Chains. Unpublished manuscript, September 1992.

17. L. Vandenberghe and S. Boyd. Semidefinite programming. *SIAM Review*, 1995. To appear.

18. G. A. Watson. Characterization of the subdifferential of some matrix norms. *Linear Algebra and Appl.*, 170:33–45, 1992.

The Quadratic Assignment Problem with a Monotone Anti-Monge and a Symmetric Toeplitz Matrix: Easy and Hard Cases *

Rainer E. Burkard[1], Eranda Çela[1], Günther Rote[1] and Gerhard J. Woeginger[2]

[1] Technische Universität Graz, Institut für Mathematik B, Steyrergasse 30, A-8010 Graz, Austria
[2] Eindhoven University of Technology, Department of Mathematics and Computer Science, P.O. Box 513, 5600 MB Eindhoven, The Netherlands

Abstract. The Anti-Monge–Toeplitz QAP (AMT-QAP) is a restricted version of the Quadratic Assignment Problem (QAP), with a monotone Anti-Monge matrix and a symmetric Toeplitz matrix. The following problems can be modeled via the AMT-QAP: (P1) The "Turbine Problem", i. e. the assignment of given masses to the vertices of a regular polygon such that the distance of the gravity center of the resulting system to the polygon's center is minimized. (P2) The Traveling Salesman Problem on symmetric Monge matrices. (P3) The linear arrangement of records with given access probabilities in order to minimize the average access time.
We identify conditions on the Toeplitz matrix that lead to polynomially solvable cases of the AMT-QAP. In each of these cases an optimal permutation can be given without regarding the numerical problem data. We show that the Turbine Problem is NP-hard and consequently, that the AMT-QAP is NP-hard.

1 Introduction

Given two $n \times n$ matrices $A = (a_{ij})$ and $B = (b_{ij})$, $1 \le i, j \le n$, with real entries, the *Quadratic Assignment Problem* (QAP) in the Koopmans-Beckmann form [8] consists in finding the permutation π which minimizes

$$\sum_{i=1}^{n} \sum_{j=1}^{n} a_{\pi(i)\pi(j)} b_{ij}. \tag{1}$$

Here, π ranges over the set S_n of all permutations of $\{1, 2, \ldots, n\}$. The QAP with matrices A and B will be abbreviated by QAP(A, B). The QAP has many applications e.g. in location theory, scheduling, manufacturing, parallel and distributed computing, and statistical data analysis. The QAP is well known to be NP-hard, as it contains other well studied NP-hard problems as special cases,

---

* This research has been supported by the Spezialforschungsbereich F 003 "Optimierung und Kontrolle", Projektbereich Diskrete Optimierung.

e.g. *the Traveling Salesman Problem*. Also from the practical point of view, the QAP is a very hard problem; currently, solving instances of size $n \geq 20$ is considered intractable (see Clausen and Perregård [6]). For more information on the QAP, the reader is referred to the surveys by Lawler [10], Burkard [3], and Pardalos, Rendl and Wolkowicz [11].

1.1 The Anti-Monge–Toeplitz QAP: Definitions

We investigate a restricted version of the QAP, the *Anti-Monge–Toeplitz QAP* (AMT-QAP), where matrix A is a *monotone Anti-Monge matrix* and matrix B is a *symmetric Toeplitz matrix*.

Definition 1. An $n \times n$ matrix $A = (a_{ij})$ is called an *Anti-Monge matrix* if it satisfies the inequality $a_{ij} + a_{rs} \geq a_{is} + a_{rj}$ for all $1 \leq i < r \leq n$ and $1 \leq j < s \leq n$ (this inequality is called the *Anti-Monge inequality*). Matrix A is called *monotone* if $a_{ij} \leq a_{i,j+1}$ and $a_{ij} \leq a_{i+1,j}$ for all i, j, i. e. the entries in every row and in every column are in non-decreasing order.

Matrices M for which $-M$ is an Anti-Monge matrix are called *Monge matrices*.

Simple examples of Anti-Monge matrices are *sum matrices* and *product matrices*. A matrix A is a *sum matrix* (respectively, a *product matrix*) if there exist real numbers r_i and c_i, $1 \leq i \leq n$, such that $a_{ij} = r_i + c_j$ (respectively, $a_{ij} = r_i \cdot c_j$) holds for all $1 \leq i, j \leq n$. Every sum matrix is an Anti-Monge matrix. A product matrix is a monotone Anti-Monge matrix, if the numbers r_i and c_j are non-negative and sorted in increasing order.

Definition 2. An $n \times n$ matrix $B = (b_{ij})$ is a *Toeplitz matrix*, if there exists a function $f: \{-n+1, \ldots, n-1\} \to \mathbb{R}$ such that $b_{ij} = f(i-j)$, for $1 \leq i, j \leq n$. The Toeplitz matrix B is said to be *generated* by function f.

1.2 The Anti-Monge–Toeplitz QAP: Applications

Our interest in the Anti-Monge–Toeplitz QAP arose from the "Turbine Problem" initially introduced by Bolotnikov [2]. In the manufacturing of turbines, n given positive masses (=blades) m_i, $1 \leq i \leq n$, have to be placed on the vertices of a regular polygon (=turbine rotor) in a balanced way, which means that the center of gravity of the resulting system should be as close to its rotational center (=rotational axis) as possible. It can be shown that this objective leads to a quadratic assignment problem (see Section 3 for more details).

$$\min_{\phi \in S_n} \sum_{i=1}^{n} \sum_{j=1}^{n} m_{\phi(i)} m_{\phi(j)} \cos \frac{2(i-j)\pi}{n}$$

W.l.o.g. we may assume that $m_1 \leq m_2 \leq \cdots \leq m_n$. Then in the QAPabove, the matrix $A = (a_{ij}) = (m_i m_j)$ is a monotone Anti-Monge matrix. The matrix $B = (b_{ij}) = \left(\cos \frac{2(i-j)\pi}{n}\right)$ is a symmetric Toeplitz matrix generated by the function $f(x) = \cos \frac{2\pi x}{n}$. Thus, the Turbine Problem is a special case of the AMT-QAP.

There are also two other optimization problems that can be formulated as AMT-QAPs: The *Traveling Salesman Problem* with a symmetric Monge distance matrix (Supnick [12]) and a *Data Arrangement Problem* initially introduced by Burkov, Rubinstein and Sokolov [5]. These two applications of the AMT-QAP are described in detail in Section 4.

1.3 The Anti-Monge–Toeplitz QAP: Negative Results

Despite its simple combinatorial structure, the AMT-QAP can be shown to be NP-hard. In fact, we provide two NP-hardness results for special cases of the AMT-QAP which imply NP-hardness for the general problem.

First, we show NP-hardness for the Turbine Problem, and hence solve an "old" open complexity question. Secondly, the QAP(A, B) with a monotone Anti-Monge matrix A and a Toeplitz matrix $B = (b_{ij})$, defined by $b_{ij} = (-1)^{i+j}$ is shown to be NP-hard. Thus, the Anti-Monge–Toeplitz QAP is NP-hard even in the case that the generating function of the Toeplitz matrix B is periodic and takes only ± 1-values.

1.4 The Anti-Monge–Toeplitz QAP: Positive Results

We identify two polynomially solvable versions of the AMT-QAP. Both versions can be solved very easily: One can give an optimal permutation π even if the corresponding matrices A and B are not explicitly given. The information about the structure of the problem alone suffices to solve it, even without knowing the numerical values of the matrix entries. More generally we introduce the *constant permutation property*.

Definition 3. An $n \times n$ matrix B has the *constant permutation property* with respect to a class of $n \times n$ matrices $\mathcal{A}$, if there is a permutation $\pi^B \in S_n$ which solves the problem QAP(A, B) for all matrices $A \in \mathcal{A}$.

A class of matrices $\mathcal{B}$ has the constant permutation property with respect to a class of matrices $\mathcal{A}$ if each matrix $B \in \mathcal{B}$ has the constant permutation property with respect to class $\mathcal{A}$.

We identify conditions on the symmetric Toeplitz matrix B which guarantee that B has the constant permutation property with respect to monotone Anti-Monge matrices. The first of these conditions involves *benevolent* matrices.

Definition 4. A function $f: \{-n+1, \ldots, n-1\} \to \mathbb{R}$ is called *benevolent* if it fulfills the following three properties.

(BEN1) $f(-i) = f(i)$ for all $1 \leq i \leq n - 1$.
(BEN2) $f(i) \leq f(i+1)$ for all $1 \leq i \leq \lfloor \frac{n}{2} \rfloor - 1$.
(BEN3) $f(i) \leq f(n - i)$ for all $1 \leq i \leq \lceil \frac{n}{2} \rceil - 1$.

A matrix is called benevolent if it is a Toeplitz matrix generated by a benevolent function.

Example 1. Let us give the example of a benevolent function $f: \{-7, -6, \ldots, 0, 1, \ldots 7\}$ defined by

$$f(0) = 1.5, \quad f(1) = 0.5, \quad f(2) = 1, \quad f(3) = 1.25, \quad f(4) = 2,$$

$$f(5) = 2.5, \quad f(6) = 2.5, \quad f(7) = 1.5$$

The graph of this function is shown in Figure 1.

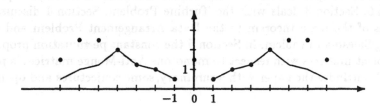

Fig. 1. The graph of the function f

By property (BEN1), a benevolent matrix is symmetric.

The permutation π^* which is optimal for every QAP with a monotone Anti-Monge matrix A and a benevolent matrix B is given below.

Definition 5. The permutation $\pi^* \in S_n$ is defined by $\pi^*(i) = 2i - 1$ for $1 \leq i \leq \lceil \frac{n}{2} \rceil$, and $\pi^*(n + 1 - i) = 2i$ for $1 \leq i \leq \lfloor \frac{n}{2} \rfloor$.

Note that π^* starts with the odd numbers in increasing order followed by the even numbers in decreasing order. We adopt the following notation for permutation ϕ:

$$\phi = \langle \phi(1), \phi(2), \ldots, \phi(n) \rangle$$

With this notation we have: $\pi^* = \langle 1, 3, 5, 7, 9, \ldots, 10, 8, 6, 4, 2 \rangle$.

Our main result states that every benevolent matrix B has the constant permutation property with respect to monotone Anti-Monge matrices. A sketch of the proof is given in Section 2. The reader is referred to [4] for a complete proof.

Theorem 6 (Main theorem). *The permutation π^* solves $QAP(A, B)$ when A is a monotone Anti-Monge matrix and B is a benevolent matrix.*

Theorem 6 generalizes and unifies a number of known results on the Data Arrangement Problem and on the Traveling Salesman Problem. For example, we give a very simple proof for the polynomial solvability of the TSP on symmetric Monge matrices, whereas the original proof given by Supnick in 1957 [12] is quite involved. Further, the main theorem leads to a generalization of polynomial results on the Data Arrangement Problem obtained in the late 1960s and early 1970s. Finally, Theorem 6 yields that the *maximization version of the Turbine Problem* is solvable in polynomial time, see Section 3.

Further, we investigate certain periodic extensions of benevolent matrices, the so called *k-benevolent matrices*, which fulfill the constant permutation property with respect to monotone Anti-Monge matrices.

1.5 Organization of the Paper

In Section 2, we give a sketch of the rather technical and complex proof of Theorem 6. Section 3 deals with the Turbine Problem. Section 4 discusses applications of the main theorem to the Data Arrangement Problem and to the Traveling Salesman Problem. In Section 5 the constant permutation property of *k*-benevolent matrices with respect to monotone Anti-Monge matrices is proven. Section 6 concludes the paper with a summary, some conjectures and open questions.

2 The Main Result: Sketch of the Proof

In this section we consider only $n \times n$ matrices, for some fixed $n \geq 3$. W.l.o.g., we assume that all entries in the matrices A and B of $QAP(A, B)$ are nonnegative and that all Toeplitz matrices have 0-entries on the diagonal. The proof of the main theorem follows the simple idea described below.

2.1 Idea of the Proof

We want to prove that π^* solves $QAP(A, B)$ when A is a non-negative monotone Anti-Monge matrix and B is a benevolent matrix with 0-entries on the diagonal. Actually, we prove an even stronger statement. Consider the relaxation of $QAP(A, B)$ where the columns and the rows of matrix A may be permuted *independently of each other*, the columns according to some permutation ϕ and the rows according to some permutation ψ:

$$\min_{\phi, \psi \in S_n} \sum_{i=1}^{n} \sum_{j=1}^{n} a_{\phi(i)\psi(j)} b_{ij} \tag{2}$$

This problem is called *independent-QAP(A, B)*.

We show that the double sum in (2) is minimized by $\phi = \psi = \pi^*$. This trivially guarantees that π^* is an optimal solution of $QAP(A, B)$. Thus, Theorem 6 can be derived as a corollary of the following theorem:

Theorem 7. *The pair (π^*, π^*) solves the independent-QAP(A, B) when A is a non-negative monotone Anti-Monge matrix and B is a benevolent matrix with 0-entries on the diagonal.*

In the proof of Theorem 7, we exploit the fact that the set $\mathcal{A}$ of non-negative monotone Anti-Monge matrices is a cone, and likewise, the set $\mathcal{B}$ of benevolent matrices with 0-entries on the diagonal is a cone, too. Then, it is sufficient to prove the theorem for the extreme rays of these cones, as shown by Observation 11. It turns out that these extreme rays have a simple structure, which makes the proof tractable.

The following lemmas, whose simple proofs are omitted in this version of the paper, give the generators of the extreme rays of the cones $\mathcal{A}$ and $\mathcal{B}$, respectively.

Lemma 8. *The monotone Anti-Monge matrices with non-negative entries form a cone $\mathcal{A}$. The extreme rays of this cone are generated by the $0-1$ matrices $R^{(pq)} = \left(r_{ij}^{(pq)}\right)$, $1 \leq p, q \leq n$, defined by $r_{ij}^{(pq)} = 1$ for $n - p + 1 \leq i$ and $n - q + 1 \leq j$, and $r_{ij}^{(pq)} = 0$ otherwise.*

In a pictorial setting, matrix $R^{(pq)}$ has a $p \times q$ block of one entries in the lower right corner and zero entries elsewhere.

Lemma 9. *The benevolent functions $f: \{-n+1, \ldots, n-1\} \to \mathbb{R}$ with $f(0) = 0$ form a cone. The extreme rays of this cone are the functions g^α, $\lfloor \frac{n}{2} \rfloor + 1 \leq \alpha \leq n-1$, and h^β, $1 \leq \beta \leq \lfloor \frac{n}{2} \rfloor$, defined in Definition 10.*

Accordingly, the benevolent matrices with 0-entries on the diagonal form a cone $\mathcal{B}$, whose extreme rays are the Toeplitz matrices generated by the functions g^α, $\lfloor \frac{n}{2} \rfloor + 1 \leq \alpha \leq n-1$, and h^β, $1 \leq \beta \leq \lfloor \frac{n}{2} \rfloor$.

Definition 10. The benevolent functions g^α, $\lfloor \frac{n}{2} \rfloor + 1 \leq \alpha \leq n - 1$, and h^β, $1 \leq \beta \leq \lfloor \frac{n}{2} \rfloor$, which map $\{-n+1, \ldots, n-1\}$ into $\{0, 1\}$ are defined by

$$g^\alpha(x) = \begin{cases} 1, & \text{for } x \in \{-\alpha, \alpha\}, \\ 0, & \text{for } x \notin \{-\alpha, \alpha\}. \end{cases} \qquad h^\beta(x) = \begin{cases} 1, & \text{for } \beta \leq |x| \leq n - \beta, \\ 0, & \text{otherwise.} \end{cases}$$

Example 2. For $n = 6$ the two functions $g^4, h^2: \{-5, -4, \ldots, 4, 5\} \to \{0, 1\}$ generate the following Toeplitz matrices.

$$B(g^5) = \begin{pmatrix} 0&0&0&0&1&0 \\ 0&0&0&0&0&1 \\ 0&0&0&0&0&0 \\ 0&0&0&0&0&0 \\ 1&0&0&0&0&0 \\ 0&1&0&0&0&0 \end{pmatrix} \qquad B(h^2) = \begin{pmatrix} 0&0&1&1&1&0 \\ 0&0&0&1&1&1 \\ 1&0&0&0&1&1 \\ 1&1&0&0&0&1 \\ 1&1&1&0&0&0 \\ 0&1&1&1&0&0 \end{pmatrix}$$

Now, the proof of Theorem 7 can be reduced to proving the optimality of (π^*, π^*) for the independent-QAP(A, B) with $A = R^{(pq)}$ for some $1 \leq p, q \leq n$, and B being a Toeplitz matrix generated by some function g^α or h^β, where $\lfloor \frac{n}{2} \rfloor + 1 \leq \alpha \leq n - 1$ and $1 \leq \beta \leq \lfloor \frac{n}{2} \rfloor$. This follows from an elementary observation:

Observation 11 *Assume that the problems independent-QAP(A_i, B), $i = 1, 2$, are both solved by the pair of permutations (π_0, ψ_0). Then for any two real numbers $k_1, k_2 \geq 0$, the problem independent-QAP$(k_1 A_1 + k_2 A_2, B)$ is also solved by (π_0, ψ_0).*

Of course, an analogous statement holds for the linear combinations of the second matrix B. Concluding, our job is reduced to the proof of the following lemma:

Lemma 12. *If $1 \leq p, q \leq n$ and B is a Toeplitz matrix generated by g^α, $\lfloor \frac{n}{2} \rfloor + 1 \leq \alpha \leq n - 1$, or h^β, $1 \leq \beta \leq \lfloor \frac{n}{2} \rfloor$, then the pair of permutations (π^*, π^*) is an optimal solution of the independent-QAP$(R^{(pq)}, B)$.*

2.2 The Independent-QAP(A, B) for $A = R^{(pq)}$

Due to the structure of the matrices $R^{(pq)}$, the independent-QAP$(R^{(pq)}, B)$, can be formulated as a *selection* problem, which is then more convenient to handle. Indeed, the objective function of the independent-QAP$(R^{(pq)}, B)$ is the sum of $p \cdot q$ entries of matrix B. These entries lie in p rows and q columns. The selected rows are those which are mapped to the last p rows $n - p + 1, \ldots, n$ of A by the permutation ϕ, and the selected columns are those which are mapped to the last q columns $n - q + 1, \ldots, n$ of A by the permutation ψ. Hence, the independent-QAP$(R^{(pq)}, B)$ amounts to selecting p rows and q columns from matrix B such that the total sum of all pq selected entries is minimized. Then, in the language of the selection problem, our goal is to prove the following lemma:

Lemma 13 (The optimal selection lemma). *Let $1 \leq p, q \leq n$. Let B be a Toeplitz matrix generated by g^α, $\lfloor \frac{n}{2} \rfloor + 1 \leq \alpha \leq n - 1$, or h^β, $1 \leq \beta \leq \lfloor \frac{n}{2} \rfloor$. Suppose that p rows and q columns of the matrix B have to be selected such that the total sum of all pq selected entries of B is minimized. Then, it is optimal to select the last p elements of the sequence $1, n, 2, n - 1, 3, \ldots$ as row indices and the last q elements of this sequence as column indices.*

It is easy to see that the following holds:

Claim 14 *Lemmas 12 and 13 are equivalent.*

So we finally have reduced the proof of Theorem 7 to the proof of the optimal selection lemma. The optimal solution claimed by this lemma selects the p rows from $p1$ to $p2$, and the q columns from q_1 to q_2, where $p_1 := \lceil \frac{n-p}{2} \rceil + 1$, $p_2 := n - \lfloor \frac{n-p}{2} \rfloor$, $q_1 := \lceil \frac{n-q}{2} \rceil + 1$ and $q_2 := n - \lfloor \frac{n-q}{2} \rfloor$.

2.3 Proof of the Optimal Selection Lemma - Sketch

Lemma 13 is proved by splitting it in two parts. The cases where the Toeplitz matrix B is generated by one of the functions g^α or one of the functions h^β are considered separately.

Lemma 15. *For any $1 \leq p, q \leq n$ and any $\lfloor \frac{n}{2} \rfloor + 1 \leq \alpha \leq n - 1$, the independent-QAP$(A, B)$ with $A = R^{(pq)}$ and B being the benevolent matrix generated by g^α, is solved to optimality by the pair of permutations (π^*, π^*).*

In other words, selecting p rows and q columns from B according to Lemma 13 minimizes the sum of the selected elements.

Sketch of the proof. We first show that the value of the objective function must be at least $p + q - 2\alpha$, and we then show that our claimed solution achieves the value $\max\{0, p + q - 2\alpha\}$. □

Lemma 16. *For any $1 \leq p, q \leq n$ and any $1 \leq \beta \leq \lfloor \frac{n}{2} \rfloor$, the independent-QAP$(A, B)$ with $A = R^{(pq)}$ and B being the benevolent matrix generated by h^β, is solved to optimality by the pair of permutations (π^*, π^*).*

In other words, selecting p rows and q columns from B according to Lemma 13 minimizes the sum of the selected elements.

Sketch of the proof. First, the sum z^* of the selected entries is computed, when the rows and the columns are selected as specified by Lemma 13. Notice that each column of B has a very simple structure: If the row indices are arranged in a circular sequence $1, 2, \ldots, n, 1, 2, \ldots$, then the 1-entries in this column form an interval of length $\gamma := n - 2\beta + 1$. Moreover, let us denote $N := \max\{0, p + q + \gamma - n\}$ and $k := \min\{p, q, \gamma, \lfloor N/2 \rfloor\}$. With these notations it can be shown that $z^* = N(N - k)$. The key remark is that the p rows selected according to Lemma 13 are exactly the rows with the p smallest sums of entries in the selected columns[3].

Finally, a couple of technical lemmas and auxiliary results are involved in showing that z^* is a lower bound for the sum of the selected entries in an arbitrary selection. □

3 The Turbine Problem

Hydraulic turbine runners as used in electricity generation consist of a cylinder around which a number of blades are welded at regular spacings. Due to imprecisions in the manufacturing process, the weights of these blades differ slightly, and it is desirable to locate the blades around the cylinder in such a way that the distance between the center of mass of the blades and the axis of the cylinder is minimized. This problem was initially introduced by Bolotnikov [2] in 1978, who formulated it as a QAP. The QAP formulation of the Turbine Problem as described below is due to Laporte and Mercure [9].

The places at regular spacings on the cylinder are modeled by the vertices $v_1, \ldots, v_n$ of a regular n-gon on the unit circle in the Euclidean plane, The masses of the n blades are given by the positive reals $0 < m_1 \leq m_2 \leq \cdots \leq m_n$. The goal is to find a permutation $\phi \in S_n$ that minimizes the distance of the gravity center of the whole mass system from the origin. This distance equals the Euclidean norm of the vector

$$\sum_{i=1}^{n} m_{\phi(i)} \begin{pmatrix} \sin \frac{2i\pi}{n} \\ \cos \frac{2i\pi}{n} \end{pmatrix}.$$

---

[3] It is perhaps astonishing that the formula $z^* = N(N - k)$ is completely symmetric in p, q, and γ.

Then, minimizing the Euclidean norm of this vector is equivalent to minimizing the expression

$$\sum_{i=1}^{n}\sum_{j=1}^{n} m_{\phi(i)} m_{\phi(j)} \cos \frac{2(i-j)\pi}{n}. \tag{3}$$

This is a quadratic assignment problem $QAP(A, B)$ with matrix $A = (a_{ij})$ defined by $a_{ij} = m_i \cdot m_j$, and matrix $B = (b_{ij})$ defined by $b_{ij} = \cos \frac{2(i-j)\pi}{n}$. Thus, A is a monotone Anti-Monge matrix and B is a symmetric Toeplitz matrix generated by the function $f(i) = \cos(2\pi i/n)$. The function f is not benevolent, and hence the Turbine Problem does not fall under Theorem 6. Notice that the function $g(i) = -\cos(2\pi i/n)$ which generates the matrix $-B$ is benevolent.

Several heuristics for this problem have been proposed and tested by different authors[4], but no fast (polynomial time) exact solution algorithm has been derived till today. We show that the problem is in fact NP-hard.

Theorem 17. *The Turbine Problem, i. e. the minimization of* (3) *over all permutations* $\phi \in S_n$, *is an NP-hard problem.*

Proof. Omitted in this version of the paper. The proof consists of a reduction from the NP-complete EVEN-ODD PARTITION problem, (cf. Garey and Johnson [7]). □

Since, on the other hand, matrix $-B$ is benevolent, Theorem 6 applies and $QAP(A, -B)$ is easy to solve. This corresponds to the *maximization* of (3). In the context of the Turbine Problem this means that the goal is to get the center of gravity of the mass system as far away from the origin as possible.

Corollary 18. *The maximization version of the Turbine Problem, i. e. the maximization of (3) over all permutations* $\phi \in S_n$, *is solved to optimality by permutation* π^*.

4 Two Further Applications of the Main Theorem

In this section we apply the main theorem to the Traveling Salesman Problem (P2) and to the Data Arrangement Problem (P3), getting short proofs for known results on these problems.

4.1 The TSP on Symmetric Monge Matrices

The *Traveling Salesman Problem* (TSP) consists in finding a shortest closed tour through a set of cities with given distance matrix. One of the first results on polynomially solvable special cases of this NP-hard problem was derived by Supnick in 1957. We derive the result of Supnick as a corollary of Theorem 6. For this we need a definition and two elementary observations. A matrix A is called a *circulant* if there is a function $g: \{0, \ldots, n-1\} \to \mathbb{R}$ such that $a_{ij} = g((i-j) \bmod n)$. In other words, A is a Toeplitz matrix generated by a function f with $f(i) = f(i-n)$ for $i = 1, \ldots, n-1$.

---

[4] See [4] for related references.

Observation 19 *For a sum matrix A and a circulant matrix B, the objective function value of $QAP(A, B)$ is independent of the permutation π. In other words, $QAP(A, B)$ is solved by any permutation $\pi \in S_n$.*

Observation 20 *Assume that $QAP(A_1, B)$ and $QAP(A_2, B)$ are both solved by permutation π_0. Then, for any positive reals $k_1, k_2 \geq 0$, the problem $QAP(k_1 A_1 + k_2 A_2, B)$ is also solved by π_0.*

Proposition 21 (Supnick [12], 1957). *For every instance of the TSP with a symmetric Monge distance matrix $D = (d_{ij})$ an optimal tour is given by*

$$\pi^*(1) \rightarrow \pi^*(2) \rightarrow \cdots \rightarrow \pi^*(n) \rightarrow \pi^*(1). \tag{4}$$

Proof. Let $\Delta = 2 \max_{1 \leq i,j \leq n} \{|d_{ij}|\}$ and define a sum matrix $S = (s_{ij})$ by $s_{ij} = (i+j)\Delta$. Moreover, let us define a symmetric Toeplitz matrix B by its generating benevolent function $f(1) = f(-1) = f(n-1) = f(-n+1) = -1$ and $f(i) = 0$ for $i \notin \{-n+1, -1, 1, n-1\}$. The proof results from the main theorem in three easy steps.

Firstly, $QAP(S - D, B)$ is solved by π^*: $S - D$ is a monotone Anti-Monge matrix and the function f is benevolent. Thus, Theorem 6 applies.

Secondly, $QAP(-S, B)$ is solved by π^*: Since $-S$ is a sum matrix and B is a circulant, Observation 19 applies.

Finally, by adding $S-D$ and S we get, by Observation 20, that $QAP(-D, B)$, and equivalently $QAP(D, -B)$, is solved by π^*.

Now, matrix $-B$ is the adjacency matrix of an undirected cycle on n vertices. Hence, $QAP(D, -B)$ exactly corresponds to the TSP with distance matrix D and the optimal solution π^* of $QAP(D, -B)$ exactly corresponds to the optimal tour (4) for the TSP. $\qquad\square$

4.2 Data Arrangement in a Linear Storage Medium

Consider a set of n records $r_1, \ldots, r_n$ to be referenced repetitively. The reference is to record r_i with probability p_i and different references are independent. W.l.o.g. the records are numbered such that $p_1 \leq p_2 \leq \cdots \leq p_n$. The goal is to place these records into a linear array of storage cells, like a magnetic tape, such that the expected distance between consecutively referenced records is minimized, i. e. one wishes to minimize

$$\sum_{i=1}^{n} \sum_{j=1}^{n} p_{\phi(i)} p_{\phi(j)} d_{ij} \tag{5}$$

where d_{ij} is the distance between the records placed at storage cells i and j respectively. In the case that the distance d_{ij} is given as $d_{ij} = f(i - j)$, we obtain a straightforward corollary of our main theorem:

Corollary 22. *If $d_{ij} = f(i - j)$ where f is a benevolent function, then the data arrangement problem is solved by the permutation π^*.*

Corollary 22 generalizes and unifies a number of results on polynomially solvable cases of the Data Arrangement Problem obtained in the late 1960s and early 1970s (see [4] for related references).

5 Periodic Toeplitz matrices

Let us return to well solvable special cases of the AMT-QAP. First, we give a generalization of our main Theorem 6 to matrices B with a certain periodic structure. Then, we show that the AMT-QAP with a Toeplitz matrix generated by a periodic function is in general NP-hard.

5.1 Toeplitz Matrices Generated by k-Benevolent Functions

In this section we extend the benevolent functions in a periodic way, and we show that the resulting Toeplitz matrices have the constant permutation property.

Definition 23. Let $k \geq 1$, $n' \geq 2$ and $n = kn'$. A function $f: \{-n+1, \ldots, n-1\} \to \mathbb{R}$ is called k-*benevolent* if it fulfills the following four properties.

(i) $f(i) \leq f(i+1)$, for $0 \leq i \leq \lfloor \frac{n'}{2} \rfloor - 1$.
(ii) $f(i) = f(n' - i)$, for $0 \leq i \leq \lceil \frac{n'}{2} \rceil - 1$.
(iii) $f(i) = f(i + jn')$, for $0 \leq i \leq n' - 1$, $1 \leq j \leq k - 1$.
(iv) $f(-i) = f(i)$, for $0 \leq i \leq n - 1$.

A Toeplitz matrix generated by a k-benevolent function is called a k-benevolent matrix.

Notice that Properties (i), (ii), and (iv) are similar to the properties of benevolent functions for the range $\{-n'+1, \ldots, n'-1\}$. Notice, moreover, that there exist benevolent matrices which are not 1-benevolent, whereas any 1-benevolent matrix is also benevolent.

Example 3. Let $n = 10$, $k = 2$, $n' = 5$. Define a function $f: \{-9, -8, \ldots, 0, 1, \ldots, 9\} \to \mathbb{R}$, fulfilling properties (i)–(iv), by the following equalities: $f(0) = 1$, $f(1) = 2$, $f(2) = 3$. Figure 2 represents its graph. The Toeplitz matrix B

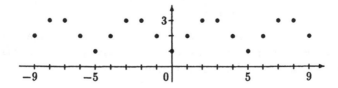

Fig. 2. *The graph of the function f in Example 3.*

generated by this function consists of $k \times k = 4$ identical submatrices B' of size $n' \times n' = 5 \times 5$ each and looks as follows.

$$B = \left(\begin{array}{c|c} B' & B' \\ \hline B' & B' \end{array} \right) \qquad \text{where} \quad B' = \begin{pmatrix} 1\,2\,3\,3\,2 \\ 2\,1\,2\,3\,3 \\ 3\,2\,1\,2\,3 \\ 3\,3\,2\,1\,2 \\ 2\,3\,3\,2\,1 \end{pmatrix}$$

Clearly, two columns whose indices are congruent modulo $n' = 5$ are identical, and the same holds for the rows.

In general, a k-benevolent Toeplitz matrix B consists of $k \times k$ identical submatrices of size $n' \times n'$. These are the submatrices B_{uv}, $1 \leq u, v \leq k$, obtained by selecting in B n' rows with indices in N_u and n' columns with indices in N_v, where

$$N_t := \{(t-1)n' + 1, \ldots, tn'\}, \quad \text{for } 1 \leq t \leq k.$$

It is easy to see that the matrices B_{uv}, $1 \leq u, v \leq k$, are identical and benevolent. According to the main theorem, each of these matrices has the constant permutation property with respect to monotone Anti-Monge matrices and π^* is an optimal solution of the corresponding QAP. Recall that $\pi^* = \langle 1, 3, 5, 7, 9, \ldots, 8, 6, 4, 2 \rangle$, and regard π^* as a permutation of $\{1, \ldots, n'\}$ throughout this section. We show that the k-benevolent matrices have the constant permutation property, too. The corresponding optimal solution of the AMT-QAP is denoted by $\pi^{(k)}$. In terms of $\pi^* \in S_{n'}$, the permutation $\pi^{(k)} \in S_n$ (with $n = kn'$) is given as follows

$$\pi^{(k)}((u-1)n' + i) = k\pi^*(i) - (u-1), \quad \text{for } 1 \leq u \leq k, \ 1 \leq i \leq n'. \qquad (6)$$

For example, for $k = 4$, $n' = 5$, $n = kn' = 20$,

$$\pi^{(4)} = \langle \underbrace{4, 12, 20, 16, 8}, \ \underbrace{3, 11, 19, 15, 7}, \ \underbrace{2, 10, 18, 14, 6}, \ \underbrace{1, 9, 17, 13, 5} \rangle.$$

The sequence $\langle \pi^{(k)}(1), \pi^{(k)}(2), \ldots \rangle$ is naturally divided into $k = 4$ blocks with $n' = 5$ elements each. The first block, corresponding to $u = 1$ in (6), is obtained from $\pi^* = \langle 1, 3, 5, 4, 2 \rangle$ by multiplying every element by $k = 4$. Each successive block is obtained from the previous one by subtracting one from each entry.

Theorem 24. *The permutation $\pi^{(k)}$ solves QAP(A, B) when A is a monotone Anti-Monge matrix and B is a k-benevolent matrix.*

As previously, we can show the constant permutation property even for the independent-QAP. Again, w.l.o.g. we can assume that matrix A has non-negative entries and that matrix B has 0-entries on the diagonal. Theorem 24 follows then from the following theorem.

Theorem 25. *The pair $(\pi^{(k)}, \pi^{(k)})$ solves independent-QAP(A, B) when A is a non-negative monotone Anti-Monge matrix and B is a k-benevolent matrix with 0-entries on the diagonal.*

As in Section 2, we restrict our attention to the matrices $A = R^{(pq)}$ which are the extreme rays of the cone of non-negative monotone Anti-Monge matrices. Thus, our final goal in this subsection is to prove the following lemma:

Lemma 26. *For any $1 \leq p, q \leq n$, the independent-QAP$(R^{(pq)}, B)$, where B is a k-benevolent matrix with 0-entries on the diagonal, is solved to optimality by the pair of permutations $(\pi^{(k)}, \pi^{(k)})$.*

Sketch of the proof. This problem can be seen as a selecting problem: Select p rows and q columns from matrix B such that the total sum of all pq selected entries is minimized. Due to the special structure of matrix B we may impose the following structure on the selected set of rows and columns, respectively.

Claim 27 *There is an optimal selection of p rows and q columns, where the number p_u of selected rows in each block N_u is either $\lfloor p/k \rfloor$ or $\lceil p/k \rceil$, and the number q_u of selected columns in each block N_u is either $\lfloor q/k \rfloor$ or $\lceil q/k \rceil$.*

Let us denote $R_p := p \bmod k$. Then R_p of the values p_u must be equal to $\lceil p/k \rceil$, and the remaining $k - R_p$ of the values p_u are equal to $\lfloor p/k \rfloor$. Similarly, $R_q := q \bmod k$ of the values q_v are equal to $\lceil q/k \rceil$, and $k - R_q$ of them are equal to $\lfloor q/k \rfloor$. The entries of B which lie in the selected rows and columns can be summed separately for each block B_{uv}, $(1 \leq u, v \leq k)$. All blocks B_{uv} are identical to a certain $n' \times n'$ benevolent Toeplitz matrix B'. By Lemma 13 we know how to optimally select a given number p' of rows and a given number q' of columns from B' if we want to minimize the overall sum of the selected entries. It remains then only to check that the permutation $\pi^{(k)}$ indeed selects the optimal set of p_u rows out of each block of rows N_u and the optimal set of q_v columns out of each block of columns N_v, as specified by Lemma 13. $\qquad\square$

5.2 Toeplitz Matrices Generated by General Periodic Functions

The simplest non-trivial periodic functions f for generating a Toeplitz matrix B have period $n' = 2$ and thus only two values: $f(0) = f(i)$ for all even i and $f(1) = f(i)$ for all odd i. These two values form a chess-board pattern in the matrix B. The case $f(0) \leq f(1)$ leads to a k-benevolent function and hence to the constant permutation property. It can be shown that the other case, $f(0) > f(1)$, leads to an NP-hard problem. Consequently, the AMT-QAP with a periodic Toeplitz matrix is in general NP-hard. W.l.o.g. we can assume that $f(0) = 1$ and $f(1) = -1$. In this case $B = (b_{ij})$ is given by $b_{ij} = (-1)^{i+j}$.

Theorem 28. *The QAP is NP-hard even if A is a $(2k) \times (2k)$ monotone Anti-Monge matrix and $B = (b_{ij})$ is a $(2k) \times (2k)$ symmetric Toeplitz matrix with $b_{ij} = (-1)^{i+j}$.*

Proof. The proof, omitted in this version of the paper, is done by a reduction from the NP-complete EquiPartition problem (cf. Garey and Johnson [7]). $\qquad\square$

6 Conclusions

We investigated the AMT-QAP, a restricted version of the quadratic assignment problem $QAP(A, B)$ where A is a monotone Anti-Monge matrix and B is a symmetric Toeplitz matrix. We have shown that the TSP on symmetric Monge matrices, the Turbine Problem and a Data Arrangement Problem are instances of the AMT-QAP. By proving that the Turbine Problem is NP-hard, we have shown that even this apparently simple version of the QAP is NP-hard. We conjecture that the Turbine Problem is even *strongly* NP-hard. In particular, we propose the following one-dimensional version of the Turbine Problem.

Let $a_1, a_2, \ldots$ be the fixed sequence $1, -1, 2, -2, 3, -3, \ldots$.

Problem: WEIGHTED PARTITION
Instance: n positive integers $x_1, x_2, \ldots, x_n$.
Question: Is there a permutation $\phi \in S_n$ such that $\sum_{i=1}^{n} a_{\phi(i)} x_i = 0$?

The numbers a_i are the positions to which the masses x_i have to be assigned. Using some ideas from the NP-hardness proof for the Turbine Problem, we can show that this problem is also NP-hard, but we do not know whether it is NP-hard in the strong sense. For the sequence $a_i = (-1)^i$ we get just the EQUIPARTITION problem (see [7]), which can be solved in pseudo-polynomial time.

A related problem which arose in the context of loading cargo on a truck or on an airplane, was considered by Amiouny et al. [1]. This problem consists of packing boxes of given lengths and weights inside an interval along a one-dimensional axis, so that the center of gravity gets as close as possible to a given target point. For the special case when all boxes have the same length and the target point is in the middle of the interval, we get our WEIGHTED PARTITION problem.

Our main result is the identification of an easy special case of the AMT-QAP: If A is a *monotone Anti-Monge* matrix and B is a *benevolent* matrix, then a fixed permutation π^* is an optimal solution of $QAP(A, B)$. Thus, in this case $QAP(A, B)$ is trivial in the sense that an optimal solution of an instance of the problem can be given independently on its numerical data. More generally, we introduced matrices with the *constant permutation property*: A matrix B has the constant permutation property with respect to a class of matrices $\mathcal{A}$, if there exists a permutation π^B that solves $QAP(A, B)$ for all matrices $A \in \mathcal{A}$. *Deriving a characterization of all Toeplitz matrices that have this property* with respect to monotone Anti-Monge matrices is an open problem whose complete solution is currently out of sight. As a first step toward the solution of this problem, we have identified two classes of Toeplitz matrices which have the constant permutation property with respect to monotone Anti-Monge matrices. These are the *benevolent* and *k-benevolent* matrices. Another class of Toeplitz matrices which have this property are the Toeplitz matrices with bandwidth 1, see [4]. As for Toeplitz matrices of larger bandwidth, it can be shown that they do not have the constant permutation property with respect to monotone Anti-Monge matrices. However, the computational complexity of the AMT-QAP with a Toeplitz matrix of limited bandwidth remains an open question.

As a "negative" result, it is shown that the Anti-Monge–Toeplitz QAP remains NP-hard even when considering only Toeplitz matrices generated by a function of period two taking only ±1 values.

Thus, there is a "thin" borderline between "easy" and "hard" cases of this restricted version of the QAP, as well as between Toeplitz matrices with and without the constant permutation property with respect to monotone Anti-Monge matrices. It is an open question whether these two borderlines coincide.

Acknowledgement. First, we thank Bettina Klinz for several discussions which motivated our interest in k-benevolent matrices. Moreover, we thank N. Metelski and M. Rubinstein for providing us with references to the Russian literature. Finally, we would like to thank an anonymous referee for helpful suggestions on the first version of this paper.

References

1. Amiouny, S.V., Bartholdi, J.J. III, Vande Vate, J.H., Zhang, J.: Balanced loading, Operations Research **40** (1992) 238–246
2. Bolotnikov, A.A.: On the best balance of the disk with masses on its periphery, (in Russian), Problemi Mashinostroenia **6** (1978) 68–74
3. Burkard, R.E.: Locations with spatial interactions: The quadratic assignment problem, Discrete Location Theory (Mirchandani, P.B., Francis, R.L., eds.), John Wiley, New York (1990) 387–437
4. Burkard, R.E., Çela, E., Rote, G., Woeginger, G.J.: The quadratic assignment problem with an Anti-Monge and a Toeplitz matrix: easy and hard cases, Technical report SFB-34 (1995) Institut für Mathematik B, Technische Universität Graz file://ftp.tu-graz.ac.at/pub/papers/math/sfb34.ps.gz
5. Burkov, V.N., Rubinstein, M.I., Sokolov, V.B.: Some problems in optimal allocation of large-volume memories, (in Russian), Avtomatika i Telemekhanika **9** (1969) 83–91
6. Clausen, J., Perregård, M.: Solving Large Quadratic Assignment Problems in Parallel, Computational Optimization and Applications (1994) (to appear)
7. Garey, M.R., Johnson, D.S.: Computers and Intractability: A Guide to the Theory of NP-Completeness, Freeman, San Francisco (1979)
8. Koopmans, T.C., Beckmann, M.J.: Assignment problems and the location of economic activities, Econometrica **25** (1957) 53–76
9. Laporte, G., Mercure, H.: Balancing hydraulic turbine runners: a quadratic assignment problem, European J. Oper. Res. **35** (1988) 378–382
10. Lawler, E.L.: The quadratic assignment problem: a brief review, Combinatorial Programming: Methods and Applications, (Roy, B., ed.), Dordrecht, Holland (1975) 351–360
11. Pardalos, P., Rendl, F., Wolkowicz, H.: The Quadratic Assignment Problem: A Survey and Recent Developments, Proceedings of the DIMACS Workshop on Quadratic Assignment Problems (Pardalos, P., Wolkowicz, H., eds.), DIMACS Series in Discrete Mathematics and Theoretical Computer Science **16** (1994) 1–42
12. Supnick, F.: Extreme Hamiltonian lines, Annals of Math. **66** (1957) 179–201

On Optimizing Multiplications of Sparse Matrices

Edith Cohen[1]

AT&T Bell Laboratories, Murray Hill, NJ 07974

Abstract. We consider the problem of predicting the nonzero structure of a product of two or more matrices. Prior knowledge of the nonzero structure can be applied to optimize memory allocation and to determine the optimal multiplication order for a chain product of sparse matrices. We adapt a recent algorithm by the author and show that the essence of the nonzero structure and hence, a near-optimal order of multiplications, can be determined in near-linear time in the number of nonzero entries, which is much smaller than the time required for the multiplications. An experimental evaluation of the algorithm demonstrates that it is practical for matrices of order 10^3 with 10^4 nonzeros (or larger). A relatively small pre-computation results in a large time saved in the computation-intensive multiplication.

1 Introduction

Large sparse matrices occur in the solutions of many practical problems. Many matrix operations that are computation intensive or infeasible on the full representation of the matrix become feasible when sparsity is exploited. Indeed, extensive literatures exists on more efficient storage schemes and algorithms for matrix operations on sparse matrices. (see, e.g.,[7] [8] [4]). Since sparse matrices are easier to manipulate, much effort was made on minimizing *fill-in* (nonzeros introduced during some computation) when there is freedom in the choice of operations. (for example, minimizing fill-in in Gaussian eliminations.) It is also common to represent matrices as products of sparser matrices. To exploit sparsity, it is of value to be able to determine the nonzero structure of an output matrix of a matrix operation in advance (see Gilbert and Ng [5]). Many matrix computations have an initial phase that predicts the nonzero structure of the output matrix, followed by the actual numerical computation. Prior knowledge of the nonzero structure is used to improve memory allocation, data structure setup, and running time.

We consider the problem of determining the nonzero structure of products of sparse matrices. The intention is to predict the structure using a low cost procedure, before performing the computation-intensive matrix multiplication. The nonzero structure can serve as a guide for efficient memory allocation for the output matrix when performing the actual multiplication. The nonzero structure can also be used to determine the *optimal order* of multiplications (that is, the order that minimizes the total number of operations) when computing a chain product of three or more matrices. Consider the product $A_1 \cdots A_k$ of $k \geq 3$ sparse matrices. The multiplication can be performed in many orders. The number of operations can vary significantly for different orders. The nonzero structure is used to determine the least-expensive order in which to perform the multiplications. For dense matrix multiplications, the optimal order is determined solely by the dimensions of the matrices, and it is well known that it can be computed efficiently by dynamic programming [9]. However, when using multiplications schemes that utilize sparsity, the number of operations varies with the order even for square matrices.

Throughout the paper we assume *no cancellation*. That is, the inner product of two vectors with overlapping nonzero entries never cancels to zero. This is a very common assumption in the mathematical programming literature. The justification is that cancellations are unlikely when computing over real numbers. Furthermore, even when (due to a singular structure) cancellation occurs, rounding errors may prevent the resulting computed value from being zero.

Even with the no cancellation assumption it is costly to determine the full nonzero structure (exact locations of all nonzero entries) of a matrix product. This operation amounts to performing a boolean matrix multiplication (transitive closure computation). Known boolean matrix multiplication algorithms require the same number of operations to predict the structure as to perform the actual multiplication. (To be precise, in practice each "operation" in a real matrix multiplication is a floating point multiply-add. These are much more expensive than boolean and-or operations and hence it is still much cheaper to perform a boolean multiplication. However, the cost of boolean sparse matrix multiplication still grows quadratically with the average number of nonzeros in each row.)

We consider a relaxed goal of determining the number of nonzeros in each row and column of the product. This partial knowledge is as valuable as the full structure for optimizing memory allocation, data structure setup, and determining the optimal order of computing chain products. We introduce an algorithm where each operation is a simple compare-store operation on small integers and also the number of operations is linear in the number of nonzeros.

The obvious method of quickly estimating row and column densities in a matrix product assumes random distributions of nonzero entries. This assumption, however, often does not hold in practice. The method we propose, based on an adaptation of a recent reachability-set size estimation algorithm [1], obtains accurate estimates for products of *arbitrary* matrices. We also provide better estimators and analysis. We estimate the densities of all rows and columns of the product matrix in time linear in the number of nonzero entries of the input matrices. We introduce an algorithm that for chain products of $k > 3$ matrices, proposes a multiplication-order that (nearly) minimizes the number of operations. The algorithm combines the estimation technique with dynamic programming and runs is $O(kZ + k^3N)$, where Z is the total number of nonzeros in the input matrices and N is the maximum dimension (number of rows or columns) of an input matrix.

We experimentally tested our method on products of 3 randomly-generated large sparse matrices. We introduced correlations between nonzero locations that model patterns arising in "real" data. We also experimented with data obtained in an Information Retrieval application. For these matrices, there was a significant difference in the costs of the two multiplication orders. Nevertheless, methods that assume random nonzero distribution are not powerful enough to determine the less-costly order. We concluded that even for matrices of dimension 10^3 with 2×10^4 nonzeros it is cost-effective to use our algorithm prior to performing the multiplication. The number of operations performed by the estimation routine was small in comparison to the gain obtained by selecting the less-costly order. The potential savings are even larger for longer chain products and larger matrices.

In Section 2 we provide background material and review graph representation of matrices and sparse matrix multiplication. In Section 3 we present estimation algorithms for number of nonzeros in rows and columns of a product. In Section 4 we present algorithms for determining a near-optimal order for computing a chain product. Section 5 contains experimental performance study. Section 6 contains a summary.

2 Preliminaries

Extensive literature exists on storage schemes and implementing matrix operations on sparse matrices (see, e.g., [8, 11]). We review some relevant material. We denote by $[A]_{i,j}$ the entry at the ith row and jth column of a matrix A, by $[A]_{i.}$ the ith row, by $[A]_{.j}$ the jth column, and by $|A|$ the number of nonzero entries in the matrix A. We use the term *size* of a column, a row, or a matrix to refer to the number of nonzero entries.

2.1 Graph representation of matrices and matrix-products

We represent a matrix A of dimension $n_1 \times n_2$ by a bipartite graph G_A, with n_1 nodes $\{v_1, \ldots, v_{n_1}\}$ at the lower level and n_2 nodes $\{v'_1, \ldots, v'_{n_2}\}$ at the upper level. For $1 \leq i \leq n_1$ and $1 \leq j \leq n_2$, there is a directed edge from v_i to v'_j if and only if $[A]_{ij} \neq 0$. It is easy to see that the outdegree of v_i (resp., indegree of v'_j) is the size of the ith row (resp., jth column) of A.

We represent a product $A = A_1 \cdots A_k$ as a layered graph with $(k+1)$ layers as follows. Suppose A_i is of dimensions $n_i \times n_{i+1}$ $(1 \leq i \leq k)$. For $1 \leq i \leq k$, the ith layer has n_i nodes and the edges between the ith and the $(i+1)$st layers are as is the bipartite graph G_{A_i}.

For $1 \leq r < r' \leq k$, consider the subproduct $B = A_r A_{r+1} \cdots A_{r'}$. B is of dimension $n_r \times n_{r'+1}$. The following proposition is immediate.

Proposition 1. *Assuming no cancellation*

1. *The number of nonzero entries in the ith row of B equals the number of nodes in layer $r' + 1$ that the ith node in layer r reaches via directed paths.*

2. *The number of nonzero entries in the ith column of B is equal to the number of nodes in layer r that can reach the ith node in layer $r' + 1$.*

2.2 Storage schemes of sparse matrices

A good storage scheme that exploits sparsity should be both compact (memory usage of the order of the number of nonzero entries) and facilitate an efficient computation of the applicable matrix operations. The simplest representation is an array or a list of all nonzero entries, where for each entry we store its contents and location. Common representations that allow for efficient access to the content of specified rows or columns are the *sparse row-wise* and *sparse column-wise* formats. The sparse row-wise representation consists of an array of indexes to the contents of the rows of the matrix. Each row is stored as a linked list of the nonzero entries in the row. For each nonzero in the row the list contains a record of the content and the column number. A column-wise format is similar to row-wise when the roles of rows and columns are reversed. For efficient sparse multiplication of two matrices AB it suffices that either the matrix A is represented in column-wise format or B is in row-wise format. Our algorithm for estimating column and row sizes of products of arbitrary matrices can be implemented efficiently even when each matrix in the product is represented as a large (unsorted) array of all nonzero locations.

2.3 Sparse matrix multiplication

Dense matrix multiplication algorithms perform a cubic number of operations in the dimension of the matrices (some [10, 2] are faster). When the matrices have a relatively small number of nonzero entries, the sparsity can be exploited to significantly reduce the amount of storage and running time. Consider the product of the matrices $A^{n_1 \times n}$ and $B^{n \times n_2}$. Every entry $[AB]_{i_1, i_2}$ in the product is expressed as the sum of products

$$[AB]_{i_1, i_2} = \sum_i [A]_{i_1, i} [B]_{i, i_2} .$$

Sparse multiplication considers the product of $[A]_{i_1, i}$ and $[B]_{i, i_2}$ only if both are nonzero. The total number of such pairs and a good measure for the number of operations performed by a sparse multiplication of (appropriately-represented) matrices is

$$\sum_{i=1}^{n} |[A]_{.i}||[B]_{i.}| = \sum_{i=1}^{n} |\{j|[A]_{ji} \neq 0\}||\{j|[B]_{ij} \neq 0\}| . \tag{1}$$

The sum over the columns of A (rows of B), of the products of the size of the ith column of A and the size of the ith row of B. Hence, the cost of multiplying $A = A_1 \cdots A_i$ and $B = B_1 \cdots B_j$ can be deduced from the number of nonzero entries in each column of A and each row of B. To see this recall that the product of A and B can be expressed as the sum of n outer products: Sum,

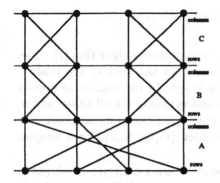

Fig. 1. The layered graph of the product ABC

for $1 \leq i \leq n$, the outer product of the ith column of A and the ith row of B (see, e.g., [6]). For our purposes we use Equation 1 as a measure of the number of operations performed by the sparse-multiplication algorithm. Note that this is an optimistic measure of the true cost. We comment that a significant part of the cost of multiplying large matrices in practice is the transfer of data between slow and fast memory. In light of this, our cost measure of number of operations, although common, is rather simplistic. To justify it, we note that the same memory management procedures applied in the multiplication itself can also be carried out in our structure-prediction algorithms. Hence, we expect the number of operations performed to be a valid measure for comparing the costs of determining the nonzero structure (and the optimal order of multiplications) and of performing the multiplication (in the optimal or arbitrary order).

2.4 Example

Consider the following two nonzero structures (x denotes a nonzero entry).

$$T_1 = \begin{pmatrix} x & x & 0 & 0 \\ x & x & 0 & 0 \\ 0 & 0 & x & x \\ 0 & 0 & x & x \end{pmatrix} \qquad T_2 = \begin{pmatrix} x & 0 & x & 0 \\ 0 & x & 0 & x \\ 0 & x & x & 0 \\ x & 0 & 0 & x \end{pmatrix}$$

Let B and C be matrices with nonzero structure in the form of T_1. Let A have the structure of T_2. The goal is to compute the product ABC. See Figure 1 for the graph representation of the product. The product ABC is evaluated using two matrix multiplications. There are two possible orders of evaluation, either $(AB)C$ (where AB is the first multiplication) or $A(BC)$ (where BC is the first multiplication). All three matrices have the same column and row sizes (size 2). However, the evaluation order $A(BC)$ requires 32 scalar multiplication operations and $(AB)C$ requires 48 operations. Note that knowledge of the row sizes of the product BC and the column sizes of AB is needed in order to determine the costs of the second multiplications.

3 Estimating column sizes in a product matrix

The algorithm is based on a reachability-set size estimation algorithm [1]. The size estimation algorithm in [1] produces estimates on the sizes of the reachability sets for all nodes in an arbitrary directed graph. In the context of matrix products, we designate a subset of the nodes and estimate for all other nodes, the number of ancestors or descendents they have from the designated set. We utilize the layered structure of the matrix product graphs to obtain a simpler specialized algorithm.

Consider a matrix product $A_1 \cdots A_k$ represented as a $k+1$ layer graph (as in Subsection 2.1). Let A_i be of dimension $n_i \times n_{i+1}$ $(1 \le i \le k)$. For $1 \le i \le k+1$, denote by V_i the nodes in layer i. We have $|V_i| = n_i$. We use an adapted single application of the estimation algorithm of [1] to produce for each $1 < i \le k$ and each column of the product matrix $A'_i = A_1 \cdots A_i$, an estimate on the size of the column. Each column of the product A'_i corresponds to a node $v \in V_{i+1}$. The size of the column equals the number of V_1-nodes that can reach v. The size of each row in each of the subproducts $A_i \cdots A_k$ $(1 \le i \le k)$ can be estimated using a symmetric procedure.

The estimation algorithm gets as a parameter the *number of rounds r*. Both estimate quality and running time increase with r, and hence, it should be set according to the desired performance. We state a column-size estimation algorithm. (The row-size estimation algorithm performs essentially a symmetric computation on the graph with the layers in reverse order.) Each round amounts to first assigning independent random numbers (*keys*) to all V_1 nodes (the first layer). Following that, for all other nodes, the least key of a V_1 node that reaches them is determined. This is performed in time linear in the size of the layered graph, by traversing the layers sequentially and propagating the smallest keys. The intuition, is that for every node $v \in V_i$ $(i > 1)$, the distribution and the expected value of the key of the minimum-ranked V_1-node that reaches v depends solely on the number of V_1 nodes that reach v. Each round provides a single sample for every such node v. The estimation algorithm applies a "reversed" strategy to deduce the size from the least-key samples. The accuracy and confidence of the estimate increase with the number of samples (rounds). Note that the rounds may be performed independently on a parallel platform. The algorithm assumes in Step 2 that the matrices are represented in a column-wise format. That is, every node $v \in V_i$ has an associated list of its neighbors from layer V_{i-1}. A alternate statement for Step 2 that is suitable for matrices represented as an arbitrary list of their nonzero entries is also provided.

3.1 Estimation algorithm for column-sizes

1. Assign for each node in V_1 an r-vector that consists of random samples from the Exponential distribution with parameter $\lambda = 1$ (another value of λ or the Uniform distribution on $[0, 1]$ could be used instead).

2. For layers $i = 2, \ldots, k+1$ do as follows:
   - For each node $v \in V_i$:
     Assign to v the r-vector that equals the coordinate-wise minima of the r-vectors of all the V_{i-1} neighbors of v.
3. At this point, every node in the graph has an r-vector associated with it. To estimate the number of nonzeros of a column in A'_j, apply an *estimator* to the r-vector of the corresponding node from layer $j + 1$. For a vector $(a_1, \ldots, a_r)$ we apply the estimator

$$\frac{r-1}{\sum_{i=1}^{r} a_i}$$

($r - 1$ divided by the sum of the coordinates. If $\lambda \neq 1$, divide by λ.)

When the matrices are provided as unsorted lists of nonzero locations (edges) we replace Step 2 by the following computation:
For layers $i = 2, \ldots, k+1$ do as follows:
- For each node $v \in V_i$ initialize all entries in the r-vector of v to $+\infty$:
- For each edge (v, u) where $v \in V_{i-1}$ and $u \in V_i$: (nonzero entry in A_i)
  Assign to u the r-vector that equals the coordinate-wise minimum of u's current vector and the r-vector of v.

3.2 Time bounds

Obtaining the r-vectors requires $r n_1$ random samples and $r \sum_{i=1}^{k} |A_i|$ compare and store operations. Computing each estimate amounts to $r - 1$ additions and a division. Hence, estimating the column sizes of all submatrices A'_i requires $\sum_{i=2}^{k+1} n_i$ divisions and $(r-1) \sum_{i=2}^{k+1} n_i$ additions. The total number of operations is dominated by the term $r \sum_{i=1}^{k} |A_i|$.

3.3 Accuracy and time tradeoffs

We analyze the quality of each estimate $\hat{z}$ on the number of nonzeros z (in some row or column). Let the r.v. $M^{(z)}$ be the minimum key for a (product) column (or row) of size z. $M^{(z)}$ distributes like the minimum of z independent Exponential r.v.'s. The sum $S^{(r,z)}$ of r independent r.v.'s drawn from $M^{(z)}$ has density and distribution functions (see, e.g. [3])

$$g_{r,z}(x) = z\frac{(zx)^{r-1}}{(r-1)!}e^{-zx} \ , \ G_{r,z}(x) = 1 - e^{-zx}\left(1 + \sum_{i=1}^{r-1}\frac{(zx)^i}{i!}\right) \ , x \geq 0 \ .$$

We estimate z with the estimator $\hat{z} = (r-1)/S^{(r,z)}$. Consider the r.v. $y = z/\hat{z}$. The density and distribution functions of y are independent of z and are given by

$$f_r(y) = \frac{(r-1)^r}{(r-1)!}y^{r-1}e^{-(r-1)y} \ , \ F_r(y) = 1 - e^{-(r-1)y}\left(1 + \sum_{i=1}^{r-1}\frac{((r-1)y)^i}{i!}\right) \ , y \geq 0 \ .$$

We obtain the following expressions for the relative error

$$\text{Prob}\{\hat{z} \geq z(1+\epsilon)\} = \text{Prob}\{y \leq 1/(1+\epsilon)\} = 1 - e^{-(r-1)/(1+\epsilon)}\left(1 + \sum_{i=1}^{r-1} \frac{((r-1)/(1+\epsilon))^i}{i!}\right)$$

$$\text{Prob}\{\hat{z} \leq z(1-\epsilon)\} = \text{Prob}\{y \geq 1/(1-\epsilon))\} = e^{-(r-1)/(1-\epsilon)}\left(1 + \sum_{i=1}^{r-1} \frac{((r-1)/(1-\epsilon))^i}{i!}\right)$$

Further analysis establishes the asymptotic behavior

$$\text{Prob}\left\{\frac{|z - \hat{z}|}{z} \geq \epsilon\right\} = \exp(-\Omega(\epsilon^2 r)) \ .$$

(the probability that an estimate is ϵ fraction off the true value decreases exponentially with $r\epsilon^2$.) The bias of the estimator $\hat{z}$ is $E(1/y)-1 = \int_0^\infty \frac{f_r(y)}{y}dy-1 = 0$ (the estimator is unbiased). The variance is $E((1-1/y)^2) = \frac{1}{r-2}$. The expected relative error is $E(|1 - 1/y|) = \frac{2(r-1)^{(r-2)}}{(r-2)!e^{r-1}} \approx \sqrt{2/(\pi(r-2))}$. (using Stirling's formula.)

4 Near-optimal order for computing chain-products

We present algorithms that utilize column and row-size estimates to determine a near-optimal order of multiplications for evaluating a *chain product* (product of 3 or more matrices). The cost of multiplying two matrices A, B can be computed from the sizes C_A of the columns of A and R_B of the rows of B and is given in Equation 1. These sizes are likely to be readily available with the representation of the matrices. Consider computing a product ABC of three matrices. There are two possible orders to compute the product. The first multiplication is either AB or BC. The second is either $(AB)C$ or $A(BC)$ accordingly. The costs of the two possible first multiplications are $C_A^T R_B$ and $C_B^T R_C$ (obtained using Equation 1). To compute the costs of the two possible second multiplications, we use *estimates* $\hat{C}_{AB}$ on the column sizes of the matrix AB and $\hat{R}_{AB}$ on the row sizes of the matrix BC. Our estimates for the cost of the second multiplication, $\hat{C}_{AB}^T R_C$ and $C_A^T \hat{R}_{BC}$, are obtained by applying Equation 1 to the *estimated* sizes. The algorithm of Section 3 produces estimates for the column sizes of (AB) using $r(|A| + |B|)$ operations and for the row sizes of (BC) using $r(|B| + |C|)$ operations. An optimal order of multiplications (relative to the estimates) is obtained by comparing the two quantities

$$C_A^T R_B + \hat{C}_{AB}^T R_C \text{ and } C_B^T R_C + C_A^T \hat{R}_{BC} \ .$$

4.1 Algorithm for longer products

We present an algorithm that computes a near-optimal multiplication order for
a chain product $A_1 \cdots A_k$ of $k > 3$ matrices. For $1 \leq i \leq k$, A_i has dimensions
$n_i \times n_{i+1}$. We apply a two-stage algorithm. The first stage consists of using
several applications of the algorithm of Section 3 to compute estimated column
and row sizes for all the subproducts. The second stage is a dynamic programming algorithm that utilizes these estimates and produces an optimal order of
multiplication with respect to these estimates. We elaborate on the two stages.

1. For $j > i > 1$ we produce estimates $C_{i,j}$ for the column sizes of each subproduct
   $A_i \cdots A_j$. For $i < j < k$ we produce estimates $R_{i,j}$ for the row sizes of each
   subproduct. These estimates are used in the dynamic programming stage and are
   obtained as follows:
   - For $i = 1, \ldots, k - 2$:
     Apply the column-size estimation algorithm (Section 3)
     on the product $A_i \cdots A_{k-1}$.
     This provides us with estimates $C_{i,i+1}, \ldots, C_{i,k-1}$ of the column sizes of the
     subproducts $A_i \cdots A_j$ (for $i < j \leq k - 1$).
   - For $i = 3, \ldots, k$:
     Apply the row-size estimation algorithm (Section 3)
     on the product $A_2 \cdots A_i$.
     This provides us with estimates $R_{2,i}, \ldots, R_{i-i,i}$ of the row sizes of the
     subproducts $A_j \cdots A_i$ (for $2 \leq j < i$).
   For $1 \leq i \leq k$, let $C_{i,i}$ (resp., $R_{i,i}$) be the vector whose entries are the column
   (resp., row) sizes of the matrix A_i.

2. We use dynamic programming to compute an optimal order of multiplications with respect to the *estimated* row and column sizes of the subproducts.
   The dynamic programming computation is performed in phases, where the
   ℓth phase computes optimal multiplication order for all subproducts of size
   ℓ. In the last phase $\ell = k$ and the algorithms produces the optimal order for
   computing the whole product. For $1 \leq i < j \leq k$ denote by $c(i, j)$ the cost
   of computing the subproduct $A_i \cdots A_j$ using the least-cost ordering. $c(1, k)$
   is the cost of the least-cost order for computing $A_1 \cdots A_k$.
   - For $1 \leq i \leq k$, let $c(i, i) \leftarrow 0$.
   - Compute the cost of all subproducts of size 2:
     For $i = 1, \ldots, k - 1$, let $c(i, i + 1) \leftarrow C_{i,i}^T R_{i+1,i+1}$
     (apply Equation 1 to the column sizes of A_i and the row sizes of A_{i+1}).
   - For $\ell = 3, \ldots, k$: (compute costs of all subproducts of size ℓ)
     For $j = 1, \ldots, k - \ell + 1$: (compute cost of $A_j \cdots A_{j+\ell-1}$)

$$c(j, j + \ell - 1) = \min_{j \leq i < j + \ell - 1} \left(c(j, i) + c(i + 1, j + \ell - 1) + C_{j,i}^T R_{i+1, j+\ell-1} \right) .$$

4.2 Time bounds and accuracy

The number of operations (per round) performed in the first stage of the algorithm is:

$$\text{for } k = 3 \quad |A_1| + 2|A_2| + |A_3|$$
$$\text{for } k \geq 4 \quad |A_1| + |A_k| + k(|A_2| + |A_{k-1}|) + (k+1)\sum_{i=3}^{k-2} |A_i|$$

The second stage of the algorithm requires $\sum_{i=2}^{k} n_i(i-1)(k+1-i)$ operations. It follows that the asymptotic cost for r rounds of the estimation method is

$$O(kr \sum_{i=1}^{k} |A_i| + k^3 N)$$

where $N = \max_i n_i$ is the maximum dimension of the matrices. Since estimates on column and row sizes are unbiased, so are the estimates on multiplication costs. As for convergence, in the worst case it is the same as a single row (see Subsection 3.3). The worst case occurs when many rows (or columns) in the appropriate subproduct have similar nonzero patterns (and hence size estimates are biased the same way). Generally, however, relative errors "cancel out" and convergence is much faster.

5 Experimental evaluation

We experimented with products of 2 and 3 randomly generated square matrices of various patterns and sizes and with products of Document-Term matrices that arise in Information Retrieval. We studied the performance of our algorithm for the following two objectives.

1. There are 2 evaluation orders for the matrix product ABC: Either multiply (AB) first and then $(AB)C$, or compute (BC) first and then $A(BC)$. We estimate the costs of the two evaluation orders to determine the less costly one. We study the dependence of the accuracy of our estimates on the number of rounds. We explored the effectiveness of the estimation procedure by comparing the cost of estimation to the cost of multiplication and to the possible saving by selecting the less costly order.
2. We studied the distribution of the relative error of estimates on individual column and row sizes of products of two matrices. The tradeoffs of accuracy and cost of obtaining the estimate are evaluated. We also conclude that the estimates obtained experimentally using pseudo random numbers have the distribution predicted by the theory (Subsection 3.3).

5.1 Computing and comparing costs

We denote by $c(M_1 * M_2)$ the cost according to Equation 1 (the number of floating-point multiply-add operations), of a sparse multiplication of the two matrices M_1 and M_2. In our study we used r rounds, for several values of r, to estimate the multiplication costs $c(AB * C)$ and $c(A * BC)$. The number of operations per round for estimating $c(AB * C)$ (resp., $c(A * BC)$) is $|A| + |B|$ (resp., $|B| + |C|$). Each operation in the estimation procedure is a compare-store on numbers of size at most $O(\log N)$, where N is the largest dimension of the matrices. There is a much smaller number of other operations that is subsumed by the compare-store operations. Hence, we measure the cost of each round by $c_r = |A| + 2|B| + |C|$ and the total cost by rc_r.

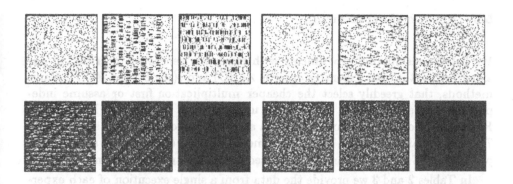

Fig. 2. Visualization of matrices in experiments E_0, E_1

We measure the effectiveness of multiplication-costs estimation by comparing its cost rc_r to the cost $\min\{c(A * B) + c(AB * c), c(B * C) + c(A * BC)\}$ of multiplication and to the gain $|c(A * B) + c(AB * C) - c(B * C) - c(A * BC)|$ by selecting the less costly order. We note that on most computing platforms, an "operation" of the multiplication algorithm is significantly slower that an "operation" of the estimation algorithm. The exact relation is platform-dependent. Our results, however, support the use of the estimation scheme even if the two units are comparable.

5.2 Generating the test matrices

We selected the matrices A, B, and C randomly according to some patterns. To generate the matrix A we used a small set of column densities (probability for an entry to be nonzero.). The matrix A is generated column by column. To generate a column we first select a probability p uniformly at random from the set of densities. Each entry in the column is determined to be zero or nonzero independently with probability p. The matrices B and C are generated using a "block like" pattern that also introduces correlations between the nonzero locations in B and C. Each column of C and each row of B were generated independently according to the following procedure. A *block-type* that consists of a set of rows (resp., columns for C) is selected uniformly at random from a set of block-types. Each block-type has two different probabilities $\{p_{in}, p_{out}\}$ associated with it. The probability p_{in} (resp., p_{out}) is the probability of an entry inside (resp., outside) the block is nonzero. Entries that are inside the block have higher probability of being nonzero ($p_{in} > p_{out}$). In Table 1 we list the parameters used in each of 9 experiments: The dimension of the matrices, the number and size of blocks, the probabilities to be inside and outside the block, and the set of densities used for selecting the columns of matrix A. Figure 2 visualizes the structures of subproducts of ABC in experiments E_0 and E_1.

5.3 Results of experiments

We list the results of 9 experiments with randomly generated matrices. In our experiments, the least expensive order is $A(BC)$ and $c(A*B) < c(B*C)$. Naive methods, that greedily select the cheaper multiplication first or assume independent nonzero locations, fail. We also used matrices that arise in Information Retrieval, where the matrix A relates a set of documents D_1 and set of terms T_1, B is a relation between T_1 and documents D_2, and C relates D_2 and T_2. The product ABC relates D_1 and T_2. We used square matrices of dimension 5000.

In Tables 2 and 3 we provide the data from a single execution of each experiment. In Table 2 we list the number of nonzeros in each matrix and subproduct and the (easy to compute) multiplication costs $c(A*B)$ and $c(B*C)$. In Table 3 we list the cost c_r of a single round of the estimation algorithm, the exact costs of the two multiplications $c(A*BC)$ and $c(AB*C)$, and the estimates to these costs using $r = \{5, 10, 15, 20\}$ rounds of the estimation algorithm. In experiments 1–7, the number of operations required for a sufficiently good estimate of the multiplication costs is small compared to the gap between the best and the lesser multiplication orders. The number of rounds needed for a sufficiently good estimate decreases for larger matrices. For experiments 4-7, only 5 rounds suffice. We studied the accuracy of the multiplication-cost estimates as a function of the number of rounds in different experiments. Figure 4 contains plots of the ratio of the estimated cost of multiplication and the actual cost. It plots the dependence of this ratio on the number of rounds performed. Each plot contains the accuracy curve for 10 repetitions of the experiment with $r = [2, \ldots, 25]$ rounds. We included plots for the product $(AB)C$ (cost of multiplying the product AB by the matrix C). As expected, the accuracy of the multiplication-cost estimates increases with input size and for a larger number of rounds. The cost of multiplication and the potential savings of selecting the less-costly order grow much faster than input size (and the cost of estimating multiplication-costs).

5.4 Estimates of sizes of rows and columns

Prior knowledge of (approximate) sizes of the columns or rows in a matrix product can be used to optimize memory and time usage by the multiplication procedure. For example, by allocating memory and setting up data structures of appropriate sizes. We included here histograms that plot the distribution of the estimate accuracies for various numbers of rounds. Figure 3 provides plots of the distribution functions and histograms for the accuracy (ratio of estimate to the value estimated) of estimates on the column sizes of the product AB in 10 repetitions of experiment 5 (hence, $50,000$ estimates). The histograms plot the accuracy distribution for $r = \{5, 10, 15, 20\}$ rounds. Observe that even for only 5 rounds, it is very unlikely that the estimate is over 3 times or below 0.4 times the actual value.

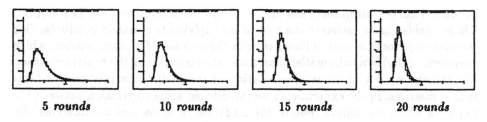

5 *rounds*   10 *rounds*   15 *rounds*   20 *rounds*

Fig. 3. Exp.5:Histograms for accuracy of estimates on AB column sizes

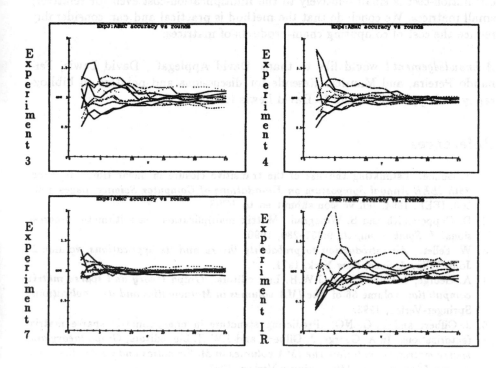

Fig. 4. Accuracy vs. Rounds for estimating $c(AB * C)$ (10 repetitions)

6 Summary

Multiplications of sparse matrices arise often in practice, both as stand alone operations and as part of more complex computation, such as solving linear programs. We considered predicting the nonzero structure of a product of 2 or more sparse matrices prior to performing the multiplication. In particular, predicting the size (number of nonzero elements) of each columns and row of the product. Prior knowledge of the nonzero structure may improve memory requirements and running time of the multiplication and be used to determine the least-costly order to multiply 3 or more sparse matrices. Naive approaches that predict the structure assume that nonzero entries are spread randomly. This assumption is often not justified. We adapted a reachability-set size estimation method [1] to produce accurate estimates of row and column sizes of arbitrary matrix products. These enabled us to estimate the cost of multiplying two matrix products. Our estimation procedure runs in linear time in the number of nonzero entries, which (asymptotically) is much smaller than the cost of computing the matrix product. We experimented with an implementation of the method on products of large sparse matrices. In the experiments we introduced some correlation between the matrices which we believe reflects the situation in some real applications (for example, matrices that arise in Information Retrieval). We observed that the estimation-cost is small relatively to the multiplication-cost even for relatively small matrices. We conclude that the method is practical and can considerably reduce the cost of computing chain-products of matrices.

Acknowledgement I would like to thank David Applegate, David Lewis, Fernando Pereira, and Mauricio Resende for discussions and pointers to bibliography and applications, and to David Lewis for the IR data.

References

1. E. Cohen. Estimating the size of the transitive closure in linear time. In *Proc. 35th IEEE Annual Symposium on Foundations of Computer Science*, pages 190–200. IEEE, 1994. full version submitted to JCSS.
2. D. Coppersmith and S. Winograd. Matrix multiplication via arithmetic progressions. *J. Symb. Comput.*, 9:251–280, 1990.
3. W. Feller. *An introduction to probability theory and its applications*, volume 2. John Wiley & Sons, New York, 1971.
4. A. George, J. Gilbert, and J.W.H. Liu, editors. *Graph theory and sparse matrix computation*, volume 56 of *The IMA volumes in Mathematics and its Applications*. Springer-Verlag, 1993.
5. J. Gilbert and E. G. NG. Predicting structure in nonsymmetric sparse matrix factorizations. In A. George, J. Gilbert, and J.W.H. Liu, editors, *Graph theory and sparse matrix computation The IMA volumes in Mathematics and its Applications*, volume 56, pages 107–140. Springer-Verlag, 1993.
6. G. Golub. *Matrix Computations*. The Johns Hopkins U. Press, Baltimore, MD, 1989.
7. A. Jennings and J. J. McKeown. *Matrix computations*. John Wiley & Sons, New York, second edition, 1992.
8. S. Pissanetzky. *Sparse matrix technology*. Academic Press, New York, 1984.
9. R. Sedgewick. *Algorithms*. Addison-Wesley Publishing Co., Reading, MA, 1988.
10. V. Strassen. Gaussian elimination is not optimal. *Numerische Mathematik*, 14(3):345–356, 1969.
11. R. P. Tewarson. *Sparse matrices*. Academic Press, New York, 1973.

Exp.	dimension	# blocks	block-size	p_{in}	p_{out}	A densities
E_0	256	8	15	0.8	0.02	{0.03}
E_1	512	8	60	0.25	0.01	{0.02}
1	1,000	20	50	0.3	0.005	{0.0150.0350.0200.0100.020}
2	1,000	10	30	0.4	0.005	{0.0150.0030.0200.0100.020}
3	1,000	10	60	0.2	0.005	{0.01}
4	5,000	100	40	0.5	0.0015	{0.0040.0040.0050.0050.004}
5	5,000	50	60	0.4	0.001	{0.0040.0030.0050.0050.003}
6	10,000	100	60	0.4	0.001	{0.0020.0010.0030.0030.002}
7	10,000	500	20	0.7	0.001	{0.002}

Table 1. Parameter choices for the experiments

Exp.	$|A|$	$|B|$	$|C|$	$|AB|$	$|BC|$	$|ABC|$	$c(A*B)$	$c(B*C)$
E_0	2,042	4,317	4,242	23,780	32,489	63,975	34,325	111,465
E_1	5,245	9,906	10,047	83,937	98,419	255,678	101,590	202,547
1	19,377	19,697	19,700	316,806	198,822	971,975	381,391	388,186
2	13,923	16,753	16,916	185,205	218,316	944,800	233,781	612,263
3	9,948	16,572	16,800	147,979	209,317	866,913	164,809	369,591
4	110,421	137,176	137,524	2,830,071	1,946,403	20,254,461	3,028,574	4,276,973
5	100,294	144,729	144,908	2,669,364	1,773,636	18,678,073	2,900,704	6,117,018
6	219,762	339,549	339,583	7,099,568	6,589,297	75,692,609	7,458,357	15,343,737
7	199,865	240,035	240,155	4,685,728	3,929,670	53,909,513	4,799,504	5,760,134
IR	82,670	85,793	85,155	5,787,707	1,174,978	12,305,303	8,870,366	2,412,967

Table 2. Matrix sizes

Exp.	c_r ($\times 10^3$)	cost: ($\times 10^3$)	exact	5 rnds.	10 rnds.	15 rnds.	20 rnds.
E_0	14.9	$c(AB*C)$	582	382	465	520	489
		$c(A*BC)$	257	237	248	265	268
E_1	35.1	$c(AB*C)$	1,709	1,887	1,613	1,563	1,652
		$c(A*BC)$	1,009	822	929	870	930
1	78.5	$c(AB*C)$	6,256	7,937	6,934	6,321	6,317
		$c(A*BC)$	3,860	5,168	4,163	4,235	4,246
2	64.3	$c(AB*C)$	6,456	7,760	7,416	6,465	6,647
		$c(A*BC)$	3,058	3,956	3,307	3,345	3,331
3	59.9	$c(AB*C)$	3,274	3,963	3,489	3,288	3,310
		$c(A*BC)$	2,222	2,690	2,222	2,287	2,286
4	522.3	$c(AB*C)$	88,001	93,130	95,477	88,911	83,709
		$c(A*BC)$	42,964	41,033	40,830	42,984	42,927
5	534.7	$c(AB*C)$	112,208	115,820	121,008	111,493	105,116
		$c(A*BC)$	35,510	33,486	33,106	35,031	352,821
6	1,238.4	$c(AB*C)$	319,546	298,115	306,531	313,904	322,776
		$c(A*BC)$	144,820	141,311	143,162	143,812	146,296
7	920.1	$c(AB*C)$	112,476	106,704	108,683	112,046	112,839
		$c(A*BC)$	78,607	74,261	76,961	76,949	78,113
IR	339.4	$c(AB*C)$	127,798	92,766	110,700	104,433	116,506
		$c(A*BC)$	68,892	45,018	50,595	58,875	65,785

Table 3. Estimation cost and results

Continuous Relaxations for Constrained Maximum-Entropy Sampling

Kurt M. Anstreicher[1], Marcia Fampa[2*], Jon Lee[3**], Joy Williams[3***]

[1] Dept. of Management Sciences, University of Iowa, Iowa City, IA 52242
[2] Dept. of Systems Engineering and Computer Sciences, COPPE, Federal University of Rio de Janeiro, CP 68511, 21945 Rio de Janeiro, RJ, Brazil
[3] Dept. of Mathematics, University of Kentucky, Lexington, KY 40506-0027

Abstract. We consider a new nonlinear relaxation for the Constrained Maximum Entropy Sampling Problem – the problem of choosing the $s \times s$ principal submatrix with maximal determinant from a given $n \times n$ positive definite matrix, subject to linear constraints. We implement a branch-and-bound algorithm for the problem, using the new relaxation. The performance on test problems is far superior to a previous implementation using an eigenvalue-based relaxation.

1 Introduction

Let n be a positive integer. For $N := \{1, \ldots, n\}$, let $Y_N := \{Y_j \mid j \in N\}$ be a set of n random variables, with joint-density function $g_N(\cdot)$. Let s be an integer satisfying $0 < s \leq n$. For $S \subset N$, $|S| = s$, let $Y_S := \{Y_j \mid j \in S\}$, and denote the marginal joint-density function of Y_S by $g_S(\cdot)$. The *entropy* of S is defined by

$$h(S) := -E[\ln g_S(Y_S)].$$

Let m be a nonnegative integer, and let $M := \{1, 2, \ldots m\}$. The *Constrained Maximum-Entropy Sampling Problem* is then the problem of choosing a subset S, with $|S| = s$, so as to maximize the entropy of S, while satisfying the linear constraints

$$\sum_{j \in S} a_{ij} \leq b_i, \quad i \in M.$$

The concept of entropy was introduced, in physics, by Rudolph Clausius. Somewhat later, Boltzmann [4] formalized the definition mathematically, as he laid the foundations of statistical mechanics. Shannon [16] and Blackwell [3] further developed and popularized entropy in the contexts of information theory and statistics, respectively. Constrained maximum-entropy sampling problems

---

* Visiting the Dept. of Management Sciences, University of Iowa, supported by a Research Fellowship from CNPq-Brasilia-Brazil.
** Supported in part by NSF grant DMI-9401424.
*** Supported in part by NSF grant DMI-9401424 and by the U.K. Center for Computational Sciences.

are fundamental in the area of statistical design. They arise in problems of environmental, geological, and meteorological monitoring; see, for example, [5], [6], [10], [9], [12], [17], and [18]. Moreover, the classical vertex-packing problem of combinatorial optimization can be modeled as an unconstrained (that is, $M = \emptyset$) maximum-entropy sampling problem; see Ko, Lee, and Queyranne [14].

Entropy is attractive for several reasons. First, it provides a robust *model-independent* measure of the information available in a set of random variables. Second, for a particular second-order linear-model, maximizing the entropy coincides with minimizing the generalized variance of the parameter estimates (see [10]). Finally, for the important case in which Y_N has a joint Gaussian distribution with covariance matrix C, $h(S)$ has an easily computable form:

$$h(S) = k_s + \alpha \operatorname{ldet} C[S, S],$$

where $\operatorname{ldet} := \ln \det$, α and k_s are constants, and $C[S, T]$ is the submatrix of C with rows indexed by S and columns indexed by T. Hence $C[S, S]$ is the principal submatrix of C indexed by S. Therefore, in the Gaussian case, we can recast our problem as:

$$
\begin{aligned}
\text{CMESP:} \quad \max \quad & \operatorname{ldet} C[\underline{x}, \underline{x}] \\
\text{s.t.} \quad & Ax \leq b \\
& e^T x = s \\
& x_j \in \{0, 1\}, \ j \in N,
\end{aligned}
$$

where $e \in \Re^n$ is the vector of ones, and $\underline{x} \subset N$ denotes the support of x; $\underline{x} = \{j \in N \mid x_j = 1\}$.

Computational approaches to the constrained maximum-entropy sampling problem have concentrated on the Gaussian case. Early approaches focused on the unconstrained case ($M = \emptyset$), using greedy and interchange heuristics. Ko, Lee and Queyranne developed the first algorithm to find optimal solutions and prove their optimality. Their method is of the branch-and-bound variety. Branching is accomplished by fixing indices into and out of S. If f elements F are fixed into S, and u elements U are fixed out of S, then one needs to choose $s' := s - f$ elements S' from the $n' := n - f - u$ elements $N' := N \setminus (F \cup U)$, so as to maximize the *conditional* entropy of S'. The latter is easy to work with since the resulting conditional distribution of $Y_{N'}$ is joint Gaussian, with covariance matrix

$$C_{N'|F} := C[N', N'] - C[N', F] C^{-1}[F, F] C[F, N'].$$

In other words, the conditional covariance matrix $C_{N'|F}$ is simply the Schur complement of $C[F, F]$ in $C[N \setminus U, N \setminus U]$. Ko, Lee and Queyranne observed that an upper bound on the objective in CMESP, for given sets F and U, is then

$$\operatorname{ldet} C[F, F] + \sum_{i=1}^{s'} \ln \lambda_i \left(C_{N'|F} \right), \tag{1}$$

where $\lambda_1 \left(C_{N'|F} \right) \geq \ldots \geq \lambda_{n'} \left(C_{N'|F} \right)$ are the eigenvalues of $C_{N'|F}$. In the unconstrained case, the bounds provided by (1) are sharp enough to allow the exact solution of moderate size, real-world problems. Unfortunately, however, there is no obvious way to incorporate constraint information using the bounds of (1), and therefore this method will not be very effective when $M \neq \emptyset$.

Lee [15] extended the bounds based on (1) to incorporate constraints, by using a Lagrangian approach. Minimizing the Lagrangian requires some kind of nonlinear programming algorithm. Each iteration of the minimization method requires the calculation of an eigensystem, to obtain an objective value and gradient of the Lagrangian function. (The objective value is the unconstrained eigenvalue bound applied to a symmetrically-scaled conditional covariance matrix, where the scaling is based on multipliers for the constraints. A gradient, or subgradient, is then obtained from the eigenvectors). A major shortcoming of this method is that the number of nonlinear programming iterations needed to approximately minimize the Lagrangian increases as the number of constraints increases.

Notation. For a vector $v \in \Re^n$, let diag(v) denote the diagonal matrix with diag$(v)_{ii} := v_i$ for each i. For any matrix V, let Diag(V) denote the diagonal matrix whose diagonal components are those of V. For two square matrices U and V, let $U \circ V$ denote the Hadamard product of V and U: $(U \circ V)_{ij} := u_{ij} v_{ij}$, and let $V^{(2)} := V \circ V$. For symmetric matrices U and V we write $U \preceq V$, or $V \succeq U$, to mean that $V - U$ is positive semidefinite, and $U \prec V$, or $V \succ U$, to mean that $V - U$ is positive definite. For u and v both in $\Re^n$, $v \geq 0$, and $V = \text{diag}(v)$, we use the notation V^u to denote the diagonal matrix having $(V^u)_{ii} := v_i^{u_i}$, $i = 1, \ldots, n$.

2 New Bounds Based on Nonlinear Relaxation

Define constants $d_j > 0$ and $p_j \geq 1$ for $j \in N$, and let $D := \text{diag}(d)$, $P := \text{diag}(p)$. For nonnegative $x \in \Re^n$, let $X := \text{diag}(x)$. Define

$$f(x) := \text{ldet } M(x), \quad \text{where} \quad M(x) := X^{p/2}(C - D)X^{p/2} + D^x . \tag{2}$$

Notice that if x is 0/1-valued, then $f(x) = \text{ldet } C[\underline{x}, \underline{x}]$. So we consider the relaxation

$$\begin{aligned} \text{NLP:} \quad & \max \ f(x) \\ & \text{s.t.} \quad Ax \leq b \\ & \qquad e^T x = s \\ & \qquad 0 \leq x \leq e. \end{aligned}$$

The relaxation NLP is motivated by the observation that if C is diagonal (the case of independent random variables), then setting $D = C$ yields a linear objective-function. In the unconstrained case ($M = \emptyset$), NLP then becomes a trivial linear program that has a 0/1 optimal solution solving the original maximum-entropy sampling problem. In general, for NLP to be computationally useful, it must be a tractable problem. In particular, $f(\cdot)$ should be concave on the feasible region of NLP. We will next show that this is the case for appropriate choices of D and p. We assume throughout that C is positive definite.

Lemma 1. *Assume that $p_i \geq 1$ and $0 < d_i \leq \exp(p_i)$ for $i \in N$. Then $M(x)$ is positive definite for $0 \leq x \leq e$.*

Proof. $M(0) = I$ is positive definite, and therefore $M(x)$ is positive definite for all x in some neighborhood of $x = 0$. For $x > 0$, $X^{p/2}CX^{p/2}$ is certainly positive definite, and it is straightforward to show that $d_i^{x_i} \geq d_i x_i^{p_i}$ for any $0 < d_i \leq \exp(p_i)$, $0 \leq x_i \leq 1$. It follows that $M(x) \succeq X^{p/2}CX^{p/2}$ for $x > 0$, and therefore $M(x)$ is positive definite for $0 \leq x \leq e$. $\qquad\square$

Lemma 1 implies that $M(x)$ is nonsingular for $0 \leq x \leq e$. Define $l(x) \in \Re^n$ by $l_i(x) := p_i - \ln(d_i^{x_i})$, $i = 1, \ldots, n$, and let $L(x) = \mathrm{diag}(l(x))$. Note that the conditions of Lemma 1 then imply that $l(x) \geq 0$ for $0 \leq x \leq e$. Using straightforward but laborious matrix calculus, the Hessian of $f(x)$, for $x > 0$, can be shown to be:

$$\nabla^2 f(x) = X^{-1} P \left[D^x \, \mathrm{Diag}(M^{-1}(x)) - I \right] X^{-1} \tag{3}$$
$$+ X^{-1} L(x) \left[D^x \, \mathrm{Diag}(M^{-1}(x)) - D^x (M^{-1}(x))^{(2)} D^x \right] L(x) X^{-1}.$$

Theorem 2. *Assume that $D \succeq C$, $p_j \geq 1$ and $0 < d_j \leq \exp(p_j - \sqrt{p_j})$, $j \in N$. Then $f(\cdot)$ is concave for $0 < x \leq e$.*

Proof. By construction we have $M(x) \preceq D^x$ for $0 \leq x \leq e$, which implies (see [13, Section 7.7])

$$M^{-1}(x) \succeq D^{-x}. \tag{4}$$

From (4) it follows that

$$D^x M^{-1}(x) D^x \succeq D^x. \tag{5}$$

Then (4) and (5) together imply (see [13, Section 7.7])

$$D^x M^{-1}(x) D^x \circ (M^{-1}(x) - D^{-x}) \succeq D^x \circ (M^{-1}(x) - D^{-x}),$$

which can be re-written as

$$D^x (M^{-1}(x))^{(2)} D^x - D^x \, \mathrm{Diag}(M^{-1}(x)) \succeq D^x \, \mathrm{Diag}(M^{-1}(x)) - I. \tag{6}$$

But (4) also implies that

$$D^x \, \mathrm{Diag}(M^{-1}(x)) = D^x \circ M^{-1}(x) \succeq D^x \circ D^{-x} = I, \tag{7}$$

so $[D^x \, \mathrm{Diag}(M^{-1}(x)) - I]$ is a positive semidefinite, diagonal matrix. Next, the conditions $d_i \leq \exp(p_i - \sqrt{p_i})$ and $p_i \geq 1$ together imply that $l_i(x) \geq \sqrt{p_i}$ for $0 \leq x_i \leq 1$, and therefore

$$L(x) \left[D^x \, \mathrm{Diag}(M^{-1}(x)) - I \right] L(x) \succeq P \left[D^x \, \mathrm{Diag}(M^{-1}(x)) - I \right]. \tag{8}$$

Then (6) and (8) together imply that

$$L(x)[D^x (M^{-1}(x))^{(2)} D^x - D^x \, \mathrm{Diag}(M^{-1}(x))]L(x) \succeq P \left[D^x \, \mathrm{Diag}(M^{-1}(x)) - I \right]$$

and the theorem follows immediately from (3). $\qquad\square$

From Theorem 2, any D with $D \succeq C$ can be used to obtain a concave $f(\cdot)$. Next we show that for any fixed $D \succeq C$, the best bound from NLP will be obtained using the smallest p satisfying the conditions of Theorem 2. In several subsequent proofs we will use the well-known fact that if V is a nonsingular matrix, then

$$\frac{\partial}{\partial V_{ij}} \operatorname{ldet} V = (V^{-1})_{ij}. \tag{9}$$

Lemma 3. *Assume that p and d satisfy the conditions of Theorem 2, and $p' \geq p$. Let $f'(\cdot)$ be defined as in (2), but using p' in place of p. Then $f'(x) \geq f(x)$ for all $0 < x \leq e$.*

Proof. We will consider $f(x)$ to be a function of p, and show that for any $0 < x \leq e$, and every i,

$$\frac{\partial}{\partial p_i} f(x) \geq 0. \tag{10}$$

Note that we can write

$$f(x) = \sum_{j=1}^{n} p_j \ln(x_j) + \operatorname{ldet}(C - D + X^{-p}D^x),$$

and therefore

$$\frac{\partial}{\partial p_i} f(x) = \ln(x_i) + \frac{\partial}{\partial p_i} \operatorname{ldet}(C - D + X^{-p}D^x). \tag{11}$$

Using the chain rule and (9),

$$\frac{\partial}{\partial p_i} \operatorname{ldet}(C - D + X^{-p}D^x) = (C - D + X^{-p}D^x)_{ii}^{-1} \frac{d}{dp_i} d_i^{x_i} x_i^{-p_i}$$

$$= -[\ln(x_i)]d_i^{x_i} x_i^{-p_i}(C - D + X^{-p}D^x)_{ii}^{-1}$$

$$= -[\ln(x_i)]d_i^{x_i} M_{ii}^{-1}(x). \tag{12}$$

Substituting (12) into (11) results in

$$\frac{\partial}{\partial p_i} f(x) = \ln(x_i)\left(1 - d_i^{x_i} M_{ii}^{-1}(x)\right). \tag{13}$$

But (7) implies that $d_i^{x_i} M_{ii}^{-1}(x) \geq 1$ for every i, and then (13) implies (10). □

Note that in the original problem CMESP, scaling the matrix C by a constant $\gamma > 0$ simply adds $s \ln(\gamma)$ to the objective, for any feasible x. Such a scaling can also be applied in NLP. In particular, define

$$f_\gamma(x) = \operatorname{ldet} M_\gamma(x) - s \ln(\gamma), \tag{14}$$

$$\text{where } M_\gamma(x) = \gamma X^{p/2}(C - D)X^{p/2} + (\gamma D)^x.$$

Then $f(x) = f_\gamma(x)$ for all x satisfying $e^T x = s$, $x_j \in \{0, 1\}$, $j \in N$. Therefore, we can replace $f(x)$ with $f_\gamma(x)$ in the nonlinear programming relaxation. The use of such a scaling is important because, in general, the bound obtained from NLP is dependent on the scaling factor γ. For any $D \succeq C$, and scaling factor γ, Lemma 3 implies that the best bound is obtained using

$$p_j(\gamma) = \begin{cases} 1 & \text{if } \gamma d_j \leq 1, \\ \left(1 + \sqrt{1 + 4\log(\gamma d_j)}\right)^2 / 4 & \text{if } \gamma d_j > 1. \end{cases} \tag{15}$$

In particular, if $\gamma D \preceq I$, then $p = e$ gives the best possible bound. In the next two lemmas we show that to get the best bound from NLP, it is never desirable to have $\gamma D \prec I$, or $\gamma D \succ I$.

Lemma 4. *Assume that* $C \preceq D \preceq I$, *and* $p = e$. *Then* $f_\gamma(x) \geq f(x)$ *for all* $0 \leq x \leq e$, $e^T x = s$, *and* $0 < \gamma \leq 1$.

Proof. It suffices to show that

$$\frac{d}{d\gamma} f_\gamma(x) \leq 0,$$

for all $0 < \gamma \leq 1$, where from (14) we have

$$\frac{d}{d\gamma} f_\gamma(x) = \frac{d}{d\gamma} \operatorname{ldet} M_\gamma(x) - \frac{s}{\gamma}. \tag{16}$$

Instead of working directly with $M_\gamma(\cdot)$, it is convenient in the analysis to consider scaling the rows and columns of C and D individually. Consequently let $w > 0$ be a vector in $\Re^n$, $W := \operatorname{diag}(w)$, and define

$$M_w(x) := X^{e/2} W^{e/2} (C - D) W^{e/2} X^{e/2} + (WD)^x$$
$$= W^{e/2} \left(X^{e/2} (C - D) X^{e/2} + W^{x-e} D^x \right) W^{e/2}.$$

Immediately then

$$\operatorname{ldet} M_w(x) = \sum_{j=1}^{n} \ln(w_j) + \operatorname{ldet}(X^{e/2}(C - D)X^{e/2} + W^{x-e} D^x).$$

Using the chain rule, and (9), we obtain

$$\frac{\partial}{\partial w_i} \operatorname{ldet} M_w(x) = \frac{1}{w_i} + (X^{e/2}(C - D)X^{e/2} + W^{x-e} D^x)_{ii}^{-1} d_i^x (x_i - 1) w_i^{x_i-2}$$
$$= \frac{1}{w_i} \left(1 + (x_i - 1)[M_w^{-1}(x)]_{ii}(w_i d_i)^{x_i}\right).$$

Next, the analog of (7) for $M_w(x)$ implies that $[M_w^{-1}(x)]_{ii}(w_i d_i)^{x_i} \geq 1$, and therefore

$$\frac{\partial}{\partial w_i} \operatorname{ldet} M_w(x) \leq \frac{x_i}{w_i}.$$

Finally, $M_\gamma(x) = M_w(x)$ for $w = \gamma e$, so

$$\frac{d}{d\gamma} \operatorname{ldet} M_\gamma(x) = \sum_{i=1}^n \frac{\partial}{\partial w_i} \operatorname{ldet} M_w(x)|_{w=\gamma e} \le \frac{s}{\gamma}, \qquad (17)$$

where we are using the fact that $e^T x = s$. The proof is completed by combining (16) and (17). □

Lemma 5. *Suppose that $C \preceq D$, and $D \succeq I$. Then $f_\gamma(x) \ge f(x)$ for any $0 < x \le e$, $e^T x = s$, and $\gamma \ge 1$, where $p = p(\gamma)$ is given by (15) for all γ.*

Proof. It suffices to show that

$$\frac{d}{d\gamma} f_\gamma(x) \ge 0,$$

for $\gamma \ge 1$. As in the proof of Lemma 4 consider a scaling vector w, where now $w \ge e$, and let

$$
\begin{aligned}
M_w(x) &:= X^{p(w)/2} W^{e/2}(C-D)W^{e/2} X^{p(w)/2} + (WD)^x \\
&= X^{p(w)/2} W^{e/2} \left(C - D + X^{-p(w)} W^{x-e} D^x\right) W^{e/2} X^{p(w)/2},
\end{aligned}
$$

where $p(w)$ is defined by

$$p_i(w) := \left(1 + \sqrt{1 + 4\ln(w_i d_i)}\right)^2 / 4, \quad i = 1, \dots, n. \qquad (18)$$

Immediately then

$$\operatorname{ldet} M_w(x) = \sum_{j=1}^n \ln(w_j) + \sum_{j=1}^n p_j(w) \ln(x_j) + \operatorname{ldet}(C - D + X^{-p(w)} W^{x-e} D^x).$$

Using the chain rule and (9), we obtain

$$\frac{\partial}{\partial w_i} \operatorname{ldet} M_w(x) = \frac{1}{w_i} + \ln(x_i)\frac{\partial p_i(w)}{\partial w_i} \qquad (19)$$

$$+ (C - D + X^{-p(w)} W^{x-e} D^x)_{ii}^{-1} d_i^{x_i} \frac{\partial}{\partial w_i} \left(x_i^{-p_i(w)} w_i^{x_i-1}\right).$$

A straightforward differentiation next obtains

$$\frac{\partial}{\partial w_i} x_i^{-p_i(w)} w_i^{x_i-1} = x_i^{-p_i(w)} w_i^{x_i-1} \left(\frac{x_i - 1}{w_i} - \ln(x_i)\frac{\partial p_i(w)}{\partial w_i}\right). \qquad (20)$$

Substituting (20) into (19), and simplifying, results in

$$\frac{\partial}{\partial w_i} \operatorname{ldet} M_w(x) = \frac{1}{w_i} + \ln(x_i)\frac{\partial p_i(w)}{\partial w_i} \qquad (21)$$

$$+ [M_w^{-1}(x)]_{ii}(w_i d_i)^{x_i} \left(\frac{x_i - 1}{w_i} - \ln(x_i)\frac{\partial p_i(w)}{\partial w_i}\right).$$

Next, for $p(w)$ given as in (18), it is easy to show that

$$\frac{\partial p_i(w)}{\partial w_i} = \frac{2\sqrt{p_i(w)}}{w_i(2\sqrt{p_i(w)} - 1)}. \tag{22}$$

Using (22), we obtain

$$\frac{x_i - 1}{w_i} - \ln(x_i)\frac{\partial p_i(w)}{\partial w_i} = \frac{1}{w_i}\left(x_i - 1 - \ln(x_i)\frac{2\sqrt{p_i(w)}}{2\sqrt{p_i(w)} - 1}\right)$$

$$= \frac{1}{w_i(2\sqrt{p_i(w)} - 1)}(2\sqrt{p_i(w)}[x_i - 1 - \ln(x_i)] + 1 - x_i)$$

$$\geq 0, \tag{23}$$

where the inequality follows from $x_i \leq 1$, and $\ln(x_i) \leq x_i - 1$ for $x_i > 0$. Finally, the analog of (7) for $M_w(x)$ implies that $[M_w^{-1}(x)]_{ii}(w_i d_i)^{x_i} \geq 1$, and this fact combined with (21) and (23) gives

$$\frac{\partial}{\partial w_i}\text{ldet } M_w(x) \geq \frac{x_i}{w_i}.$$

The remainder of the proof is similar to that of Lemma 4. □

The above results suggests several reasonable strategies for choosing D and γ. For example:

1. Let ρ be the maximum eigenvalue of C, $D = \rho I$, $\gamma = 1/\rho$.
2. Let $D_0 = \text{Diag}(C)$. Let $\rho > 0$ be the smallest number such that $\rho D_0 - C$ is positive semidefinite. (Then ρ is the maximum eigenvalue of $D_0^{-1/2}CD_0^{-1/2}$.) Let $D = \rho D_0$, and choose $\gamma \in [1/d_{\max}, 1/d_{\min}]$.
3. Choose D to minimize some function of the eigenvalues of $D - C$. For example, find D that approximately minimizes the trace of $D - C$, or the maximum eigenvalue of $D - C$, subject to $D - C$ being positive semidefinite. Then choose $\gamma \in [1/d_{\max}, 1/d_{\min}]$.

The advantage of strategy 2 over 1 is that the same work is required (a single maximum-eigenvalue computation), but in the case where C is diagonal and $M = \emptyset$, using 2 results in the solution of NLP being a solution to the original discrete problem. The problems for finding D from 3 are more involved, but turn out to be examples of *semidefinite programming* problems, and can be efficiently solved using recent advances in interior-point theory.

Bounds based on NLP can also be applied to a "complementary problem," using the observation that for any S,

$$\text{ldet } C[S, S] = \text{ldet } C + \text{ldet } C^{-1}[\bar{S}, \bar{S}],$$

where $\bar{S} = N \setminus S$. As a result, we can treat C^{-1} as the covariance matrix and calculate bounds for the problem in which we seek to choose $\bar{s} = n - s$ indices.

In general the bounds obtained from the original and complementary problem can differ substantially from one another.

To illustrate the NLP bounds we consider an example from [14]. The original data set consists of covariate data from 52 acid rain monitoring sites in the United States. After fixing 36 sites, we are left with the (unconstrained) problem of choosing s sites from the remaining 16 so as to maximize the conditional entropy. In the figure below we illustrate the gap between the NLP value and the true discrete solution value for $s = 1$ to 15. For each s we compute bounds based on the original and complementary problems, using a D that approximately minimizes the trace of $D - C$ (or $D - C^{-1}$). For comparison we also plot the gaps from the eigenvalue bounds (1). (In the figure "EIG" refers to the bounds from (1), "NLP-Tr" refers to the bounds based on the original problem, using C, and "INV-Tr" refers to the bounds based on the complementary problem, using C^{-1}.) Clearly the combination of the two NLP bounds (based on the original and complementary problems) is significantly sharper than the eigenvalue bound for these problems.

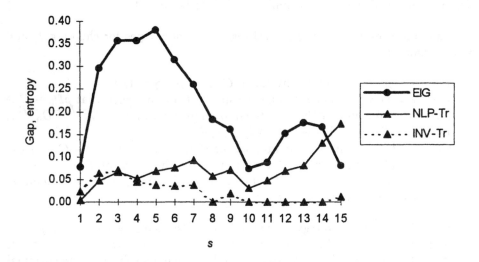

Fig. 1. Gaps for EIG versus NLP-Tr bounds on $52/16/s$ problems

3 Solving the Relaxation

We use a "SUMT", or "long-step path following," approach to approximately solve NLP; see for example [8] or [11]. For simplicity, assume that NLP has an interior-point solution, that is, an x with $e^T x = s$, $0 < x < e$, $Ax < b$. (If such an x does not exist then there are inequalities in NLP that hold as equalities for all feasible solutions. Such constraints are identified by first solving a "Phase I" linear-program, and then changed from inequalities to equalities in NLP.) For scalar $\mu > 0$, define the logarithmic barrier function

$$F(x, \mu) := -f(x) - \mu \sum_{j=1}^{n} \log(x_j) - \mu \sum_{j=1}^{n} \log(1 - x_j) - \mu \sum_{i=1}^{m} \log(b_i - a_i^T x),$$

where a_i^T is the ith row of A. For a fixed μ, we approximately minimize $F(\cdot, \mu)$ on $\{x \mid e^T x = s\}$ using damped Newton-steps, and then reduce μ and repeat the process. This solution procedure suits our application well for a number of reasons. First, it is well known that if $x^*(\mu)$ is the exact minimizer of $F(\cdot, \mu)$ on $\{x \mid e^T x = s\}$, then

$$f(x^*(\mu)) \le v_{\text{NLP}} \le f(x^*(\mu)) + \epsilon, \tag{24}$$

where v_{NLP} denotes the solution value in NLP, and $\epsilon = (2n + m)\mu$. Of course the exact minimizer cannot be obtained, but for any $\epsilon > (2n + m)\mu$, sufficiently accurate minimization of $F(\cdot, \mu)$ produces a "dual solution" which provides an upper bound as in (24). Such bounds are essential in our application, since we use NLP as a relaxation of CMESP, and therefore require an *upper* bound for the optimal value in NLP. In practice we set a desired tolerance ϵ, decrease μ until $\mu \le \epsilon/(4n + 2m)$, and then for this final value of μ continue minimizing $F(\cdot, \mu)$ until the dual solution verifying (24) is generated. In our application highly accurate approximations of the solution value in NLP are *not* required. For example, the use of $\epsilon = .001$ will usually make the inaccuracy in solving NLP substantially less than the difference between v_{NLP} and the solution value in the original problem CMESP. By using such a large ϵ we can economize substantially on the time required to generate a bound from NLP. In addition, when a lower bound for CMESP is available, the dual solution that generates the upper bound in NLP can potentially be used to *fix* variables in NLP at 0/1 values. In the branch-and-bound context where NLP is ultimately applied (see Sect. 4), this is an important feature, since a relaxation that is close to fathoming may be able to fix a number of variables, and in so doing improve the bounds for successive subproblems. Another useful observation is that if ever $f(x)$ is found to exceed a lower bound for CMESP, then computations in NLP could immediately be terminated since fathoming will be impossible.

In addition to the problem of approximately solving NLP, we need to consider the problem of finding an appropriate D when the third strategy of the previous section is employed. For example, we consider the problem of finding $D \succeq C$ that approximately minimizes the trace of $D - C$. We also use a long-step path following strategy for this semidefinite programming problem; see [2] for details.

4 The Branch-and-Bound Algorithm

The complete algorithm is of a branch-and-bound variety. Subproblems are described by the indices that are fixed into the solution and those indices that remain eligible. We maintain a set of active subproblems whose relaxations have been solved. A subproblem is removed from the active set, and its bound is compared to the current lower bound for possible fathoming. If we cannot fathom, then we choose an eligible index j and create two subproblems: one in which j is fixed into the solution and one in which j is fixed out of the solution. For each of these two subproblems we do the following:

1. If there is a unique 0/1-valued solution to the cardinality constraint, then we check feasibility. If the solution is infeasible, then we fathom by infeasibility, discarding the subproblem. If the solution is feasible, we check the objective value of the solution, update the lower bound if possible, and discard the subproblem.

2. Otherwise, we solve a linear program to determine a relative-interior solution of the feasible region of the continuous relaxation. In doing so, we identify the constraints that must always be satisfied as equations.

3. If the dimension of the relative interior is -1, then we fathom by infeasibility, discarding the subproblem.

4. If the dimension of the relative interior is 0, then we check for integrality; fathoming by infeasibility if the solution is fractional; checking the objective value and possibly updating the lower bound if the solution is integer valued. In either case, we discard the subproblem.

5. If the dimension of the relative interior is positive, we determine a basis for the null-space of the tight-constraint matrix, and use it in the calculation of an upper bound and associated continuous solution, and then fathom by bounds if possible.

6. If the subproblem was not fathomed by bounds, we use the continuous solution as the vector of objective function coefficients in a linearization of the problem. We solve the linearization by either linear programming or integer linear-programming. If we discover an integer solution to the linearization, we check its objective value and update the lower bound if possible. In practice, we use integer linear-programming for the linearization of the first subproblem, to make sure that we have a reasonable lower bound. For subsequent subproblems, we use linear programming.

5 Computational Results

We coded the branch-and-bound algorithm in C. Our code makes use of the matrix-algebra library LAPACK 2.0 [1], which is written in FORTRAN 77, and the linear-programming library CPLEX 3.0 [7], which is written in C. The upper-bounding routines require the calculation of Schur complements, matrix inverses, linear-equation solves, null-space bases, eigensystems, etc., which we carry out using a combination of LAPACK routines, and specialized routines of our own construction. Linear programs in steps 2 and 6 of the branch-and-bound algorithm are solved using CPLEX.

Table 1. Number of bounds computed

	Bounds	C0	C1	C2	C3	C4	C5	C6	C7	C8	C9	Mean	MAD
A0	EIG	241	266	782	1161	2178	328	1161	956	4837	565	1247.5	904.0
	NLP-Id	79	146	84	213	80	300	237	70	161	723	209.3	127.2
	NLP-Di	65	88	66	85	70	110	79	46	165	107	88.1	23.5
	NLP-Tr	57	62	58	73	46	84	69	46	113	65	67.3	14.0
A2	EIG	227	246	960	4344	3667	319	903	4835	3182	513	1919.6	1669.9
	NLP-Id	77	128	108	914	207	389	211	549	76	655	331.4	236.3
	NLP-Di	65	68	52	382	101	79	57	135	56	103	109.8	59.5
	NLP-Tr	57	56	52	286	55	59	57	87	56	63	82.8	41.5
A5	EIG	197	553	907	1618	3833	144	1236	728	5632	505	1535.3	1295.4
	NLP-Id	71	534	110	288	131	258	276	101	206	617	259.2	135.6
	NLP-Di	47	80	60	128	95	82	86	47	148	107	88.0	25.2
	NLP-Tr	43	58	52	100	53	52	72	43	78	65	61.6	13.7
A10	EIG	258	643	919	1023	4769	244	574	1819	2790	477	1351.6	1064.6
	NLP-Id	744	606	122	198	251	434	160	275	124	507	342.1	184.5
	NLP-Di	64	86	82	90	125	94	54	95	104	105	89.9	14.7
	NLP-Tr	54	54	52	80	69	56	52	79	56	65	61.7	9.2
A15	EIG	37	147	85	85	289	217	67	143	347	47	146.4	82.9
	NLP-Id	63	231	65	81	91	329	45	57	79	97	113.8	66.5
	NLP-Di	29	67	29	39	41	91	29	29	53	29	43.6	16.0
	NLP-Tr	29	49	29	29	29	53	29	29	29	29	33.4	7.0

We have begun to solve a variety of constrained and unconstrained problems. Here, we report results using ten different covariance matrices (C0-C9), and five sets of constraints (A0/A2/A5/A10/A15). The covariance matrices and constraints are the same ones used in [15]. The covariance matrices are order-30 matrices that were derived from an order-63 matrix used by [12]. The order-63 matrix was derived from sulfate data obtained from a network of 63 environmental monitoring stations in the United States. Each order-30 matrix that we used corresponds to the conditional covariance matrix arising from fixing a particular choice of 33 stations. We solved all problems with $s = 15$, so our problems are network expansion problems in which we seek to augment a network of 33 stations with the optimal choice of 15 more stations from a set of 30 potential stations. (In general we have found that for a given n, CMESP problems are most difficult when s is about $n/2$.) Each constraint set Am contains m constraints. Using all combinations of the ten matrices and five sets of constraints, we worked with fifty distinct problems. We solved each of the fifty problems using each of four possible upper-bounding methods. One method was the eigenvalue method

Table 2. Wall seconds

	Bounds	C0	C1	C2	C3	C4	C5	C6	C7	C8	C9	Mean	MAD
A0	EIG	112	116	335	502	942	139	503	417	2092	245	540.4	390.7
	NLP-Id	16	22	15	33	13	45	41	13	19	106	32.3	19.2
	NLP-Di	16	12	12	13	12	21	15	8	17	18	14.5	2.9
	NLP-Tr	20	20	17	25	13	29	23	14	36	22	21.8	5.0
A2	EIG	137	148	511	2766	2399	221	490	2887	1881	314	1175.3	1046.3
	NLP-Id	26	34	33	229	62	112	63	143	21	180	90.4	60.7
	NLP-Di	20	18	14	102	32	28	16	42	15	31	31.8	16.1
	NLP-Tr	31	22	20	134	23	27	25	39	22	29	37.0	19.9
A5	EIG	149	367	537	992	2414	110	762	479	3622	326	975.6	820.1
	NLP-Id	25	158	36	86	42	80	85	30	57	182	78.1	40.1
	NLP-Di	20	25	23	49	30	27	28	13	43	35	29.3	7.9
	NLP-Tr	20	26	22	47	22	23	34	19	35	36	28.6	7.7
A10	EIG	225	572	659	897	4273	226	411	1330	1924	441	1096.0	847.9
	NLP-Id	250	207	48	68	99	145	63	87	39	174	117.9	60.8
	NLP-Di	21	29	29	30	41	33	17	29	33	38	30.0	4.9
	NLP-Tr	27	27	23	41	30	27	24	37	24	34	29.3	5.0
A15	EIG	76	214	157	142	401	236	148	225	477	86	216.3	94.8
	NLP-Id	16	49	18	24	23	66	14	17	19	27	27.4	12.1
	NLP-Di	8	18	8	13	11	24	8	8	14	8	11.8	4.3
	NLP-Tr	11	20	11	13	11	22	11	12	12	13	13.6	2.9

(see [15]). We also used the NLP bounds with the three different choices for choosing D and γ suggested in Sect. 2. (For each method for choosing D and γ, we computed the NLP bound for the original subproblem as well as for its "complementary problem," as described in Sect. 2.)

We ran our code on a 50MHZ HP 9000/715, a very modest workstation. In Tables 1 and 2, we report statistics (number of upper bounds calculated, and total solution time) for the optimal solution of these problems. In the tables, "EIG" refers to eigenvalue-based bounds and "NLP" refers to NLP-based bounds. "Id" (identity), "Di" (diagonal), and "Tr" (trace) refer to the three methods for choosing D and γ suggested in Sect. 2. The last two columns report the mean, and mean absolute deviation, of the values in each row. Clearly the average performance using the NLP bounds is far superior to that when the eigenvalue bounds are used. In general the use of the "Di" strategy results in fewer bound evaluations than "Id," and "Tr" results in fewer still. However, on these problems the reduction in the number of bounds computed by the "Tr" strategy is often not worth the additional time required to implement this strategy. It is also worth noting that the relative variation (MAD/Mean) for both statistics is lower when the NLP bounds are used, especially with the "Di" and "Tr" strategies.

These preliminary results, using the complete branch-and-bound algorithm with the NLP-based bounds, are very encouraging. In general, the bounds generated using the NLP approach appear to be much stronger than the eigenvalue bounds used in [14] and [15]. The only downside is that for unconstrained problems, bounds based on NLP are more time consuming to compute than the eigenvalue bounds (1), for subproblems of the same size. However, the time to compute the eigenvalue bounds grows much more quickly than the time to compute the NLP bounds, as the number of constraints increases. For the solution of all fifty problems considered here, the use of NLP-Di reduces the total number of bounds computed by a factor of about 15, and the total time required by a factor of about 34, compared to the use of the eigenvalue bounds.

The code is still in the process of development. We plan on using dual solutions to fix variables, in the hopes of reducing the number of bounds calculated. We also plan to experiment with terminating a given upper-bound calculation once some cutoff value is exceeded (the cutoff would be chosen above the current lower bound), so as to not waste time calculating an accurate upper bound once it is determined that it is impossible to fathom by bounds. However, there are potential drawbacks in terminating the calculation of upper bounds prior to convergence, since (i) some useful branching strategies depend on the upper bounds, and (ii) our lower-bounding strategy depends on the quality of the continuous solutions. Finally, we are developing a parallel implementation for the Convex Exemplar.

References

1. E. Anderson, Z. Bai, C. Bischof, J. Demmel, J. Dongarra, J. Du Croz, A. Greenbaum, S. Hammarling, A. McKenney, S. Ostrouchov and D. Sorensen (1992), LAPACK *Users' Guide*. SIAM, Philadelphia.
2. K.M. Anstreicher and M. Fampa (1996), A long-step path following algorithm for semidefinite programming problems. Working paper, Dept. of Management Sciences, University of Iowa, Iowa City.
3. D. Blackwell (1951), Comparison of experiments. In *Proceedings of the 2nd Berkeley Symposium,* 93-102, University of California Press, Berkeley.
4. L. Boltzmann (1877), Beziehung zwischen dem zweiten Haupstatz der Wärmetheorie und der Wahrscheinlichkeitsrechnung resp. den Sätzen über das Wärmegleichgewicht (Complexionen-Theorie). *Wien Ber.* 76^2, p. 373.
5. W.F. Caselton, L. Kan and J.V. Zidek (1991), Quality data network designs based on entropy. In *Statistics in the Environmental and Earth Science*, P. Guttorp and A. Walden, eds., Griffin, London.
6. W.F. Caselton and J. Zidek (1984), Optimal monitoring network designs. *Stat. Prob. Lett.* **2**, 223-227.
7. CPLEX Optimization, Inc. (1994), *Using the* CPLEX *Callable Library.*
8. D. den Hertog, C. Roos, and T. Terlaky (1992), On the classical logarithmic barrier method for a class of smooth convex programming problems. *JOTA* **73**, 1-25.
9. V. Fedorov, S. Leonov, M. Antonovsky and S. Pitovranov (1987), The experimental design of an observation network: software and examples. WP-87-05, International Institute for Applied Systems Analysis, Laxenburg, Austria.

10. V. Fedorov and W. Mueller (1989), Comparison of two approaches in the optimal design of an observation network. *Statistics* **20**, 339-351.
11. A.V. Fiacco and G.P. McCormick (1968), *Nonlinear Programming, Sequential Unconstrained Minimization Techniques.* John Wiley, New York; reprinted as *Classics in Applied Mathematics Vol. 4*, SIAM, Philadelphia, 1990.
12. P. Guttorp, N.D. Le, P.D. Sampson and J.V. Zidek (1992), Using Entropy in the redesign of an environmental monitoring network. Technical Report #116, Department of Statistics, The University of British Columbia, Vancouver, British Columbia.
13. R.A. Horn and C.R. Johnson (1985), *Matrix Analysis.* Cambridge University Press, Cambridge.
14. C.-W. Ko, J. Lee, and M. Queyranne (1995), An exact algorithm for maximum entropy sampling. *Operations Research* **43**, 684-691.
15. J. Lee (1995), Constrained maximum-entropy sampling. Working paper, Dept. of Mathematics, University of Kentucky, Lexington, KY.
16. C.E. Shannon (1948), The mathematical theory of communication. *Bell Systems Technical Journal* **27**, 379-423, 623-656.
17. M.C. Shewry and H.P. Wynn (1987), Maximum entropy sampling. *Journal of Applied Statistics* **46**, 165-170.
18. S. Wu and J.V. Zidek (1992), An entropy based review of selected NADP/NTN network sites for 1983-86. *Atmospheric Environment* **26A**, 2089-2103.

A Submodular Optimization Problem with Side Constraints

David Hartvigsen

Department of Management, College of Business Administration, University of Notre Dame, Notre Dame, IN 46556-0399

Abstract. In this paper we consider the general problem of optimizing over the intersection of a submodular base polyhedron and an affine space. An example is the following flow problem defined on a capacitated network: we wish to send a commodity from locations in a producing country to locations in a number of client countries so as to simultaneously maximize profits and send to each client country a prescribed proportion of the total sent. We present an algorithm for the general problem and analyze its complexity.

1 Introduction

Let V be a finite set and let $f : 2^V \to \Re$ be a submodular function; that is,

$$f(X) + f(Y) \geq f(X \cup Y) + f(X \cap Y) \quad \forall X, Y \subseteq V.$$

Without loss of generality, we assume that $f(\emptyset) = 0$. For convenience in stating our results, we additionally assume $f(V) = 0$, although this is not necessary (see discussion below).

The *base polyhedron of f* is defined to be

$$B(f) \equiv \{x : x(X) \leq f(X) \; \forall X \subseteq V, x(V) = f(V)\} \text{ where } x(X) \equiv \sum_{i \in X} x_i.$$

If we let $w \in \Re^{|V|}$, then the *submodular optimization problem* is the following:

$$\max \; wx \; \text{ subject to } x \in B(f).$$

This problem generalizes, for example, the problem of finding a maximum weight base of a matroid as well as the problem of finding max flows in a network; hence, it has many applications. This problem, and its variants, have been widely studied. The reader is referred, for example, to [3], [6], [7], and [10].

We are interested in a generalization of $B(f)$. Let A be a $p \times |V|$ matrix of reals with linearly independent rows. We define the *constrained base polyhedron of f* as follows:

$$B(f, A) \equiv \{x : x(X) \leq f(X) \; \forall X \subseteq V, x(V) = f(V), Ax = 0\}$$

(The idea is that $B(f, A)$ is a base polyhedron intersected with an affine space. To simplify notation, we have assumed that the base polyhedron and the affine space both contain the origin, hence our assumption that $f(V) = 0$.) It is not difficult to see that with even one row in A, $B(f, A)$ need not be a base polyhedron see discussion in Section 3). In this paper we study the following *constrained submodular optimization problem*:

$$\max wx \quad \text{subject to } x \in B(f, A). \tag{P}$$

An example is the following network flow problem. We wish to send a commodity from locations in a producing country to locations in a number of client countries so as to simultaneously do the following: maximize profits; stay within the transportations capacities; and send to each client country a prescribed proportion of the total sent. The "proportion constraint" is modeled with the constraints $Ax = 0$. Further details are given in Section 3, along with other examples.

This paper contains an algorithm for the constrained submodular optimization problem called Algorithm: CSO. The algorithm "combinatorializes" the structure of f through the use of a simple oracle (which computes "exchange capacities"; see the next section). Roughly speaking, this algorithm is efficient when the number of rows in A is "small." By efficient, we mean that the time complexity of the algorithm is always independent of the numbers in the range of f and dependent in some way on the numbers in w and the description at hand for A. With two quick definitions, we can make this precise.

For $\alpha \in Z_+$, a class of problems of type (P) is called α-*bounded* if the number of rows in the matrices A occurring in these problems is at most α. A matrix A is called *combinatorial* if the space $\{x : x(V) = 0, Ax = 0\}$ is given in the form of a basis of vectors whose entries are in $\{0, \pm 1\}$.

Theorem 1. *Algorithm: CSO has the following properties:*

(1.1) *The algorithm is finite even if A, w and the range of f contain irrational numbers.*

For the remainder, assume w and A are rational and let $\mathbf{C}$ be a class of problems (P).

(1.2) *If $\mathbf{C}$ is α-bounded, then the algorithm can be implemented in (weakly) polynomial time (i.e., the complexity is polynomial in $|V|$, α, and the size of the numbers in A and w).*

(1.3) *If $\mathbf{C}$ is α-bounded and the matrices A are combinatorial, then the algorithm can be implemented in strongly polynomial time (i.e., the complexity is polynomial in $|V|$ and α).*

The paper is organized as follows. In Section 2 we present the main algorithm. Section 3 contains some examples of constrained submodular optimization problems and how the algorithm works on some special cases. The last section contains a description of the proof of Theorem 1 and statements of a few of the main results upon which its proof depends.

2 Main Algorithm

In this section we present our main algorithm. First we need a few definitions. Recall that V denotes our ground set.

Let $D = (V, E)$ be a digraph with node set V. The *incidence matrix* B of D has a row for each node in V and a column for each arc in A such that

$$B_{ij} = \begin{cases} +1 & \text{if arc } j \text{ is incident into node i} \\ -1 & \text{if arc } j \text{ is incident out of node i} \\ 0 & \text{otherwise} \end{cases}$$

If D contains exactly one arc that goes from node u to node v, then we denote its (one column) incidence matrix $\lambda(u, v)$.

D is called a *bipartite forest* if it satisfies the following two conditions:

(2.1) There exists a partition of V into V_1 and V_2 such that every arc in A has its tail in V_1 and its head in V_2;

(2.2) The undirected graph underlying D is acyclic.

Note that a bipartite forest need not span V. The incidence matrix for a bipartite forest is called a *bipartite forest matrix* or, simply, a *b-f matrix*.

Let B be a $|V| \times m$ matrix. A $|V| \times m'$ matrix consisting of a subset of columns of B is called a *submatrix* of B. If B is a b-f matrix and $y \in \Re^m$, then the pair (B, y) is called a *feasible pair* if the following conditions are satisfied:

(2.3) $y \geq 0$; $\sum_{i=1}^{m} y(i) = 1$;

(2.4) $By \in \{x : x(V) = 0, Ax = 0\}$;

(2.5) There does not exist a pair (B', y') that satisfies (2.3) and (2.4), where B' is a submatrix of B.

Note that if there is no matrix $Ax = 0$ in the problem, then the feasible pairs all have the form $(\lambda(u, v), 1)$. Also note that if (B, y) is a feasible pair, then the sum of the absolute values of the entries of By is 2.

Consider a problem (P) and an $x \in B(f, A)$. A feasible pair (B, y) is called an *augmenting pair for* x if, for some $\delta > 0$, $x + \delta By \in B(f, A)$ and $w(x + \delta By) > w(x)$. We say that a b-f matrix B is *augmenting for* x if there exists an augmenting pair (B, y) for x. This notion suggests the following idea for an algorithm for problem (P):

Step 1 Let 0 be the initial feasible solution.

Iteratively apply the following two steps:

Step 2 Find an augmenting pair for the current feasible solution. If no such pair exists, then done.

Step 3 Use this augmenting pair to obtain a new feasible solution with as large an objective value as possible.

Unfortunately, this algorithm runs into the same sort of problem that the augmenting path algorithm of Ford and Fulkerson [5] ran into. Namely, the number of augmenting pairs used can depend on the numbers in the range of f. (This follows easily from the network example given in the next section.) Edmonds and Karp [4] and Dinic [2] solved this problem for max flows by augmenting along paths of minimum length. In a somewhat analogous fashion, we introduce the following notion and then our main algorithm.

The *weight* of a b-f matrix B is max $\{wBy : (B, y)$ is feasible$\}$. If y' achieves this max, then we call (B, y') a *feasible pair associated with* B. (Note that the associated feasible pair depends on w.)

Algorithm Constrained Submodular Optimization (CSO)
Input A problem (P).
Output An optimal solution to (P).
Step 1 Perturb the weights w. Set $x_0 := 0$ and $i := 1$.
Step 2 Find a maximum weight b-f matrix B_i that is augmenting for x_{i-1}. Let (B_i, y_i) be an associated feasible pair. If there is no augmenting b-f matrix, output x_{i-1}; done.
Step 3 Find $\delta_i = \max \{\delta > 0 : x_{i-1} + \delta B_i y_i \in B(f, A)\}$. Set $x_i := x_{i-1} + \delta_i B_i y_i$; set $i := i + 1$; and go to Step 2.

By "perturb the weights w" in Step 1, we mean to replace w with a new weight vector w' such that $w'(V_1) \neq w'(V_2)$ for all $V_1, V_2 \subseteq V$, where $V_1 \neq V_2$, and such that the unique optimal solution for (P) with w' is also an optimal solution for (P) with w. This can be accomplished by taking $w'(j) := w(j) + \varepsilon^j$ for all $j \in V$, where $V = \{1, \ldots, n\}$, and some small number ε. The effect of this in implementing the algorithm is to add $|V|$ bits to each $w(j)$ and a factor of $|V|$ to the overall complexity. To simplify notation we assume in the remainder of the paper that the weights w are already perturbed.

The details of how to implement the algorithm remain to be given. In particular, we have not described how to find (B_i, y_i) in Step 2, nor how to find δ_i in Step 3. This is discussed in Section 4.

Finally, we note that the questions of complexity are dealt with by using a standard oracle for computing "exchange capacities" (e.g., see [6]). That is , we assume for $x' \in B(f)$ and $u, v \in V$, $u \neq v$, that we have an oracle for computing

$$c(x', u, v) \equiv \max \{\delta > 0 : x' + \delta \lambda(u, v) \in B(f)\}.$$

3   Examples

This section contains a somewhat informal discussion of examples of the constrained submodular optimization problem and how *Algorithm: CSO* behaves on these examples. We consider the submodular optimization problem, the constrained problem with one special constraint, a network flow problem, a matching problem, and a matroid problem.

Let us first consider the case that there is no system $Ax = 0$ in (P) and hence we have a submodular optimization problem. Trivially, such problems are α-bounded for all α. Such problems are also combinatorial since any maximal independent set of vectors of the form $\lambda(u,v)$, $u,v \in V$, is a basis for $\{x : x(V) = 0\}$. Hence *Algorithm: CSO* is strongly polynomial for such problems by Theorem 1. It is well known that the submodular optimization problem can be solved by a straightforward greedy algorithm: the components of V are considered in decreasing order of weight and, as each component is considered, its value is increased as much as possible while maintaining feasibility (e.g., see [6]). Let us consider in more detail how *Algorithm: CSO* solves this problem.

As we observed in Section 2, all feasible pairs for this problem have the form $(\lambda(u,v), 1)$ hence all augmenting b-f matrices have the form $\lambda(u,v)$ and their weight is $w(v) - w(u)$. The algorithm begins with the feasible solution zero. In the iterative part of the algorithm we have a feasible solution, say z, and we find the pair $u,v \in V$ for which $w(v) - w(u)$ is a maximum and $c(z,u,v) > 0$. We then find a new feasible solution $z := z + c(z,u,v)\lambda(u,v)$. Our main theorem in the next section shows that the values $w(v) - w(u)$ are strictly decreasing from one iteration to the next. Hence, in this case, the algorithm iterates at most $|V|^2$ times and is strongly polynomial.

We next consider a well-known example of a submodular optimization problem (e.g., see [6]). Let $N = (V, E)$ be a network with arc capacities $c \in \Re_+^{|E|}$. A vector $y \in \Re^{|E|}$ such that $0 \le y \le c$ is called a *flow*. Let M denote the incidence matrix for N. A vector $x \in \Re^{|V|}$ is called *feasible for N and c* if there exists a flow y such that $x = My$. The polyhedron $P(N,c) \equiv \{x : x \text{ is feasible for } N \text{ and } c\}$ is well known to be a base polyhedron for the submodular function f defined by $f(X)$ equals the sum of the capacities of the arcs incident into $X \subseteq V$, where $f(\oslash) = f(V) = 0$. Hence $P(N,c) = B(f)$. Given node weights w, the submodular optimization problem is to find a maximum weight feasible vector. Consider the iterative part of *Algorithm:CSO* with feasible solution z. We find the pair of nodes u,v for which $w(v) - w(u) > 0$ is a maximum and for which there exists an augmenting path from u to v with respect to z. We then send a max flow from u to v, with respect to z, to get the next feasible solution.

Next, let us consider the constrained submodular optimization problem where the matrix A contains a single constraint $ax = 0$ whose coefficients are in $\{0, \pm 1\}$ and are both positive and negative. Let $T_1 = \{a(j) : a(j) = +1, j \in V\}$ and let $T_2 = \{a(j) : a(j) = -1, j \in V\}$. The vectors in $B(f,a)$ are those vectors $x \in B(f)$ such that $x(T_1) = x(T_2)$.

First, we observe that $B(f,a)$ is not a submodular base polyhedron. To see this, consider the well known result (e.g., see Theorem 3.27 in [6]) that says if $x_1, x_2 \in B(f')$ for some submodular f', then there exist $z_j \in B(f')$, $\varepsilon_j \in \Re_+$, and $\lambda(u_j, v_j)$, for $j = 1, \ldots, p$ such that $z_1 = x_1$, $z_p = x_2$, and $z_{j+1} = z_j + \varepsilon_j \lambda(u_j, v_j)$ for $j = 1, \ldots, p-1$. Furthermore, the progression $z_1, z_2, \ldots$ is monotonic. Clearly this is not possible for $x_1 = 0$ and any vector $x_2 \in B(f,a)$ for which $x_2(T_1) \neq 0$.

Next, it is not difficult to see that feasible pairs have either the form $(\lambda(u,v), 1)$ or $(B, (\frac{1}{2}, \frac{1}{2}))$ such that B contains two columns of the form $\lambda(u_1, v_1)$, $\lambda(u_2, v_2)$, where either u_1, $u_2 \in V\backslash(T_1 \cup T_2)$ and $v_1 \in T_1$, $v_2 \in T_2$, or B is the negative of such a matrix. From this it follows that a is combinatorial. Since these problems are 1-bounded, Theorem 1 tells us that *Algorithm: CSO* is strongly polynomial for this class. However, finding the values δ_i in the algorithm is no longer a straightforward application of our exchange oracle (e.g., if we specialize this problem to networks, as described above, we cannot augment by solving a simple max flow problem).

Let us next consider a generalized max flow problem. Let $N = (V, E)$ be a network with arc capacities c and incidence matrix M, as above. Let $S_1, \ldots, S_p$, $T_1, \ldots, T_q$ be disjoint subsets of V with $S = S_1 \cup \cdots \cup S_p$ and $T = T_1 \cup \cdots \cup T_q$. Let $b_1, \ldots, b_p, d_1, \ldots, d_q \in \Re_+$ satisfy $\sum_{i=1}^{p} b_i = \sum_{i=1}^{q} d_i = 1$. Let us now call a vector x *feasible* if

(3.1) there exists a flow y such that $x = My$;

(3.2) $x(j) \leq 0$ for $j \in S$, $x(j) \geq 0$, for $j \in T$, $x(j) = 0$ for $j \in V\backslash(S \cup T)$;

(3.3) $\sum_{j \in S_i} x(j) = b_i \sum_{j \in S} x(j)$ for $i = 1, \ldots, p$;
$\sum_{j \in T_i} x(j) = d_i \sum_{j \in T} x(j)$ for $i = 1, \ldots, q$.

(3.2) says, roughly, that y is a flow from S to T. (3.3) says that, for a feasible vector, the flow out of each S_i equals b_i times the total flow; and the flow into each T_i equals d_i times the total flow. Vectors that satisfy (3.1) and (3.2) can easily be seen to be a submodular base polyhedron and the additional constraints in (3.3) can be described by a suitable side constraint matrix A. So let $B(f, A)$ denote the polyhedron of feasible solutions to such a problem.

We may interpret the problem (P) defined on $B(f, A)$ as follows. We have a set of source nodes at which a commodity may be purchased for a specified cost w. And we have a set of sinks at which the commodity may be sold at a specified price w. We wish to buy some of the commodity, move it through the network without violating the capacities, and sell it so as to maximize our profits, subject to a side condition. The additional condition is an equity consideration: if F is the total amount of the commodity bought, then we wish to buy a fraction b_i of F from S_i and we wish to sell a fraction d_i of F to T_i, for each i. Work on related problems occurs in [1], [8], [9], and [16].

If $p = q = |S_1| = |T_1| = 1$, then, of course, we have the classical max flow problem. If we have only $p = q = 1$, then the problem is equivalent to a special min cost flow problem (where the node weights for $S \cup T$ are moved onto new arcs incident with a supersource and a supersink). If we have $|S_1| = \cdots = |S_p| = |T_1| = \cdots = |T_q| = 1$, then we have, essentially, what is referred to in the literature as a *parametric flow problem*. Strongly polynomial algorithms exist for this problem (see [13] and [14]). This special case is an example of a problem (P) for which $\{x : x(V) = 0, Ax = 0\}$ has dimension 1.

If $p = 1$ and $d_1 = \cdots = d_q$, then one can see that the problem becomes combinatorial. If, in addition, $1 + q \leq \alpha$, for some fixed α, then we obtain an α-bounded and combinatorial class of problems for which *Algorithm: CSO* is strongly polynomial, by Theorem 1. If $p = 1$, $|S_1| = 1$, and $d_1 = \cdots = d_q$, then the algorithm is again strongly polynomial, regardless of the size of q

(although, in this case, A is combinatorial but not α-bounded; we must use the main theorem in the next section to prove strong polynomiality). In fact, we have a stronger result. For this special case, it is not difficult to see that Steps 2 and 3 of *Algorithm: CSO* can be implemented as a transshipment problem and as a parametric flow problem, respectively. It follows from Theorem 3, in the next section, that this special case can be solved in strongly polynomial time for any class defined by an upper bound on $|S| + |T|$, which is a class of α-bounded such problems (that are not necessarily combinatorial).

Another example of a constrained submodular optimization problem comes from matching theory. Let $G = (S, T, F)$ be an undirected bipartite graph on node set $S \cup T$. A *2-matching* is an assignment of numbers from $\{0, 1, 2\}$ to each edge so that the sum of the values on the edges incident with each node is at most 2. The *cardinality* of a 2-matching is the sum of the numbers on all edges. The problem of finding a maximum cardinality (and maximum weight, in general) 2-matching in a general graph is a well-studied variant of the classical matching problem (see [11]; it is sometimes called the *fractional matching problem*).

Let us consider this problem with a side constraint. Let T_1, T_2 be a partition of T. Then a (T_1, T_2)-*2-matching* is a 2-matching such that the sum of the values on the edges incident with T_1 equals the sum of the values on the edges incident with T_2. The problem of finding a maximum cardinality (T_1, T_2)-2-matching can be modeled as a constrained submodular optimization problem with a single constraint and can be shown to be solvable in strongly polynomial time using *Algorithm: CSO*.

Let us finally consider the case of matroids. If $M = (V, \mathcal{I})$ is a matroid, then the convex hull of incidence vectors of bases is a base polyhedron. The submodular optimization problem is thus to find a maximum weight base (see [3]). Our algorithm has the following interpretation when applied to this problem (with the $f(0) = 0$ assumption relaxed). Let $b \in \mathcal{I}$ be an arbitrary base. Consider every ordered pair of elements (u, v), for which $w(v) - w(u) > 0$, in decreasing order of $w(v) - w(u)$. For each pair, if $b \setminus \{u\} \cup \{v\}$ is a base, then set $b := b \setminus \{u\} \cup \{v\}$. So in $O\left(|V|^2\right)$ steps, this algorithm transforms an arbitrary base into one with maximum weight.(Of course the greedy algorithm accomplishes this starting from 0 in $O(|V|)$ steps.) We note that if we add side constraints to this problem (i.e., hyperplanes that contain b) the integrality of the optimal solution may be lost.

4   Supporting Results

We begin this section by stating a decomposition theorem (Theorem 2) for vectors contained in a constrained base polyhedron. The proof of this result is somewhat technical and is omitted. We next present a theorem (Theorem 3) that states that the weights of the b-f matrices used in *Algorithm: CSO* are decreasing. We include a proof of this result that relies upon the preceding decomposition theorem. We then use Theorem 3 to prove (1.1) of Theorem 1.

Two hurdles remain in completing the proof of Theorem 1: we must show that Steps 2 and 3 of *Algorithm:CSO* can be implemented in the claimed complexity. This is accomplished with two subroutines, one for each step. We include in this section a statement of the first of these (with the analysis of its precise complexity omitted). The second subroutine requires a bit more machinery and is omitted. We mention that it is closely based upon the algorithms of McCormick and Ervolina [12] and Radzik [14] for the maximum mean weight cut problem. The claimed strongly polynomial complexity in (1.3) is obtained by using the algorithm of Tardos [15] in the implementation of the subroutine for Step 2.

With a quick definition, we can present our next result. Let $u, v \in \Re^n$. We say v *conforms to* u if $v(j) \neq 0$ implies $u(j)v(j) > 0$.

Theorem 2. *Let $x^* \in B(f, A)$. Then there exists a collection of feasible pairs $(B_1, y_1), \ldots, (B_p, y_p)$ such that*

(4.1) $x^* = \sum_{j=1}^p \alpha_j B_j y_j$ *where* $\alpha_j > 0$ *and* $B_j y_j$ *conforms to* x^*, *for* $j = 1, \ldots, p$;

(4.2) $\delta B_j y_j \in B(f, A)$, *for* $j = 1, \ldots, p$ *and some* $\delta > 0$.

Theorem 3. *Suppose in two consecutive iterations of* Algorithm: CSO *the augmenting pairs (B_i, y_i) and (B_{i+1}, y_{i+1}) are used. Then $wB_i y_i > wB_{i+1} y_{i+1}$.*

Proof. Suppose this is not the case, that is, suppose $wB_i y_i < wB_{i+1} y_{i+1}$. Note that we have a strict inequality here because $B_i y_i \neq B_{i+1} y_{i+1}$ and the weights w are perturbed.)

Let $B(f', A)$ denote the constrained base polyhedron obtained from $B(f, A)$ by subtracting (feasible solution) x_{i-1} from every vector in $B(f, A)$. Consider $z = \delta_i B_i y_i + \delta_{i+1} B_{i+1} y_{i+1}$. Clearly, $z \in B(f', A)$.

Apply Theorem 2 to z in $B(f', A)$ to get feasible pairs (B'_j, y'_j), $j = 1, \ldots, p$, that satisfy (4.1) and (4.2). In particular, we have $z = \sum_{j=1}^p \alpha_j B'_j y'_j$, using the notation of Theorem 2. Thus we have

$$w(\delta_i B_i y_i + \delta_{i+1} B_{i+1} y_{i+1}) = w(\sum_{j=1}^p \alpha_j B'_j y'_j).$$

We also have $\delta_i + \delta_{i+1} \geq \sum_{j=1}^p \alpha_j$, since each vector $B'_j y'_j$ conforms to z. It follows that

$$\frac{\delta_i wB_i y_i + \delta_{i+1} wB_{i+1} y_{i+1}}{\delta_i + \delta_{i+1}} \leq \frac{\sum_{j=1}^p \alpha_j wB'_j y'_j}{\sum_{j=1}^p \alpha_j}$$

Because *Algorithm: CSO* chooses a maximum augmenting pair in Step 2, it follows that

$$wB'_j y'_j \leq wB_i y_i < wB_{i+1} y_{i+1}, \text{ for } j = 1, \ldots, p.$$

(The second inequality is by assumption.) But a weighted average of $wB_i y_i$ and $wB_{i+1} y_{i+1}$ cannot be less than a weighted average of $wB'_j y'_j$, for $j = 1, \ldots, p$, which is what we are claiming above. This is a contradiction and the result follows. $\square$

Proof of (1.1). First, observe that if the algorithm is in Step 2, and if x_{i-1} is not optimal, then there exists a b-f matrix B_i that is augmenting for x_{i-1}. To see this, form the polyhedron $B(f', A)$ described in the proof of Theorem 3. Next, apply the decomposition described in Theorem 2 to the vector from 0 to the optimal solution. Clearly, at least one of the b-f matrices in the decomposition has positive weight and is, therefore, augmenting for x_{i-1}.

Second, observe that the number of b-f matrices for a problem (P) is finite. Hence, the number of different weights that b-f matrices can have is bounded. It follows from Theorem 3 that *Algorithm: CSO* is finite, even if w, A, and the range of f contain irrational numbers. $\square$

The following subroutine is used for Step 2 in *Algorithm: CSO*.

Subroutine Max weight b-f matrix

Input $x_{i-1} \in B(f, A)$ for a problem (P).

Output A maximum weight b-f matrix that is augmenting for x_{i-1} plus an associated feasible pair (B_i, y_i); or the fact that none exists.

Step 1 If this is the first call of this subroutine, then find a matrix C with independent columns such that $\{x : x(V) = 0, Ax = 0\} = \{y : y = Cz\}$.

Step 2 Form a matrix B whose columns are those vectors $\lambda(u, v)$ such that $c(x_{i-1}, u, v) > 0$. Let m denote the number of columns in B.

Step 3 Form a vector $w'(j) := w(v) - w(u)$ where j corresponds to column $\lambda(u, v)$ in B, for $j = 1, \ldots, m$.

Step 4 Solve the following linear program (with variable vectors y and z):

$$\max \; w'x$$

$$\text{s.t. } By = Cz$$
$$1 \cdot y = 1 \qquad \qquad \text{(LP)}$$
$$y \geq 0$$

If there is no optimal solution, then done; output "no augmenting pair." Otherwise, let (y^*, z^*) be a basic optimal solution.

Step 5 Let y' denote the vector of nonzero entries in y^* (i.e., delete the zero entries) and let B' denote the corresponding submatrix of columns of B. If B' contains no two columns of the form $\lambda(a, b)$, $\lambda(b, c)$, then done; output (B', y'). Otherwise, note that $\lambda(a, c)$ is a column of B (but not necessarily B'). Let $y'_{ab}, y^*_{ab}, y'_{bc}, y^*_{bc}, y^*_{ac}$ be the entries of y' and y^* that correspond to $\lambda(a, b)$, $\lambda(b, c)$, and $\lambda(a, c)$, respectively.

If $y'_{ab} = y'_{bc}$, change y^* as follows:

$$y^*_{ab} := 0$$
$$y^*_{bc} := 0$$
$$y^*_{ac} := y'_{ab}$$

If $y'_{ab} > y'_{bc}$, change y^* as follows:

$$y^*_{ab} := y'_{ab} - y'_{bc}$$
$$y^*_{bc} := 0$$
$$y^*_{ac} := y'_{bc}$$

If $y'_{ab} < y'_{bc}$, change y^* as follows:

$$y^*_{ab} := 0$$
$$y^*_{bc} := y'_{bc} - y'_{ab}$$
$$y^*_{ac} := y'_{ab}$$

Repeat Step 5.

Proposition 4. Subroutine:Max weight b-f matrix *works*.

Proof. Observe that the columns of B, formed in Step 2, are precisely the candidates for columns in B_i. Indeed, matrix C is chosen so that every augmenting pair yields a feasible solution to (LP).

Conversely, observe that the matrix B' constructed in the first pass through Step 5 is an incidence matrix for a digraph, say D, on V. Because (y^*, z^*) is basic, the columns of B' are independent, which is equivalent (from network theory) to the graph underlying D being acyclic. It remains to check if D is a bipartite forest. This is what we make sure of in the remainder of Step 5. With each pass through Step 5 we reduce the number of dipaths of length 2 in D by one, while simultaneously producing a new solution (y^*, z^*) with the same objective value in (LP) and while maintaining an acyclic underlying graph. Since there are a linear number of dipaths of length 2 in D, the algorithm terminates with a maximum weight augmenting pair. $\square$

References

1. Brown, J.R.: The sharing problem. Operations Research **27** (1979) 324–340.
2. Dinic, E.A.: Algorithm for a solution of a problem of maximum flow in a network with power estimation. Soviet Math. Dokl. **2** (1970) 1277–1280.
3. Edmonds, J.: Submodular functions, matroids, and certain polyhedra. In: Combinatorial Structures and their Applications. (R.K. Guy, H. Hanoni, N. Sauer, and J. Schonheim, eds.), Gordon and Breach, New York, 1970, 69–87.
4. Edmonds, J., Karp, R.M.: Theoretical improvements in algorithmic efficiency for network flow problems. J. of ACM **19** (1972) 248–264.
5. Ford, L.R., Fulkerson, D.R.: Maximal flow through a network. Canadian J. of Math. **8** (1956) 399–404.
6. Fujishige, S.: Submodular Functions and Optimization. North Holland, Amsterdam, 1991.
7. Fujishige, S.: Submodular systems and related topics. Math. Programming Study **22** (1984) 113–131.

8. Hall, N.G., Vohra, R.V.: Towards equitable distribution via proportional equity constraints. Math. Prog. **58** (1993) 287–294.
9. Ichimori, T., Ishii, H., Nishida, T.: Optimal sharing. Math. Prog. **23** (1982) 341–348.
10. Lovasz, L.: Submodular functions and convexity. In: Mathematical Programming–The State of the Art. (A. Machem, M. Grotschel, and B. Korte, eds.), Springer, Berlin (1983) 235–257.
11. Lovasz, L., Plummer, M.D.: Matching Theory. North Holland, Amsterdam, 1986.
12. McCormick, S.T., Ervolina, T.R.: Computing maximum mean cuts. Discrete Applied Math. **52** (1994) 53–70.
13. Megiddo, N.: Applying parallel computation algorithms in the design of serial algorithms. J. of ACM **30** (1983) 852–865.
14. Radzik, T.: Parametric flows, weighted means of cuts, and fractional combinatorial optimization. In: Complexity in Numerical Optimization (P. Pardalos, ed.), World Scientific (1993) 351–386.
15. Tardos, E.: A strongly polynomial algorithm to solve combinatorial linear programs. Operations Research **34** (1986) 250–256.
16. Zimmerman, U.: Duality for balanced submodular flows. Disc. Appl. Math. **15** (1986) 365–376.

Convexity and Steinitz's Exchange Property (Extended Abstract)

Kazuo Murota

Research Institute for Mathematical Sciences, Kyoto University, Kyoto 606-01, Japan

Abstract. A theory of "convex analysis" is developed for functions defined on integer lattice points. We investigate the class of functions which enjoy a variant of Steinitz's exchange property. It includes linear functions on matroids, valuations on matroids, and separable concave functions on the integral base polytope. It is shown that a function ω has the exchange property if and only if it can be extended to a concave function $\overline{\omega}$ such that the maximizers of ($\overline{\omega}$+any linear function) form an integral base polytope. A Fenchel-type min-max theorem and discrete separation theorems are given, which contain, e.g., Frank's discrete separation theorem for submodular functions, and also Frank's weight splitting theorem for weighted matroid intersection.

1 Introduction

The analogy between convex/concave functions and submodular/supermodular functions attracted research interest in the 80's. Fujishige [8] formulated Edmonds' intersection theorem into a Fenchel-type min-max duality theorem. Frank [6] showed a separation theorem for a pair of submodular/supermodular functions, with integrality assertion for the separating hyperplane in the case of integer-valued functions. These theorems can also be regarded as being equivalent to Edmonds' intersection theorem. A precise statement, beyond analogy, about the relationship between convex functions and submodular functions was made by Lovász [12]. Namely, a set function is submodular if and only if the so-called Lovász extension of that function is convex. This penetrating remark also established a direct link between the duality for convex/concave functions and that for submodular/supermodular functions. The essence of the duality for submodular/supermodular functions is now recognized as the discreteness (integrality) assertion in addition to the duality for convex/concave functions.

In spite of these developments, our understanding of the relationship between convexity and submodularity seems to be only partial. In convex analysis, a convex function is minimized over a convex domain of definition, which can be described by a system of inequalities in (other) convex functions. In the polyhedral approach to matroid optimization, a linear function is optimized over a (discrete) domain of definition, which is described by a system of inequalities involving submodular functions. The relationship between convexity and submodularity we have understood so far is concerned only with the domain of definitions and not with the objective functions. In the literature, however, we can find a number of results on the optimization of nonlinear functions over the base polytope of a submodular system. In particular, the minimization of a separable convex function over a base polytope has been considered by Fujishige [7] and Groenevelt [10], and the submodular flow problem with a separable convex objective function has been treated by Fujishige [9]. Our present knowledge does not help us understand these results in relation to convex analysis.

Quite independently of the developments in the theory of submodular functions, Dress and Wenzel [3], [4] have recently introduced the concept of a valuated matroid as a quantitative generalization of matroid. A matroid $(V, \mathcal{B})$, defined in terms of the family of bases $\mathcal{B} \subseteq 2^V$, is characterized by the simultaneous exchange (a variant of Steinitz's exchange) property:

For $X, Y \in \mathcal{B}$ and $u \in X - Y$ there exists $v \in Y - X$ such that $X - u + v \in \mathcal{B}$ and $Y + u - v \in \mathcal{B}$.

A valuation of $(V, \mathcal{B})$ is a function $\omega : \mathcal{B} \to \mathbb{R}$ which enjoys the quantitative extension of this exchange property:

(MV) For $X, Y \in \mathcal{B}$ and $u \in X - Y$ there exists $v \in Y - X$ such that $X - u + v \in \mathcal{B}$, $Y + u - v \in \mathcal{B}$ and $\omega(X) + \omega(Y) \leq \omega(X - u + v) + \omega(Y + u - v)$.

It has turned out recently that the valuated matroids afford a nice combinatorial framework to which the optimization algorithms for matroids can be generalized. Variants of greedy algorithms work for maximizing a matroid valuation, as has been shown by Dress–Wenzel [3] as well as by Dress–Terhalle [1, 2] and Murota [13]. The weighted matroid intersection problem has been extended by Murota [14] to the valuated matroid intersection problem.

This direction of research can be extended further as follows. Let us say that $B \subseteq \mathbb{Z}^V$ is an integral base set if it is a nonempty set that satisfies:

(BAS) For $x, y \in B$ and for $u \in \mathrm{supp}^+(x - y)$, there exists $v \in \mathrm{supp}^-(x - y)$ such that $x - \chi_u + \chi_v \in B$ and $y + \chi_u - \chi_v \in B$,

where $\mathrm{supp}^+(x - y) = \{u \in V \mid x(u) > y(u)\}$, $\mathrm{supp}^-(x - y) = \{v \in V \mid x(v) < y(v)\}$ and χ_u denotes the characteristic vector of $u \in V$. The following folk theorem states that (BAS) characterizes the set of the integral points of an integral base polytope.

Theorem 1. Let B be a finite nonempty subset of $\mathbb{Z}^V$. B satisfies (BAS) if and only if there exists an integer-valued supermodular function $\mu : 2^V \to \mathbb{Z}$ with $\mu(\emptyset) = 0$ such that $B = \mathbb{Z}^V \cap \mathbf{B}(\mu)$ and $\overline{B} = \mathbf{B}(\mu)$, where $\overline{B}$ denotes the convex hull of B, and $\mathbf{B}(\mu) = \{x \in \mathbb{R}^V \mid x(X) \geq \mu(X) \ (\forall X \subset V), x(V) = \mu(V)\}$.

We then consider a function $\omega : B \to \mathbb{R}$ on a finite integral base set B such that:

(EXC) For $x, y \in B$ and $u \in \mathrm{supp}^+(x - y)$ there exists $v \in \mathrm{supp}^-(x - y)$ such that $x - \chi_u + \chi_v \in B$, $y + \chi_u - \chi_v \in B$ and $\omega(x) + \omega(y) \leq \omega(x - \chi_u + \chi_v) + \omega(y + \chi_u - \chi_v)$.

We call such ω an M-concave function, where M stands for Matroid.

M-concave functions arise naturally in combinatorial optimization; for example, a linear function on a matroid, a separable concave function on the integral base polytope of a submodular system, and the maximum cost of a network flow, illustrated below. It is remarked that a general concave function on a base polytope does not satisfy (EXC) when restricted to $\mathbb{Z}^V$.

Example 1. Let $G = (V, A)$ be a directed graph with a vertex set V and an arc set A. Assume further that we are given an upper capacity function $\bar{c} : A \to \mathbb{Z}$ and a lower capacity function $\underline{c} : A \to \mathbb{Z}$. A feasible (integral) flow φ is a function $\varphi : A \to \mathbb{Z}$ such that $\underline{c}(a) \leq \varphi(a) \leq \bar{c}(a)$ for each $a \in A$. Its boundary $\partial\varphi : V \to \mathbb{Z}$ is defined by

$$\partial\varphi(v) = \sum\{\varphi(a) \mid a \in \delta^+ v\} - \sum\{\varphi(a) \mid a \in \delta^- v\} , \qquad (1)$$

where $\delta^+ v$ and $\delta^- v$ denote the sets of the out-going and in-coming arcs incident to v, respectively. Then

$$B = \{\partial\varphi \mid \varphi : \text{ feasible flow }\}$$

is known to satisfy (BAS). See, e.g., [9].

Suppose further that we are given a family of convex functions $f_a : \mathbb{Z} \to \mathbb{R}$ indexed by $a \in A$, where we call $f : \mathbb{Z} \to \mathbb{R}$ convex if its piecewise linear extension $\tilde{f} : \mathbb{R} \to \mathbb{R}$ is a convex function. Define $\Gamma(\varphi) = \sum\{f_a(\varphi(a)) \mid a \in A\}$. Then the function $\omega : B \to \mathbb{R}$ defined by

$$\omega(x) = -\min\{\Gamma(\varphi) \mid \varphi : \text{ feasible flow with } \partial\varphi = x\} \qquad (x \in B)$$

is M-concave, satisfying (EXC) (see [16] for the proof). $\qquad\square$

Theorem 1 shows that (BAS) is (cryptomorphically) equivalent to sub/supermodularity. With the correspondence between convexity and submodularity in mind, we may say that (BAS) prescribes a certain "convexity" of the domain of definition of the function ω. The main theme of this paper is to demonstrate that the property (EXC) can be interpreted as "concavity" of the objective function in the context of combinatorial optimization. Three central questions are the following:

– We know a pair of "conjugate" characterizations of the base polytope of a sub/supermodular system, namely, the exchange property (BAS) for the points in the polytope and the sub/supermodularity for (the inequalities describing) the faces of the polytope. The property (EXC) is a quantitative generalization of (BAS). Then what is the generalization of sub/supermodularity that corresponds to (EXC)? An answer is given in Theorem 11.

$$
\begin{array}{ccc}
\text{[Domain]} & & \text{[Function]} \\
\text{(BAS)} & \Longrightarrow & \text{(EXC)} \\
\Updownarrow & & \Updownarrow \\
\text{sub/supermodularity} & \Longrightarrow & ?
\end{array}
\qquad (2)
$$

- Can an M-concave function be extended to a concave function in the usual sense, just as a submodular function can be extended to a convex function through the Lovász extension? Theorem 9 gives a positive answer to this.
- Is there any duality for M-convex/concave functions that corresponds to the duality for convex/concave functions? The main concern here will be the discreteness (integrality) assertion for such pair of integer-valued functions. This amounts to a generalization of the potential characterization of the optimality due to Iri–Tomizawa [11] and the weight splitting theorem of Frank [5] for the weighted matroid intersection.

2 Properties of M-concave Functions

Let $B \subseteq \mathbb{Z}^V$ be a finite integral base set (satisfying (BAS) by definition). We are concerned with an M-concave function $\omega : B \to \mathbb{R}$, which by definition satisfies (EXC). Convention: $\omega(x) = -\infty$ for $x \notin B$.

For $p : V \to \mathbb{R}$ we define $\omega[p] : B \to \mathbb{R}$ by

$$\omega[p](x) = \omega(x) + \langle p, x \rangle$$

with

$$\langle p, x \rangle = \sum_{v \in V} p(v)x(v) \ .$$

Theorem 2. $\omega[p]$ *satisfies* (EXC).

We introduce the notation

$$\omega(x, u, v) = \omega(x - \chi_u + \chi_v) - \omega(x) \qquad (x \in B; u, v \in V) \tag{3}$$

to represent the information about the local behavior of ω in the neighborhood of x. The following fact is fundamental, showing the local optimality implies the global optimality. (This is a straightforward extension of the result of [3], [4].)

Theorem 3 [16]. *Let* $x \in B$. *Then* $\omega(x) \geq \omega(y)$ $(\forall y \in B)$ *if and only if*

$$\omega(x, u, v) \leq 0 \qquad (\forall u, v \in V) \ . \tag{4}$$

An upper estimate of $\omega(y) - \omega(x)$ can be given in terms of the local information $\omega(x, u, v)$. For $x, y \in B$ we consider a bipartite graph $G(x, y) = (V^+, V^-; \hat{A})$, where $(V^+, V^-) = (\text{supp}^+(x - y), \text{supp}^-(x - y))$ is the vertex bipartition and

$$\hat{A} = \{(u, v) \mid u \in V^+, v \in V^-, x - \chi_u + \chi_v \in B\}$$

is the arc set. Each arc (u, v) is associated with "arc weight" $\omega(x, u, v)$. We define

$$\hat{\omega}(x, y) = \max \left\{ \sum_{(u,v) \in \hat{A}} \omega(x, u, v)\lambda_{uv} \ \middle| \ \lambda_{uv} \geq 0 \ ((u, v) \in \hat{A}), \right.$$

$$\left. \sum_{v:(u,v) \in \hat{A}} \lambda_{uv} = x(u) - y(u) \ (u \in V^+), \quad \sum_{u:(u,v) \in \hat{A}} \lambda_{uv} = y(v) - x(v) \ (v \in V^-) \right\} .$$

It is known that such $\lambda \in \mathbb{R}^A$ exists, so that $\widehat{\omega}(x, y)$ is defined to be a finite value. We have the following lemma, called "upper-bound lemma" in [16].

Theorem 4 [16]. *For $x, y \in B$, $\omega(y) \leq \omega(x) + \widehat{\omega}(x, y)$.*

Just as the maximizers of a concave function form a convex set, the family of the maximizers of ω, denoted argmax (ω), enjoys a nice property.

Lemma 5. *If $\omega : B \to \mathbb{R}$ has the property (EXC), then argmax (ω) is an integral base set.*

This lemma implies furthermore that argmax $(\omega[p])$ is an integral base set for each $p : V \to \mathbb{R}$, since $\omega[p]$ also satisfies (EXC) by Theorem 2. This turns out to be a key property for the M-concavity (EXC) as follows.

Theorem 6. *Let $\omega : B \to \mathbb{R}$ be a function on a finite integral base set $B \subseteq \mathbb{Z}^V$. Then ω satisfies (EXC) if and only if argmax $(\omega[p])$ is an integral base set for each $p : V \to \mathbb{R}$.*

3 Conjugate Function and Concave Extension

In line with the standard method in convex analysis [18], we introduce the concept of conjugate function. For a function $g : B \to \mathbb{R}$ in general we define $g^\circ : \mathbb{R}^V \to \mathbb{R}$ by

$$g^\circ(p) = \min\{\langle p, x \rangle - g(x) \mid x \in B\} . \tag{5}$$

We call g° the concave conjugate function of g. Since $|B|$ is finite, g° is a polyhedral concave function, taking finite values for all p. Furthermore we define $\hat{g} : \mathbb{R}^V \to \mathbb{R}$ by

$$\hat{g}(b) = \inf\{\langle p, b \rangle - g^\circ(p) \mid p \in \mathbb{R}^V\} . \tag{6}$$

Obviously, $\hat{g}$ is a concave function, which we call the concave closure of g. By a standard result from convex analysis we see

$$\hat{g}(b) = \begin{cases} \max\{\sum_{y \in B} \lambda_y g(y) \mid b = \sum_{y \in B} \lambda_y y, \ \lambda \in \Lambda(B)\} & (b \in \overline{B}) \\ -\infty & (b \notin \overline{B}) \end{cases} \tag{7}$$

where $\lambda = (\lambda_y \mid y \in B) \in \mathbb{R}^B$, $\overline{B}$ denotes the convex hull of B, and $\Lambda(B) = \{\lambda \in \mathbb{R}^B \mid \sum_{y \in B} \lambda_y = 1, \ \lambda_y \geq 0 \ (y \in B)\}$. Define

$$\text{argmax}\,(g) = \{x \in B \mid g(x) \geq g(y) \ (\forall y \in B)\} , \tag{8}$$

$$\text{argmax}\,(\hat{g}) = \{b \in \overline{B} \mid \hat{g}(b) \geq \hat{g}(c) \ (\forall c \in \overline{B})\} , \tag{9}$$

and denote by $\overline{\text{argmax}\,(g)}$ the convex hull of argmax (g).

Lemma 7. (1) $\hat{g}(x) \geq g(x)$ for $x \in B$.

(2) $\max\{\hat{g}(b) \mid b \in \overline{B}\} = \max\{g(x) \mid x \in B\}$.

(3) $\text{argmax}\,(\hat{g}) = \overline{\text{argmax}\,(g)}$.

We reveal a precise relationship between the M-concavity (EXC) and the ordinary concavity. By Lemma 7(1) we know that $\hat{\omega} : \overline{B} \to \mathbb{R}$ is a concave function such that $\hat{\omega}(x) \geq \omega(x)$ for $x \in B$. The M-concavity (EXC) guarantees the equality here as follows.

Lemma 8. *If $\omega : B \to \mathbb{R}$ has the property (EXC), then $\hat{\omega}(x) = \omega(x)$ for $x \in B$.*

Theorem 9 (Extension Theorem). *Let $\omega : B \to \mathbb{R}$ be a function on a finite integral base set $B \subseteq \mathbb{Z}^V$. Then ω satisfies (EXC) if and only if it can be extended to a concave function $\overline{\omega} : \overline{B} \to \mathbb{R}$ such that $\mathrm{argmax}\,(\overline{\omega}[p])$ is an integral base polytope for each $p : V \to \mathbb{R}$. Here, $\overline{\omega}[p] : \overline{B} \to \mathbb{R}$ is defined by*

$$\overline{\omega}[p](b) = \overline{\omega}(b) + \langle p, b \rangle = \overline{\omega}(b) + \sum_{v \in V} p(v)b(v).$$

Proof. "only if": We can take $\overline{\omega} = \hat{\omega}$, which is an extension of ω by Lemma 8 and meets the requirement by $\mathrm{argmax}\,(\hat{\omega}[p]) = \mathrm{argmax}\,((\omega[p])\hat{\,}) = \overline{\mathrm{argmax}\,(\omega[p])}$ and Theorem 6.

"if": Obviously we have $\max(\overline{\omega}[p]) \geq \max(\omega[p])$, since $\overline{\omega}[p](x) = \omega[p](x)$ for $x \in B$. On the other hand, $\mathrm{argmax}\,(\overline{\omega}[p])$ contains an integral point, which belongs to $\mathbb{Z}^V \cap \overline{B} = B$. Therefore we have $\max(\overline{\omega}[p]) = \max(\omega[p])$ and $\mathbb{Z}^V \cap \mathrm{argmax}\,(\overline{\omega}[p]) = \mathrm{argmax}\,(\omega[p])$. Since $\mathrm{argmax}\,(\overline{\omega}[p])$ is an integral base polytope by the assumption, it follows from Theorem 6 that ω satisfies (EXC). $\square$

4 Supermodularity in Conjugate Function

In Theorem 1 we have seen that the exchange property (BAS) of B is equivalent to the supermodularity of the function μ describing the face of the polytope $\overline{B}$. As the property (EXC) for ω can be regarded as a quantitative extension of (BAS) for B, it is natural to seek for an extension of this correspondence between the exchangeability and the sub/supermodularity (see (2)). Theorem 11 below says that (EXC) for ω is equivalent to "local supermodularity" of the concave conjugate function ω°.

4.1 Exchangeability (BAS) and Supermodularity

We reformulate Theorem 1 into a form that is suitable for our subsequent extension. We assume $B \subseteq \mathbb{Z}^V$ is a finite nonempty set such that $B = \mathbb{Z}^V \cap \overline{B}$.

We define $\psi^\circ : \mathbb{R}^V \to \mathbb{R}$ by

$$\psi^\circ(p) = \min\{\langle p, x \rangle \mid x \in B\} \ . \tag{10}$$

Note that ψ° is the concave conjugate function of $\psi \equiv 0$ (on B) in the sense of (5). Obviously, $\psi^\circ(p)$ is concave, $\psi^\circ(0) = 0$, and positively homogeneous, i.e., $\psi^\circ(\lambda p) = \lambda \psi^\circ(p)$ for $\lambda > 0$.

Suppose B satisfies (BAS). We first observe that the function $\mu : 2^V \to \mathbb{R}$ defined by $\mu(X) = \psi^\circ(\chi_X)$ $(X \subseteq V)$ is supermodular. In fact, we have

$$\mu(X) = \min\{\langle \chi_X, x \rangle \mid x \in B\} = \min\{x(X) \mid x \in B\}$$

and this is how the supermodular function μ in Theorem 1 is constructed. Secondly, the value of $\psi^\circ(p)$ at arbitrary p can be expressed as a linear combination of $\psi^\circ(\chi_X)$ ($X \subseteq V$). In fact, the greedy algorithm for minimizing a linear function over the base polytope, say $\mathbf{B}(\mu)$, of the supermodular system $(2^V, \mu)$ shows

$$\min\{\langle p, x\rangle \mid x \in \mathbf{B}(\mu)\} = \sum_{j=1}^{n}(p_j - p_{j+1})\mu(V_j) \ , \tag{11}$$

where, for given $p \in \mathbb{R}^V$, the elements of V are indexed as $\{v_1, v_2, \cdots, v_n\}$ (with $n = |V|$) in such a way that

$$p(v_1) \geq p(v_2) \geq \cdots \geq p(v_n) \ ; \tag{12}$$

we put $p_j = p(v_j)$, $V_j = \{v_1, v_2, \cdots, v_j\}$ for $j = 1, \cdots, n$, and $p_{n+1} = 0$. Noting $\overline{B} = \mathbf{B}(\mu)$ we obtain

$$\psi^\circ(p) = \sum_{j=1}^{n}(p_j - p_{j+1})\psi^\circ(\chi_{V_j}) \ . \tag{13}$$

Conversely, suppose $h = \psi^\circ : \mathbb{R}^V \to \mathbb{R}$ defined from B by (10) satisfies the two conditions:

(C1) [supermodularity] $\mu(X) = h(\chi_X)$ is supermodular,

(C2) [greediness] $h(p) = \sum_{j=1}^{n}(p_j - p_{j+1})h(\chi_{V_j})$ (under convention (12)).

Then B satisfies (BAS) by Theorem 1. Note also that such h is necessarily concave by a result of Lovász [12]. The above observations are summarized in the following theorem.

Theorem 10. *Let $B \subseteq \mathbb{Z}^V$ be a finite nonempty set with $B = \mathbb{Z}^V \cap \overline{B}$. Then B satisfies (BAS) if and only if ψ° satisfies (C1) and (C2).*

4.2 Exchangeability (EXC) and Supermodularity

We now consider the concave conjugate function

$$\omega^\circ(p) = \min\{\langle p, x\rangle - \omega(x) \mid x \in B\} \tag{14}$$

of $\omega : B \to \mathbb{R}$ defined on a finite integral base set $B \subseteq \mathbb{Z}^V$. As opposed to ψ°, ω° is not a positively homogeneous function though it is concave.

Since $\omega^\circ(p)$ is a concave function, we can think of its subdifferential in the ordinary sense in convex analysis. Namely, the subdifferential of ω° at $p_0 \in \mathbb{R}^V$, denoted $\partial\omega^\circ(p_0)$, is defined by

$$\partial\omega^\circ(p_0) = \{b \in \mathbb{R}^V \mid \omega^\circ(p) - \omega^\circ(p_0) \leq \langle p - p_0, b\rangle \ (\forall p \in \mathbb{R}^V)\} \ .$$

Using this we define a positively homogeneous concave function $\hat{L}(\omega^\circ, p_0)$: $\mathbb{R}^V \to \mathbb{R}$ by

$$\hat{L}(\omega^\circ, p_0)(p) = \inf\{\langle p, b\rangle \mid b \in \partial\omega^\circ(p_0)\} , \tag{15}$$

which we call the localization of ω° at p_0 (provided $\partial\omega^\circ(p_0) \neq \emptyset$). Note that

$$\omega^\circ(p) \leq \omega^\circ(p_0) + \hat{L}(\omega^\circ, p_0)(p - p_0) \tag{16}$$

and that this holds with equality in a neighborhood of p_0.

The following theorem allows us to say that the exchange property (EXC) is nothing but "a collection of local supermodularity," just as the exchange property (BAS) corresponds to supermodularity on the boolean lattice 2^V.

Theorem 11. *Let $\omega : B \to \mathbb{R}$ be a function on a finite integral base set $B \subseteq \mathbb{Z}^V$. Then ω satisfies (EXC) if and only if the localization $\hat{L}(\omega^\circ, p_0)$ of ω° satisfies (C1) and (C2) at each point p_0.*

Proof. We have $\hat{L}(\omega^\circ, p_0)(p) = \min\{\langle p, x\rangle \mid x \in \text{argmax}\,(\omega[-p_0])\}$. By Theorem 10, $\hat{L}(\omega^\circ, p_0)$ satisfies (C1) and (C2) if and only if $\text{argmax}\,(\omega[-p_0])$ satisfies (BAS). The latter condition for all p_0 is equivalent to (EXC) by Theorem 6. $\square$

The following theorem shows the "global supermodularity" of ω° on the vector lattice $\mathbb{R}^V$, in which the join $p \vee q \in \mathbb{R}^V$ and the meet $p \wedge q \in \mathbb{R}^V$ of $p, q \in \mathbb{R}^V$ are defined by

$$(p \vee q)(v) = \max(p(v), q(v)), \qquad (p \wedge q)(v) = \min(p(v), q(v)) .$$

Theorem 12. *If $\omega : B \to \mathbb{R}$ satisfies (EXC), then*

$$\omega^\circ(p) + \omega^\circ(q) \leq \omega^\circ(p \vee q) + \omega^\circ(p \wedge q) \qquad (p, q \in \mathbb{R}^V) . \tag{17}$$

Proof. Since (16) holds with equality in a neighborhood of p_0, the supermodularity (C1) of $\hat{L}(\omega^\circ, p_0)$, shown in Theorem 11, means the validity of (17) in a local sense. This in turn implies (17) for any p, q. $\square$

5 Duality

Using the standard Fenchel duality framework of convex analysis, we derive a min-max duality formula for a pair of M-convex/concave functions. The content of the min-max relation lies in the integrality assertion that both the primal (maximization) problem and the dual (minimization) problem have the integral optimum solutions when the given functions are integer-valued. This min-max formula is a succinct unified statement of the two groups of more or less equivalent theorems, (i) Edmonds' polymatroid intersection theorem, Fujishige's Fenchel-type duality theorem, and Frank's discrete separation theorem for a pair of sub/supermodular functions and (an extension of) (ii) Iri–Tomizawa's potential characterization of the optimality for the independent assignment problem, Fujishige's generalization thereof to the independent flow problem and Frank's weight splitting theorem for the weighted matroid intersection problem. The min-max formula can be reformulated also as discrete separation theorems, which are distinct from Frank's.

Let B_1 and B_2 be finite integral base sets. For $\omega : B_1 \to \mathbb{R}$ and $\zeta : B_2 \to \mathbb{R}$, we define the conjugate functions

$$\omega^\circ(p) = \min\{\langle p, x \rangle - \omega(x) \mid x \in B_1\}, \quad \zeta^\bullet(p) = \max\{\langle p, x \rangle - \zeta(x) \mid x \in B_2\}$$

and also the concave/convex closure functions

$$\hat{\omega}(b) = \inf\{\langle p, b \rangle - \omega^\circ(p) \mid p \in \mathbb{R}^V\}, \quad \check{\zeta}(b) = \sup\{\langle p, b \rangle - \zeta^\bullet(p) \mid p \in \mathbb{R}^V\} \ .$$

Convention: $\omega(x) = -\infty$ $(x \notin B_1)$, $\zeta(x) = +\infty$ $(x \notin B_2)$.

We consider a primal-dual pair of problems and a relaxation:

[**Primal**] Maximize $\omega(x) - \zeta(x)$ $(x \in B_1 \cap B_2)$,
[**Dual**]   Minimize $\zeta^\bullet(p) - \omega^\circ(p)$ $(p \in \mathbb{R}^V)$,
[**Primal, relaxed**] Maximize $\hat{\omega}(b) - \check{\zeta}(b)$ $(b \in \overline{B_1} \cap \overline{B_2})$.

The following identity is known as the Fenchel duality in convex analysis [18]:

$$\max\{\hat{\omega}(b) - \check{\zeta}(b) \mid b \in \overline{B_1} \cap \overline{B_2}\} = \inf\{\zeta^\bullet(p) - \omega^\circ(p) \mid p \in \mathbb{R}^V\} \ , \qquad (18)$$

which holds true independently of (EXC). Here we assume the convention that the maximum taken over an empty family is equal to $-\infty$. With this convention, the above formula implies that $\overline{B_1} \cap \overline{B_2} \neq \emptyset$ if the infimum on the right-hand side is finite.

Combining (18) with the obvious inequalities: $\omega(x) \leq \hat{\omega}(x)$ $(x \in B_1)$, $\zeta(x) \geq \check{\zeta}(x)$ $(x \in B_2)$ (cf. Lemma 7(1)), we obtain the following weak duality.

Lemma 13. *For any functions* $\omega : B_1 \to \mathbb{R}$ *and* $\zeta : B_2 \to \mathbb{R}$,

$$\max\{\omega(x) - \zeta(x) \mid x \in B_1 \cap B_2\}$$
$$\leq \max\{\hat{\omega}(b) - \check{\zeta}(b) \mid b \in \overline{B_1} \cap \overline{B_2}\} = \inf\{\zeta^\bullet(p) - \omega^\circ(p) \mid p \in \mathbb{R}^V\} \ .$$

Naturally, we are interested in whether the equality holds in the weak duality above. The next theorem shows that this is indeed the case if ω and $-\zeta$ enjoy (EXC). Remember that B_1 and B_2 are assumed to be finite, whereas the more general case will be mentioned in Conclusion.

Theorem 14 (Fenchel-type Duality). *Let* $\omega : B_1 \to \mathbb{R}$ *and* $\zeta : B_2 \to \mathbb{R}$ *be such that* ω *and* $-\zeta$ *satisfy* (EXC).
   (1) $\max\{\omega(x) - \zeta(x) \mid x \in B_1 \cap B_2\} = \inf\{\zeta^\bullet(p) - \omega^\circ(p) \mid p \in \mathbb{R}^V\}$.
To be more precise,
   (P1) *If* $\inf\{\zeta^\bullet(p) - \omega^\circ(p) \mid p \in \mathbb{R}^V\} \neq -\infty$, *then* $B_1 \cap B_2 \neq \emptyset$,
   (P2) *If* $B_1 \cap B_2 \neq \emptyset$, *the maximum and the infimum are finite and equal, and the infimum is attained by some* $p \in \mathbb{R}^V$.
   (2) *If* ω *and* ζ *are integer-valued, the infimum can be taken over integral vectors, i.e.,* $\max\{\omega(x) - \zeta(x) \mid x \in B_1 \cap B_2\} = \inf\{\zeta^\bullet(p) - \omega^\circ(p) \mid p \in \mathbb{Z}^V\}$, *and the infimum is attained by some* $p \in \mathbb{Z}^V$ *if it is finite.*

Before giving the proof, we observe from Lemma 13 that the essence of the first half of Theorem 14 lies in the integrality of the relaxed primal problem. The proof of Theorem 14 relies on Frank's discrete separation theorem for a pair of sub/supermodular functions and a recent theorem of the present author.

Theorem 15 (Discrete Separation Theorem [6]). *Let* $\rho : 2^V \to \mathbb{R}$ *and* $\mu :$ $2^V \to \mathbb{R}$ *be submodular and supermodular functions, respectively, with* $\rho(\emptyset) = \mu(\emptyset) = 0$. *If* $\mu(X) \leq \rho(X)$ $(X \subseteq V)$, *there exists* $x^* \in \mathbb{R}^V$ *such that*

$$\mu(X) \leq x^*(X) \leq \rho(X) \qquad (X \subseteq V) . \tag{19}$$

Moreover, if ρ *and* μ *are integer-valued, there exists such* $x^* \in \mathbb{Z}^V$.

Theorem 16 (M-concave Weighted Intersection [16]). *Assume that* $\omega_1 :$ $B_1 \to \mathbb{R}$ *and* $\omega_2 : B_2 \to \mathbb{R}$ *satisfy* (EXC) *and let* $x^* \in B_1 \cap B_2$. *Then*

$$\omega_1(x^*) + \omega_2(x^*) \geq \omega_1(x) + \omega_2(x) \qquad (\forall x \in B_1 \cap B_2)$$

if and only if there exists $p^* \in \mathbb{R}^V$ *such that*

$$\omega_1[-p^*](x^*) \geq \omega_1[-p^*](x) \quad (\forall x \in B_1), \quad \omega_2[p^*](x^*) \geq \omega_2[p^*](x) \quad (\forall x \in B_2) .$$

Moreover, if ω_1 *and* ω_2 *are integer-valued, there exists such* $p^* \in \mathbb{Z}^V$.

Remark. When ω_1 and ω_2 are affine functions, Theorem 16 above agrees with the optimality criterion for the weighted (polymatroid) intersection problem. When $B_1, B_2 \subseteq \{0,1\}^V$, on the other hand, it reduces to the optimality criterion [14, I, Theorem 4.2] for the valuated matroid intersection problem. If, in addition, ω_1 is affine and $\omega_2 = 0$, this criterion recovers Frank's weight splitting theorem for the weighted matroid intersection problem, which is in turn equivalent to Iri–Tomizawa's potential characterization of the optimality for the independent assignment problem.

We now prove (P1) of Theorem 14(1). Recall Theorem 1 and let μ_1 be the supermodular function describing B_1 and ρ_2 be the submodular function describing B_2. We have $\mu_1(\emptyset) = \rho_2(\emptyset) = 0$. We also introduce (cf. (10))

$$\psi_1^\circ(p) = \min\{\langle p, x \rangle \mid x \in B_1\}, \qquad \psi_2^\bullet(p) = \max\{\langle p, x \rangle \mid x \in B_2\} .$$

Lemma 17.

$$\inf\{\zeta^\bullet(p) - \omega^\circ(p) \mid p \in \mathbb{R}^V\} \neq -\infty \tag{20}$$
$$\Longleftrightarrow \psi_2^\bullet(p) \geq \psi_1^\circ(p) \qquad (p \in \mathbb{R}^V) \tag{21}$$
$$\Longleftrightarrow \rho_2(X) \geq \mu_1(X) \quad (X \subseteq V), \qquad \rho_2(V) = \mu_1(V). \tag{22}$$

Moreover, (20) $\Longleftrightarrow \inf\{\zeta^\bullet(p) - \omega^\circ(p) \mid p \in \mathbb{Z}^V\} \neq -\infty.$

Proof. Since $|\omega^\circ(p) - \psi_1{}^\circ(p)| \leq \max_{x \in B_1} |\omega(x)|$, $|\zeta^\bullet(p) - \psi_2{}^\bullet(p)| \leq \max_{x \in B_2} |\zeta(x)|$, and $\psi_1{}^\circ(p)$ and $\psi_2{}^\bullet(p)$ are positively homogeneous, we have $\inf\{\zeta^\bullet(p) - \omega^\circ(p) \mid p \in \mathbb{R}^V\} \neq -\infty \iff \inf\{\psi_2{}^\bullet(p) - \psi_1{}^\circ(p) \mid p \in \mathbb{R}^V\} \neq -\infty \iff \psi_2{}^\bullet(p) \geq \psi_1{}^\circ(p)$ $(p \in \mathbb{R}^V)$. By Theorem 10 it suffices to consider the last inequality for $p = \chi_X$ $(X \subseteq V)$. A straightforward calculation using (13) shows this is further equivalent to (22). $\qquad\square$

Theorem 15 shows that (22) implies $B_1 \cap B_2 \neq \emptyset$. $\qquad$ [End of proof of (P1)]

Next, we prove (P2) of Theorem 14(1). By Lemma 13 we see that (P2) is equivalent to the existence of $x^* \in B_1 \cap B_2$ and $p^* \in \mathbb{R}^V$ such that $\omega(x^*) - \zeta(x^*) = \zeta^\bullet(p^*) - \omega^\circ(p^*)$. Put $\omega_1 = \omega$ and $\omega_2 = -\zeta$ and denote by x^* a common base that maximizes $\omega_1(x) + \omega_2(x)$. By Theorem 16 we have $\omega_1[-p^*](x^*) = \max\{\omega_1[-p^*](x) \mid x \in B_1\}$, $\omega_2[p^*](x^*) = \max\{\omega_2[p^*](x) \mid x \in B_2\}$ for some $p^* \in \mathbb{R}^V$. This implies $\omega(x^*) - \zeta(x^*) = \omega_1(x^*) + \omega_2(x^*) = \omega_1[-p^*](x^*) + \omega_2[p^*](x^*) = \max_{x \in B_1} \omega_1[-p^*](x) + \max_{x \in B_2} \omega_2[p^*](x) = \max_{x \in B_1}(-\langle p^*, x \rangle + \omega(x)) + \max_{x \in B_2}(\langle p^*, x \rangle - \zeta(x)) = \zeta^\bullet(p^*) - \omega^\circ(p^*)$.

The second half of Theorem 14 follows from the second half of Theorem 16 that guarantees the existence of integral p^*. $\qquad$ [End of proof of Theorem 14]

The min-max identity of Theorem 14 yields a pair of separation theorems, one for the primal pair (ω, ζ) and the other for the dual (conjugate) pair $(\omega^\circ, \zeta^\bullet)$. We emphasize these separation theorems do not exclude the case of $B_1 \cap B_2 = \emptyset$.

Theorem 18 (Primal Separation Theorem). *Let* $\omega : B_1 \to \mathbb{R}$ *and* $\zeta : B_2 \to \mathbb{R}$ *be such that* ω *and* $-\zeta$ *satisfy* (EXC). *If* $\omega(x) \leq \zeta(x)$ $(x \in B_1 \cap B_2)$, *there exist* $\alpha^* \in \mathbb{R}$ *and* $p^* \in \mathbb{R}^V$ *such that* $\omega(x) \leq \alpha^* + \langle p^*, x \rangle \leq \zeta(x)$ $(x \in \mathbb{Z}^V)$[1].

Moreover, if ω *and* ζ *are integer-valued, there exist such* $\alpha^* \in \mathbb{Z}$ *and* $p^* \in \mathbb{Z}^V$.

Theorem 19 (Dual Separation Theorem). *Let* $\omega : B_1 \to \mathbb{R}$ *and* $\zeta : B_2 \to \mathbb{R}$ *be such that* ω *and* $-\zeta$ *satisfy* (EXC). *If* $\omega^\circ(p) \leq \zeta^\bullet(p)$ $(p \in \mathbb{R}^V)$, *there exist* $\beta^* \in \mathbb{R}$ *and* $x^* \in B_1 \cap B_2$ *such that* $\omega^\circ(p) \leq \beta^* + \langle p, x^* \rangle \leq \zeta^\bullet(p)$ $(p \in \mathbb{R}^V)$.

Moreover, if ω *and* ζ *are integer-valued, there exists such* $\beta^* \in \mathbb{Z}$.

Remark. The dual separation theorem for $\omega = 0$ and $\zeta = 0$ reduces to Frank's discrete separation theorem (Theorem 15) for sub/supermodular functions. In fact, the assumption reduces to (21), which is equivalent to (22), and we have $\beta^* = 0$.

Finally we schematically summarize the relationship among the duality theorems. It is emphasized that the "equivalence" relies on Lemmas 8 and 17.

---

[1] This is a short-hand expression for: $\omega(x) \leq \alpha^* + \langle p^*, x \rangle$ $(x \in B_1)$, $\alpha^* + \langle p^*, x \rangle \leq \zeta(x)$ $(x \in B_2)$.

Primal separation
(Theorem 18)
$\Updownarrow$

$$\begin{cases} \text{(P1)} \iff & \text{Frank's discrete separation} \\ & \text{(Theorem 15)} \\ \text{(P2)} \iff & \text{M-concave weighted intersection} \\ & \text{(Theorem 16)} \end{cases}$$

Min-max duality
(Theorem 14)

$\Updownarrow$

Dual separation
(Theorem 19)

6 Induction through Networks

We show that an M-concave function can be transformed into another M-concave function through a network. This is an extension of the well-known fact in matroid theory that a matroid can be transformed through a bipartite graph into another matroid.

Let $G = (V, A; V^+, V^-)$ be a directed graph with a vertex set V, an arc set A, a set V^+ of entrances and a set V^- of exits such that $V^+, V^- \subseteq V$ and $V^+ \cap V^- = \emptyset$. Also let $\bar{c} : A \to \mathbb{Z}$ be an upper capacity function, $\underline{c} : A \to \mathbb{Z}$ be a lower capacity function, and $w : A \to \mathbb{R}$ be a weight function. Suppose further that we are given a finite set $B^+ \subseteq \mathbb{Z}^{V^+}$ and a function $\omega^+ : B^+ \to \mathbb{R}$.

A flow is a function $\varphi : A \to \mathbb{Z}$. Its boundary $\partial\varphi : V \to \mathbb{Z}$ is defined by (1). We denote by $(\partial\varphi)^+$ (resp. $(\partial\varphi)^-$) the restriction of $\partial\varphi$ to V^+ (resp. V^-). A flow φ is called feasible if

$$\underline{c}(a) \le \varphi(a) \le \bar{c}(a) \qquad (a \in A) , \tag{23}$$

$$\partial\varphi(v) = 0 \qquad (v \in V - (V^+ \cup V^-)) , \tag{24}$$

$$(\partial\varphi)^+ \in B^+ . \tag{25}$$

We assume throughout that a feasible flow exists.

Define $\tilde{B} \subseteq \mathbb{Z}^{V^-}$ and $\tilde{\omega} : \tilde{B} \to \mathbb{R}$ by

$$\tilde{B} = \{(\partial\varphi)^- \mid \varphi : \text{feasible flow}\} , \tag{26}$$

$$\tilde{\omega}(x) = \max\{\langle w, \varphi \rangle_A + \omega^+((\partial\varphi)^+) \mid \varphi : \text{feasible flow with } (\partial\varphi)^- = x\}, \tag{27}$$

where $\langle w, \varphi \rangle_A = \sum_{a \in A} w(a)\varphi(a)$.

Lemma 20. *If B^+ satisfies* (BAS), *then $\tilde{B}$ of* (26) *satisfies* (BAS).

Theorem 21. *If ω^+ satisfies* (EXC), *then $\tilde{\omega}$ of* (27) *satisfies* (EXC).

The above theorem, the main result of this section, has important consequences.

Theorem 22. *If $\omega_1 : B_1 \to \mathbb{R}$ and $\omega_2 : B_2 \to \mathbb{R}$ satisfy (EXC), then the supremum convolution $\omega_1 \square \omega_2 : B_1 + B_2 \to \mathbb{R}$ satisfies (EXC), where*

$$(\omega_1 \square \omega_2)(x) = \sup\{\omega_1(x_1) + \omega_2(x_2) \mid x_1 + x_2 = x, x_1 \in B_1, x_2 \in B_2\} .$$

Proof. Let V_1 and V_2 be disjoint copies of V and consider a bipartite graph $G = (V^+, V^-, A)$ with $V^+ = V_1 \cup V_2$, $V^- = V$ and $A = \{(v_1, v) \mid v \in V\} \cup \{(v_2, v) \mid v \in V\}$, where $v_i \in V_i$ is the copy of $v \in V$ $(i = 1, 2)$. Take $\bar{c}$ sufficiently large, $\underline{c}$ sufficiently small, $w \equiv 0$, $B^+ = B_1 \times B_2$, and $\omega^+(x_1, x_2) = \omega_1(x_1) + \omega_2(x_2)$. Then $\omega_1 \square \omega_2 = \tilde{\omega}$, where $\tilde{\omega}$ is the induced M-concave function. $\square$

As a special case of Theorem 21, a valuated matroid can be induced by matchings in a bipartite graph. Let $G = (V^+, V^-, A)$ be a bipartite graph, $w : A \to \mathbb{R}$ a weight function, and $\mathbf{M}^+ = (V^+, B^+)$ a matroid with valuation $\omega^+ : B^+ \to \mathbb{R}$. Then $\tilde{B} = \{\partial^- M \mid M \text{ is a matching with } \partial^+ M \in B^+\}$ forms the base family of a matroid, provided $\tilde{B} \neq \emptyset$. Here $\partial^+ M \subseteq V^+$ and $\partial^- M \subseteq V^-$ denote the sets of vertices incident to M. Define $\tilde{\omega} : \tilde{B} \to \mathbb{R}$ by

$$\tilde{\omega}(X) = \max\{w(M) + \omega^+(\partial^+ M) \mid M : \text{matching}, \partial^+ M \in B^+, \partial^- M = X\} . \tag{28}$$

Theorem 23. *$\tilde{\omega}$ of (28) is a valuation of $(V^-, \tilde{B})$.*

Let $\mathbf{M}_1 = (V, B_1)$ and $\mathbf{M}_2 = (V, B_2)$ be matroids with valuations $\omega_1 : B_1 \to \mathbb{R}$ and $\omega_2 : B_2 \to \mathbb{R}$. Let $\mathbf{M}_1 \vee \mathbf{M}_2 = (V, B_1 \vee B_2)$ denote the union of $\mathbf{M}_1$ and $\mathbf{M}_2$, where $B_1 \vee B_2$ is defined to be the family of the maximal elements of $\{X_1 \cup X_2 \mid X_1 \in B_1, X_2 \in B_2\}$. Define $\omega_1 \vee \omega_2 : B_1 \vee B_2 \to \mathbb{R}$ by

$$(\omega_1 \vee \omega_2)(X) = \max\{\omega_1(X_1) + \omega_2(X_2) \mid X_1 \cup X_2 = X, X_1 \in B_1, X_2 \in B_2\} .$$

Theorem 24 (Union). *$\omega_1 \vee \omega_2$ is a valuation of $\mathbf{M}_1 \vee \mathbf{M}_2$.*

Proof. Let V_1 and V_2 be disjoint copies of V, and U be a set of size $= \text{rank}\,\mathbf{M}_1 + \text{rank}\,\mathbf{M}_2 - \text{rank}\,(\mathbf{M}_1 \vee \mathbf{M}_2)$. Consider a bipartite graph $G = (V^+, V^-, A)$ with $V^+ = V_1 \cup V_2$, $V^- = V \cup U$ and $A = \{(v_1, v) \mid v \in V\} \cup \{(v_2, v) \mid v \in V\} \cup \{(v_2, u) \mid v \in V, u \in U\}$, where $v_i \in V_i$ is the copy of $v \in V$ $(i = 1, 2)$. Let $\tilde{\omega}$ be the valuation induced on V^- from the valuation ω^+ on V^+ defined by $\omega^+(X_1 \cup X_2) = \omega_1(X_1) + \omega_2(X_2)$ $(X_i \in B_i$ $(i = 1, 2))$. Then $(\omega_1 \vee \omega_2)(X) = \tilde{\omega}(X \cup U)$ for $X \subseteq V$. $\square$

For a matroid $\mathbf{M} = (V, B)$ its truncation to rank k is given by $\mathbf{M}_k = (V, B_k)$ with $B_k = \{X \subseteq V \mid |X| = k, \exists B : X \subseteq B \in B\}$. For a valuation $\omega : B \to \mathbb{R}$ of $\mathbf{M}$, define $\omega_k : B_k \to \mathbb{R}$ by

$$\omega_k(X) = \max\{\omega(B) \mid X \subseteq B \in B\} \qquad (X \in B_k) .$$

Theorem 25 (Truncation [15]). *ω_k is a valuation of $\mathbf{M}_k$, where $k \leq \text{rank}\,\mathbf{M}$.*

Proof. Let V' be a copy of V, and U be a set of size $= \operatorname{rank} \mathbf{M} - k$. Consider a bipartite graph $G = (V^+, V^-, A)$ with $V^+ = V'$, $V^- = V \cup U$ and $A = \{(v', v) \mid v \in V\} \cup \{(v', u) \mid v \in V, u \in U\}$, where $v' \in V'$ is the copy of $v \in V$. Let $\tilde{\omega}$ be the valuation induced on V^- from $\omega^+ = \omega$ on V^+. Then $\omega_k(X) = \tilde{\omega}(X \cup U)$ for $X \subseteq V$. □

The proof of Theorem 21 relies on the optimality criterion (Theorem 26 below) for the submodular flow problem with an objective function satisfying (EXC). In addition to the network $(G = (V, A; V^+, V^-), \underline{c}, \overline{c}, w)$ suppose we are given a pair of M-concave functions $\omega^+ : B^+ \to \mathbb{R}$ and $\omega^- : B^- \to \mathbb{R}$, where $B^+ \subseteq \mathbb{Z}^{V^+}$ and $B^- \subseteq \mathbb{Z}^{V^-}$ are finite integral base sets. [Do not confuse B^- with $\tilde{B}$, and ω^- with $\tilde{\omega}$.] The generalized submodular flow problem reads:

[Problem P] Maximize $\langle w, \varphi \rangle_A + \omega^+((\partial\varphi)^+) + \omega^-((\partial\varphi)^-)$
   subject to (23), (24), (25) and

$$(\partial\varphi)^- \in B^- . \tag{29}$$

The following theorem refers to a "potential" function $q : V \to \mathbb{R}$. We denote by $q^+ : V^+ \to \mathbb{R}$ and $q^- : V^- \to \mathbb{R}$ the restrictions of q and define

$$\omega^+[q^+](x) = \omega^+(x) + \langle q^+, x \rangle \qquad (x \in B^+) ,$$
$$\omega^-[q^-](x) = \omega^-(x) + \langle q^-, x \rangle \qquad (x \in B^-) .$$

Also we use the notation $w_q(a) = w(a) - q(\partial^+ a) + q(\partial^- a)$ $(a \in A)$, where $\partial^+ a$ and $\partial^- a$ denote the initial vertex and the terminal vertex of $a \in A$.

Theorem 26 [16]. (1) *A flow* $\varphi : A \to \mathbb{Z}$ *with* (23), (24), (25) *and* (29) *is optimal for* Problem P *if and only if there exists a "potential" function* $q : V \to \mathbb{R}$ *such that* (i)–(ii) *below hold true.*
   (i) *For each* $a \in A$,

$$w_q(a) < 0 \implies \varphi(a) = \underline{c}(a), \qquad w_q(a) > 0 \implies \varphi(a) = \overline{c}(a) .$$

   (ii) $(\partial\varphi)^+$ *maximizes* $\omega^+[q^+]$ *and* $(\partial\varphi)^-$ *maximizes* $\omega^-[q^-]$.
Moreover, if ω^+ *and* ω^- *are integer-valued, there exists integer-valued such* q.

   (2) *Let* q *be a potential that satisfies* (i)–(ii) *above for some (optimal) flow* φ. *A flow* φ' *with* (23), (24), (25) *and* (29) *is optimal if and only if it satisfies* (i)–(ii) *(with* φ *replaced by* φ').

7 Conclusion

In this paper we have restricted ourselves to M-concave functions defined on base polytopes (bounded base polyhedra). The boundedness assumption is not essential but only for the sake of simplicity in presentation. All the results can be extended mutatis mutandis; for instance, the first part of the Fenchel-type duality (Theorem 14) in the general case reads as follows.

Theorem 27. *Let* $\omega : B_1 \to \mathbb{R}$ *and* $\zeta : B_2 \to \mathbb{R}$ *be such that* ω *and* $-\zeta$ *satisfy* (EXC), *where* B_1 *and* B_2 *are nonempty (possibly unbounded) subsets of* $\mathbb{Z}^V$ *satisfying* (BAS). *If* $B_1 \cap B_2 \neq \emptyset$ *or* $\zeta^\bullet(p) - \omega^\circ(p) \neq +\infty$ *for some* $p \in \mathbb{R}^V$, *then*

$$\sup\{\omega(x) - \zeta(x) \mid x \in B_1 \cap B_2\} = \inf\{\zeta^\bullet(p) - \omega^\circ(p) \mid p \in \mathbb{R}^V\} \ .$$

The author thanks András Frank, Satoru Fujishige, Satoru Iwata, András Sebő, and Akiyoshi Shioura for valuable discussions.

References

1. Dress, A. W. M., Terhalle, W.: Well-layered maps and the maximum-degree $k \times k$-subdeterminant of a matrix of rational functions. Appl. Math. Lett. **8** (1995) 19–23
2. Dress, A. W. M., Terhalle, W.: Well-layered maps — A class of greedily optimizable set functions. Appl. Math. Lett. **8** (1995) 77–80
3. Dress, A. W. M., Wenzel, W.: Valuated matroid: A new look at the greedy algorithm. Appl. Math. Lett. **3** (1990) 33–35
4. Dress, A. W. M., Wenzel, W.: Valuated matroids. Adv. Math. **93** (1992) 214–250
5. Frank, A.: A weighted matroid intersection algorithm. J. Algorithms **2** (1981) 328–336
6. Frank, A.: An algorithm for submodular functions on graphs. Ann. Disc. Math. **16** (1982) 97–120
7. Fujishige, S.: Lexicographically optimal base of a polymatroid with respect to a weight vector. Math. Oper. Res. **5** (1980) 186–196
8. Fujishige, S.: Theory of submodular programs: A Fenchel-type min-max theorem and subgradients of submodular functions. Math. Progr. **29** (1984) 142–155
9. Fujishige, S.: Submodular Functions and Optimization. Ann. Disc. Math. **47**, North-Holland, 1991
10. Groenevelt, H.: Two algorithms for maximizing a separable concave function over a polymatroid feasible region. Working Paper, Grad. School Management, Univ. Rochester, 1995.
11. Iri, M., Tomizawa, N.: An algorithm for finding an optimal "independent assignment". J. Oper. Res. Soc. Japan **19** (1976) 32–57
12. Lovász, L.: Submodular functions and convexity. In "Mathematical Programming — The State of the Art" (A. Bachem, M. Grötschel and B. Korte, eds.), Springer, 235–257, 1983
13. Murota, K.: Finding optimal minors of valuated bimatroids. Appl. Math. Lett. **8** (1995) 37–42
14. Murota, K.: Valuated matroid intersection, I: optimality criteria, II: algorithms. SIAM J. Disc. Math. **9** (1996) No.3 (to appear)
15. Murota, K.: Matroid valuation on independent sets. Report 95842-OR, Inst. Disc. Math., Univ. Bonn, 1995
16. Murota, K.: Submodular flow problem with a nonseparable cost function. Report 95843-OR, Inst. Disc. Math., Univ. Bonn, 1995
17. Murota, K.: Convexity and Steinitz's exchange property. Report 95848-OR, Inst. Disc. Math., Univ. Bonn, 1995
18. Rockafellar, R. T.: Convex Analysis. Princeton Univ. Press, 1970

On Ideal Clutters, Metrics and Multiflows

Beth Novick[1] and András Sebő[2]

[1] Clemson University, Clemson, South Carolina 29634-1907, USA
[2] CNRS, Laboratoire Leibniz, Grenoble, France

Abstract. "Binary clutters" contain various objects studied in Combinatorial Optimization, such as paths, Chinese Postman Tours, multiflows and one-sided circuits on surfaces. Minimax theorems about these can be generalized in terms of ideal binary clutters. Seymour has conjectured a characterization of these, and the goal of the present work is to study this conjecture in terms of multiflows in matroids. Seymour's conjecture is equivalent to the following:

Let F be a binary clutter. Then the Cut Condition is sufficient in the underlying matroid, for all $F \in \mathcal{F}$ as demand-set, to have a multiflow, if and only if it implies the so called K_5, F_7 and R_{10}-conditions.

These three conditions are applications of the general "Metric Condition" to particular $0 - 1$ *bipartite weightings*. In this paper we prove the following weakening of this conjecture:

The Cut Condition is sufficient for all $F \in \mathcal{F}$ as demand-set, to have a multiflow, if and only if it implies the Metric Condition for every bipartite $0 - 1$ weighting.

A special case of this result has been stated as a conjecture in Robertson and Seymour's "Graph Minors" volume, (1991, "Open Problem 11 (A. Sebő)"). Using Lehman's theorem on minimal non-ideal clutters, we sharpen the properties of minimally non-ideal clutters for the *binary* and *graphic* special cases.

1 Introduction

A *clutter* is a family of bts of a finite ground set N, none of which contains any other. We will suppose that every $e \in N$ is contained in at least one set of the family. A clutter $\mathcal{H}$ is *ideal* (has the max-flow-min-cut property) if its *blocking polyhedron*, that is the polyhedron $\{x \in \mathbf{R}_+^n : x(H) \geq 1 \text{ for all } H \in \mathcal{H}\}$, has only integer vertices. Clearly, a clutter is ideal precisely when the set of vertices of its blocking polyhedron are exactly the characteristic vectors of its blocker. The *blocking clutter* or *blocker* of the clutter $\mathcal{A} \subseteq 2^N$, denoted by $\mathcal{B} = b(\mathcal{H})$, is defined to be the family of minimal elements of $\{B \subseteq N : |B \cap A| \geq 1 \text{ for all } A \in \mathcal{A}\}$. It is easy to see $b(b(\mathcal{A})) = \mathcal{A}$, [7]. The notation $\mathcal{A}$ and $\mathcal{B}$ will automatically mean that these are clutters and they are the blocker of each other. We denote the ground set of a specific clutter $\mathcal{A}$ by $N(\mathcal{A})$. When it causes no confusion, we will speak interchangeably about a subset of N and its incidence vector considered as a member of $GF(2)^n$, where $n := |N|$; similarly, a family of subsets and a 0-1 matrix are interchangeable, as well as the mod 2 sum of vectors and their "symmetric difference", denoted by $\triangle$. dPe linear independence, rank, span, orthogonality etc. is usually understood over over $GF(2)^n$, when we understand it over the reals we will mention it. The sign "$\equiv$" means congruence modulo 2.

A *binary* clutter is the family of (inclusionwise) minimal supports of elements of affine subspaces (shifts of linear subspaces) of vector spaces over $GF(2)$. For results on binary spaces and binary clutters see [17], [21], [3], [4]. It is easy to see that the blocker of a binary clutter $\mathcal{H}$ is $b(\mathcal{H}) = \{K \subseteq N : |K \cap H| \equiv 1 \mod 2 \text{ for all } H \in \mathcal{H}, K \text{ minimal}\}$. It follows that $b(\mathcal{H})$ is also a binary clutter (see [17]). The result of *deleting* or *contracting* $e \in N$ in a binary clutter $\mathcal{H}$ is denoted by $\mathcal{H}\backslash e$, and $\mathcal{H}/e$ respectively, and defined by $\mathcal{H}\backslash e := \{H \in \mathcal{H} : e \notin H\}$, and $\mathcal{H}/e :=$ the minimal elements of $\{H - \{e\} : H \in \mathcal{H}\}$. $\mathcal{H}\backslash X/Y$ denotes the the result of deleting the elements of X and then contracting those of Y and is called a *minor*. It is easy to see that $b(\mathcal{H} \backslash v) = b(\mathcal{H})/v$ and that the classes of ideal and binary clutters are minor-closed. Throughout the sequel *clutter* will mean "binary clutter."

Binary clutters generalize various objects studied in combinatorial optimization such as paths, Chinese postman tours and one-sided circuits on surfaces. Seymour, in his works [21] and [20] shows how good characterization theorems about these special cases can be extended to binary clutters, and conjectures the following. $\mathcal{F}_7$ will denote the clutter consisting of the lines of Fano plane; $\mathcal{K}_5$ the set of odd circuits of the complete graph K_5.

Seymour's Conjecture : *A binary clutter is ideal if and only if it contains none of $\mathcal{K}_5$, $b(\mathcal{K}_5)$ or $\mathcal{F}_7$ as minors.*

Compare the three excluded minors of this conjecture to the infinite set of minimal non-ideal clutters (see some infinite classes in [6], [14], [12]).

The collection of all odd circuits of an undirected graph and its blocker are binary clutters. More generally, a *signed* graph is the pair (G, R), where $G = (V, E)$ is an undirected graph, and $R \subseteq E(G)$. We define the pair of binary clutters

$$\mathcal{A}(G, R) := \{A : |A \cap R| \text{ is odd}, A \text{ a circuit of } G\}.$$

$$\mathcal{B}(G, R) := \{B = R \Delta Q : Q \text{ is a cocycle of } G, B \text{ minimal non-empty}\}.$$

$\mathcal{A}(G, R)$ is called an *odd circuit* clutter, and $\mathcal{B}(G, R)$, the blocker of $\mathcal{A}(G, R)$, a *signing* clutter. The class of odd circuits clutters of graphs is minor closed. Signed graphs are studied by Gerards in [3], [4], where the various particular cases have a topological meaning. The approach of the present work is different: we see the idealness of binary clutters as a problem about metrics and multiflows in matroids; exploiting this tool we get a characterization of idealness which sharpens Lehman's theorem (also using it) for binary clutters.

It is straightforward to extend the definition of signed graph to *signed* matroid: a pair (M, R), where $R \subseteq N(M)$. In this context $\mathcal{A}(M, R)$ $(\mathcal{B}(M, R))$ is $\mathcal{A}(G, R)$ $(\mathcal{B}(G, R))$ with G replaced by M in the above definition and is, again, called an *odd circuit* clutter (a *signing* clutter). It is clear that $R \in \mathcal{B}(M, R)$. The generalization will be useful in the context of multiflows, since *every* binary clutter is an odd circuit clutter for some M and some R, see (2.4).

The *core* of a binary clutter $\mathcal{A}$ is the family of its minimal cardinality elements; the family will be denoted by $core\,(\mathcal{A})$, and the cardinality of its elements by $r := r(\mathcal{A})$; $s := s(\mathcal{A}) := r(b(\mathcal{A}))$. Let $M_{\mathcal{A}}$ (and $M_{\mathcal{B}}$) be a matrix whose rows are the incidence vectors of the members of core $(\mathcal{A})$ (core $(b(\mathcal{A}))$ respectively). The $n \times n$ identity matrix will be denoted by I, whereas J is the $n \times n$ matrix of all 1's. We state Lehman's theorem on minimal non-ideal clutters, making obvious simplifications due to the binary case (the degenerate projective plane is not a binary clutter) :

Theorem 1.1 [Lehman [22]] *If $\mathcal{A}$ is a minimal non-ideal binary clutter, then the unique non-integer vertex of its blocking polyhedron is the constant vector $1/r$, $rs \geq n + 1$,and $M_{\mathcal{A}}$, $M_{\mathcal{B}}$ are $n \times n$ matrices whose row and column sums are r and s respectively, and the rows of these matrices can be ordered so that* $M_{\mathcal{A}}M_{\mathcal{B}}^{\mathsf{T}} = J + (rs - n)I = M_{\mathcal{B}}^{\mathsf{T}}M_{\mathcal{A}}.$ □

The key statement is that the row and column sums are r and s respectively. As Seymour [22] observes, this provides sufficient information for a coNP-characterization of idealness. Still, the matrix equations of Theorem 1.1 also express combinatorial facts that will be important for us. Let us look more into these for later use. Let $\alpha = rs - n + 1$. By Theorem 1.1 each $A \in$ core $(\mathcal{A})$ is paired to a unique $B \in$ core $(\mathcal{B})$ with $|A \cap B| = \alpha$. The sets A and B will be called *partners* of each other.

Besides the equation $M_{\mathcal{A}}M_{\mathcal{B}}^{\mathsf{T}} = J + (rs - n)I$ which provides us with the cardinalities of the intersections we have $M_{\mathcal{B}}^{\mathsf{T}}M_{\mathcal{A}} = J + (rs - n)I$, expressing the following:

(1.1) *[22] Let $\mathcal{B}_v := \{B \in \text{core}\,(\mathcal{B}) : \text{the partner of } B \text{ contains } v\}$. Then for all $v \in N$, $\mathcal{B}_v/v$, is a partition of $N \setminus v$. Out of the r members of $\mathcal{B}_v$ exactly α contain v.*

If $\mathcal{A}$ and $\mathcal{B}$ are blockers of each other and the corresponding matrices $M_{\mathcal{A}}$, $M_{\mathcal{B}}$ have row and column sums equal to r and s respectively and satisfy the equations $M_{\mathcal{A}}M_{\mathcal{B}}^{\mathsf{T}} = J+(rs-n)I = M_{\mathcal{B}}^{\mathsf{T}}M_{\mathcal{A}}$, then we say that the clutters are *partitionable*.

A partitionable binary clutter is not ideal, and it is also not necessarily minimal non-ideal. It is easy to see that the binary clutter $\mathcal{A}'$ generated by the core of $\mathcal{A}$ (that is, $\mathcal{A}'$ consists of the set of minimal supports of vectors obtained by summing an odd number of rows of A), is also partitionable. It will be called the *normalization* of $\mathcal{A}$. If $\mathcal{A}' = \mathcal{A}$, then $\mathcal{A}$ will be said to be *normalized*. It is easy to see that by normalization, the core, r, s, and the core of the blocker do not change; the normalization is the minimal binary clutter for which the core is the same as originally.

The fact that $M_{\mathcal{A}}$ and $M_{\mathcal{B}}^{\mathsf{T}}$ commute can be summarized in a simple "polarity" relation between the core of a minimal non-ideal clutter and the set of elements $\{1, \ldots, n\}$. An analogous notion for antiblocking (minimal imperfect graphs), defined by Tucker [23], has also been useful.

We define the *polar* of the partitionable binary clutter $\mathcal{A}$ to be the family of (inclusionwise) minimal supports of vectors in the affine space generated by the *columns* of A.

(1.2) *The polar A^* of a partitionable binary clutter is also a binary clutter which is partitionable. The binary clutter A^{**} is the normalization of A. The normalization of $b(A^*)$ is B^*.* □

The proof is straightforward from Theorem 1.1. The analogous equality for "antiblocking" plays an important role in Gasparian's recent proof [2] of Lovász's perfect graph theorem.

Note that the three known minimal non-ideal clutters are normalized and self-polar.

In Section 2 we prove some basic relations between binary clutters, metrics and multiflows in underlying matroids, which are used in the following two sections to prove the main results. Section 3 provides a coNP characterization for the Cut Condition to be sufficient for a class of multiflow problems whose demand-edges are members of a binary clutter, and Section 4 sharpens this characterization. The property we characterize is a reformulation of idealness, and is interesting for its own sake. It sharpens the application of Lehman's theorem to binary clutters, using Theorem 1.1 and the above mentioned consequences. In Section 5 we make some additional comments about the graphic special case of our results and the most important graphic special case of Seymour's conjecture.

2   Clutters, Metrics and Multiflows

Prior to presenting our main results, we will need to introduce some matroidal tools and to clarify some powerful connections between ideal binary clutters, on the one hand, and multiflows and metrics in matroids, on the other. Such is the topic of the present section.

For us, *matroid* will mean "binary matroid" — one representable over $GF(2)$ — and will be a pair $M = (N, C)$, where C is the set of its circuits. The linear space generated by C is called the *cycle-space* of M. (Its members are the cycles; the circuits are the inclusionwise minimal cycles.) The linear space orthogonal to C is called the *cocycle space*, its elements are the *cocycles*, or cuts, its (inclusionwise) minimal nonempty elements are the *cocircuits*. The matroid $M^* = (N, C^*)$, where C^* is the set of cocircuits of M, is the *dual matroid* of M. If the columns of a matrix represent a binary matroid, then the rows generate the cocycle space of the matroid. The ground-set N of a matroid M will be referred to as $N(M)$. The rank function of the matroid is denoted by $r := r(M)$. For more on the basic notions and simple facts related to binary clutters and matroids, see the introductions of [17], [21], [19], [20], [3], [4].

Three Matroids Associated with each Binary Clutter (see more details in [15])

Let $\mathcal{H}$ be a binary clutter. We define the *down matroid* of $\mathcal{H}$ to be $M_0(\mathcal{H}) := (N, C_0)$, where C_0 consists of the minimal non-empty subsets of N that can be written as the sum of an even number of elements of $\mathcal{H}$. The *up matroid* of $\mathcal{H}$ is defined by $M_1(\mathcal{H}) := (N, C_1)$, where C_1 is the set of minimal non-empty elements of the linear space generated by $\mathcal{H}$. Proofs of (2.1), (2.2) and (2.3), below, are straightforward and hence omitted (see also [15]).

(2.1) If $\emptyset \neq \mathcal{H} \neq \{\emptyset\}$, then $\mathcal{C}_0$ generates a subspace of corank $= r(\mathcal{C}_1) - r(\mathcal{C}_0) = 1$ of $\mathcal{C}_1$, and $\mathcal{H} = \mathcal{C}_1 \backslash \mathcal{C}_0$. $\qquad\square$

One gets M_1 from M_0 by "undeleting" and contracting an element, and the matroid one gets in the intermediate step is uniquely determined:

(2.2) If $\emptyset \neq \mathcal{H} \neq \{\emptyset\}$, there exists a uniquely determined matroid M_2, and $t \in S(M_2)$ such that $M_2(\mathcal{A}) \backslash t = M_0(\mathcal{A})$, $M_2(\mathcal{A})/t = M_1(\mathcal{A})$. $\qquad\square$

M_2 will be called the *port matroid* of $\mathcal{H}$.

Given our assumption that $\emptyset \neq \mathcal{H} \neq \{\emptyset\}$, M_0, M_1, M_2 are uniquely determined by $\mathcal{H}$. [3] Conversely, the pair (M_0, M_1) or the pair (M_2, t), $t \in N(M_2)$ uniquely determines $\mathcal{H}$.

(2.3) Let $\mathcal{A} \subseteq 2^N$ be a binary clutter, and let $\mathcal{B}$ be its blocker. For $e \in N$, if $\mathcal{A} \backslash e \neq \emptyset$ and $\mathcal{A}/e \neq \{\emptyset\}$, then $M_i(\mathcal{A} \backslash e) = M_i(\mathcal{A}) \backslash e$ and $M_i(\mathcal{A}/e) = M_i(\mathcal{A})/e$, $(i = 0, 1, 2)$. Moreover, the matroids $M_i(\mathcal{A})$, $M_i(\mathcal{B})$ $(i = 0, 1, 2)$ relate as follows:

> (i) $M_2(\mathcal{A}) \backslash t = M_0(\mathcal{A})$, $M_2(\mathcal{A})/t = M_1(\mathcal{A})$.
> (ii) $M_2(\mathcal{B}) = M_2^*(\mathcal{A})$.
> (iii) $M_0(\mathcal{B}) = M_1^*(\mathcal{A})$.

$\qquad\square$

We make the convention that the down, up and port matroids of the clutter $\mathcal{A} = \emptyset$ is any triple of matroids $M_0 = (N, \mathcal{C}_0)$, $M_1 = (N, \mathcal{C}_1)$, $M_2 = (N \cup \{t\}, \mathcal{C}_2)$, such that t is a coloop in M_2 and $M_1 = M_2/t$. Accordingly, the down, up and port matroid of $\mathcal{B} = \{\emptyset\}(= b(\mathcal{A}))$ is any triple where t is a loop in M_2 and $M_0 = M_2 \backslash t$. It follows given this convention, that the claims of (2.3) hold for the blocking pair of clutters $\emptyset$ and $\{\emptyset\}$ as well.

The following is not difficult to show:

(2.4) Every binary clutter $\mathcal{A}$ is equal to $\mathcal{A}(M, R)$ for some uniquely determined binary matroid M, and arbitrary $R \in b(\mathcal{A})$; the same is true for $\mathcal{B}(M, R)$; $\mathcal{A}(M, R)$ and $\mathcal{B}(M, R)$ are the blocker of each other. Furthermore,

(i) If $\mathcal{A}$ is an arbitrary binary clutter, then $\mathcal{A} = \mathcal{A}(M, R)$, $\mathcal{B} = \mathcal{B}(M, R)$ with $M = M_1(\mathcal{A})$, and arbitrary $R \in b(\mathcal{A})$;

(ii) $\mathcal{A} \backslash e = \mathcal{A}(M \backslash e, R)$; $\mathcal{A}/e = \mathcal{A}(M/e, R')$, where $R' \in \mathcal{B}$, $e \notin R'$, or if $e \in R'$ for all $R' \in \mathcal{B}$, then $\mathcal{A}/e = \{\emptyset\}$. $\qquad\square$

The fact that the sets $\mathcal{A} = \mathcal{A}(M, R)$, $\mathcal{B} = \mathcal{B}(M, R)$, and $e \in N(M)$ do not change if we replace R by any $R' \in \mathcal{B}$, (that is, by $R \triangle D$ where D is an arbitrary cocycle), is used in proving 2.4 (ii).

Note that a binary clutter $\mathcal{A}$ is a (graphic) odd circuit clutter if and only if $M_1(\mathcal{A})$ is graphic. The graph obtained by removing, that is *deleting*, a set of edges Y from a graph G then shrinking to a single node, that is *contracting* each of the edges $X \in E(G)$, where $X \cap Y = \emptyset$, is denoted $G/X \backslash Y$. Since the process corresponds to the same minor taking in the graphic matroid,

$$\mathcal{A}(G/X \backslash Y, R) = (\mathcal{A}(G, R))/X \backslash Y.$$

---

[3] These statements are equivalent to the well-known fact due to Lehman[10] that a binary clutter uniquely determines the underlying matroid.

Ideal Clutters, Multiflows and Metrics. Let $M = (E, \mathcal{C})$ be a matroid. $m : E \mapsto \mathbf{R}_+ \cup \{\infty\}$ is called a *metric*, if for every circuit $C \in \mathcal{C}$ and every $e \in C$: $m(e) \leq m(C \backslash e)$. Metrics on M form a cone, whose extreme rays are called *primitive* metrics. A *cut* metric is a function 1 on a cocycle, 0 elsewhere; if the cocycle is a cocircuit, it is easy to see that the cut metric is primitive. A function $d : E \mapsto \mathbf{N}$ is called a *distance function* if for some $F \subseteq E$ it is defined in the following way: for $e \in F$, $d(e) := 1$, and for $e \notin F$, $d(e) := \min\{|C \backslash e| : C \in \mathcal{C}, \{e\} = C \backslash F\}$. It is easy to see that a distance function is a metric and that it is finite if and only if $N(M) \backslash F$ does not contain a cocircuit. This metric will be denoted by $[M, F]$. For instance, $[M(K_5), E(K_{2,3})]$ is 1 on a $K_{2,3}$ subgraph of K_5, and 2 otherwise.

We define now some other basic examples of metrics. Let L be a line of F_7; C_3 is any 3-element subset of any 4-element circuit of R_{10}; $R_{10} := M_0(\mathcal{K}_5)$. All of $[F_7, F_7 - L]$, $[K_5, K_{2,3}]$, $[R_{10}, R_{10} - C_3]$ can be checked to be primitive (follows from (3.1) below).

Besides these, Papernov's multiflow problem (H_6, R) is also a useful example serving often as a counterexample: H_6 is the graph (unique up to isomorphism) one gets from K_5 by uncontracting an edge ab so that no series edges occur; R consists of three edges ab, $x_1 x_2$ and $y_1 y_2$ forming a matching of H_6, and such that a is adjacent to both x_1 and x_2 and b is adjacent to both y_1 and y_2. The metric $[H_6, H_6 \backslash R]$ is primitive.

We define an $[M, F]$-*metric* to be a function m on the elements N of an arbitrary matroid such that contractit elements with m-value 0 and deleting some of the $e \in N$ with $m(e) = m(C \backslash e)$ for some circuit C, we get the matroid M, with the metric $[M, F]$ (up to isomorphism). Clearly, $[M, F]$-metrics are metrics. Cut-metrics are $[M, \emptyset]$ metrics where M is the matroid having one coloop element. A metric $m : N(M) \mapsto \mathbf{Z}_+$ is *bipartite* if $\sum_{e \in C} m(e)$ is even for all circuits C of M. All the examples above are bipartite metrics. A binary matroid is called *bipartite* if every circuit is even and *Eulerian* if every cocircuit is even.

A *multiflow problem*, (M, R, c), on a matroid $M = (N, \mathcal{C})$ is defined by a set of *demands* $R \subseteq N$, and a function $c : N \mapsto \mathbf{R}_+$. A *multiflow* is a function $f : \mathcal{C} \mapsto \mathbf{Q}_+$, so that if $f(C) > 0$, then $|C \cap R| = 1$, and the sum of the f-values of circuits containing a given $e \in N$ is at most $c(e)$, moreover equality holds here for $e \in R$. (M, R) will stand for the class of multiflow problems in M with demands R. For basic facts about multiflows in matroids we refer to [20], for their connection to binary clutters see [20] or [4]. For metrics in matroids we refer to [16]. If $e \in R$, $c(e)$ is called the *demand* of e, if $e \in N \backslash R$ it is called the *capacity* of e.

The connection between metrics and multiflows is provided by the following statement, well-known for graphs, which is also easy from linear programming duality (Farkas' lemma) for matroids (see [16]).

Metric Criterion *For $R \subseteq N(M)$ and $c : N(M) \mapsto \mathbf{R}_+$ there exists a multiflow if and only if the following Metric Condition is satisfied: for every primitive metric m, $\sum_{e \in R} m(e) c(e) \leq \sum_{e \in N(M) \backslash R} m(e) c(e)$.* ◻

The Metric Condition specialized to a class μ of metrics will be called the

μ Condition. The Metric Condition specialized the incidence vectors of cocycles is called *Cut Condition* (see [20]). The matroid M is called *R-flowing* ([20]) for $R \subseteq N(M)$, if for arbitrary $c : N(M) \mapsto \mathbf{R}_+$ for which the Cut Condition is satisfied there exists a multiflow.

Let us illustrate these definitions on H_6 with the above introduced notation: let c be 1 everywhere except on ab, and $c(ab) := 2$. It is easy to see that *the Cut Condition and the $[K_5, K_{2,3}]$-Condition are satisfied, but the $[H_6, H_6 \backslash R]$-condition is not satisfied*. This is the only violated metric condition, and $A(H_6, R)$ is not minimally non-ideal ! However, the edges adjacent to a form a tight cut, switching on which the following $[K_5, K_{2,3}]$ Condition is violated: $m(ab) := 0$, $m(e) := 1$ for $e \in H_6 \backslash R'$ and $m(e) := 2$ for $e \in R'$, where R' is the set of demands after the switching.

By the polarity of cones, primitive metrics provide the facets of the "multicommodity cone":

(2.5)   *The matroid M is R-flowing $(R \subseteq N(M))$, if and only if the Cut Condition implies the m-condition for every primitive metric m, where $m(e) = m(C \backslash e)$ for $e \in R$.*

A cocycle for which the Cut Condition holds with equality will be called a *tight cocycle*. It is easy to see that *switching* on a tight cocycle (that is interchanging the edges in R and those which are not in R), we get an equivalent multiflow problem which has an (integer) solution if and only if the original problem has one. If the Cut Condition holds then a tight cocycle is the disjoint union of tight cocircuits, and a sequence of switchings can be replaced by just one, on a tight cocycle. (Note that the emptyset is a tight cocycle.) The example of H_6 (see above) shows that the condition proving the non-existence of a flow may become simpler after switching on a tight cut.

Seymour pointed out the connection between the cut condition and idealness. (for instance in [20]). The following statement includes this connection plus the above remarks about the up, down matroid and primitive metrics. The proof relies only on an understanding of these notions, we omit it here:

(2.6)   *Let A be a binary clutter, $B = b(A)$. The following statements are equivalent:*
(i) A is ideal.
(ii) $M_0^*(A)$ is A-flowing for all $A \in \mathcal{A}$.
(iii) $M_1(A)$ is B-flowing for all $B \in \mathcal{B}$.
(iv) For every multiflow problem (M, R), where $M = M_1(A)$ and $R \in \mathcal{B}$, if $c : N(M) \mapsto \mathbf{Z}_+$ satisfies the Cut Condition, then for every metric it also satisfies the Metric Condition. $\square$

3   The Sufficiency of Distance Functions

Throughout this section and the next we will suppose that A is minimal non-ideal and binary with $b(A) := B$, and we will use the notations r, s, core, M_A, M_B, introduced in Section 1. If $R \in$ core $(\mathcal{B})$, let $\overline{R} = N(A) - R$. Let $\underline{1}$ denote the vector, of appropriate dimension, of all 1's.

We next prepare for the main result of this section, Theorem (3.1).

(3.1) *If A is partitionable, then with arbitrary $R \in core\,(B)$, $m := [M_1(A), \overline{R}]$ is a primitive metric.*

Indeed, by definition, m is a metric. Clearly, if $A \in core\,(A)$ and $A \cap R = \{e\}$, then m satisfies the equality $m(e) = m(A \setminus e)\ (= r - 1)$. But according to Lehman's theorem (Theorem 1.1), out of the n linearly independent vectors $A \in core\,(A)$, $n - 1$ satisfy $|A \cap R| = 1$. Hence for $n - 1$ linearly independent $A_1, \ldots, A_{n-1} \in core\,(A)$ (all but one member of core (A)), the inequality $m(A \setminus e) - m(e) \geq 0$ is satisfied with equality, where $\{e\} = A \cap R$. Note that the coefficient vectors of these equalities arise by switching the sign $A_1, \ldots, A_{n-1}$ (from $+$ to $-$) for all $e \in R$. But clearly, a set of vectors is linearly independent if and only if after switching the signs of some coordinates it is linearly independent. Hence m satifies $n - 1$ linearly independent equations, which means that it is an extreme ray of the cone of metrics. □

Such a metric will be called *partitionable*; if A is in addition minimal non-ideal, then we will call the metric *minimal non-ideal* as well.

The most natural converse of (3.1) is not true, as the example of $[H_6, \overline{R}]$ (see Section 2) shows. Still, with some care, (3.1) can be reversed:

(3.2) *Let $M = (N, C)$ be a binary matroid, and $R^* \subseteq N$. If M is not R^*-flowing, then there exists $R \in B(M, R^*)$ (that is, $R = R^* \triangle C$, C cocycle of M, R minimal) and $c : N \mapsto \mathbf{R}_+$ such that for the multiflow problem $(M, R = R^* \triangle C, c)$, there exists a violated minimal non-ideal metric condition.*

Proof. Suppose M is not R^*-flowing. Then according to (2.6) $A(M, R^*)$ is not ideal, whence there exist $X, Y \subseteq N(M)$, $X \cap Y = \emptyset$ such that $Q := A(M, R^*) \setminus X/Y$, $R := b(Q)$ is a minimally non-ideal pair of clutters. Denote the respective minima of cardinalities by q and r. Denote the cardinality of $N(Q) = N(R)$ by n. Let $R \in core\,(R)$ be arbitrary, and $C := R^* \triangle R$. Define $m(e) := 0$ if $e \in Y$, $m(e) := 1$ if $e \in N(M) \setminus (R \cup X \cup Y)$; if $e \in R \cup X$, then $m(e) := \min\{|C \setminus e| : C \in C, \{e\} = C \cap (R \cup X)\}$. Define now $c(e) := 0$ if $e \in X$, $c(e)$ to be arbitrarily big if $e \in Y$, and $c(e) := 1$ otherwise. (M, R, c) satisfies the Cut Condition, sincoevery cut C of M, $R \triangle C \in B(M, R^*)$, so $c(R \triangle C) \geq c(R)$, that is, $0 \leq c(R \triangle C) - c(R) = c(C \setminus R) - c(C \cap R)$; but the Metric Condition is violated, since $\sum_{e \in R} m(e)c(e) = (q - 1)r > n - r = |N(M) \setminus (X \cup Y \cup R)| = \sum_{e \in N(M) \setminus R} m(e)c(e)$. □

In the example of $[H_6, \overline{R}]$, C is the tight cut mentioned in Section 2; this example shows that (3.2) cannot be sharpened so that for (M, R^*) itself (and some capacity function) there exists a violated minimal non-ideal metric condition. A special case of the following theorem was stated as a conjecture in Robertson and Seymour's "Graph Minors" volume, (1991, "Open Problem 11 (A. Sebő)").

Theorem 3.1 *Let A be a binary clutter, B its blocker. Then $M_1(A)$ is B-flowing for all $B \in B$, if and only if for every capacity function for which the cut condition is satisfied, every m-condition where m is a distance function, is also satisfied. Furthermore, m can be restricted to be a minimal non-ideal metric.*

Proof. The only if part is easy, using results in Section 2. Indeed, suppose $(M_1(\mathcal{A}), B)$ is B-flowing for all $B \in \mathcal{B}$. Let $c : S \mapsto \mathbf{R}_+$ be such that the Cut Condition is satisfied. Then there exists a multiflow, and by (the easy part of) the Metric Criterion, the Metric Condition is satisfied for every metric m, in particular also for distance functions, as claimed.

To prove the if part, suppose $B \in \mathcal{B}$ and $(M_1(\mathcal{A}), B)$ is not B-flowing. Then by (3.2) there exists $R \in \mathcal{B}(= \mathcal{B}(M_1(\mathcal{A}), B)$ so that for some $c : S \mapsto \mathbf{R}_+$ there exists a violated minimally non-ideal metric condition for $(M_1(\mathcal{A}), R, c)$. $\qquad\square$

4 Bipartite Distance Functions

The next result, Theorem 4.1, relies on the application of Lehman's theorem to binary clutters, and not at all on the theory of metrics and multiflows; yet, through Theorem 3.1 it will yield a surprising statement on the primitive metrics of binary matroids.

(4.1) *If $\mathcal{A}$ is minimal non-ideal and binary then the cardinalities of all elements of $\mathcal{A}$ have the same parity.*

Proof. We prove that $\underline{1}$ is the $\bmod\, 2$ sum of incidence vectors of some $\mathcal{B}' \subseteq \mathrm{core}(\mathcal{B})$. We will be done then, since $|A| = \chi_A^T \underline{1} \equiv \chi_A^T \sum_{B \in \mathcal{B}'} \chi_B = \sum_{B \in \mathcal{B}'} |A \cap B|$, where $|A \cap B|$ is odd for all $A \in \mathcal{A}$. So for all $A \in \mathcal{A}$, $|A|$ has the same parity as $|\mathcal{B}'|$, as claimed.

We show now using Theorem 1.1 that $\sum_{B \in \mathcal{B}_v} \chi_B$, where $\mathcal{B}_v$ is the set defined in (1.1), is $\underline{1}$, so $\mathcal{B}' = \mathcal{B}_v$ will do. We actually know from (1.1) the exact value of $w := \sum_{B \in \mathcal{B}_v} \chi_B$: $w(u) = 1$ if $u \in N \setminus v$ and $w(v) = \alpha$. But using Theorem 1.1 again, α is also equal to $|A \cap B|$ for some $A \in \mathcal{A}$ and $B \in B$ (namely for any pair of associates), so it is odd. $\qquad\square$

We get as a consequence:

(4.2) *Minimal non-ideal metrics are bipartite, that is, if $\mathcal{A}$ is minimal non-ideal and $R \in B$, then $m := [M_1(\mathcal{A}), \overline{R}]$ is a bipartite metric.*

Proof. Let C be a cycle in $M_1(\mathcal{A})$. If C is also in $M_0(\mathcal{A})$ then (4.1) implies that $|C|$ is even. Since $|R \cap C|$ is odd, and hence also $|C \setminus R|$ is even, it follows that:

$$\sum_{e \in C} m(e) = \sum_{e \in R \cap C} m(e) + \sum_{e \in C \setminus R} m(e) = |R \cap C|(s-1) + |C \setminus R| \equiv 0$$

Similarly, if $C \in M_1(\mathcal{A}) \setminus M_0(\mathcal{A})(= \mathcal{A})$, then $|R \cap C|$ is odd; according to (4.1) $|C| \equiv s$; these two observations together imply $|C \setminus R| \equiv s - 1$, and we deduce:

$$\sum_{e \in C} m(e) = |R \cap C|(s-1) + |C \setminus R| \equiv (s-1) + |C \setminus R| \equiv 0 .$$

$\qquad\square$

With the help of (4.2) Theorem 3.1 can be sharpened:

Theorem 4.1   Let A be a binary clutter, B its blocker. The Cut Condition is sufficient for $(M_1(A), B)$ $(B \in B)$ to have a multiflow, if and only if for every capacity function for which the cut condition is satisfied, every m-condition where m is a bipartite distance function is also satisfied.

Theorem 4.1 is an immediate consequence of Theorem 3.1 and (4.2). One can actually deduce further structural properties from Lehman's theorem in the particular case of binary clutters:

(4.3)   Let Q and R be a minimally non-ideal pair of binary clutters, $M :=$ $M_1(Q)$, and $R \in$ core (R). Let Q be the partner of R and $C := Q \cap R$, $|C|$ is odd, $|C| \geq 3$. Consider the multiflow problem $(M, R, \underline{1})$.
(i) M is an Eulerian matroid, and either M or M/R (in both cases $M - R$) is bipartite.
(ii) There exist exactly $n - 1$ non-empty tight cuts each of which meets C in $|C| - 1$ elements.
(iii) If K is a tight cut, then it is an (inclusionwise) minimal cut, furthermore $K \setminus R$ is a minimal cut in $M - R$.

Proof.   By Theorem 1.1 $|C| = \alpha > 1$. and $\alpha = |Q \cap R|$ is odd, since Q and R are a blocking pair of binary clutters.

By (4.1), every member of R has the same parity, so $M_0(R)$ is bipartite, and then $M_0^*(R) = M_1(Q) = M$ is Eulerian. Applying (4.1) to Q now, we get that all members of Q have the same parity. If this parity is even, then $M = M_1(Q)$ is bipartite; if it is odd, then, since $|R \cap Q|$ is odd for all $Q \in Q$, we get that M/R is bipartite. The proof of (i) is finished.

Let now the different minimum cardinality members of R be $R = R_0, R_1, \ldots,$ R_{n-1}; according to Theorem 1.1 this is a complete list. Therefore, since $B \in R$ is of minimum cardinality if and only if $R\Delta B$ is a tight cut, the complete list of tight cuts is $R\Delta R_i$ $(i = 1, \ldots, n - 1)$. Since $Q \cap R = C$, but $|Q \cap R_i| = 1$ if $i = 1, \ldots, n - 1$, $|(R\Delta R_i) \cap C| = |C| - |R_i \cap C| \geq |C| - |R_i \cap Q| = |C| - 1$, and (ii) is proved.

Now (ii) implies that any pair of non-empty tight cuts have $|C| - 2 \geq 1$ common demands, so they are not disjoint. If K is a tight cut which is not minimal, then, since the matroid is binary, it is the disjoint union of at least two tight cuts, which is impossible, because we have just noticed that tight cuts are not disjoint. If $K - R$ is he disjoint union of the cuts $C_1, \ldots, C_k$ of $M - R$, then there exist $R_1, \ldots, R_k$ such that $C_i \cup R_i$ is a cut $(i = 1, \ldots, k)$. Since K is a minimal cut, it follows that $R_1 \cup \ldots \cup R_k = R$ (not necessarily disjoint union). Now

$$0 \leq \sum_{i=1}^{k} (|C_i| - |R_i|) \leq |K \setminus R| - |K \cap R| = 0,$$

so there is equality throughout. It follows that the R_i are disjoint and $C_i \cup R_i$ $(i = 1, \ldots, k)$ are tight cuts, so by what has already been proved, $i = 1$, and (iii) is proved.   □

5 Summary

In this section we summarize the results and illustrate them in the graphic special case. We reformulate Seymour's conjecture, putting it into a form which helps compare it with what has been proved. We then show what remains to be proved for showing this conjecture for graphs This special case contains many important multiflow problems in graphs [4], so deserves special attention.

Conjecture Let $M = (N, C)$ be a binary matroid, and $R \subseteq N$. Then either M is R-flowing, or there exists $R' \in \mathcal{B}(M, R)$ (that is, $R' = R \triangle C$, where C is a cut and R' minimal with this property) and $c : S \mapsto \mathbb{Z}$ so that the multiflow problem (M, R', c) satisfies the Cut Condition, but some $[F_7, F_7 - L]$, $[K_5, K_{2,3}]$, $[R_{10}, R_{10} - C_3]$ condition is violated.

For graphs, only one of these conditions can be violated, the $[K_5, K_{2,3}]$ condition. The proof of the equivalence of this and Seymour's conjecture is not difficult, we omit it here. (It uses (2.6).) Let us restate Theorem 4.1 using (4.3) in the special case of graphs:

Theorem 5.1 Let $G = (V, E)$ be a graph, and $R \subseteq E$. Then either G is R-flowing, or there exists a $c : E \mapsto \mathbb{Z}$ (in fact $0 - 1 - \infty$, suppose there are no 0-weight edges [contract them]) so that possibly after switching on a cocycle there exists a minimal non-ideal metric m for which the metric condition is not satisfied, but the Cut Condition is satisfied. This metric m has the following properties:
Let Q be the partner of R and $C := Q \cap R$, $|C|$ is odd, $|C| \geq 3$. Consider the multiflow problem $(M, R, \underline{1})$.
(i) G is an Eulerian graph, $G - R$ is connected and bipartite and either all demand edges join two points in different classes or all of them join two points in the same class.
(ii) There exist exactly $n - 1$ non-empty tight cuts each of which meets C in $|C| - 1$ elements.
(iii) If K is a tight cut, then it is an (inclusionwise) minimal cut, furthermore $K \setminus R$ is a minimal cut in $M - R$.

In the graphic case Theorem 3.1 becomes a coNP-characterization theorem where the negative result is particularly easy to check: *either the Cut Condition is satisfied for every $B \in \mathcal{B}$, or if not, there exists a $B \in \mathcal{B}$ and a $0 - 1$-weighting of the edges for which the length of the shortest paths violates the metric criterion.*

In order to prove Seymour's conjecture for graphs, we need to prove that the only metrics with the properties listed in Theorem 5.1 are $[K_5, K_{2,3}]$-metrics.

If G is a graph and $R \subseteq E(G)$, let us call a path R-geodesic if it occurs as a subset of a shortest path in $\overline{R}$ between the endpoints of an $uv \in R$.

Consider the union of three disjoint even R geodesics with the same endpoints a and b at distance d. (It follows from Theorem 1.1 and $\alpha \geq 3$ that these geodesics exist.) If the set of pairwise distances between the mid-points c_1, c_2, c_3 is also d, and the minimum paths between c_i and c_j ($i \neq j$) are R-geodesics, we will call this disjoint union of R geodesics a *kite*.

Let G be a graph and $R \subseteq E(G)$.
–If there exists an R-kite, $\mathcal{A}(G, R)$ has a $\mathcal{K}_5$-clutter minor. (In (H_6, R), see the definition in Section 2, there exists a kite !)

The following also seems to be plausible:
–If there exists no R-kite, for all $R' \in \mathcal{B}(G, R)$ (including R) the Cut Condition is sufficient to have a flow.

References

1. D. Avis, "On the Extreme Rays of the Metric Cone," *Canadian Journal of Mathematics*, **32**, (1980), 126-144.

2. G. Gasparian, Minimal Imperfect Graphs: a Simple Approach, to appear in Combinatorica.

3. A.M.H Gerards, Graphs and Polyhedra: Binary Spaces and Cutting Planes, PhD. Thesis, Tilburg University (1988).

4. A.M.H Gerards, Multicommodity Flows and Polyhedra, CWI Quarterly, Volume 6, Number 3 (1993)

5. W.G. Bridges and H.J. Ryser, "Combinatorial Designs and Related Systems," *Journal of Algebra*, 13, (1969), 432-446.

6. G. Cornuéjols and B. Novick, "Ideal 0-1 Matrices," *The Journal of Combinatorial Theory, series B*, **60**, 1, January 1994, pp. 145-157.

7. J. Edmonds and D.R. Fulkerson, "Bottleneck Extrema," *Journal of Combinatorial Theory B* (8), (1970), 299-306.

8. A.V.Karzanov, "Sums of Cuts and Bipartite Metrics," *European Journal of Combinatorics*, 11, (1990), 473-484.

9. D.R. Fulkerson, "Blocking Polyhedra," in : *Graph Theory and Its Applications* (B. Harris, eds.), Academic Press, New York, (1970), 112.

10. A. Lehman, "A solution of the Shannon switching game", *Journal of SIAM*, 12, 4, (1964), 687-725.

11. M. Lomonosov, "Combinatorial Approaches to Multiflow Problems," *Discrete Applied Mathematics*, 11, (1985), 1-94.

12. C. Lütolf and F. Margot, "A Catalog of Minimally Nonideal Matrices," document, (1995).

13. M. Lomonosov and A. Sebő, "On the Geodesic-Structure of Graphs: a Polyhedral Approach to Metric Decomposition," Technical Report, Bonn, Institut für Ökonometrie und Operations Research, No. 93793.

14. B.Novick, "Ideal 0,1 Matrices," Ph.D. dissertation, Carnegie Mellon University (1990).

15. B.Novick, A. Sebő, "On integer multiflows and metric packings in matroids", IPCO 4, E. Balas and J. Clausen (Eds.), Lecture Notes in Computer Science, 920, Springer Verlag (1995).
    top

16. A. Sebő, "Cographic multicommodity flow problems: an epilogue", DIMACS Vol 1, Polyhedral Combinatorics (W. Cook and P. D. Seymour AMS, ACM eds., 1991).

17. P.D.Seymour, "The Forbidden Minors of Binary Clutters," *The Journal of the London Mathematical Society*, (12) (1976), 356-360.

18. P.D.Seymour, "A note on the production of matroid minors," *Journal of Combinatorial Theory B*, (22) (1977), 289-295.
19. P.D.Seymour, "Decomposition of Regular Matroids," *Journal of Combinatorial Theory B*, (28), (1980), 305-359.
20. P.D.Seymour, "Matroids and Multicommodity Flows," *European Journal of Combinatorics* (1981), 257-290.
21. P.D.Seymour, "The Matroids with the Max-Flow Min-Cut Property," *Journal of Combinatorial Theory B*, (23) (1977), 189-222.
22. P.D.Seymour, "On Lehman's Width-Length Characterization,", *DIMACS 1*, W. Cook and P. Seymour eds., (1990), 75-78.
23. A. Tucker, "Uniquely Colorable Perfect Graphs," Discr. Math., **44**, (1984), 187–194

A Supermodular Relaxation
for Scheduling with Release Dates

Michel X. Goemans[*]

Dept. of Mathematics, Room 2-382, M.I.T., Cambridge, MA 02139.[**]

Abstract. We consider the scheduling problem of minimizing a weighted sum of completion times under release dates. We present a relaxation which is a supermodular polyhedron. We show that this relaxation is precisely the projection of a time-indexed relaxation introduced by Dyer and Wolsey.

1 Introduction

We consider the 1-machine scheduling problem with release dates and in which the objective is to minimize a weighted sum of completion times. In the classical scheduling notation, this corresponds to $1|r_j| \sum_j w_j C_j$.

We use the following notation. Job j has a processing time p_j, and a release date r_j. Its completion time in a schedule is given by C_j. Let $N = \{1, \cdots, n\}$ denote the set of jobs to be processed. For any subset S of jobs, let $p(S)$ denote the sum of the processing times of jobs in S, let $p^2(S)$ (not to be confused with $p(S)^2 = p(S) * p(S)$) denote $\sum_{j \in S} p_j^2$, and let $r(S) = \min_{j \in S} r_j$.

In this paper, we consider the scheduling problem $1|r_j| \sum_j w_j C_j$ from a polyhedral point-of-view. For a survey of the study of scheduling polyhedra, we refer the reader to Queyranne and Schulz [10]. Let P denote the convex hull of the set of feasible completion times over all possible schedules, i.e. $P = conv\{(C_1, \cdots, C_n) :$ there exists a feasible schedule with completion time C_j for job $j\}$. When all jobs are available at time 0, i.e. $r_j = 0$ for all j, Queyranne [9] has completely characterized P in terms of linear inequalities:

$$P = \left\{ C : \sum_{j \in S} p_j C_j \geq h(S) \text{ for all } S \subseteq N \right\},$$

where $h(S) = \frac{1}{2}(p^2(S) + p(S)^2)$. Moreover, Queyranne has shown that $h(S)$ is supermodular, meaning that $h(S) + h(T) \leq h(S \cup T) + h(S \cap T)$ for all S and T. This implies that P is a supermodular polyhedron and that the greedy algorithm can be used to optimize over P when there are no release dates. For a discussion

---

[*] Part of this work was supported by NSF contract 9302476-CCR, a Sloan Foundation Fellowship, and ARPA Contract N00014-95-1-1246.

[**] Email:goemans@math.mit.edu

of supermodularity and related topics, see Lovász [5], Fujishige [3], or Nemhauser and Wolsey [6].

When there are release dates, one possible way to strengthen the supermodular constraints is to "shift" the entire schedule by the smallest release date in the set S. This implies that the following polyhedron is a relaxation of P

$$Q = \{C : \sum_{j \in S} p_j C_j \geq l(S) \text{ for all } S \subseteq N\},$$

where $l(S) = p(S)r(S) + \frac{1}{2}(p^2(S) + p(S)^2)$. However, it is easy to see that $l(S)$ is not necessarily supermodular. We may, however, construct an equivalent formulation of Q by replacing $l(S)$ by

$$g(S) = \max_{\{\text{Partitions } S_1, \cdots, S_k \text{ of } S\}} \sum_{i=1}^{k} l(S_i).$$

The function $g(S)$ is called the upper Dilworth truncation of $l(S)$ (see Lovász [5]). One of the results of this paper is to show that $g(S)$ is supermodular and therefore Q defines a supermodular polyhedron. This means that one can optimize over P by using the greedy algorithm, and we show how the greedy algorithm can be implemented in $O(n \log n)$ time. A very similar situation appears in [11]. We also give necessary and sufficient conditions under which this supermodular inequality defines a facet of P.

There are typically three types of scheduling formulations (see [10]); the first ones use the completion times as variables, the second ones have linear ordering variables indicating whether job j precedes job k in the schedule, and the third ones are time-indexed and have one variable for each job and each unit of time. Dyer and Wolsey [2] present several types of relaxations for $1|r_j|\sum_j w_j C_j$, the strongest ones being time-indexed. We actually show that one of their time-indexed relaxations is equivalent to Q; by projecting the time-indexed variables, one obtains Q. The way we derive time-indexed variables out of Q is quite interesting and may be useful for other types of scheduling problems. We should point out that the equivalence between these two relaxations gives another proof of the supermodularity of $g(S)$.

This research was actually performed back in 1989 and was motivated by the preprints of the papers of Queyranne [9] and Dyer and Wolsey [2]. Very recently, Hall, Shmoys and Wein [4] used time-indexed formulations to derive the first approximation algorithms for several scheduling problems with a weighted sum of completion times as objective. Subsequently, Schulz [12] also showed that approximation algorithms for these problems can be obtained from relaxations based solely on completion times. This motivated us to write these results on the relationship between time-indexed relaxations and completion times relaxations for $1|r_j|\sum_j w_j C_j$.

2   Supermodular relaxation

Instead of computing the upper Dilworth truncation of $l(S)$, we will directly introduce the function $g(S)$ and then its relationship with $l(S)$ will become clear later on.

Let S be a subset of jobs. Consider the schedule in which we first schedule all jobs in S in order of non-decreasing release dates (ties are broken arbitrarily) and as early as possible, and then schedule all other jobs in any order. We refer to this schedule as the S-schedule (it is only well specified for the jobs in S, but this will not cause any problem). Some of the jobs in S will be scheduled at their release dates in the S-schedule, and the others will start strictly after their release dates. This leads to a partition of S into $S_1, \cdots, S_k$, such that (i) exactly one job in each S_i is scheduled at its release date, and (ii) all jobs in S_i are scheduled before the jobs in S_{i+1}. We refer to $S_1, \cdots, S_k$ as the *canonical decomposition* of S. Observe that the canonical decomposition does not depend on how ties are broken in the definition of the S-schedule when there are several jobs with the same release date. We will refer to the S-schedule corresponding to the canonical decomposition of S by a superscript (S).

For any subset S of jobs, let

$$g(S) = \sum_{j \in S} p_j C_j^{(S)}.$$

An expression of $g(S)$ in terms of its canonical decomposition will be given later. This will relate the function g to the function l discussed in the introduction.

Theorem 1. *For any S, the inequality*

$$\sum_{j \in S} p_j C_j \geq g(S) \tag{1}$$

is a valid inequality for P.

Proof. We need to prove that $z(C) = \sum_{j \in S} p_j C_j$ is minimized by the S-schedule. Consider any schedule and assume that the start time of some job in S can be decreased slightly without violating feasibility. Then $z(C)$ decreases, and therefore we can assume that all jobs in S are scheduled as early as possible. Furthermore, we can assume that all jobs not in S are scheduled after all the jobs in S, and therefore do not interfere with the jobs in S.

Now, if the jobs in S are not scheduled according to non-decreasing release dates then there exist two consecutive jobs j_1 and j_2 such that $C_{j_1} < C_{j_2}$ and $r_{j_1} > r_{j_2}$. By interchanging j_1 and j_2, we obtain a new feasible schedule with $C'_{j_2} = C_{j_2} - p_{j_1}$ and $C'_{j_1} = C_{j_1} + p_{j_2}$. Observe that $p_{j_1} C_{j_1} + p_{j_2} C_{j_2} = p_{j_1} C'_{j_1} + p_{j_2} C'_{j_2}$, and therefore $z(C) = z(C')$. Repeating this argument and the first part of the proof, we derive that the S-schedule must minimize $g(S)$.

Our next goal is to show that $g(S)$ is supermodular, but this will require some preliminaries. First, for any schedule, we define the indicator function $I_S(t)$ to be 1 at time t if some job in S is being processed at time t, and 0 otherwise. To avoid any confusion, if a job of S is being processed up to time t and no other job of S is started at time t, we let $I_S(t)$ to be 0. For simplicity, we denote $I_{\{j\}}(t)$ by $I_j(t)$. The indicator function will be very useful not only to prove supermodularity of $g(S)$ but also to relate our formulation to a time-indexed relaxation introduced by Dyer and Wolsey [2] (which in turn can be used to prove the supermodularity of $g(S)$ as well).

Here are two elementary properties of indicator functions.

Lemma 2. *1. If $S_1, \cdots, S_k$ is a partition of S then $I_S(t) = \sum_{i=1}^{k} I_{S_i}(t)$. In particular, $I_S(t) = \sum_{j \in S} I_j(t)$.*

2. For any S, $\int_0^\infty I_S(t)dt = p(S)$.

Proof. 1. is obvious. 2. follows from the fact that the total processing time of jobs in S is $p(S)$.

The following lemma will be very useful for the analysis.

Lemma 3. *For any schedule and any set S,*

$$\sum_{j \in S} p_j C_j = \frac{1}{2} p^2(S) + \int_0^\infty t I_S(t) dt = \frac{1}{2} p^2(S) + \int_0^\infty \left[p(S) - \int_0^\tau I_S(t) dt \right] d\tau.$$

Note that the value $\sum_{j \in S} p_j C_j$ depends only on when the machine is processing a job of S, but does not depend on which specific job is being processed.

Proof. We first claim that, for any j,

$$p_j C_j = \frac{p_j^2}{2} + \int_0^\infty t I_j(t) dt.$$

If we then sum over all $j \in S$, we obtain the first equality by using Lemma 2, part 1.

To prove the claim, it suffices to observe that $I_j(t)$ is equal to 1 on the interval $[C_j - p_j, C_j)$ and 0 otherwise. As a result, the integral on the right-hand-side is equal to

$$\int_{C_j - p_j}^{C_j} t\, dt = \frac{C_j^2}{2} - \frac{(C_j - p_j)^2}{2} = p_j C_j - \frac{p_j^2}{2},$$

and the claim follows.

The second equality follows from simple calculus:

$$\int_0^\infty t I_S(t) dt = \int_0^\infty \int_0^t I_S(t) d\tau dt$$

$$= \int_0^\infty \int_\tau^\infty I_S(t) dt d\tau$$

$$= \int_0^\infty \left[p(S) - \int_0^\tau I_S(t) dt \right] d\tau,$$

using Lemma 2, part 2.

To prove the supermodularity of $g(S)$, we will only be dealing with the indicator functions corresponding to S-schedules for all S. Let $J_S(t)$ denote $I_S^{(S)}(t)$, namely the indicator function for the set S associated with the S-schedule. We refer to $J_S(t)$ as the canonical indicator function. Although this is not needed to prove supermodularity, we first give an expression for $g(S)$ in terms of its canonical decomposition.

Lemma 4. *Let $S_1, \cdots, S_k$ be the canonical decomposition of S. Then*

$$g(S) = \frac{1}{2}p^2(S) + \sum_{i=1}^{k} p(S_i)\left(r(S_i) + \frac{1}{2}p(S_i)\right) = \sum_{i=1}^{k} l(S_i).$$

Proof. By definition, $g(S) = \sum_{j \in S} p_j C_j^{(S)}$. For $j \in S_i$, the definition of the canonical decomposition implies that $C_j^{(S)} = C_j^{(S_i)}$, and thus

$$g(S) = \sum_{i=1}^{k} \sum_{j \in S_i} p_j C_j^{(S_i)}.$$

By Lemma 3,

$$\sum_{j \in S_i} p_j C_j^{(S_i)} = \frac{1}{2}p^2(S_i) + \int_0^\infty \left[p(S_i) - \int_0^\tau J_{S_i}(t)dt\right] d\tau.$$

But $J_{S_i}(t)$ is 1 if $t \in [r(S_i), r(S_i) + p(S_i))$ and 0 otherwise. Therefore,

$$\int_0^\infty \left[p(S_i) - \int_0^\tau J_{S_i}(t)dt\right] d\tau = p(S_i)\left[r(S_i) + \frac{p(S_i)}{2}\right],$$

which proves the result.

This implies that g is the upper Dilworth truncation of l (since there is a schedule achieving $g(S)$).

We now give some properties of the canonical indicator function.

Lemma 5 (Monotonicity). *For $S \subseteq T$, and for any t, $J_S(t) \le J_T(t)$.*

Proof. Let $S_1, \cdots, S_k$ and $T_1, \cdots, T_l$ be the canonical decompositions of S and T respectively. In the S-schedule, the machine is busy from $r(S_i)$ to $r(S_i) + p(S_i)$ for every i. In the T-schedule, independently of whether a job is being processed right before $r(S_i)$, the machine must be continuously busy at least between $r(S_i)$ and $r(S_i) + p(S_i)$ since all the jobs in S_i were released on or after $r(S_i)$. In other words, whenever the S-schedule is processing a job in S, the T-schedule must be processing a job in T, proving the claim.

Lemma 6. *For $j \notin S$,*

$$\int_0^t [J_{S \cup \{j\}}(\tau) - J_S(\tau)] \, d\tau = \min \left(p_j, \int_{\min(r_j, t)}^t (1 - J_S(\tau)) \, d\tau \right).$$

Proof. If $r_j > t$ then both sides are 0. So assume that $r_j \leq t$. The integral on the right-hand-side represents the amount of idle time between r_j and t in the S-schedule. In the $(S \cup \{j\})$-schedule, these periods of idle time will be used to process some job in $S \cup \{j\}$, up to p_j units. This proves the result.

A function f is submodular if for all S and T, $f(S) + f(T) \geq f(S \cup T) + f(S \cap T)$. So, f is submodular iff $-f$ is supermodular. An alternate definition for submodularity (see, for example, [6]) is that for any $T \subset S$ and for any $j \notin T$, we have that

$$f(S \cup \{j\}) - f(S) \leq f(T \cup \{j\}) - f(T).$$

Let $f_S(t) = \int_0^t J_S(\tau) d\tau$.

Lemma 7. *For any t, $f_S(t)$ is submodular.*

Proof. By Lemma 6, for any $j \notin S$, we have that

$$f_{S \cup \{j\}}(t) - f_S(t) = \min \left(p_j, \int_{\min(r_j, t)}^t (1 - J_S(\tau) d\tau \right).$$

The same expression with S replaced by T can also be written.
Using Lemma 5, it then follows that

$$f_{S \cup \{j\}}(t) - f_S(t) \leq f_{T \cup \{j\}}(t) - f_T(t),$$

since the integrand on the right-hand-side is no greater for S than for T. This shows submodularity.

We are now ready to prove:

Theorem 8. *The function $g(S)$ is supermodular.*

Proof. From Lemma 3, $g(S)$ can be rewritten as:

$$g(S) = \frac{1}{2} p^2(S) + \int_0^\infty [p(S) - f_S(t)] \, d\tau.$$

Using the fact that $p(S)$ and $p^2(S)$ are modular (i.e. both supermodular and submodular) and the submodularity of $f_S(t)$ proved in Lemma 7, we derive the supermodularity of g.

We now consider the following linear program:

$$Z_R = \text{Min} \sum_{j \in N} w_j C_j$$

subject to:

(R)
$$\sum_{j \in S} p_j C_j \geq g(S) \qquad\qquad S \subseteq N.$$

Because of Theorem 1, (R) is a valid relaxation for the scheduling problem $1|r_j| \sum_j w_j C_j$ and therefore Z_R is a lower bound on the optimum value Z^*. If we make a change of variables and replace $p_j C_j$ by x_j for all j, we can rewrite (R) as

$$Z_R = \text{Min} \sum_{j \in N} \frac{w_j}{p_j} x_j$$

subject to:

(R')
$$\sum_{j \in S} x_j \geq g(S) \qquad\qquad S \subseteq N.$$

Because of Theorem 8, this is a *supermodular polyhedron* (see [3]). This implies that its optimum solution can be computed by the greedy algorithm. More precisely, first order the jobs in non-increasing order of w_j/p_j, i.e. assume that $\frac{w_1}{p_1} \geq \frac{w_2}{p_2} \geq \cdots \geq \frac{w_n}{p_n}$. Let $N_j = \{1, 2 \ldots, j\}$. Then, let $\tilde{x}_j = g(N_j) - g(N_{j-1})$ for $j = 1, \cdots, n$, where $g(\emptyset) = 0$. Then $\tilde{x}$ is an optimum solution for (R'). As a result, an optimum solution to (R) is given by $\tilde{C}_j = \frac{1}{p_j}(g(N_j) - g(N_{j-1}))$ for $j = 1, \cdots, n$, assuming that $\frac{w_1}{p_1} \geq \frac{w_2}{p_2} \geq \cdots \geq \frac{w_n}{p_n}$.

Computing $\tilde{C}_j$ for all j can be done in $O(n \log n)$ time, although we will need some data structures for this purpose. First, we can sort the $\frac{w_j}{p_j}$ in $O(n \log n)$ time. Then we need to be able to compute $g(N_j)$ efficiently for all j. A naive implementation would take $O(n^2)$ time. Here is a brief sketch of how to implement this computation in $O(n \log n)$ time. For a given j, we keep track of the disjoint intervals (corresponding to the canonical decomposition of N_j) in which $J_{N_j}(t) = 1$. For this purpose, we use a balanced binary search tree (such as a red-black tree, see [1]). When considering $j+1$, we start by finding the first interval I whose right endpoint is greater than r_{j+1}. This requires one SEARCH operation (for the terminology see [1]). The addition of job $j+1$ will either create a new interval (if $r_{j+1} + p_{j+1}$ is no greater than the left endpoint of I) or will cause several consecutive intervals, say q, to merge (because we need p_{j+1} units of idle time to accommodate job $j+1$). This can be determined by performing at most q SUCCESSOR operations, and then performing either one INSERT operation or $q-1$ DELETION operations. This gives $O(q)$ operations, each taking $O(\log n)$ time. However, the number of intervals decreases by $q-1$, and therefore this can be charged to the time when these intervals were created. This shows that maintaining the intervals over all j takes $O(n \log n)$ time. As we update these intervals, it is very easy to compute $g(N_{j+1}) - g(N_j)$ in $O(q)$ time.

3 Equivalence with a relaxation of Dyer and Wolsey

In [2], Dyer and Wolsey introduce 5 different relaxations of $1|r_j|\sum_j w_j C_j$ that they denote by $(A), (B), (C), (D)$ and (E). They show that their values satisfy $Z_A \leq Z_B \leq Z_C \leq Z_D \leq Z_E$, i.e. E is the strongest one. Relaxation (D) is a time-indexed formulation using two types of variables: $y_{j\tau} = 1$ if job j is being processed at time τ, and t_j represents the start time of job j. For simplicity, we add p_j to t_j and replace the resulting expression by C_j; this gives an equivalent relaxation.

$$Z_D = \text{Min} \sum_j w_j C_j$$

subject to:

(D)
$$\sum_j y_{j\tau} \leq 1 \qquad \qquad \tau = 0, 1, \cdots, T$$

$$\sum_\tau y_{j\tau} = p_j \qquad \qquad j \in N \qquad\qquad (2)$$

$$C_j = \frac{p_j}{2} + \frac{1}{p_j} \sum_\tau \left(\tau + \frac{1}{2}\right) y_{j\tau} \quad j \in N$$

$$0 \leq y_{j\tau} \qquad \qquad j \in N, \tau = r_j, r_j + 1, \cdots, T,$$

where T is an upper bound on the makespan of any schedule. In this relaxation, the release dates are assumed to be integral, and we will keep this assumption for the rest of this paper. Because of the equalities (2), the expression for C_j can take many different equivalent forms; we selected the easiest for the forthcoming analysis. Observe that the number of variables of this formulation is only pseudo-polynomial.

Theorem 9. $Z_D = Z_R$.

Proof. The most interesting part of the proof is to show that $Z_D \leq Z_R$. Consider the optimal solution $\tilde{C}$ given by the greedy algorithm for (R). We will construct variables $y_{j\tau}$ such that $\tilde{C}$ and y are feasible in (D), showing that $Z_D \leq Z_R$.

Assume that $\frac{w_1}{p_1} \geq \frac{w_2}{p_2} \geq \cdots \geq \frac{w_n}{p_n}$. Define $y_{j\tau} = J_{N_j}(\tau) - J_{N_{j-1}}(\tau)$ for every job j and every integral τ. We claim that $\tilde{C}$ and y are feasible in (D). First, $y_{j\tau} \geq 0$ because of Lemma 5. Moreover, $y_{j\tau}$ as defined will be 0 if $\tau < r_j$ or if $\tau \geq C_j^{(N_j)}$ and thus certainly if $\tau \geq T$. Equality (2) follows from Lemma 2 part 2, and the fact that $J_S(\tau)$ is constant over $[t, t+1)$ for any integer t:

$$\sum_\tau y_{j\tau} = \sum_\tau \left(J_{N_j}(\tau) - J_{N_{j-1}}(\tau)\right)$$

$$= \int_0^\infty \left(J_{N_j}(\tau) - J_{N_{j-1}}(\tau)\right) d\tau = p(N_j) - p(N_{j-1}) = p_j.$$

Furthermore, we have that

$$\sum_j y_{j\tau} = \sum_{j=1}^{n} \left(J_{N_j}(\tau) - J_{N_{j-1}}(\tau) \right) = J_N(\tau) \le 1,$$

for all τ. Finally, we need to verify the expression for $\tilde{C}_j$:

$$\frac{p_j}{2} + \frac{1}{p_j} \sum_\tau (\tau + \tfrac{1}{2}) y_{j\tau} = \frac{p_j}{2} + \frac{1}{p_j} \sum_\tau \int_\tau^{\tau+1} t \left(J_{N_j}(t) - J_{N_{j-1}}(t) \right) dt$$

$$= \frac{p_j}{2} + \frac{1}{p_j} \int_0^\infty t \left(J_{N_j}(t) - J_{N_{j-1}}(t) \right) dt$$

$$= \frac{p_j}{2} + \frac{1}{p_j} \left(g(N_j) - \tfrac{1}{2} p^2(N_j) - g(N_{j-1}) + \tfrac{1}{2} p^2(N_{j-1}) \right)$$

$$= \frac{1}{p_j} \left(g(N_j) - g(N_{j-1}) \right)$$

$$= \tilde{C}_j,$$

the third equality following from Lemma 3.

The second part of the proof is to show that $Z_D \ge Z_R$. For this purpose, we show that if y and C are feasible in (D) then C is also feasible in (R). Consider any set S. Then,

$$\sum_{j \in S} p_j C_j = \sum_{j \in S} p_j \left[\frac{p_j}{2} + \frac{1}{p_j} \sum_\tau \left(\tau + \frac{1}{2} \right) y_{j\tau} \right]$$

$$= \frac{p^2(S)}{2} + \sum_{j \in S} \sum_\tau \left(\tau + \frac{1}{2} \right) y_{j\tau}$$

$$= \frac{p^2(S)}{2} + \sum_\tau \left(\tau + \frac{1}{2} \right) \sum_{j \in S} y_{j\tau}.$$

Now, to y, we can easily associate a preemptive schedule: between τ and $\tau + 1$, $y_{j\tau}$ units of time are used to process job j. Because of the constraints on y in (D), every job is processed for a total of p_j units of time and the machine never works on two jobs at the same time. Let I' be the indicator function for this preemptive schedule. Then the above expression can be rewritten as:

$$\sum_{j \in S} p_j C_j = \frac{p^2(S)}{2} + \int_0^\infty t I_S'(t) dt.$$

But, the integral only depends on the times at which the preemptive schedule is busy processing some job in S, but not on which job is being processed at any such time. As a result, the expression on the right is minimized by a non-preemptive schedule. But, over the non-preemptive schedules, the expression is minimized for the S-schedule (see Lemma 3 and Theorem 1). This shows that $\sum_{j \in S} p_j C_j \ge g(S)$.

The proof actually shows a stronger result. When T is ∞, the projection of relaxation (D) onto the space of the C_j's is precisely (R).

From the proof, we can also derive a stronger result for the problem $1|r_j, p_j = 1|\sum_j w_j C_j$ with unit processing times. Indeed, the preemptive schedule derived in the first part of the proof is not preemptive if the processing times are all equal to 1. This shows that the value Z_R is precisely equal to the optimum value in the case of unit processing times, and that the feasible region defined by (R) is precisely equal to P. This was already known; see Queyranne and Schulz [11].

We should point out that Dyer and Wolsey [2] indicate that, after elimination of the variables C_j, (D) becomes a transportation problem. This has several consequences. First, we could have restricted our attention to *extreme* points of this transportation problem in the second part of the proof above, which have the property that $y_{j\tau} \in \{0, 1\}$. This would have avoided dealing with fractional y's.

Also, as indicated in [2], the relaxation (D) can be solved in $O(n \log n)$ time by exploiting the special cost structure of this transportation problem. Dyer and Wolsey refer the reader to Posner [8] for details. This gives the same time bound as the one we obtained for (R).

Finally, the fact that (D) becomes a transportation problem after elimination of the C_j's can be used to give an alternate proof of the supermodularity of g. Clearly,

$$g(S) = \min \left\{ \sum_{j \in S} p_j C_j \text{ subject to } \sum_{j \in T} p_j C_j \geq g(T) \text{ for all } T \subseteq N \right\}.$$

By Theorem 9, $g(S)$ can then be expressed by the following transportation problem:

$$g(S) = \frac{p^2(S)}{2} + \text{Min} \sum_{j \in S} \sum_{\tau} \left(\tau + \frac{1}{2} \right) y_{j\tau}$$

subject to:

$$\sum_j y_{j\tau} \leq 1 \qquad \tau = 0, 1, \cdots, T$$

$$\sum_{\tau} y_{j\tau} = p_j \qquad j \in S$$

$$0 \leq y_{j\tau} \qquad j \in S, \tau = r_j, r_j + 1, \cdots, T.$$

From an interpretation by Nemhauser et al. [7] of a result of Shapley [13], it then follows that the value of the above transportation problem as a function of S is supermodular.

This also shows that the value of Z_R (or Z_D) as a function of the set S of jobs is supermodular. But this follows from the fact that Z_R is obtained by optimizing over a supermodular polyhedron.

4 Additional results

The separation problem for the inequalities (1) can be solved efficiently. Indeed, we only need to separate over $\sum_{j\in S} p_j C_j \geq l(S)$, and this can be done by trying all n possible values for $r(S)$ and then applying a separation routine of Queyranne [9] for the problem without release dates. The overall separation routine can be implemented in $O(n^2)$ time.

We can also give necessary and sufficient conditions under which the inequalities (1) define facets of P. We should point out that the condition that all the release dates in S should be identical is not necessary as claimed in [11]. We say that S is inseparable if it is its own canonical decomposition. Certainly, the condition that S is inseparable is necessary for the inequality (1) to define a facet. For a given inseparable set S, we say that a schedule is S-tight if $I_S(t) = 1$ for all $t \in [r(S), r(S) + p(S))$. The proof of the following theorem is omitted.

Theorem 10. *Let $S = \{1, \cdots, i\}$ be an inseparable set with $r_1 \leq r_2 \leq \cdots \leq r_i$. Then $\sum_{j\in S} p_j C_j \geq g(S)$ defines a facet of P if and only if, for every $j \in \{2, \cdots, i\}$, there exists an S-tight schedule for which j precedes a job k with $k < j$.*

Proof. We start with the "only if" part. If, for some j, there is no S-tight schedule for which j precedes some job k with $k < j$ then we claim that every S-tight schedule starts with the jobs $\{1, 2, \cdots, j - 1\}$ (in any order). Indeed, assume there is a schedule in which a job $l > j - 1$ precedes a job $k \leq j - 1$. Then if we schedule j just in front of l and continue the schedule as before, we obtain a feasible schedule (since $r_l \geq r_j$). This gives a S-schedule violating the hypothesis (since j precedes k). Hence, every S-tight schedule is also $\{1, \cdots, j - 1\}$-tight and therefore the set of feasible schedules satisfying (1) at equality are included in the face defined by the set $\{1, \cdots, j - 1\}$.

For the "if" part, assume that any feasible schedule satisfying (1) at equality also satisfy $\sum_l a_l C_l = b$. For every j, consider an S-tight schedule in which j precedes k with $k < j$. If, in any schedule, we interchange two consecutive jobs such that the release date of the first one is at least the release date of the second one, then the schedule remains feasible. This means that, by the interchange of consecutive jobs (and possibly redefining k), we can assume that in our schedule j immediately precedes k. Now, if we interchange j and k, C_j increases by p_k and C_k decreases by p_j. But the equality $\sum_i a_i C_i = b$ must still be satisfied, implying that $a_k p_j = a_j p_k$, or $\frac{a_j}{p_j} = \frac{a_k}{p_k}$. Since, for every $2 \leq j \leq i$, we can find such a k with $k < j$, this proves that $\frac{a_1}{p_1} = \frac{a_2}{p_2} = \cdots = \frac{a_i}{p_i}$. Moreover, any job $j > i$ can be delayed arbitrarily, implying that $a_j = 0$ for $j > i$. This therefore shows that the equality must be $\sum_{j\in S} p_j C_j = g(S)$.

For example, the condition of the theorem is satisfied if $r_1 + p(S_{j-2}) \geq r_j$ for $j = 2, \cdots, i - 1$.

The results in this paper can also be translated to the problem with deadlines instead of release dates by simply inverting the time. For the problem

$1|d_j| \sum w_j C_j$, we can therefore introduce valid inequalities of the form $\sum_{j \in S} p_j C_j \leq k(S)$ for all S. If we now have simultaneously release dates and deadlines, we can consider the relaxation defined by $k(S) \geq \sum_{j \in S} p_j C_j \geq g(S)$ for all S. However, in the case of unit processing times $(1|r_j, d_j, p_j = 1| \sum w_j C_j)$, these inequalities are not sufficient to describe the convex hull of feasible completion times. This was also observed in [11]. The problem however is simply an assignment problem and can therefore be solved efficiently. Moreover, from Hall's theorem, we can easily derive that this scheduling problem is feasible if and only if $k(S) \geq g(S)$ for all S.

Acknowledgments

The author would like to thank Andreas Schulz for stimulating discussions and for convincing him to write these results. The results presented here resulted from discussions many years ago with Leslie Hall and Tim Magee on the papers of Maurice Queyranne and of Martin Dyer and Laurence Wolsey.

References

1. T.H. Cormen, C.E. Leiserson, and R.L. Rivest, *Introduction to algorithms*, McGraw Hill (1990).
2. M.E. Dyer and L.A. Wolsey, "Formulating the single machine sequencing problem with release dates as a mixed integer program", *Discrete Applied Mathematics*, 26, 255–270 (1990).
3. S. Fujishige, "Submodular functions and optimization", *Annals of Discrete Mathematics*, 47, North-Holland (1991).
4. L.A. Hall , D.B. Shmoys and J. Wein, "Scheduling to minimize average completion time: Off-line and on-line algorithms", *Proceedings of the 7th Annual ACM-SIAM Symposium on Discrete Algorithms*, 142–151, 1996.
5. L. Lovász, "Submodular functions and convexity", in: A. Bachem, M. Grötschel and B. Korte (eds.), *Mathematical Programming: The State of the Art, Bonn, 1982*, Springer, Berlin, 235–257, 1983.
6. G.L. Nemhauser and L.A. Wolsey, *Integer and Combinatorial Optimization*, John Wiley & Sons, New York (1988).
7. G.L. Nemhauser, L.A. Wolsey and M.L. Fisher, "An analysis of approximations for maximizing submodular set functions — I", *Mathematical Programing*, 14, 265–294 (1978).
8. M.E. Posner, "A sequencing problem with release dates and clustered jobs", *Management Science*, 32, 731–738 (1986).
9. M. Queyranne, "Structure of a simple scheduling polyhedron", *Mathematical Programming*, 58, 263–285 (1993).
10. M. Queyranne and A. S. Schulz, "Polyhedral approaches to machine scheduling", Preprint 408/1994, Department of Mathematics, Technical University of Berlin, Berlin, Germany, 1994.
11. M. Queyranne and A. S. Schulz, "Scheduling units jobs with compatible release dates on parallel machines with nonstationary speeds", Proceedings of the 4th Integer Programming and Combinatorial Optimization Conference, E. Balas and

J. Clausen, eds., Lecture Notes in Computer Science, **920**, Springer-Verlag, 307–320 (1995).

12. A.S. Schulz, "Scheduling to minimize total weighted completion time: Performance guarantees of LP-based heuristics and lower bounds", these proceedings, 1996.

13. L.S. Shapley, "Complements and substitutes in the optimal assignment problem", *Naval Research Logistics Quaterly*, **9**, 45–48 (1962).

Scheduling to Minimize Total Weighted Completion Time: Performance Guarantees of LP–Based Heuristics and Lower Bounds

Andreas S. Schulz

Technische Universität Berlin, Fachbereich Mathematik (MA 6-1),
Straße des 17. Juni 136, D–10623 Berlin, Germany,
schulz@math.tu–berlin.de

Abstract. There has been recent success in using polyhedral formulations of scheduling problems not only to obtain good lower bounds in practice but also to develop provably good approximation algorithms. Most of these formulations rely on binary decision variables that are a kind of assignment variables. We present quite simple polynomial–time approximation algorithms that are based on linear programming formulations with completion time variables and give the best known performance guarantees for minimizing the total weighted completion time in several scheduling environments. This amplifies the importance of (appropriate) polyhedral formulations in the design of approximation algorithms with good worst–case performance guarantees.

In particular, for the problem of minimizing the total weighted completion time on a single machine subject to precedence constraints we present a polynomial–time approximation algorithm with performance ratio better than 2. This outperforms a $(4 + \epsilon)$–approximation algorithm very recently proposed by Hall, Shmoys, and Wein that is based on time–indexed formulations. A slightly extended formulation leads to a performance guarantee of 3 for the same problem but with release dates. This improves a factor of 5.83 for the same problem and even the 4–approximation algorithm for the problem with release dates but without precedence constraints, both also due to Hall, Shmoys, and Wein.

By introducing new linear inequalities, we also show how to extend our technique to parallel machine problems. This leads, for instance, to the best known approximation algorithm for scheduling jobs with release dates on identical parallel machines. Finally, for the flow shop problem to minimize the total weighted completion time with both precedence constraints and release dates we present the first approximation algorithm that achieves a worst–case performance guarantee that is linear in the number of machines. We even extend this to multiprocessor flow shop scheduling.

The proofs of these results also imply guarantees for the lower bounds obtained by solving the proposed linear programming relaxations. This emphasizes the strength of linear programming formulations using completion time variables.

1 Introduction

An important goal in real–world scheduling is to optimize "average" performance. One of the classic criteria is the average weighted completion time, or equivalently, the total weighted completion time. Approximation algorithms for scheduling problems with this objective have very recently achieved much attention, see [PSW95, HSW96]. In this paper, we present new ρ–approximation algorithms for a variety of NP–hard scheduling problems to minimize the total weighted completion time. That is, these algorithms have a polynomial running time and produce schedules whose objective function value is not worse than ρ times the optimal one. Here, ρ is a constant in case of a single machine or identical parallel machines, and a sublinear or a linear function in the number of machines in case of uniform parallel machines or a (multiprocessor) flow shop environment, respectively.

In particular, Hall, Shmoys, and Wein [HSW96] showed very recently that an optimal fractional solution of the linear relaxation of a time–indexed integer programming formulation of the single–machine total weighted completion time problem with precedence constraints can be rounded to a schedule of objective function value at most 4 times as large. Unfortunately, because of the time indices this formulation is only pseudo–polynomial in the input size. Since their algorithm relies on solving the LP relaxation this algorithm is only pseudo–polynomial as well. However, Hall, Shmoys, and Wein present a new linear programming formulation of polynomial size, by relying on time intervals instead of single points in time, which leads to an algorithm guaranteed to produce a schedule with total weighted completion time at most $(4 + \epsilon)$ of the optimal value, for any fixed $\epsilon > 0$. When incorporating release dates, their algorithm produces a solution at most a factor of $(3 + 2\sqrt{2} + \epsilon)$ worse than the optimal one. If there are release dates but no precedence constraints, they are able to turn it into a 4–approximation algorithm. Up to now, these have been the best bounds obtained; see [HSW96] for the short history.

Our work is strongly motivated by the work of Hall, Shmoys, and Wein (but is not based on their results), and the observations Maurice Queyranne and the author made comparing different polyhedral formulations of scheduling problems, see [QS94]. We also refer the reader to [QS94] for an overview on polyhedral approaches to scheduling.

We present much simpler algorithms using the optimal solution to a linear programming relaxation in completion time (or natural date) variables that achieve a performance ratio of $\left(2 - \frac{2}{n+1}\right)$ for the single–machine problem subject to precedence constraints, and of 3 for the single–machine problem subject to precedence constraints and release dates (Sect. 2). This not only significantly improves the best performance guarantees known but provides new insight into the strength of several related LP relaxations of these problems. In particular, it implies that both the linear programming relaxation in time–indexed variables as well as in linear ordering variables can be used to obtain an approximation algorithm with the same performance ratio for the first problem. Moreover, by proposing new linear programming relaxations for identical parallel machine problems (Sect. 3) as well as uniform parallel machine problems (Sect. 4) we extend our method to these models as well. This leads, for instance, to a $(4 - \frac{1}{m})$-approximation algorithm for identical machines and release dates. This algorithm is slightly better than the $(4 + \epsilon)$-approximation algorithm that is the best previously known [HSW96] and is much simpler. We also extend these techniques to the problem of minimizing the total weighted completion time subject to both release dates and precedence constraints in a flow shop and develop a simple $(2m + 1)$-approximation algorithm in this case. Finally, we present a $3m$-approximation algorithm for multi–stage parallel processor flow shop scheduling to minimize the total weighted completion time (Sect. 5).

Tables 1 – 5 summarize the results of this paper, and compare the bounds with the best previously known, if any.

Single Machine	Known	New
Precedence Constraints	$4 + \epsilon$	$2 - \frac{2}{n+1}$
Release Dates	4	3
Precedence Constraints and Release Dates	$3 + 2\sqrt{2} + \epsilon$	3

Table 1: Summary of results for the minimization of the total weighted completion time on a single machine. The "Known" column lists the best known previous bounds, while the "New" column lists new results from this paper. All the best known bounds are due to Hall, Shmoys, and Wein [HSW96]. Note that for the models with precedence constraints this were the best known bounds for the minimization of the total (unweighted) completion time as well.

Identical Parallel Machines	Known	New
	$\frac{\sqrt{2}+1}{2}$	$(3 - \frac{1}{m})(1 - \frac{1}{n+1})$
Release Dates	$4 + \epsilon$	$4 - \frac{1}{m}$

Table 2: Summary of results for the minimization of the total weighted completion time on identical parallel machines. The first known bound is due to Kawaguchi and Kyan [KK86]; the second is due to Hall, Shmoys, and Wein [HSW96]. Although the former is better than ours, we have included it since it shows what can be achieved with our method. Indeed, our approach has the advantage of being able to handle quite naturally additional restrictions like release dates and precedence constraints. In addition, at the same time it gives the same guarantee for the LP lower bound. Here, m denotes the number of machines.

Uniform Parallel Machines	Known	New
	$4 + \epsilon$	$\min\{2 + (m - 1)\frac{\max s_i}{\sum s_i}, 2\frac{\sum s_i}{\max s_i}\}$
Release Dates	$4 + \epsilon$	$\min\{3 + (m - 1)\frac{\max s_i}{\sum s_i}, 1 + 2\frac{\sum s_i}{\max s_i}\}$

Table 3: Summary of results for the minimization of the total weighted completion time on uniform parallel machines. Here, s_i denotes the speed of machine i. Both best known bounds are due to Hall, Shmoys, and Wein [HSW95]. In the first case our bound is not worse than $1 + \sqrt{2m - 1}$. Hence it is always better than the known one for $m \leq 5$. In case of release dates our bound is not worse than $2 + \sqrt{2m - 1}$. Therefore it is always better than the known one for $m \leq 2$. Moreover, our algorithm and its analysis are directly accessible. The result of Hall, Shmoys, and Wein relies on a generalization of Hochbaum and Shmoys' dual approximation approach for minimizing the makespan on uniform parallel machines (see [HS88]).

Flow Shop	Known	New
	$-$	$2m - \frac{2m}{n+1}$
Precedence Constraints	$-$	$2m - \frac{2m}{n+1}$
Release Dates	$-$	$2m + 1$
Precedence Constraints and Release Dates	$-$	$2m + 1$

Table 4: Summary of results for the minimization of the total weighted completion time in a flow shop environment. The "$-$" indicates the lack of a relevant result. To the best of our knowledge, the only known bound is for minimizing the total completion time, i. e., for unit weights without additional restrictions. It is m and achieved by the shortest processing time rule, see Gonzalez and Sahni [GS78].

Multiprocessor Flow Shop	Known	New
	$-$	$3m$
Release Dates	$-$	$3m + 1$

Table 5: Summary of results for the minimization of the total weighted completion time in a multi–stage flow shop with identical parallel machines. Here, m is the number of stages.

In each case we give a polynomial solvable LP relaxation of the problem under consideration and work with an optimal LP solution. Let $w(\text{LP})$ be its value, and $w(\text{OPT})$ be the optimal value of the scheduling problem. We use the optimal LP solution to create a schedule with value $w(\text{HEURISTIC}) \leq \rho\, w(\text{LP})$, for some $\rho > 1$. Since $w(\text{LP}) \leq w(\text{OPT}) \leq w(\text{HEURISTIC})$, it follows that

$$w(\text{HEURISTIC}) \quad \leq \quad \rho\, w(\text{OPT}) \quad \text{(performance guarantee of primal heuristic)},$$

$$w(\text{LP}) \quad \geq \quad \frac{1}{\rho}\, w(\text{OPT}) \quad \text{(performance guarantee of dual heuristic)}.$$

That is, each time we present a ρ–approximation algorithm we also give a guarantee for the quality of the lower bound obtained by solving the associated LP relaxation.

Due to space limitations some details are omitted from this paper. A complete version can be obtained from the author [Sch95b] (see also [Sch95a]).

The power of linear programming formulations in completion time variables has further been proved by a number of results that have been obtained after a preliminary version of this paper appeared. Hall, Schulz, Shmoys, and Wein [HSSW96] extended this approach to scheduling of precedence–constrained jobs on parallel machines as well as to preemptive scheduling. In the combined journal version of our papers [HSSW96], we will report on these results, including a 7–approximation algorithm for $P|r_j, \text{prec}| \sum w_j C_j$, a 3–approximation algorithm for $P|p_j = 1, r_j, \text{prec}| \sum w_j C_j$, a 2–approximation algorithm for $1|\text{pmtn}, r_j, \text{prec}| \sum w_j C_j$, and a 3–approximation algorithm for $P|\text{pmtn}, r_j, \text{prec}| \sum w_j C_j$. Subsequently, Chakrabarti, Phillips, Schulz, Shmoys, Stein, and Wein [CPS$^+$96] turned the algorithm for $P|r_j, \text{prec}| \sum w_j C_j$ into a 5.33–approximation algorithm. Moreover, LP formulations in completion time and certain binary variables can also be used to obtain the first constant–factor approximation algorithms for scheduling with communication delays so as to minimize the total weighted completion time [MSS96].

Improved approximation algorithms for some of the problems that only involve release dates have also been obtained subsequently by the use of a general on–line framework; see [CPS+96] for details.

2 Single Machine with Precedence Constraints

In this section, we present a $(2 - \frac{2}{n+1})$-approximation algorithm for the problem of minimizing the total weighted completion time on one machine subject to precedence constraints. We are given a set $N = \{1, \ldots, n\}$ of n jobs to be processed on a single machine which can execute at most one job at a time. Each job j requires a processing time $p_j > 0$ and cannot be interrupted during its execution. We denote the completion time of a job j by C_j and let $w_j \geq 0$ be its weight. Throughout, we assume, w. l. o. g., that all data are integral. Each feasible sequence has to be consistent with precedence constraints, and we are interested in minimizing the weighted sum of the completion times. This problem is well–known to be NP–hard, see [LRKB77, Law78]. In the usual classification scheme [GLLRK79] it is denoted by $1|\text{prec}| \sum w_j C_j$. We write $j \rightarrow k$ when job j has to precede job k, and $j \parallel k$ if j and k are incomparable with respect to the precedence constraints.

Since a feasible schedule is uniquely determined by the corresponding job completion times it seems to be natural to work with completion time variables. Hence a feasible schedule can be seen as a vector $C \in \mathbb{R}^N$, satisfying:

$$C_j \geq p_j \qquad \text{for all jobs } j,$$

$$C_k \geq C_j + p_k \qquad \text{for all jobs } j, k \text{ with } j \rightarrow k,$$

$$C_k \geq C_j + p_k \quad \text{or} \quad C_j \geq C_k + p_j \qquad \text{for all jobs } j \parallel k, j \neq k.$$

The difficulties clearly arise from the disjunctive constraints. Queyranne [Que93] showed that the convex hull of feasible completion time vectors in the absence of precedence constraints is completely described by the following linear inequalities:

$$\sum_{j \in A} p_j C_j \geq \frac{1}{2} \left(\left(\sum_{j \in A} p_j \right)^2 + \sum_{j \in A} p_j^2 \right) \qquad \text{for all } A \subseteq N. \tag{1}$$

Lemma 1. *Let* $C \in \mathbb{R}^N$ *be a point satisfying the inequalities* (1). *Assume,* w. l. o. g., *that* $C_1 \leq C_2 \leq \cdots \leq C_n$. *Then, for each* $j = 1, \ldots, n$,

$$\sum_{k=1}^{j} p_k \leq 2C_j - \frac{\sum_{k=1}^{j} p_k^2}{\sum_{k=1}^{j} p_k}.$$

Proof. The inequality (1) for $A = \{1, \ldots, j\}$ implies that

$$p_j C_j \geq \frac{1}{2} \left(\sum_{k=1}^{j} p_k \right)^2 + \frac{1}{2} \sum_{k=1}^{j} p_k^2 - \sum_{k=1}^{j-1} p_k C_k \ .$$

Since $C_k \leq C_j$, for $k = 1, \ldots, j-1$, we see that

$$p_j C_j \geq \frac{1}{2} \left(\sum_{k=1}^{j} p_k \right)^2 + \frac{1}{2} \sum_{k=1}^{j} p_k^2 - \sum_{k=1}^{j-1} p_k C_j \ .$$

This implies the desired inequality. $\qquad\qquad\qquad\qquad\qquad\qquad\qquad\square$

The linear programming relaxation of the scheduling problem we consider is formed by inequalities (1) and the simple precedence constraints, i. e.,

$$
\begin{aligned}
\text{minimize } & \sum_{j=1}^{n} w_j C_j \\
\text{subject to } & \sum_{j \in A} p_j C_j \geq \frac{1}{2} \left(\left(\sum_{j \in A} p_j \right)^2 + \sum_{j \in A} p_j^2 \right) \quad \text{for all } A \subseteq N, \qquad (2) \\
& C_k \geq C_j + p_k \qquad\qquad\qquad\qquad\quad \text{for } j \to k.
\end{aligned}
$$

The algorithm we analyze is based in a straightforward manner on this formulation. We call it 1–NATURAL: Find an optimal solution to the linear relaxation (2). Let C^{LP} denote this solution. Assume that $C_1^{\text{LP}} \leq C_2^{\text{LP}} \leq \cdots \leq C_n^{\text{LP}}$, where ties are broken arbitrarily. Now, schedule the jobs in this order $(1, 2, \ldots, n)$, without idle time. The output is denoted by C^{H}. Due to the simple precedence constraints incorporated in our linear relaxation, C^{H} respects the precedence constraints.

Theorem 2. *For* $1|\text{prec}| \sum w_j C_j$, *let* $w(1\text{–NATURAL})$ *be the value of the schedule produced by algorithm* 1–NATURAL. *Let* $w(\text{OPT})$ *be the value of an optimal schedule. Then*

$$w(1\text{–NATURAL}) \ \leq \ \left(2 - \frac{2}{n+1} \right) w(\text{OPT}) \ .$$

Proof. Let C^{LP} be an optimal solution to the LP relaxation (2), and let C^{H} be the output of algorithm 1–NATURAL, i. e., $C_j^{\text{H}} = \sum_{k=1}^{j} p_k$. Lemma 1 implies that

$$\sum_{j=1}^{n} w_j C_j^{\text{H}} \leq 2 \sum_{j=1}^{n} w_j C_j^{\text{LP}} - \sum_{j=1}^{n} w_j \frac{\sum_{k=1}^{j} p_k^2}{\sum_{k=1}^{j} p_k} \ .$$

Since the value of the LP solution is a lower bound on the optimal value, and

$$\sum_{j=1}^{n} w_j \frac{\sum_{k=1}^{j} p_k^2}{\sum_{k=1}^{j} p_k} = \sum_{j=1}^{n} w_j \sum_{k=1}^{j} p_k \frac{\sum_{k=1}^{j} p_k^2}{\left(\sum_{k=1}^{j} p_k\right)^2}$$

$$\geq \sum_{j=1}^{n} w_j \frac{1}{j} \sum_{k=1}^{j} p_k$$

$$\geq \frac{1}{n} \sum_{j=1}^{n} w_j C_j^{\mathrm{H}} \,,$$

the claim follows. □

Since the LP relaxation (2) can be used to obtain a lower bound on the optimal objective function value (see [QW91] for some computational experience), it is very interesting that we can also derive a bound on the ratio between the optimal value and the LP solution from the proof of Theorem 2.

Corollary 3. *For* $1|\mathrm{prec}|\sum w_j C_j$*, let* $w(\mathrm{OPT})$ *be the value of an optimal schedule, and* $w(\mathrm{LP})$ *be the value obtained from solving* (2). *Then*

$$w(\mathrm{OPT}) \leq \left(2 - \frac{2}{n+1}\right) w(\mathrm{LP}) \,.$$

Since Queyranne [Que93] showed that the separation problem associated with inequalities (1) is solvable in polynomial time, it follows from the equivalence of optimization and separation with respect to polynomial time solvability (cf., [GLS88]) that both the lower and the upper bound can be computed in polynomial time.

There exist instances showing that our analysis of the performance guarantee of algorithm 1–NATURAL is essentially tight.

Remark. In [QS94, Sch95a] affine transformations are presented mapping feasible solutions of linear programming relaxations of $1|\mathrm{prec}|\sum w_j C_j$ formulated with linear ordering variables (polynomial size) and with time–indexed variables (pseudo–polynomial size) to feasible solutions of (2), respectively. Hence the observations above imply that we can also solve one of these relaxations to optimality, transform the solution to one of (2), and then apply algorithm 1–NATURAL. This leads to approximation algorithms with performance guarantee $\left(2 - \frac{2}{n+1}\right)$, too. The algorithm based on the time–indexed formulation is, of course, only pseudo–polynomial. The same holds with respect to the lower bounds. This observation in particular implies that we have considerably improved the analysis of the quality of the time–indexed formulation performed by Hall, Shmoys, and Wein. They obtained from the interval–indexed formulation, which also is a relaxation of the time–indexed one, a polynomial–time heuristic with a performance guarantee of $4 + \epsilon$, for any fixed $\epsilon > 0$. Moreover, our algorithm seems to be more natural and simpler as they need to sort the jobs based on the time by which half of their processing has been completed in the fractional interval–indexed solution. Finally, note that we do not rely on the ellipsoid method for solving the linear programming relaxation in the linear ordering variables.

We now turn to the situation where jobs arrive over time, which is another characteristic encountered frequently in real–world problems. In addition to the setting described above, the jobs are now assumed to be released at different times, where job j is released at time $r_j \geq 0$. We show that algorithm 1–NATURAL when given the optimal solution to a slightly extended linear program produces a solution within a factor less than 3 of the optimal. We consider the following formulation:

$$
\text{minimize } \sum_{j=1}^{n} w_j C_j
$$

$$
\text{subject to } \sum_{j \in A} p_j C_j \geq \frac{1}{2}\left(\left(\sum_{j \in A} p_j\right)^2 + \sum_{j \in A} p_j^2 \right) \quad \text{for all } A \subseteq N, \tag{3}
$$

$$
C_k \geq C_j + p_k \qquad\qquad \text{for } j \to k,
$$

$$
C_j \geq r_j + p_j \qquad\qquad \text{for all } j \in N.
$$

Lemma 4. *Let C be any feasible solution to the linear relaxation (3). Assume, w. l. o. g., that $C_1 \leq \cdots \leq C_n$. Then, for each $j = 1, \ldots, n$,*

$$
\tilde{C}_j \quad \leq \quad 3 C_j \quad - \quad \frac{\sum_{k=1}^{j} p_k^2}{\sum_{k=1}^{j} p_k} \ ,
$$

where the feasible schedule $\tilde{C}$ is obtained by sequencing the jobs in order $(1, \ldots, n)$, i. e., $\tilde{C}_j = \max\{r_j + p_j, \tilde{C}_{j-1} + p_j\}$.

Proof. Notice that $\tilde{C}_j \leq \max_{k=1,\ldots,j} r_k + \sum_{k=1}^{j} p_k$. From the last class of inequalities in (3) and $C_1 \leq \cdots \leq C_n$ it follows that $\tilde{C}_j \leq C_j + \sum_{k=1}^{j} p_k$. By applying Lemma 1, we complete the proof. $\qquad\square$

Lemma 4 leads to the following results.

Theorem 5. *For $1|r_j, \text{prec}| \sum w_j C_j$, let $w(1–\text{NATURAL})$ be the value of the schedule produced by algorithm 1–NATURAL applied to an optimal solution to (3). Let $w(\text{OPT})$ be the value of an optimal schedule. Then*

$$
w(1–\text{NATURAL}) \quad \leq \quad 3\, w(\text{OPT}) \ .
$$

Corollary 6. *For $1|r_j, \text{prec}| \sum w_j C_j$, let $w(\text{OPT})$ be the value of an optimal schedule, and $w(\text{LP})$ be the value obtained from solving (3). Then*

$$
w(\text{OPT}) \quad \leq \quad 3\, w(\text{LP}) \ .
$$

Remark. There are three more linear programming relaxations of the strongly NP–hard problem to minimize the total weighted completion time on a single machine subject to release dates, i. e., $1|r_j| \sum w_j C_j$, which are at least as strong as relaxation (3). The first is also based on completion time variables and solely uses the shifted parallel inequalities (see [QS95, Goe96]); the second relies on time–indexed variables; the third uses both completion time variables and time–indexed variables, see [DW90, Goe96]. We may solve either of these relaxations to optimality, transform the solution to one of (3) (only needed for the second one), and then apply algorithm 1–NATURAL. Consequently, we obtain the same performance guarantee (a factor of 3) for both the upper and lower bounds computed by these three algorithms. Moreover, since Goemans [Goe96] observed that the polyhedron defined by the shifted parallel inequalities is supermodular, we may apply the greedy algorithm for supermodular polyhedra to solve this particular relaxation. This not only gives a $O(n \log n)$ algorithm for computing a feasible solution at most 3 times worse than the optimum as well as a lower bound with the same performance guarantee but this algorithm also neither relies on the ellipsoid method nor on any other polynomial time algorithm for linear programming (see [Sch95a, Sch95b] for details).

Since the affine image of any solution to the linear programming relaxation of $1|r_j, \text{prec}| \sum w_j C_j$ in time–indexed variables is feasible for (3), our analysis shows that the time–indexed formulation can be used to obtain a pseudo–polynomial time heuristic with worst–case performance guarantee 3. Moreover, the lower bound obtained by solving this relaxation is at least as good as the one obtained by solving (3). Hence, again we have significantly improved the analysis of the quality of the time–indexed formulation performed by Hall, Shmoys, and Wein [HSW96]. They obtained from the interval–indexed formulation a $(3 + 2\sqrt{2} + \epsilon)$–approximation algorithm.

3 Identical Parallel Machines

In this section, we show that our approach can be extended to the model in which there are identical parallel machines. Here, we are given m identical parallel machines instead of a single machine, each job must be processed by one of these machines, and may be assigned to any of these machines. The problem $P2| \sum w_j C_j$ is NP–hard, see [BCS74, LRKB77]. We again use C_j to denote the completion time of job j in a schedule (irrespective of which machine it is processed on). The convex hull of feasible completion time vectors in this setting has previously only been studied in the special case in which each job requires unit processing time, see [QS95]. We present here a new class of valid inequalities, namely

$$\sum_{j \in A} p_j C_j \geq \frac{1}{2m} \left(\sum_{j \in A} p_j \right)^2 + \frac{1}{2} \sum_{j \in A} p_j^2 \qquad \text{for all } A \subseteq N. \qquad (4)$$

Lemma 7. *The completion time vector C of every feasible schedule on m identical parallel machines satisfies inequalities (4).*

Proof. Consider any feasible schedule and its completion time vector C. Let A be an arbitrary subset of jobs. Let A_i be the set of jobs processed on machine i, for $i = 1, \ldots, m$. From the single machine case, we know

$$
\begin{aligned}
\sum_{j \in A} p_j C_j &= \sum_{i=1}^{m} \sum_{j \in A_i} p_j C_j \\
&\geq \sum_{i=1}^{m} \frac{1}{2} \left(\sum_{j \in A_i} p_j^2 + \left(\sum_{j \in A_i} p_j \right)^2 \right) \\
&= \frac{1}{2} \sum_{j \in A} p_j^2 + \frac{1}{2} \sum_{i=1}^{m} \left(\sum_{j \in A_i} p_j \right)^2 \\
&\geq \frac{1}{2} \sum_{j \in A} p_j^2 + \frac{1}{2m} \left(\sum_{j \in A} p_j \right)^2 .
\end{aligned}
$$

$\square$

It is an important observation for our purposes that the separation problem associated with the inequalities (4) can be solved in polynomial time. For brevity, we omit the description of the algorithm, which is an extension of Queyranne's algorithm [Que93] for the single–machine case. It can be found in [Sch95a, Sch95b]. We consider the following LP relaxation of the scheduling problem with release dates:

$$
\text{minimize} \quad \sum_{j=1}^{n} w_j C_j
$$

$$
\text{subject to} \quad \sum_{j \in A} p_j C_j \geq \frac{1}{2m} \left(\sum_{j \in A} p_j \right)^2 + \frac{1}{2} \sum_{j \in A} p_j^2 \quad \text{for all } A \subseteq N, \qquad (5)
$$

$$
C_j \geq r_j + p_j \qquad \text{for all } j \in N.
$$

We now define the approximation algorithm, called P–NATURAL. Let C^{LP} be an optimal solution to (5) and index the jobs such that $C_1^{\text{LP}} \leq C_2^{\text{LP}} \leq \cdots \leq C_n^{\text{LP}}$. Apply the following list scheduling rule using the list $(1, 2, \ldots, n)$ just defined. Iteratively take the next job, and schedule it on the machine on which it would start as early as possible. The resulting schedule has completion time vector C^{H} satisfying

$$
C_j^{\text{H}} \leq \max_{k=1,\ldots,j} r_k + \frac{1}{m} \sum_{k=1}^{j-1} p_k + p_j ,
$$

for each $j = 1, \ldots, n$.

Theorem 8. *For* $\text{P}|r_j| \sum w_j C_j$, *let* $w(\text{OPT})$ *be the value of an optimal schedule, and* $w(\text{P–NATURAL})$ *be the value of the schedule produced by algorithm* P–NATURAL. *Then*

$$
w(\text{P–NATURAL}) \leq \left(4 - \frac{1}{m} \right) w(\text{OPT}) .
$$

Again, we have the same worst–case guarantee for the LP lower bound obtained from solving (5). If all jobs are released at the same time, then the bound can be improved to $(3 - \frac{1}{m})(1 - \frac{1}{n+1})$. We would like to mention that after they read a preliminary version of this paper that focused on single machine and flow shop problems, Queyranne [Que95] as well as Hall, Shmoys, and Wein [HSW95] independently obtained almost the same identical parallel machine results. Moreover, in subsequent work [HSSW96, CPS$^+$96], we will show how to extend the techniques in this paper to derive approximation algorithms for identical parallel machine problems with precedence constraints and release dates. They have worst–case ratio 3 in case of unit jobs and 5.33 in general and considerably beat the previously known bounds. This can even be extended to uniform machines as will be indicated in the next section.

4 Uniform Parallel Machines

The parallel machines are now assumed to have different speeds. Let machine i have speed $s_i > 0$, $i = 1, \ldots, m$. If job j is processed on machine i its actual processing time is $\frac{p_j}{s_i}$. Perhaps surprisingly, we can use completion time variables and ignore the assignment of jobs to machines and still obtain good performance guarantees. Our approach is based on the following new valid inequalities.

Lemma 9. *The completion time vector C of every feasible schedule on m uniform machines satisfies*

$$C_j \geq \qquad\qquad r_j + \frac{p_j}{\max s_i} \qquad\qquad \text{for all jobs } j \in N, \quad (6)$$

$$\sum_{j \in A} p_j C_j \geq \frac{1}{2 \sum_{i=1}^m s_i} \left(\Big(\sum_{j \in A} p_j \Big)^2 + \sum_{j \in A} p_j^2 \right) \qquad \text{for all } A \subseteq N. \quad (7)$$

Let C^{LP} be an optimal solution to $\min\{\sum w_j C_j : C \text{ satisfies } (6) \text{ and } (7)\}$. Index the jobs such that $C_1^{\mathrm{LP}} \leq C_2^{\mathrm{LP}} \leq \cdots \leq C_n^{\mathrm{LP}}$. Apply a list scheduling method whereby the jobs are considered in the order $(1, 2, \ldots, n)$ just defined, and each job is assigned to the machine so *its completion time is as early as possible*. For the problem $Q| \,|\sum w_j C_j$, this leads to an approximation algorithm with worst–case ratio $\min\{ 2 + (m-1) \frac{\max s_i}{\sum s_i}, \, 2 \frac{\sum s_i}{\max s_i} \}$, and, for $Q|r_j|\sum w_j C_j$, to a ratio of $\min\{ 3 + (m-1) \frac{\max s_i}{\sum s_i}, \, 1 + 2 \frac{\sum s_i}{\max s_i} \}$. This can always be bounded from above by $1 + \sqrt{2m-1}$ and $2 + \sqrt{2m-1}$, respectively. Again, we can guarantee the same quality for the respective LP lower bounds. In [HSSW96], we will also show how to extend these results to models incorporating precedence constraints.

5 Flow Shop

In this section, we turn to multi–operation problems, and more precisely, to flow shop problems. Here we are given m machines and each job must be processed by all of the machines in the same order. We may think of job j as consisting of m operations with processing times $p_{1j}, p_{2j}, \ldots, p_{mj}$ associated with machines $1, 2, \ldots, m$, respectively. We denote by C_{hj} the completion time of job j on machine h, $h = 1, \ldots, m$. The problems we are going to study are $F|\text{prec}| \sum w_j C_j$ and $F|r_j, \text{prec}| \sum w_j C_j$. Here $C_j = C_{mj}$, and $j \to k$ means that, for $h = 1, \ldots, m$, the h-th operation of job j has to precede the h-th operation of job k. The problem $F2|| \sum C_j$ is strongly NP–hard [GJS76]. The previously only known approximation algorithm with a constant worst–case performance guarantee (when m is fixed) is the shortest processing time rule for $F| | \sum C_j$ which leads to a factor of m, see Gonzalez and Sahni [GS78]. Indeed, Lawler, Lenstra, Rinnooy Kan, and Shmoys shared the following opinion:

"General flow shop and job shop scheduling problems have earned a reputation for intractability. We will be mostly concerned with enumerative optimization methods for their solution and, to a lesser extent, with approximation algorithms. An analytical approach to the performance of methods of the latter type is badly needed" [LLRKS93, Page 487].

"Not much has been done in the way of worst–case analysis of *approximation algorithms* for the flow shop scheduling problem" [LLRKS93, Page 493].

The straightforward heuristic we propose is based on the following (weak) linear relaxation of the flow shop problem.

$$
\min \ \sum_{j=1}^{n} w_j C_{mj}
$$

$$
\text{s. t.} \ \sum_{j \in A} p_{hj} C_{hj} \geq \frac{1}{2} \left(\left(\sum_{j \in A} p_{hj} \right)^2 + \sum_{j \in A} p_{hj}^2 \right) \quad A \subseteq N, \quad h = 1, \ldots, m,
$$

$$
C_{hj} \geq C_{(h-1)j} + p_{hj} \qquad\qquad h = 2, \ldots, m, \tag{8}
$$

$$
C_{hk} \geq C_{hj} + p_{hk} \qquad\qquad j \to k, \quad h = 1, \ldots, m,
$$

$$
C_{1j} \geq r_j + p_{1j} \ .
$$

We call the flow shop heuristic F–NATURAL. Find an optimal solution C^{LP} to (8). Assume that $C_{m1}^{\text{LP}} \leq C_{m2}^{\text{LP}} \leq \cdots \leq C_{mn}^{\text{LP}}$, and schedule the jobs in this order, without unforced idle time. Thus, $C_{hj}^{\text{H}} = \max\{C_{(h-1)j}^{\text{H}} + p_{hj}, C_{h(j-1)}^{\text{H}} + p_{hj}\}$ where we assume, to simplify notation, that $C_{0j}^{\text{H}} = r_j$ and $C_{h0}^{\text{H}} = 0$. In particular, the solution constructed this way is a permutation schedule. Since

$$
C_{mj}^{\text{H}} \ \leq \ C_{1j}^{\text{H}} \ + \ \sum_{h=2}^{m} \sum_{k=1}^{j} p_{hk}
$$

the following theorems can be proved by multiple applications of Lemma 1.

Theorem 10. *For* F|prec| $\sum w_j C_j$, *let* w(F–NATURAL) *be the value of the schedule produced by algorithm* F–NATURAL, *and let* w(OPT) *be the value of an optimal schedule. Then*

$$w(\text{F--NATURAL}) \leq \left(2m - \frac{2m}{n+1}\right) w(\text{OPT}) .$$

In the presence of release dates we have to apply Lemma 4 for the first machine.

Theorem 11. *For* F|r_j, prec| $\sum w_j C_j$, *let* w(F–NATURAL) *be the value of the schedule produced by algorithm* F–NATURAL, *and let* w(OPT) *be the value of an optimal schedule. Then*

$$w(\text{F--NATURAL}) \leq (2m + 1) w(\text{OPT}) .$$

Again, we also derive the respective guarantees for the lower bounds obtained from solving the linear programming problem (8). In addition, our analysis carries over to *no–wait* flow shop problems (see [Sch95a, Sch95b] for details).

We close this section by giving a sketch of the results for multiprocessor flow shop scheduling. This is a generalization of the usual flow shop in which each machine is replaced by a set of identical parallel machines (cf., e. g., [RC92, HHLV]). If we combine the parallel machine inequalities (4) in a similar manner as we combined the single machine inequalities in (8) we can obtain an approximation algorithm with performance guarantee $3m$ (or $3m + 1$, in case of job release dates). Here, m is the number of stages.

6 Concluding Remarks

With this work we continue a series of recent papers on provably good approximation algorithms based on "rounding" optimal solutions to LP relaxations of scheduling problems to feasible solutions; see, in particular, [LST90, ST93, PSW95, HSW96].

Thereby, the results obtained prove that linear relaxations are a quite powerful tool for effectively producing not only lower bounds but also good feasible solutions. They also prove the strength of polyhedral formulations that are based on completion time variables. This fits nicely with the observations that, in the absence of precedence constraints, Smith's rule (cf., [Smi56]) can be obtained from this formulation, and that complete linear descriptions of the convex hull of feasible completion time vectors are also known when the additional restrictions have a special structure that admits polynomial optimization, see [QS94, QS95, Sch95a] for details.

Moreover, most of the approximation algorithms presented here are not only better than the comparable ones of Hall, Shmoys, and Wein, they seem also to be simpler, both in terms of explanation and analysis. Indeed, the techniques we use are not too sophisticated. The results simply document the strength of the LP relaxations. It is obvious that better knowledge of the underlying polyhedra, i. e., more efficiently solvable classes of valid inequalities, may not only lead to stronger lower bounds but also to better approximation algorithms.

Acknowledgements. The author is grateful to Maurice Queyranne and David Shmoys for helpful discussions and comments on the text, to Francois Margot for helpful comments, and to Leslie Hall and Joel Wein for helpful discussions.

The author has been supported by the graduate school "Algorithmische Diskrete Mathematik". The graduate school "Algorithmische Diskrete Mathematik" is supported by the Deutsche Forschungsgemeinschaft (DFG), grant We 1265/2-1.

References

[BCS74]   J. L. Bruno, E. G. Coffman Jr., and R. Sethi. Scheduling independent tasks to reduce mean finishing time. *Communications of the Association for Computing Machinery*, 17:382 – 387, 1974.

[CPS+96] S. Chakrabarti, C. A. Phillips, A. S. Schulz, D. B. Shmoys, C. Stein, and J. Wein. Improved scheduling algorithms for minsum criteria, 1996. To appear in Springer Lecture Notes in Computer Science, Proceedings of the 23rd ICALP Conference.

[DW90]   M. E. Dyer and L. A. Wolsey. Formulating the single machine sequencing problem with release dates as a mixed integer program. *Discrete Applied Mathematics*, 26:255 – 270, 1990.

[GJS76]   M. R. Garey, D. S. Johnson, and R. Sethi. The complexity of flowshop and jobshop scheduling. *Mathematics of Operations Research*, 1:117 – 129, 1976.

[GLLRK79] R. L. Graham, E. L. Lawler, J. K. Lenstra, and A. H. G. Rinnooy Kan. Optimization and approximation in deterministic sequencing and scheduling: A survey. *Annals of Discrete Mathematics*, 5:287 – 326, 1979.

[GLS88]   M. Grötschel, L. Lovász, and A. Schrijver. *Geometric Algorithms and Combinatorial Optimization*, volume 2 of *Algorithms and Combinatorics*. Springer, Berlin, 1988.

[Goe96]   M. X. Goemans. A supermodular relaxation for scheduling with release dates. This volume, 1996.

[GS78]   T. Gonzalez and S. Sahni. Flowshop and jobshop schedules: Complexity and approximation. *Operations Research*, 26:36 – 52, 1978.

[HHLV]   J. A. Hoogeveen, C. Hurkens, J. K. Lenstra, and A. Vandevelde. Lower bounds for the multiprocessor flow shop. In preparation.

[HS88]   D. S. Hochbaum and D. B. Shmoys. A polynomial approximation scheme for scheduling on uniform processors: Using the dual approximation approach. *SIAM Journal on Computing*, 17:539 – 551, 1988.

[HSSW96] L. A. Hall, A. S. Schulz, D. B. Shmoys, and J. Wein. Scheduling to minimize average completion time: Off–line and on–line algorithms, 1996. In preparation.

[HSW95]   L. A. Hall, D. B. Shmoys, and J. Wein. Personal communication, September 1995.

[HSW96]   L. A. Hall, D. B. Shmoys, and J. Wein. Scheduling to minimize average completion time: Off–line and on–line algorithms. In *Proceedings of the 7th ACM-SIAM Symposium on Discrete Algorithms*, pages 142 – 151, January 1996.

[KK86]   T. Kawaguchi and S. Kyan. Worst case bound of an LRF schedule for the mean weighted flow-time problem. *SIAM Journal on Computing*, 15:1119 – 1129, 1986.

[Law78]   E. L. Lawler. Sequencing jobs to minimize total weighted completion time subject to precedence constraints. *Annals of Discrete Mathematics*, 2:75 – 90, 1978.

[LLRKS93] E. L. Lawler, J. K. Lenstra, A. H. G. Rinnooy Kan, and D. B. Shmoys. Sequencing and scheduling: Algorithms and complexity. In S. C. Graves, A. H. G. Rinnooy Kan, and P. H. Zipkin, editors, *Logistics of Production and Inventory*, volume 4 of *Handbooks in Operations Research and Management Science*, pages 445 – 522. North–Holland, Amsterdam, The Netherlands, 1993.

[LRKB77] J. K. Lenstra, A. H. G. Rinnooy Kan, and P. Brucker. Complexity of machine scheduling problems. *Annals of Discrete Mathematics*, 1:343 – 362, 1977.

[LST90]   J. K. Lenstra, D. B. Shmoys, and É. Tardos. Approximation algorithms for scheduling unrelated parallel machines. *Mathematical Programming*, 46:259 – 271, 1990.

[MSS96]   R. H. Möhring, M. W. Schäffter, and A. S. Schulz. Scheduling with communication delays: Minimizing the average weighted completion time. Preprint, Department of Mathematics, Technical University of Berlin, Berlin, Germany, 1996.

[PSW95]   C. Phillips, C. Stein, and J. Wein. Scheduling jobs that arrive over time. In *Proceedings of the Fourth Workshop on Algorithms and Data Structures*, number 955 in Lecture Notes in Computer Science, pages 86 – 97. Springer, Berlin, 1995.

[QS94]   M. Queyranne and A. S. Schulz. Polyhedral approaches to machine scheduling. Preprint 408/1994, Department of Mathematics, Technical University of Berlin, Berlin, Germany, 1994.

[QS95]   M. Queyranne and A. S. Schulz. Scheduling unit jobs with compatible release dates on parallel machines with nonstationary speeds. In E. Balas and J. Clausen, editors, *Integer Programming and Combinatorial Optimization*, number 920 in Lecture Notes in Computer Science, pages 307 – 320. Springer, Berlin, 1995. Proceedings of the 4th International IPCO Conference.

[Que93]   M. Queyranne. Structure of a simple scheduling polyhedron. *Mathematical Programming*, 58:263 – 285, 1993.

[Que95]   M. Queyranne. Personal communication, September 1995.

[QW91]   M. Queyranne and Y. Wang. A cutting plane procedure for precedence–constrained single machine scheduling. Working paper. Faculty of Commerce, University of British Columbia, Vancouver, Canada, 1991.

[RC92]   C. Rajendran and D. Chaudhuri. A multi–stage parallel–processor flow–shop problem with minimum flowtime. *European Journal of Operational Research*, 57:111 – 122, 1992.

[Sch95a] A. S. Schulz. *Polytopes and Scheduling*. PhD thesis, Technical University of Berlin, Berlin, Germany, 1995.

[Sch95b] A. S. Schulz. Scheduling to minimize total weighted completion time: Performance guarantees of LP–based heuristics and lower bounds. Preprint 474/1995, Department of Mathematics, Technical University of Berlin, Berlin, Germany, 1995.

[Smi56]   W. E. Smith. Various optimizers for single–stage production. *Naval Research and Logistics Quarterly*, 3:59 – 66, 1956.

[ST93]   D. B. Shmoys and É. Tardos. An approximation algorithm for the generalized assignment problem. *Mathematical Programming*, 62:461 – 474, 1993.

Implementation of a Linear Time Algorithm for Certain Generalized Traveling Salesman Problems

Neil Simonetti and Egon Balas*

Carnegie Mellon University, Pittsburgh PA 15213, USA

Abstract. This paper discusses an implementation of a dynamic programming approach to the traveling salesman problem that runs in time linear in the number of cities. Optimality can be guaranteed when precedence constraints of a certain type are present, and many problems involving time windows fall into this class. Perhaps the most interesting feature of the procedure is that an auxiliary structure is built before any particular problem instance is known, reducing the computational effort required to solve a given problem instance to a fraction of what it would be without such a structure.

1 Introduction

This paper discusses an implementation of the approach proposed in [2] for solving some classes of generalized traveling salesman problems (TSP), symmetric or asymmetric, by finding a shortest path in an auxiliary digraph constructed for this purpose. The algorithm runs in time linear in the number of cities, and has a variety of uses, from solving instances of the TSP with time windows, to serving to improve a given heuristic solution to a standard TSP. At this point, the algorithm has been used on TSP problems with time windows, found in the literature [1, 3, 6, 9], has improved some standard TSP solutions generated by the Kanellakis & Papadimitriou heuristic [4], and has solved a new class of problems we call the TSP with time targets.

The main novel feature of this implementation is that an auxiliary structure is built in advance, *without prior knowledge of the problem instance*. This structure then serves as a foundation on which the algorithm runs, vastly reducing the computational effort required for solving any particular problem instance.

Section 2 gives a quick overview of the problem. Section 3 outlines how the auxiliary structure is built and used. Section 4 shows current computational results.

2 Background

The approach proposed in [2] starts from a precedence-constrained n-city TSP, symmetric or asymetric, defined on a complete graph G, directed or undirected,

* Research supported by Grant DMI-9201340 of the National Science Foundation and contract N00014-89-J-1063 of the Office of Naval Research.

with city 1 fixed as the home city where all tours must start and finish. The precedence constraints are given by:

Problem 1:

(i) a positive integer $k < n$.

(ii) an ordering $\{1, \ldots, n\}$ of the set N of cities, such that there exists an optimal permutation π of $\{1, \ldots, n\}$ (and associated tour) with the property that

(iii) for all $i, j \in N$, $j \geq i + k$ implies $\pi(i) < \pi(j)$.

The idea behind these precedence constraints is that if in the initial ordering city j comes k or more places after city i, then city j must be visited after city i in an optimal tour.

The method for solving the problem involves two steps: (1) building a special auxiliary digraph, G^*; and (2) solving a shortest path problem on G^*. G^* has $n+1$ *layers*, one layer for each position in the tour, with the home city appearing at both the beginning and end of the tour. The first and last layers of G^* each have only one node since only the home city can be at the beginning and end of the tour. If s is the name of the node in layer 1, and t is the name of the node in layer $n + 1$, then there is a 1-1 correspondence between feasible tours in G satisfying (i), (ii), and (iii) and $s - t$ paths in G^*. Furthermore, optimal tours correspond to shortest paths.

Every node in the ith layer of G^* corresponds to a unique state specifying which city is in position i, and which cities are visited in positions 1 through $i - 1$. This state can be expressed by the three elements:

1. j, the city in position i.
2. S^-, the set of cities numbered i or higher that are visited in one of the positions 1 through $i - 1$. ($S^- := \{h \in N : h \geq i, \; \pi(h) < i\}$)
3. S^+, the set of cities numbered below i that are not visited in one of the positions 1 through $i - 1$. ($S^+ := \{h \in N : h < i, \; \pi(h) \geq i\}$) (Note that this implies $|S^-| = |S^+|$.)

Nodes in G^* can be referenced by the notation (i, j, S^-, S^+). When referencing a node in a certain (possibly arbitrary) layer i, the notation will simply be (j, S^-, S^+), where the elements are dependent on i. G^* contains roughly $n(k+1)2^{k-2}$ nodes. All the arcs of G^* connect nodes of adjacent layers. There is an arc from a node in layer i to one in layer $i+1$ if the states represented by the two nodes can be part of the same tour. When this occurs, the nodes are said to be *compatible*. The cost assigned to the arc connecting node (i, j, S^-, S^+) to node $(i + 1, l, T^-, T^+)$ is the cost of the arc (j, l) in the original TSP instance. No node of G^* has an in-degree greater than k, which bounds the number of arcs at $nk(k + 1)2^{k-2}$. For further details on G^*, see [2].

One generalization of the original problem is to allow different values of k for the precedence constraint (iii) for different cities i. The conditions (i) and (iii) change as follows:

Problem 2:

(i) a family of positive integers $\{k(i) : i \in N\}$ all less than n.
(iii) for all $i, j \in N$, $j \geq i + k(i)$ implies $\pi(i) < \pi(j)$.

The following section deals with issues of implementing the algorithm for problem 2.

3 The Auxiliary Structure

We will use the following conventions when dealing with the cities of the TSP and the nodes of auxiliary structures:

"Cities" refers to vertices in G, all found in the set N.
"Nodes" refers to vertices in auxiliary structures, such as G^*.
$min(\emptyset) := \infty$; $max(\emptyset) := -\infty$.

While the auxiliary digraph G^* is problem dependent, the building blocks of G^* have a structure that allows them to be constructed before the problem instance is known. This is very important because the construction of the digraph is much more difficult than finding its shortest path. We know of no other combinatorial algorithm in which most of the extensive calculations are worked out and saved before the problem is examined.

The nodes of G^* can be partitioned into $n+1$ layers, each layer corresponding to a position in the tour. For problem 1, all the layers, except the first and last k layers, consist of identical copies of a set W_k^* of $(k + 1)2^{k-2}$ nodes, and the remaining layers are subsets of W_k^*. The compatibility test mentioned in Section 2 for determining arcs of G^* can also be applied to W_k^*. Furthermore, if we construct a layered digraph G_k^{**} by using singleton nodes for the first and last layers (call them s and t) and $n - 1$ copies of W_k^* for the remaining layers, and use the compatibility test to determine the arcs of G_k^{**} (though not their costs); then the sets of $s - t$ paths in G^* and G_k^{**} are identical. The nodes of W_k^* will be referenced using the notation (j, S^-, S^+), where the values of j, S^-, and S^+ (defined in Section 2 above) depend only on the value of i, which indicates the specific layer of G_k^{**}.

W_k^* and the set of compatible pairs from W_k^* can be built without prior knowledge of the problem, except for the value k. But since $W_h^* \subset W_m^*$ whenever $h \leq m$, by building this structure (i.e. the node set W_m^* and its set of compatible pairs) for a single "large" value of m, we can solve any problem instance with $k \leq m$ once this structure is loaded into memory.

For problem 2, all the layers of G^* are subsets of W_K^*, where $K := max\{k(i) : i \in N\}$, and the set of arcs of G^* between two layers is a subset of the compatible pairs of W_K^*. Thus G^* is still a subgraph of G_K^{**}, and W_K^* and its compatible pairs can be built beforehand. However, there may be $s - t$ paths in G_K^{**} that are not in G^*. One way to avoid examining some $s - t$ paths in G_K^{**} that are not in G^* is to replace the copy of W_K^* for layer i with a copy of $W_{m_i}^*$ for some $m_i \leq K$ which will not remove any $s - t$ path also in G^*. Call this slimmer

		Compatible	
No.	Node Label (i, j, S^-, S^+)	Predecessors	Successors
Level 1: 1:	$(i, i, \emptyset, \emptyset)$	1,3,8,20	1,2,4,9
Level 2: 2:	$(i, i+1, \emptyset, \emptyset)$	1,3,8,20	3,5,10
3:	$(i, i-1, \{i\}, \{i-1\})$	2,6,16	1,2,4,9
Level 3: 4:	$(i, i+2, \emptyset, \emptyset)$	1,3,8,20	6,7,11
5:	$(i, i+1, \{i\}, \{i-1\})$	2,6,16	8,15
6:	$(i, i-1, \{i+1\}, \{i-1\})$	4,12	3,5,10
7:	$(i, i, \{i+1\}, \{i-1\})$	4,12	8,15
8:	$(i, i-2, \{i\}, \{i-2\})$	5,7,19	1,2,4,9
Level 4: 9:	$(i, i+3, \emptyset, \emptyset)$	1,3,8,20	12,13,14
10:	$(i, i+2, \{i\}, \{i-1\})$	2,6,16	16,17
11:	$(i, i+2, \{i+1\}, \{i-1\})$	4,12	18,19
12:	$(i, i-1, \{i+2\}, \{i-1\})$	9	6,7,11
13:	$(i, i, \{i+2\}, \{i-1\})$	9	16,17
14:	$(i, i+1, \{i+2\}, \{i-1\})$	9	18,19
15:	$(i, i+1, \{i\}, \{i-2\})$	5,7,19	20
16:	$(i, i-2, \{i+1\}, \{i-2\})$	10,13	3,5,10
17:	$(i, i, \{i+1\}, \{i-2\})$	10,13	20
18:	$(i, i-1, \{i, i+1\}, \{i-2, i-1\})$	11,14	20
19:	$(i, i-2, \{i, i+1\}, \{i-2, i-1\})$	11,14	8,15
20:	$(i, i-3, \{i\}, \{i-3\})$	15,17,18	1,2,4,9

Fig. 1. Nodes of W_4^* and its compatible pairs.

auxiliary structure (which is dependant on the problem instance) G^{**}. Given the values m_i, the nodes in layer i of G^{**} are simply $W_{m_i}^*$. The arcs connecting two adjacent layers, i and $i+1$, of G^{**} are also easily found from the list of compatible pairs (u, v) of $W_K^* \times W_K^*$ by choosing the pairs (u, v) with $u \in W_{m_i}^*$ and $v \in W_{m_{i+1}}^*$. The rest of this section deals with calculating good values of m_i when choosing $W_{m_i}^*$ to represent a layer of G^{**}, and ways to avoid examining the remaining $s - t$ paths in G^{**} that are not also in G^*.

Since we have $W_h^* \subset W_m^*$ whenever $h \leq m$, there is a natural division of the nodes of W_m^* into *levels*. Let W_1^* be the first level of nodes in W_m^*, and then define the hth level of nodes ($2 \leq h \leq m$) to be those in the set $W_h^* \setminus W_{h-1}^*$. Let $L(v)$ be the level in which node v belongs for a given node $v := (j, S^-, S^+) \in W_K^*$. Figure 1 illustrates the breakdown of W_4^* into levels. The set of compatible pairs is expressed as a list of successors or predecessors.

Figure 2 shows the graphs G^*, G^{**} and G_K^{**} for the case of problem 2 where $n = 9$, $k(3) = 4$, and $k(i) = 3$ for $i \neq 3$. The paths shown in the figure correspond to the feasible tour 1-4-2-6-3-5-7-9-8-1; the infeasible tour 1-2-3-4-5-8-7-9-6-1, which extends beyond G^{**}; and the infeasible tour 1-3-2-5-7-4-8-6-9-1, which does not extend beyond G^{**}, but must be avoided because it contains nodes not in G^*.

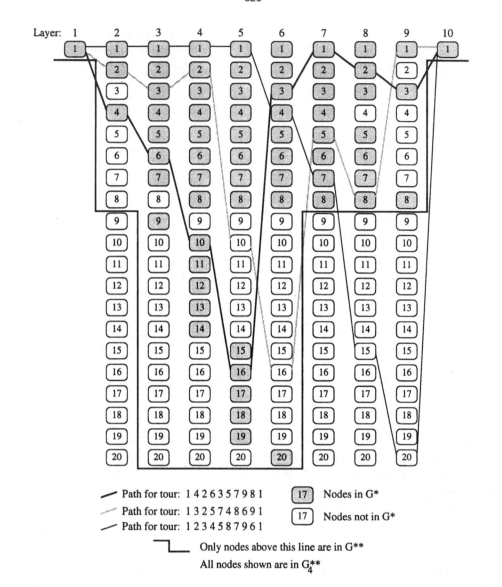

Fig. 2. G^*, G^{**}, and G^{**}_K for an example of problem 2

Proposition 3.1 *Given* $v := (j, S^-, S^+) \in W_K^*$,

$$L(v) = 1 + max\{|i - j|, \ max\{S^-\} - min\{S^+\}, \ j - min\{S^+\}\}.$$

Proof: To show that $L(v) \geq 1 + j - min\{S^+\}$, consider the following: The state associated with v implies that city j is visited before any cities in S^+, and so $\pi(min\{S^+\}) > \pi(j)$; hence if $j \geq min\{S^+\} + L(v)$, then the precedence constraint would be violated; giving $L(v) > j - min\{S^+\}$. Since $L(v)$ is integer, this is the same as $L(v) \geq 1 + j - min\{S^+\}$. The state associated with v also implies that all cities in S^- are visited before any cities in S^+, and so $\pi(min\{S^+\}) > \pi(max\{S^-\})$; hence if $max\{S^-\} \geq min\{S^+\} + L(v)$, then again the precedence constraint would be violated; giving $L(v) > max\{S^-\} - min\{S^+\}$. The state associated with v also indicates that city j is visited in the ith position. Since the precedence constraints imply that city j can only be visited in positions $j - L(v) + 1$ through $j + L(v) - 1$, we have $L(v) \geq 1 + |i - j|$, which now gives us

$$L(v) \geq 1 + max\{|i - j|, \ max\{S^-\} - min\{S^+\}, \ j - min\{S^+\}\}.$$

To show equality, we simply build a tour whose associated path contains node v. This is done by first visiting the cities of $\{1, 2, \ldots, i - 1\} \setminus S^+$ in increasing numeric order, then the cities of S^- in increasing order, then city j, then the cities of $S^+ \setminus \{j\}$ in increasing order, and finally the cities of $\{i, i+1, \ldots, n\} \setminus S^-$ in increasing order.$\square$

The *depth* of layer i is the highest level of a node in W_K^* which appears in layer i of G^*. For the example in Figure 2, $D(1) = 1$, $D(2) = 3$, $D(3) = \ldots = D(6) = 4$, and so on.

The *reach* of a city $i \in N$ is the set of layers whose depth may be affected by $k(i)$. Designate this by $R(i)$. For the example in Figure 2, $R(3) = \{3, 4, 5, 6\}$. This is explained by the following proposition.

Proposition 3.2 *Given $i \in N$,*

$$R(i) = \{i, \ldots, i + k(i) - 1\}.$$

Proof: City i cannot be assigned position $i + k(i)$ or higher, otherwise some city $h \geq i + k(i)$ would be forced into one of the positions 1 through $i + k(i) - 1$. This would imply $\pi(i) > \pi(h)$, but since $h \geq i + k(i)$, this contradicts the precedence constraint. So the reach of i can go no further than position $i + k(i) - 1$. The reach does not fall below position i because the precedence constraints that allow city i to be visited in a position less than i involve only values of $k(j)$ for $j < i$.$\square$

Proposition 3.3 *Given $i \in N$, the depth of layer i is no greater than*

$$D(i) := max_j\{k(j) : i \in R(j)\}.$$

Proof: The expression for $D(i)$ accounts for every city whose reach includes position i.$\square$

We now construct the layered digraph G^{**} using $W_{D(i)}^*$ as the node set for layer i. G^{**} is an intermediate structure between G^* and G_K^{**}. Since the complexity of solving any particular problem instance is a linear function of the number of arcs in G^{**}, the following result is of particular interest:

Theorem 3.4 *The number of arcs of G^{**} is bounded by*

$$\sum_{i=2}^{n+1} D(i-1)(D(i)+1)2^{D(i)-2}.$$

Proof: The number of nodes in layer i of G^{**} is $(D(i)+1)2^{D(i)-2}$ (shown in [2]), so we need to show only that the maximum in-degree of a node in layer i of G^{**} is $D(i-1)$. In [2], it was also shown that for a given node $v := (j, S^-, S^+)$ in layer i of G^{**}, at most one node $u := (l, T^-, T^+)$ from layer $i-1$ may be a predecessor of v for a fixed value of l, so we need to show that there are at most $D(i-1)$ candidates for l. The depth of layer $i-1$ limits the candidates for l to the set $\{(i-1) - D(i-1) + 1, \ldots, (i-1) + D(i-1) - 1\}$. Also, by the compatibility of nodes u and v, the state associated with node v must restrict l to the set $(\{1, \ldots, i-1\} \cup S^-) \setminus S^+$. Intersecting these two sets, we conclude that l must be in the set $P := (\{i - D(i-1), \ldots, i-1\} \cup S^-) \setminus S^+$. Since $|S^+| = |S^-|$, $|P| = D(i-1)$ as long as $S^+ \subset \{i - D(i-1), \ldots, i-1\}$. If this were not the case, then $min\{S^+\} \leq (i-1) - D(i-1)$ since the elements in S^+ only come from the set of cities less than i, and so, for a node $u := (l, T^-, T^+)$ to be a potential predecessor of v from layer $i-1$, we must have $min\{T^+\} \leq min\{S^+\} \leq (i-1) - D(i-1)$, which would imply

$$L(u) \leq D(i-1) \leq (i-1) - min\{T^+\} \leq max\{T^-\} - min\{T^+\}$$

contrary to Proposition 3.1.□

This improves the bound given in [2],

$$\sum_{i=2}^{n} k^*(i-1)^2 k^*(i)^2 2^{k^*(i-1)+k^*(i)-2},$$

where $k^*(i) := max\{k(i), k(j_{(i)})\}$, and $j_{(i)} := min\{j : j + k(j) \geq i + 1\}$, since the $D(i)$ are of the same order as the $k^*(i)$. Furthermore, the bound given in [2] requires an additional condition on the family $k(i)$, namely that $k(i) - k(i+1) \leq 1$ for each $i \in N$, a condition not needed for our result.

As illustrated by Figure 2, there may be paths in G^{**} which are not present in G^*. Thus we need a test for the nodes of G^{**} to prevent these paths from being considered.

The *k-threshold* for a node $v := (j, S^-, S^+) \in W_K^*$, denoted $kthresh(v)$, is the smallest value of $k(j)$ that permits the possibility of $v \in G^*$. Thus $k(j) < kthresh(v)$ implies $v \notin G^*$.

To calculate the k-threshold for a node $v := (j, S^-, S^+) \in W_K^*$, we notice that $v \in G^*$ implies that $k(j)$ is larger than the difference between j and all higher-numbered cities visited before j in the tour. The highest-numbered city visited before j is $max\{S^-\}$, unless S^- is empty, in which case no higher-numbered city can be visited before j. From this we have $k(j) > max\{0, max\{S^-\} - j\}$, so $kthresh(v) = 1 + max\{0, max\{S^-\} - j\}$.

Proposition 3.5 *Every path in G^{**} corresponding to an infeasible tour contains at least one node $v := (i, j, S^-, S^+)$, such that $kthresh(v) > k(j)$.*

Proof: Let π be the permutation for an infeasible tour T. Then there exist two cities, q and j, such that $\pi(j) > \pi(q)$ and $q \geq j + k(j)$. Let $i := \pi(j)$, (i.e. j is in the ith position). Let $v := (j, S^-, S^+)$ be the node used in layer i of G^{**} for the path corresponding to T.

If $q \geq i$, then $q \in S^-$, and so:

$$kthresh(v) = 1 + max\{0, max\{S^-\} - j\}$$
$$\geq 1 + max\{S^-\} - j \geq 1 + q - j > q - j \geq k(j).$$

If $q < i$, then $j < q < i$, and so $j \in S^+$. Since $|S^+| = |S^-|$, S^- cannot be empty, and so $max\{S^-\} \geq i$ since the elements in S^- only come from the set of cities greater than or equal to i. This gives:

$$kthresh(v) = 1 + max\{0, \; max\{S^-\} - j\} \geq 1 + i - j > 1 + q - j > q - j \geq k(j). \square$$

In Figure 2, the k-threshold of node 16 in layer 6, which is 4, is higher than $k(j)$, which is 3. (j for node 16 in layer 6 is 4.)

Once W_K^* has been built for a certain K, *any problem instance with $k(j) \leq K$* for all $j \in N$ can be solved to optimality (with a guarantee of optimality) by determining G^{**} and finding a shortest $s - t$ path in the subgraph G^* of G^{**}, where G^* is the auxiliary graph associated with the specific problem instance one wants to solve. Determinimg G^{**}, extracting the nodes and arcs of G^* from those of G^{**}, and putting the appropriate costs on the arcs does not increase the complexity of finding a shortest $s - t$ path in G^*, which remains linear in the number of arcs of G^{**}.

When applying this algorithm to time window problems, first an initial ordering based on sorting the cities by time window midpoint is constructed; then the implied precedence constraints are formed. While traversing the graph G^{**}, paths that correspond to infeasible solutions because of the time window restriction are weeded out by testing the cost at node $v := (j, S^-, S^+)$ (based on travel time) against the time window for city j. If we arrive at city j before the window opens, we pay the expense for waiting until the window opens. If we arrive at city j after the window closes, then any tour with a state associated with this node must be infeasible.

For the time window problems in [3], the objective was to minimize the total distance, not the total time, which makes a difference when tours must wait at a city for a time window to open. In this case, both the distance and the time must be kept at each node. However, this is not enough, since a tour that chooses to wait in one place to gain an advantage in distance may not be able to satisfy an important time window later.

Figure 3 illustrates this point. The route 1-2-3-4 has distance 6, but requires a time of 9 because there is a wait at city 2 This wait prevents the route from continuing to city 5 before its window closes. The route 1-3-2-4 has distance 8, but also has a total time of 8, so this route can continue through city 5. If the algorithm is not modified to keep at least the best *two* partial tours at each node in the auxiliary digraph, this solution will not be found.

In the case where the implied precedence constraints require a value of K too large to make the algorithm practical, the algorithm can still be run with a smaller value of K, but in this case, solutions found are not guaranteed to

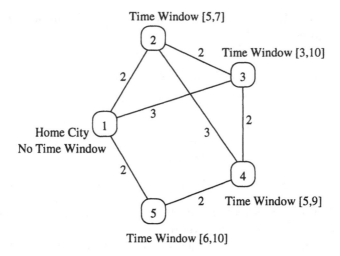

Fig. 3. Illustrating the difficulty of minimizing distance instead of time

be optimal. In such situations, using the initial sequence mentioned above, the algorithm does not perform well, but as the results show in Section 4, starting from the heuristic solution of another method, the algorithm can often improve the solution.

A variant of the time window problem, which we will call the *traveling salesman problem with time targets* (TSPTT), gives a target time for each city, rather than a window. The objective is to find a tour which minimizes the maximum deviation between the target time and actual service time over all cities. This algorithm can solve such problems by constructing windows of a fixed size d centered at each city's target time. A binary search is then used to find the smallest d, to a predetermined accuracy, that admits a feasible solution. Once the smallest such d is determined, the cheapest feasible tour is returned. Applications of this problem include delivery of perishable items (such as fresh fruit) to events without storage facilities, or routing of repair vehicles to customers with busy schedules who do not wish to wait for a potentially long period of time before being serviced.

4 Computational Experience

All of the results in this section were achieved on a Sun SparcStation 5.

Using a method similar to that given by Baker [1], Tama [10] created eight 20-city and five 26-city problems, representing one-machine scheduling problems with set-up times formulated as asymmetric TSP's with time windows, which he was unable to solve using cutting planes followed by a branch and bound algorithm, and which he kindly made available to us. We ran our algorithm on

Table 1.

	Problem Name	Required K	Solution Value (seconds)	
			$K = 13$	$K = 17$
(20-city	p192358	13	Infeasible* (1)	
problems)	p192422	10	Infeasible* (1)	
	p192433	12	Infeasible* (1)	
	p192452	10	Infeasible* (1)	
	p192572	11	Infeasible* (1)	
	p192590	12	Infeasible* (1)	
	p193489	13	Infeasible* (1)	
	p194450	9	936* (1)	
(26-city	p253574	16	Infeasible (1)	Infeasible* (7)
problems)	p253883	18	1188 (2)	1188 (42)
	p254662	16	1140 (3)	1140* (23)
	p255522	13	1373* (2)	
	p256437	13	1295* (2)	

*indicates that a guarantee of optimality (or a guarantee of the
infeasibility of the problem) was found.

these problems (using $K = 13$ and $K = 17$) with the following outcome: Of the
eight 20-city problems, 7 were *proved* to be infeasible, and the remaining 1 was
solved to optimality. Of the five 26-city problems, 1 was *proved* to be infeasible,
3 were solved to optimality, and the remaining 1 was solved without a guarantee
of optimality because the required value of K was too high. (see Table 1)

As to symmetric TSP's with time windows, thirty problems based on the RC2
problems proposed by Solomon [9] that were first studied by Potvin et al. [6] and
then Gendreau et al. [3] were run with our algorithm, using $K = 15$. The problem
name listed in Table 2 refers to the problem code used by Solomon, followed by
the route number. Neither Potvin nor Gendreau could guarantee the optimality
of a solution even if their solution was optimal. The entry for "Required K" in
Table 2 indicates what value of K would be needed for a guarantee of optimality.

Our algorithm solved 14 of these problems with a guarantee of optimality.
Of these 14, Potvin had solved 2 to optimality and Gendreau had solved 9
to optimality. Of the remaining 16 problems, our algorithm fared better than
Potvin 12 times, equalled Gendreau 2 times, and fared better than Gendreau 3
times. On the 8 problems where our algorithm had the most difficulty, we then
used the solution generated by Gendreau as the initial tour, and we improved
that solution in 5 of these 8 instances. Note that the solution we received from
Gendreau was sometimes different than the solution portrayed in the results
found in [3]. (see Table 2)

As mentioned above, the problems in Table 2 added an additional level of
difficulty in that the objective was to minimize total distance rather than total
time. For the reasons illustrated by Figure 3, we could not just keep the best
partial solution at each node, but had to keep the best several partial solutions.

Table 2.

Problem	n	Potvin [6]	Gendreau [3]	Required K	q_0	q^*	Our Numbers (seconds)
rc201.1	20	465.53	444.54	6	1	5	444.54* (1) (1)
rc201.2	26	739.79	712.91	7	1	5	711.54* (1) (1)
rc201.3	32	839.76	795.44	7	2	7	790.61* (1) (1)
rc201.4	26	803.55	793.64	7	1	4	793.63* (1) (1)
rc202.1	33	844.97	772.18	20	3	>15	772.33 (88)
rc202.2	14	328.28	304.14	13	1	11	304.14* (1) (1)
rc202.3	29	878.65	839.58	15	2	11	837.72* (25) (60)
rc202.4	28	852.73	793.03	18	14	>15	804.57 (263)
rc203.1	19	481.13	453.48	13	2	21	453.48* (5) (14)
rc203.2	33	843.22	784.16	24	2	>15	832.14 (98)
with initial tour from [3]:			789.04	32	2	>15	784.16 (70)
rc203.3	37	911.18	842.25	27	2	>15	none found
with initial tour from [3]:			842.03	36	4	>15	834.90 (136)
rc203.4	15	330.33	314.29	12	1	11	314.29* (1) (3)
rc204.1	46	923.86	897.09	39	9	>15	none found
with initial tour from [3]:			878.76	45	1	>15	878.64 (92)
rc204.2	33	686.56	679.26	28	2	>15	683.13 (167)
with initial tour from [3]:			664.14	32	4	>15	664.14 (158)
rc204.3	24	455.03	460.24	22	4	>15	466.21 (183)
rc205.1	14	381.00	343.21	7	1	8	343.21* (1) (1)
rc205.2	27	796.57	755.93	11	1	14	755.93* (1) (2)
rc205.3	35	909.37	825.06	23	3	>15	825.06 (78)
rc205.4	28	797.20	762.41	12	6	11	760.47* (3) (5)
rc206.1	4	117.85	117.85	3	1	1	117.85* (1) (1)
rc206.2	37	850.47	842.17	16	3	>15	828.06 (66)
rc206.3	25	652.86	591.20	14	1	16	574.42 (7) (37)
rc206.4	38	893.34	845.04	16	2	>15	831.67 (54)
rc207.1	34	797.67	741.53	19	2	>15	743.28 (99)
rc207.2	31	721.39	718.09	21	5	>15	702.33 (239)
rc207.3	33	750.03	684.40	23	2	>15	719.66 (108)
with initial tour from [3]:			684.40	32	1	>15	684.40 (41)
rc207.4	6	119.64	119.64	5	1	5	119.64* (1) (1)
rc208.1	38	812.23	799.19	37	1	>15	834.64 (76)
with initial tour from [3]:			804.41	37	1	>15	795.58 (65)
rc208.2	29	584.14	543.41	28	3	>15	553.43 (118)
with initial tour from [3]:			543.41	28	2	>15	543.41 (94)
rc208.3	36	691.50	660.15	35	2	>15	676.25 (155)
with initial tour from [3]:			654.27	35	7	>15	649.11 (585)

Notes:

All problems run with $K = 15$.

*indicates a guarantee of optimality.

Times listed are for the best q. If a guarantee of optimality could be achieved, a second time listed is for the optimal q.

Table 3.

Problem Name	Open-Ended K&P	Our Results (seconds)		
		$K = 10$	$K = 14$	$K = 17$
500.aa	2964	no improvement		
500.ba	7495	no improvement		
500.ca	21703	21690 (13)	21632 (290)	21618 (3641)
500.da	9950	9886 (14)	9857 (298)	9844 (3605)
500.ea	14724	no improvement		

Notes:

The Kanellakis & Papadimitriou heuristic was allowed to run for 1200 seconds.

This adds another dimension to the auxiliary graph, which we call the *thickness* of the graph, represented by the constant q, a bound on the number of partial solutions kept. Based on the problems in Table 2, the value of q required to find the best solution (q_0 in Table 2) was small, but the value of q required to guarantee a best solution for the given value of $K = 15$ (q^* in Table 2). was much larger

Five 500-city asymmetric TSP's were generated using the *genlarge* problem generator, which Repetto [8] kindly gave to us. Repetto solved these problems with an open-ended heuristic approach, where he would produce an initial tour by randomly choosing one of the nearest two neighbors, and then apply his implementation [7] of the Kanellakis & Papadimitriou heuristic [4], an adaptation to asymmetric TSP's of the Lin-Kernighan heuristic for the symmetric TSP [5]. This process would continue for 1200 seconds and the best solution found would be returned. The tours generated were used as initial tours for our algorithm, and our algorithm improved the solutions to 2 of these 5 problems. (see Table 3)

Table 4.

Number of Cities	Open-Ended K&P (seconds allowed)	$K = 8$ (seconds)	$K = 12$ (seconds)	$K = 15$ (seconds)	$K = 17$ (seconds)
200	233878 (300)	233878 (2)	233720 (19)	233720 (219)	233720 (1387)
300	272242 (600)	271597 (3)	271597 (29)	271597 (339)	271499 (2138)
400	295011 (1800)	294650 (5)	294504 (40)	294504 (449)	294443 (2928)
500	331354 (2400)	329727 (7)	329481 (52)	329371 (564)	329272 (3552)
600	351713 (3600)	351447 (9)	351318 (63)	351203 (692)	351126 (4293)
750	386920 (5500)	386457 (13)	385659 (84)	385659 (837)	385659 (5580)
1000	432558 (7500)	432329 (21)	431791 (166)	431736 (1131)	431094 (7364)

Notes:

The time shown for the Kanellakis & Papadimitriou heuristic is the running time allowed for the open-ended heuristic.

We built symmetric TSP's of various sizes using the t largest cities of the United States, for $t = 200, 300, 400, 500, 600, 750, 1000$. Population figures and coordinates were obtained from the United States Census Bureau [11]. Distances were calculated based on these coordinates, assuming a perfectly spherical earth with a radius of 6378.15 kilometers. All distances were rounded to the nearest tenth of a kilometer. Repetto then applied the above described open-ended heuristic [7] to these problems, and these tours were again used as initial tours for our algorithm. Our algorithm improved each of these solutions, usually with

Table 5.

Problem	n	Required K	Results for $K = 10$			Results for $K = 14$			Results for $K = 17$		
			d	cost	sec.	d	cost	sec.	d	cost	sec.
rc201.1	20	5	81.92*	611.10	1						
rc201.2	26	6	99.56*	875.73	1						
rc201.3	32	5	93.53*	869.68	2						
rc201.4	26	7	114.50*	891.93	2						
rc202.1	33	≈19	277.38	875.97	7	254.08	864.32	139	254.08	864.32	1337
rc202.2	14	8	118.12*	552.18	2						
rc202.3	29	8	134.62*	889.27	4						
rc202.4	28	17	244.41	925.77	6	237.26	900.66	103	273.26*	900.66	871
rc203.1	19	14	220.23	610.81	3	220.23*	610.81	56			
rc203.2	33	≈25	387.57	974.42	8	387.57	974.42	174	383.87	972.58	1976
rc203.3	37	≈27	535.94	975.11	9	482.71	957.49	221	403.46	950.40	2452
rc203.4	15	11	211.98	598.25	2	211.98*	598.25	30			
rc204.1	46	≈43	770.78	1078.49	14	658.42	1022.31	335	623.43	966.05	4101
rc204.2	33	≈30	453.11	781.41	9	434.94	793.15	221	427.36	789.36	2332
rc204.3	24	≈23	321.52	659.22	7	321.52	659.22	136	321.52	659.21	1122
rc205.1	14	3	53.03*	473.12	1						
rc205.2	27	12	186.27	816.47	5	186.27*	816.47	72			
rc205.3	35	15	240.43	965.58	7	240.43	965.58	135	240.43*	965.58	1117
rc205.4	28	11	159.55	883.81	4	159.55*	883.81	64			
rc206.1	4	3	22.08*	227.04	1						
rc206.2	37	13	202.38	902.06	8	202.38*	902.06	136			
rc206.3	25	11	145.93	697.63	3	145.93*	697.63	51			
rc206.4	38	14	204.53	941.35	8	204.53*	941.35	156			
rc207.1	34	17	234.99	906.66	7	234.99	906.66	139	234.99*	906.66	1085
rc207.2	31	≈20	241.18	798.50	7	241.18	798.50	126	241.18	798.50	1348
rc207.3	33	≈18	275.88	896.67	8	259.51	880.36	163	259.51	880.36	1793
rc207.4	6	1	12.08*	309.85	1						
rc208.1	38	≈28	449.78	936.43	10	406.04	914.56	242	375.77	899.40	2703
rc208.2	29	≈21	301.41	779.39	6	267.46	762.42	113	267.46	762.42	1072
rc208.3	36	≈24	297.89	824.90	8	285.85	818.67	177	285.85	818.67	2082

Notes:

d is the value for the minimum window size returned by the algorithm.

*indicates a guarantee of optimality (within .01).

values of K as small as 8 (see Table 4). We suspect the high rate of success on these problems comes from the tendency for cities to cluster in metropolitan areas, which would tend to imply precedence constraints of the type (iii) outlined in problems 1 and 2.

To generate instances of the traveling salesman problem with time targets (TSPTT), we used the same data studied by Potvin [6] and Gendreau [3], and used the time window midpoints for the time targets. The results are shown in Table 5. In many cases, the optimal solution was found with a value of K much smaller than that needed to guarantee optimality, which may indicate that some solutions given to problems without a guarantee of optimality are optimal. In cases where no guarantee was achieved, the exact value of K needed for a guarantee is not known. Arc costs for the problems in Table 5 generally ranged from 10 to 100 units, and 1 time unit is required to travel one distance unit.

References

1. E. Baker, "An Exact Algorithm for the Time-Constrained Traveling Salesman Problem." *Operations Research, 31,* (1983) 938-945.
2. E. Balas, "New Classes of Efficiently Solvable Generalized Traveling Salesman Problems," Management Science Research Report #MSRR-611, Graduate School of Industrial Administration, Carnegie Mellon University, March 1995.
3. M. Gendreau, A. Hertz, G. Laporte, M. Stan, "A Generalized Insertion Heuristics for the Traveling Salesman Problem with Time Windows." Publication CRT-95-07, Centre de recherche sur les transports, Montréal, January 1995.
4. P. Kanellakis, C. Papadimiriou, "Local Search for the Traveling Salesman Problem." *Operations Research, 28,* (1980) 1086-1099.
5. S. Lin, B. W. Kernighan, "An Effective Heuristic Algorithm for the Traveling Salesman Problem." *Operations Research, 21,* (1973) 495-516.
6. J.-Y. Potvin, S. Bengio, "A Genetic Approach to the Vehicle Routing Problem with Time Windows." Publication CRT-953, Centre de recherche sur les transports, Montréal, 1993.
7. B. Repetto *Upper and Lower Bounding Procedures for the Asymmetric Traveling Salesman Problem.* Ph.D. Thesis, GSIA, Carnegie Mellon University, April 1994.
8. B. Repetto, personal communication.
9. M. M. Solomon, "Algorithms for the Vehicle Routing and Scheduling with Time Windows Constraints." *Operations Research, 35,* (1987) 254-265.
10. J. Tama, personal communication.
11. United States Census Bureau, http://www.census.gov/cgi-bin/gazetteer

On Dependent Randomized Rounding Algorithms

Dimitris Bertsimas[1]   Chung-Piaw Teo[2]   Rakesh Vohra[3]

[1] Sloan School of Management and Operations Research Center, MIT
[2] Operations Research Center, MIT
[3] Fisher College of Business, Ohio State University

1   Introduction

The idea of using randomized rounding in the study of approximation algorithms was introduced by Raghavan and Thompson [23]. The generic randomized rounding technique can be described as follows:

1. Formulate and solve a continuous relaxation (in polynomial time) for a $0-1$ integer programming problem to obtain an optimal (possibly fractional) solution $\bar{x}$.

2. Devise a randomization scheme to round each variable x_i to 1.

Raghavan and Thompson [23] derive several approximation bounds for multi-commodity routing problems by rounding *independently* each variable x_i to 1 with probability $\bar{x}_i$. Goemans and Williamson [10] (for the maximum-satisfiability problem) introduce the idea to round each variable x_i independently but with probability $f(\bar{x}_i)$, for some particular nonlinear function $f(x)$. The algorithm they obtain matches the best known guarantee for the problem (originally obtained by Yannakakis [26].) Bertsimas and Vohra [4] use a nonlinear rounding function for covering problems. For the set covering problem their method matches the best known guarantee (originally obtained by Chvátal [8]). The technique has also been used to relate the bound for set-covering problems to the Vapnik-Cervonenkis (VC) dimension of the constraint matrix (Brönnimann and Goodrich [5]) to show that Chvátal's bound [8] for the set-covering problem can be improved if the VC dimension of the constraint matrix can be suitably bounded. Randomized rounding can be seen also as a generalization of deterministic rounding. Deterministic rounding techniques have been used in the the analysis of the bin-packing problem (see for instance, Fernandez de la Vega and Lueker [9]), machine scheduling (see Shmoys and Tardos [24]) and set covering (see Hochbaum [15]).

In all the above applications of randomized rounding, each variable x_i was rounded independently. Goemans and Williamson [11] in their study of maximum-cut problem introduce the idea of dependent randomized rounding of a solution of a semidefinite relaxation, in which the random variables x_i are dependent.

Their algorithm produces the best known approximation bound for this problem. Bertsimas et al. [3] apply dependent randomized rounding to obtain some of the best known bounds for multicut problems.

The heart of the rounding procedure, given a relaxation, is in the design of the randomization scheme. In a recent survey on combinatorial optimization Grötschel and Lovász [12] write:

> ... we can obtain a heuristic primal solution by fixing those variables that are integral in the optimum solution of the linear relaxation, and rounding the remaining variables "appropriately". It seems that this natural and widely used scheme for a heuristic is not sufficiently analyzed ...

This paper aims to deepen our understanding of randomized rounding and to illuminate that strong relaxations coupled with dependent randomized rounding lead to a powerful approach to study discrete optimization problems. In particular, our contributions are:

1. We establish the integrality of several classical polyhedra (min cut, uncapacitated lot-sizing, Boolean optimization, K-median for a particular metric) using dependent randomized rounding. Although the results we show are classical, the simplicity of their proofs gives promise that dependent randomized rounding can be a powerful tool to prove integrality of certain polyhedra.

2. We prove new approximation bounds for several problem classes (min-sat, maximum facility location, covering problems with forcing constraints, the joint replinishment problem, loading problems, prize covering problems).

3. Methodologically, apart from the generic dependent randomized rounding we study, we combine nonlinear and dependent rounding and also introduce a new rounding scheme that is based on rounding real vectors to $0-1$ vectors using Caratheodory's theorem.

Throughout the paper, Z_{IP} and Z_{LP} denote the optimal integral and optimal fractional solution value respectively. Z_H denotes the value returned by a heuristic H. All cost functions are assumed to be nonnegative.

2 Dependent Rounding and Integrality Proofs

In this section, we study the connection of randomized rounding and some basic combinatorial problems. In particular, with the right randomization scheme, we show that the rounding argument leads to direct integrality proofs of several well known polyhedra. The rounding idea that has been used earlier by the authors [3] for the min $s-t$ cut polyhedron. We illustrate further applications of this technique on selected problems. See the full paper for a more complete discussion.

2.1   Boolean Optimization

The quadratic optimization problem is

$$\min \left(\sum_{i,j} Q_{ij} x_i x_j + \sum_i c_i x_i \right) \quad \text{subject to } x_i \in \{0,1\}.$$

With no conditions on the Q_{ij}, the problem is NP-hard. Several researchers have focused on identifying properties of Q_{ij} that allow the problem to be solved in polynomial time. See for instance Hansen and Simeone [13]. The most general condition known is related to the notion of *sign-balancedness* in the coefficients Q_{ij} (see [13]). We explain the concept next.

We construct a graph G that has an edge between i and j if and only if $Q_{ij} \neq 0$. The edges which correspond to positive (resp. negative) Q_{ij} are called positive edges (resp. negative edges). G is sign-balanced if it does not contain any cycle with an odd number of positive edges. This implies that deleting all positive edges of G disconnects it (say $G = G_1 \cup G_2$), and $\delta(G_1, G_2)$ is the set of positive edges (see Harary [14]). The sign-balanced graph problem contains the maximum independent set problem on bipartite graphs as a special case.

Consider the following reformulation

$$\min \sum_{i,j} Q_{ij} z_{ij} + \sum_i c_i x_i$$

$$\text{subject to} \quad z_{ij} \leq \min(x_i, x_j) \quad \text{if } Q_{ij} < 0$$

$$z_{ij} \geq \max(0, x_i + x_j - 1) \quad \text{if } Q_{ij} > 0$$

$$z_{ij}, x_i \in \{0,1\}.$$

1. Generate a single random number U uniformly in $[0,1]$.

2. Starting from an optimal solution of the LP relaxation $\bar{x}, \bar{z}$, round x_i to 1 if $i \in G_1$ and $\bar{x}_i \geq U$, or $i \in G_2$ and $\bar{x}_i \geq 1 - U$.

Theorem 1. $Z_{LP} = Z_{IP}$ if G is sign-balanced.

Proof: Since $1 - U$ is also uniformly distributed in $[0,1]$, $P(x_i = 1) = \bar{x}_i$. For i, j both in G_1 or both in G_2, $P(x_i x_j = 1) = \min\{\bar{x}_i, \bar{x}_j\}$. For $i \in G_1$ and $j \in G_2$,

$$P(x_i x_j = 1) = P(U \leq \bar{x}_i, 1 - U \leq \bar{x}_j) = \max(0, \bar{x}_i + \bar{x}_j - 1).$$

At optimality, $\bar{z}_{ij} = \min(\bar{x}_i, \bar{x}_j)$ if $Q_{ij} < 0$, $\bar{z}_{ij} = \max(0, \bar{x}_i + \bar{x}_j - 1)$ if $Q_{ij} > 0$. Then $Z_{LP} \leq Z_{IP} \leq E(Z_H) = \sum_{i,j} Q(i,j) E[x_i x_j] + \sum_i c_i x_i = \sum_{i,j} Q(i,j) \bar{z}_{ij} + \sum_i c_i \bar{x}_i = Z_{LP}$. $\square$

The above was first proved by Hansen and Simeone [13] by showing that the constraint matrix is totally unimodular.

2.2 Uncapacitated Lot-Sizing

Given a time horizon T, setup costs d_i, $i = 1, \ldots, T$ and production costs c_{ij} of producing a unit in period i to satisfy a unit of demand in period j, and demands f_i, the goal of the uncapacitated lot sizing problem is to find the optimum production schedule to minimize the total setup and production cost to satisfy the demand over T.

Let y_i be a $0-1$ decision variable that indicates whether we produce during period i. Let $w_{i,j}$ be the fraction of the demand f_j in period j that is met from production in period $i \le j$. One formulation of the uncapacitated lot-sizing problem is as follows:

$$\min \sum_{i,j} f_j \, c_{ij} w_{i,j} + \sum_i d_i y_i$$

$$\text{subject to } \sum_{k=1}^{j} w_{k,j} = 1, \quad w_{ij} \le y_i, \quad 0 \le y_i \le 1.$$

It is well known that if we augment the formulation with the valid inequalities

$$w_{i,j} \le w_{i,k} \text{ if } j > k$$

and relax the $0-1$ constraints, the resulting LP is integral (see Nemhauser and Wolsey [22]). Here we prove this result using randomized rounding. Let Z_{LP} the value of this LP relaxation.

Theorem 2. $Z_{LP} = Z_{IP}$.

Proof: Let $(\overline{w}, \overline{y})$ be the optimal LP solution. Consider the following rounding method :

1. Set $r = 1$.

2. Set $y_r = 1$. Generate a random number U_r uniformly in $[0, \overline{y}_r]$.
   Let i be the index such that $\overline{w}_{r,i} \ge U > \overline{w}_{r,i+1}$.
   Set $w_{r,l}$ to 1, for all $l = r, \ldots, i$.

3. Repeat step 2 with $r \leftarrow i + 1$ until $r > T$.

We prove by induction that $P(y_i = 1) = \overline{y}_i$. Clearly $P(y_1 = 1) = \overline{y}_1 = 1$. Moreover,

$$P(y_i = 1) = \sum_{k<i} P(y_i = 1, y_{i-1} = 0, \ldots, y_{k+1} = 0, y_k = 1)$$

$$= \sum_{k<i} P(y_i = 1, y_{i-1} = 0, \ldots, y_{k+1} = 0 | y_k = 1) P(y_k = 1)$$

$$= \sum_{k<i} P(\overline{w}_{k,i-1} \ge U_k > \overline{w}_{k,i}) \overline{y}_k$$

$$= \sum_{k<i} (\overline{w}_{k,i-1} - \overline{w}_{k,i}) = \overline{w}_{i,i} = \overline{y}_i.$$

In addition, $P(w_{i,j} = 1) = P(w_{i,j} = 1, y_i = 1) = \overline{w}_{i,j}$. Hence the randomization scheme gives rise to an optimal integral solution with cost Z_{LP}, i.e., $Z_{IP} = Z_{LP}$.
□

2.3   The k-median Problem on a Cycle

Consider a cycle $C = (V, E)$ with $V = \{v_1, v_2, \ldots, v_n\}$ and $E = \{(v_1, v_2), \ldots, (v_{n-1}, v_n), (v_n, v_1)\}$. Let c be a non-negative weight function on E. Suppose each node gives rise to a unit demand. The cost of serving a unit demand at node v_i from a facility located at node v_j is equal to the distance between the two nodes defined as:

$$c_{i,j} \min \{ c(v_i, v_{i+1}) + \ldots + c(v_{j-1}, v_j),$$
$$c(v_i, v_{i-1}) + \ldots + c(v_2, v_1) + c(v_1, v_n) + \ldots c(v_{j+1}, v_j)\}.$$

The k-median problem is to locate k facilities at the nodes of the graph C in order to minimize the total cost of serving all the demands. A natural LP relaxation is as follows:

$$\min \sum_{i,j} c_{i,j} x_{i,j}$$
$$\text{subject to} \quad \sum_{j=1}^{n} x_{i,j} = 1 \quad i \in \{1, \ldots, n\}$$
$$x_{i,j} \le y_j, \quad 0 \le y_j \le 1 \quad \forall \, i, j$$
$$\sum_j y_j = k.$$

Theorem 3. $Z_{LP} = Z_{IP}$.

Proof: Let $(\overline{x}, \overline{y})$ be a optimal fractional solution to the LP relaxation. Consider the rounding heuristic:

1. Cover the interval $[0, k]$ with n non-overlapping intervals, each of length $\overline{y}_j$, $j = 1, \ldots, n$, *in that order*. Let $[y_j]$ denote the j-th subinterval.

2. Generate U uniformly in $[0, 1]$. Set y_j to 1 if one of the points in the set $S = \{U, 1 + U, \ldots, (k-1) + U\}$ falls in the j-th subinterval (of length $\overline{y}_j$).

3. Solve the assignment problem with y_j fixed.

In this way, we have selected k sites randomly. The above randomization has another simple interpretation using the circular geometry shown in Figure 1. There are k equally spaced spokes (at an angular distance of $\frac{2\pi}{k}$). The n segments of length $\frac{2\pi}{k}\overline{y}_j$, $j = 1, \ldots, n$ are arranged in an non-overlapping order to cover the circumference of the circle of radius k. Spin the wheel. When the wheel rests, y_j is rounded to 1 if there is a spoke touching the j-segment.

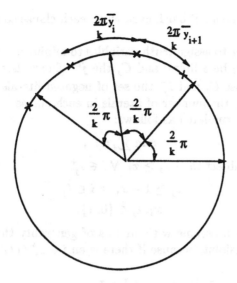

Fig. 1. Circular geometry of the randomization algorithm

For each i, the neighbours of node i are sorted in increasing distance from i. Without loss of generality, we assume that the order is $i = i_1, i_2, \ldots, i_n$. Note that the optimal solution $\bar{x}$ can be computed from $\bar{y}$:

$$\bar{x}_{i,i_j} = \begin{cases} \min\{\bar{y}_{i_j}, 1 - \sum_{\{l:l<j\}} \bar{y}_{i_l}\} & \text{if } \sum_{\{l:l<j\}} \bar{y}_{i_l} < 1 \\ 0 & \text{otherwise} \end{cases}$$

A simple consequence of the geometry in Figure 1 is that the union of the subintervals spanned by $[y_{i_1}], [y_{i_2}], \ldots, [y_{i_j}]$ forms a contiguous line segment in $[0, k]$. In this case,

$$P(x_{i,i_j} = 1) = P(S \cap [y_{i_j}] \neq \emptyset, S \cap [y_{i_l}] = \emptyset \; \forall \, l < j)$$

$$= \begin{cases} \min\{\bar{y}_{i_j}, 1 - \sum_{l:l<j} \bar{y}_{i_l}\} & \text{if } \sum_{l:l<j} \bar{y}_{i_l} < 1 \\ 0 & \text{otherwise} \end{cases}$$

Hence $E(x_{i,i_j}) = \bar{x}_{i,i_j}$. $\qquad\qquad\square$

3 Randomized Approximation Algorithms

3.1 Minimum Satisfiability

Kohli et al. [16] introduced the minimum satisfiability problem as an analog of the maximum satisfiability problem. They proved that this version of the satisfiability problem remains NP-hard, even when each clause contains at most 2 literals (min-2-sat).

The min-sat problem is to assign truth variables to minimize a weighted sum of satisfied clauses. Let x_i be a literal and C_j the jth clause. Let I_j^+ the set of unnegated literals in clause C_j and I_j^- the set of negated literals in C_j. Let K denote an upperbound on the number of literals in each clause.

The problem can be formulated as follows:

$$\min \sum_j w_j z_j$$

$$\text{subject to} \quad z_j \geq x_i \ \forall\, i \in I_j^+$$

$$z_j \geq 1 - x_i \ \forall\, i \in I_j^-$$

$$x_i, z_j \in \{0, 1\}.$$

We remark that we can assume without loss of generality that for every j the sets I_j^+ and I_j^- are disjoint, because if there is an $i \in I_j^+ \cap I_j^-$, then we will have

$$z_j \geq x_i, \quad z_j \geq 1 - x_i.$$

These inequalities force $z_j = 1$ in all feasible integer solutions and hence, in a preprocessing step we can identify all those clauses for which $I_j^+ \cap I_j^- \neq \emptyset$, set the corresponding variables z_j equal to 1 and hence reduce the problem to one for which $I_j^+ \cap I_j^- = \emptyset$.

We show that the following rounding method achieves a bound of $2(1 - \frac{1}{2^K})$ for min-sat:

1. Let $\bar{x}$ and $\bar{z}$ be an optimal LP solution. It is easy to see that

$$\bar{z}_j = \max[\max_{i \in I_j^+} \bar{x}_i, \max_{i \in I_j^-}(1 - \bar{x}_i)].$$

2. Split the x_i's into two sets A, B randomly, i.e., each x_i is assigned to the set A (resp. B) with probability $1/2$.

3. Generate U in $[0, 1]$ uniformly. For x_i in A, set $x_i = 1$ if $\bar{x}_i > U$, 0 otherwise. For x_i in B, set $x_i = 1$ if $\bar{x}_i > 1 - U$, 0 otherwise.

Theorem 4. $E(Z_H) \leq 2\left(1 - \frac{1}{2^K}\right) Z_{LP}.$

The above bound is tight as can be seen from the following example: There are K literals $x_1, \ldots, x_K$ and 2^K clauses. Clause j corresponds to the set S of literals that are unnegated. We will identify a clause by the set S of literals that appear unnegated in it. In this case the formulation becomes:

$$\min \sum_S z_S$$

$$\text{subject to} \quad z_S \geq x_i \ \forall\, i \in S, \quad z_S \geq 1 - x_i \ \forall\, i \notin S, \quad x_i, z_S \in \{0, 1\}.$$

For this example $Z_{LP} = 2^{K-1}$ and $Z_{IP} = 2^K - 1$, i.e., $Z_{IP} = 2\left(1 - \frac{1}{2^K}\right) Z_{LP}.$

3.2 The k-facility Maximization Location Problem

We are given an undirected graph $G = (V, E)$ ($|V| = n$) with profits $c_{ij} \geq 0$ for $(i, j) \in E$. The goal is to find a set $S \subseteq V$ of at most K facilities that maximizes $\sum_{i \in S} \sum_{i \in V} \max_{j \in S} c_{ij}$. The problem is NP-hard. Cornuejols et al. [7] proved that the greedy algorithm returns a solution within $1 - \frac{1}{e}$ of the optimum solution value. They analyzed their greedy heuristic with respect to the following LP relaxation:

$$\max \sum_{(i,j) \in E} c_{ij} x_{ij}$$
$$\text{subject to} \qquad \sum_{j \in V} y_j \leq k$$
$$x_{ij} \leq y_j, \quad (i,j) \in E, \ j \in V$$
$$\sum_{j=1}^{n} x_{ij} = 1, \ i = 1, \ldots, n.$$
$$x_{ij}, \ y_j \in \{0, 1\}.$$

In this section, we propose a randomized rounding heuristic for the problem. An advantage of the randomized rounding approach, over the classical greedy algorithm, is that it has the potential of delivering multiple near optimal solutions/ from a single fractional optimal solution.

The randomized heuristic is as follows:

1. Solve the LP relaxation to obtain the solution $(\overline{x}_{ij}, \overline{y}_j)$.

2. Round as follows: Choose an index j with probability y_j/k. Do this k times and set $y_j = 1$ if the index j was selected in one of the k-rounds. Set $x_{ij} = 1$ if $c_{ij} = \text{argmax} \{c_{il} : y_l = 1\}$. In this way the constraints are always satisfied.

In order to bound the performance of the heuristic we need the following proposition.

Proposition 5. Let $\frac{a_1}{b_1} \geq \frac{a_2}{b_2} \geq \ldots \geq \frac{a_n}{b_n}$ and $c_1 \geq c_2 \geq \ldots \geq c_n$ for some nonnegative numbers a_i, b_i, c_i. Then

$$\frac{\sum_{i=1}^{n} c_i a_i}{\sum_{i=1}^{n} c_i b_i} \geq \frac{\sum_{i=1}^{n} a_i}{\sum_{i=1}^{n} b_i}.$$

For every i we sort the c_{ij}'s in decreasing order. Let $r_i = \min\{j : \sum_{l=1}^{j-1} \overline{y}_l \leq 1, \ \sum_{l=1}^{j} \overline{y}_l > 1\}$.

Theorem 6.

$$E(Z_H) \geq (1 - (1 - \frac{1}{k})^k) Z_{LP} \geq (1 - \frac{1}{e}) Z_{LP}.$$

Proof: We first consider the contribution of the arcs emanating from node i to both the heuristic and the value of the LP relaxation. Without loss of generality let us assume that

$$c_{i1} \geq c_{i2} \geq \ldots \geq c_{in}.$$

Then, $Z_{LP} = \sum_{i=1}^{n} V_i$ where $V_i = c_{i1}\overline{y}_1 + c_{i2}\overline{y}_2 + \ldots + c_{ir_i} \min\{\overline{y}_{r_i}, 1 - \sum_{l=1}^{r_i-1} \overline{y}_l\}$. Letting $\overline{k}_j = \overline{y}_j$ for $j < r_i$ and $\overline{k}_{r_i} = 1 - \sum_{l=1}^{r_i-1} \overline{y}_l$, then $V_i = \sum_{j=1}^{r_i} c_{ij}\overline{k}_j$. Moreover, $E(Z_H) = \sum_{i=1}^{n} U_i$ where

$$U_i = c_{i1}\{1 - (\sum_{l=2}^{n} \overline{y}_l/k)^k\} + c_{i2}\{(\sum_{l=2}^{n} \overline{y}_l/k)^k - (\sum_{l=3}^{n} \overline{y}_l/k)^k\} + \ldots + c_{in}(\overline{y}_n/k)^k.$$

By ignoring terms for $j > r_i$ we obtain that

$$U_i \geq \sum_{j=1}^{r_i} c_{ij}\{(\sum_{l=j}^{n} \overline{y}_l/k)^k - (\sum_{l=j+1}^{n} \overline{y}_l)^k\}.$$

Since $\sum_{l=1}^{n} \overline{y}_l = k$, then $\sum_{l=j}^{n} \overline{y}_l = k - \sum_{l=1}^{j-1} \overline{y}_l$. Therefore,

$$U_i \geq \sum_{j=1}^{r_i} c_{ij}\{(1 - \sum_{l=1}^{j-1} \overline{y}_l/k)^k - (1 - \sum_{l=1}^{j} \overline{y}_l/k)^k\}$$

$$= \sum_{j=1}^{r_i-1} c_{ij}\{(1 - \sum_{l=1}^{j-1} \overline{k}_l/k)^k - (1 - \sum_{l=1}^{j} \overline{k}_l/k)^k\} +$$

$$c_{i,r_i}\{(1 - \sum_{l=1}^{r_i-1} \overline{k}_l/k)^k - (1 - \sum_{l=1}^{r_i-1} \overline{k}_l/k - \overline{y}_{r_i}/k)^k\}$$

$$\geq \sum_{j=1}^{r_i} c_{ij}\{(1 - \sum_{l=1}^{j-1} \overline{k}_l/k)^k - (1 - \sum_{l=1}^{j} \overline{k}_l/k)^k\},$$

since $\overline{k}_{r_i} \leq \overline{y}_{r_i}$. Therefore,

$$\frac{U_i}{V_i} \geq \frac{\sum_{j=1}^{r_i} c_{ij}\{(1 - \sum_{l=1}^{j-1} \overline{k}_l/k)^k - (1 - \sum_{l=1}^{j} \overline{k}_l/k)^k\}}{\sum_{j=1}^{r_i} c_{ij}\overline{k}_j}.$$

Since the sequences $\{c_{ij}\}$ and $\{\frac{(1 - \sum_{l=1}^{j-1} \overline{k}_l/k)^k - (1 - \sum_{l=1}^{j} \overline{k}_l/k)^k}{\overline{k}_j}\}$ are nondecreasing (the latter follows from the convexity of the function $f(x) = x^k$, i.e., if f is convex $\frac{f(y)-f(x)}{y-x}$ is nondecreasing), Proposition 5 applies:

$$\frac{U_i}{V_i} \geq \frac{\sum_{j=1}^{r_i}\{(1 - \sum_{l=1}^{j-1} \overline{k}_l/k)^k - (1 - \sum_{l=1}^{j} \overline{k}_l/k)^k\}}{\sum_{j=1}^{r_i} \overline{k}_j}.$$

Since $\sum_{j=1}^{r_i} \overline{k}_j = 1$, then

$$\frac{U_i}{V_i} \geq 1 - (1 - \frac{\sum_{j=1}^{r_i} \overline{k}_j}{k})^k = 1 - (1 - \frac{1}{k})^k \geq 1 - \frac{1}{e},$$

and the theorem follows. □

The above rounding technique can also be used to study a version of the multiple-type facilities location problems, where l groups of distinct facilities are to be located on the nodes of the network. By using a suitable LP relaxation for the problem, we can devise a $1 - \frac{1}{e}$ approximation bound for this class of problems. We omit the details here.

4 Combining Nonlinear and Dependent Rounding

In this section we combine nonlinear and dependent rounding to study problems of the following type:

$$\min \sum_{k=1}^{N} \sum_{j=1}^{n} c_j^k x_j^k + \sum_{k=1}^{N} f_k y_k$$

subject to
$$\sum_{j=1}^{n} a_{i,j}^k x_j^k \geq 1, \qquad i = 1, ..., m, \ k = 1, ..., N$$
$$x_j^k \leq y_k, \qquad \forall j, k$$
$$y_k, x_i^k \in \{0, 1\}.$$

We assume that $a_{i,j}^k \in \{0, 1\}$. Let A^k denote the k-th constraint matrix (a_{ij}^k). Note that $y_k = \max(x_1^k, ..., x_n^k)$.

The above generic formulation models several problem classes, including the set covering problem $(N = 1)$, and the facility location problem $(m = 1, n = N)$.

In this section, we generalize the approach followed in [4] and combine it with dependent rounding to obtain a randomized approximation algorithm for this class of problems:

1. Solve the LP relaxation to obtain solution $\overline{x}_j^k, \overline{y}_k$.

2. Generate U_k in $[0, 1]$ with distribution function $F(u) = 1 - (1 - u)^\alpha$.

3. Round x_j^k (resp. y_k) to 1 if $\overline{x}_j^k > U_k$ (resp. $\overline{y}_k > U_k$).

In preparation for the main theorem we consider a probability space defined on a ground set V. We say an event U, where $U \subset 2^V$, is an increasing event if $S \in U$ implies $T \in U$ if T contains S. For increasing events, it is well known that the following correlation inequality holds (also known as the FKG lemma, cf. Alon and Spencer [1]):

Proposition 7. *If A, B are both increasing events, then $P(A \cap B) \geq P(A)P(B)$.*

Let F_i^k denote the event that the constraint "$\sum_{j=1}^{n} a_{ij}^k x_j^k \geq 1$" is satisfied. F_i^k are clearly increasing events for each i, k. Let $\mathcal{F}$ denote the event $\{F_i^k; i = 1, \ldots, m, \ k = 1, \ldots, N\}$, i.e., the randomized procedure produces a feasible solution.

Theorem 8. *Let $\alpha = \log\left(\sum_{k=1}^{N} D_{A^k}\right)$, where D_{A^k} is the maximum column sum of the matrix A^k. Then*

$$Z_{IP} \leq E(Z_H | \mathcal{F}) \leq \frac{\log\left(\sum_{k=1}^{N} D_{A^k}\right)}{\left(1 - \frac{1}{\sum_{k=1}^{N} D_{A^k}}\right)^{\sum_{k=1}^{N} D_{A^k}}} Z_{LP}.$$

Proof: The first inequality is obvious. For the stated choice of α, we have

$$P(x_j^k = 1) = 1 - (1 - \overline{x}_j)^\alpha \leq \alpha \overline{x}_j^k.$$

Now,

$$E(Z_H | \mathcal{F}) = \sum_{j,k} c_j^k P(x_j = 1 | \mathcal{F}) + \sum_k f_k P(y_k = 1 | \mathcal{F})$$

$$= \sum_{j,k} c_j^k P(\mathcal{F} | x_j^k = 1) P(x_j^k = 1) / P(\mathcal{F}) + \sum_k f_k P(\mathcal{F} | y_k = 1) P(y_k = 1) / P(\mathcal{F}).$$

Let $E_j^k = \{i : a_{i,j}^k = 1\}$, and $\overline{E_j^k}$ its complement $\{i : a_{i,j}^k = 0\}$. Then

$$\frac{P(\mathcal{F} | x_j^k = 1)}{P(\mathcal{F})} = \frac{P(F_i^k; k = 1, \ldots, N, i \in \overline{E_j^k})}{P(\mathcal{F})}$$

$$\leq \frac{1}{P(F_i^k; i \in E_j^k, k = 1, \ldots, N)} \leq \frac{1}{\prod_{i,k : i \in E_j^k} P(F_i^k)}.$$

The last two inequalities follow from the correlation inequality. Furthermore,

$$P(F_i^k) = 1 - \prod_{l : a_{i,l}^k = 1} (1 - \overline{x}_l)^\alpha \geq 1 - e^{-\alpha \sum_l a_{il}^k \overline{x}_l}$$

$$\geq 1 - e^{-\alpha} = 1 - \frac{1}{\sum_{k=1}^{N} D_{A^k}}.$$

Hence,

$$P(\mathcal{F} | x_j^k = 1) / P(\mathcal{F}) \leq \frac{1}{\left(1 - \frac{1}{\sum_{k=1}^{N} D_{A^k}}\right)^{\sum_{k=1}^{N} D_{A^k}}}.$$

Similarly, we have

$$P(\mathcal{F}|y_k = 1)/P(\mathcal{F}) \leq \frac{1}{(1 - \frac{1}{\sum_{k=1}^N D_{A^k}})^{\sum_{k=1}^N D_{A^k}}}.$$

Thus

$$E(Z_H|\mathcal{F}) \leq \frac{\alpha(\sum_{k,j} c_j^k \overline{x}_j^k + \sum_k f_k \overline{y}_k)}{(1 - \frac{1}{\sum_{k=1}^N D_{A^k}})^{\sum_{k=1}^N D_{A^k}}} \leq \frac{\log(\sum_{k=1}^N D_{A^k})}{(1 - \frac{1}{\sum_{k=1}^N D_{A^k}})^{\sum_{k=1}^N D_{A^k}}} Z_{LP}.$$

$\square$

The above approach can also be used to derive an $O(\log T + \log N)$ bound for a class of multi-item lotsizing problems with T time periods and N number of items. See the full paper for details.

5 Vector Rounding

The rounding algorithms proposed in the previous sections essentially round fractional solutions to feasible integral solutions by introducing dependency in the rounding process. In this section, we propose a generic framework to randomly generate a feasible solution, exploiting Caratheodory's theorem:

Proposition 9. *Let P be a bounded polyhedron in $\mathcal{R}^n$ with only integral vertices. Any point x in P can be expressed as a convex combination of at most $n + 1$ integral vertices.*

Given a fractional solution $x \in P$, we can express x as a convex combination of $n + 1$ vertices, and we can round x randomly to any one of these vertices, with probability proportional to the coefficient in the convex combination. The above procedure can be executed in polynomial time (see [2]). We apply this idea to various problem classes.

5.1 Loading Problems

Magnanti and Mirchandani [18] and Chopra et al. [6] studied the following multi-commodity-one-facility network loading problem that arises in telecommunications:

Given a network G, k source-sink pairs (s_i, t_i), each with traffic flow of demand d_i. The goal is to choose ("load") a set of facilities in the edges. Each facility can support C units of flow and costs f_e. The routing cost of one unit of the i-th commodity along arc e is c_e^i. The objective is to choose a set of facilities with the least routing and loading cost.

The above network loading problem can be easily generalized to loading of arbitrary combinatorial objects, rather than directed paths between specific source-sink pairs. Let $P_1, \ldots, P_k$ be (0-1) integral polyhedra (for the network loading problem they are directed path polyhedra). For d_i a positive integer we introduce $d_i P_i = \{x^i : x^i = d_i y^i, \ y^i \in P_i\}$. Consider the following formulation of a generic loading problem:

$$\min \sum_{i=1}^{k} c^i x^i + \sum_e f_e y_e$$
$$\text{subject to} \quad x^i \in d_i P_i, \ i = 1, \ldots, k;$$
$$\sum_{i=1}^{k} x_e^i \leq C y_e,$$
$$x_j^i \in \{0, d_i\}, y_e \in \mathcal{Z}^+.$$

We apply vector rounding as follows:

1. Obtain the optimal solution $(\overline{x}^1, \ldots, \overline{x}^k, \overline{y})$ of the LP relaxation.

2. Using Caratheodory's theorem express each $\overline{x}^i$ as $\overline{x}^i / d_i = \sum_{k=1}^{dim(P_i)+1} \lambda_k z^{i,k}$, where $z^{i,k}$ are vertices of P_i.

3. For each i, select a vertex $z^{i,k}$ with probability λ_k. Return $x^i = d_i z^{i,k}$ and let $y_j = \lceil \sum_i x^i / C \rceil$.

Theorem 10. $Z_{IP} \leq E(Z_H) \leq \max_i \{ \lceil d_i / C \rceil C / d_i \} Z_{LP}$.

For the one-commodity-one-facility network loading problem, Chopra et al. [6] obtained a bound of $\frac{\lceil d_1 / C \rceil}{\lceil d_1 / C \rceil}$ using network flow techniques. The randomization gives rise to a slightly better bound. If there are multiple facilities, each of fixed capacity $C_1, \ldots, C_n$, then the same approach, applied to a suitably modified formulation, gives rise to a bound of

$$\max_{i,j} \frac{\lceil d_i / C_j \rceil}{d_i / C_j}.$$

5.2 Prize Covering Problems

In the prize covering problem, the objective is to cover as many elements as possible, using at most k combinatorial objects. The integer programming formulation is as follows:

$$\max \sum_{i=1}^{k} c^i x^i + \sum_j d_j y_j$$
$$\text{subject to} \quad x^i \in P_i, \ i = 1, \ldots, k;$$
$$\sum_{i=1}^{k} x_j^i \geq y_j, \ j = 1, \ldots, n.$$
$$x_j^i, y_j \in \{0, 1\}.$$

In this instance, we need P_i to be the convex hull of the underlying combinatorial objects. Special cases of this problem are:

1. When P_i represent matchings in a graph, the above problem, with $c^i = 0$, reduces to the max-weighted k-edge colorable subgraph problem.

2. When P_i represent cut polyhedra, the problem represents the prize collecting version of the cut-covering problem on general graphs.

We apply vector rounding as follows:

1. Obtain the optimal solution $(\overline{x}^1, ..., \overline{x}^k, \overline{y})$ of the LP relaxation.

2. Using Caratheodory's theorem express each $\overline{x}^i$ as $\overline{x}^i = \sum_{k=1}^{dim(P_i)+1} \lambda_k z^{i,k}$, where $z^{i,k}$ are vertices in P_i.

3. For each i, select a vertex $z^{i,k}$ with probability λ_k. Return $x^i = z^{i,k}$; Set $y_j = 1$ if possible.

Theorem 11. $Z_{IP} \geq E(Z_H) \geq (1 - \frac{1}{e})Z_{LP}$.

For the max-weighted k-edge colorable subgraph problem (P_i represent matchings in a graph), we obtain an $(1 - \frac{1}{e})$ approximation algorithm. When P_i represent cut polyhedra on K_5-free graphs, then the above yields an $(1 - \frac{1}{e})$ approximation algorithm for the prize-collecting version of the cut-covering problem studied by Loulou [17] and Motwani and Naor [20].

References

1. N. Alon and J. H. Spencer, *The Probabilistic Method*, Wiley, New York, 1992.

2. M. Bazaraa, J. Jarvis and H. Sherali. *Linear Programming and Network Flows*, 2nd ed. Wiley, New York, 1990.

3. D. Bertsimas, C. Teo and R. Vohra. Nonlinear relaxations and improved randomized approximation algorithms for multicut problems, *Proc. 4th IPCO Conference*, 29-39, 1995.

4. D. Bertsimas and R. Vohra. Linear programming relaxations, approximation algorithms and randomization : a unified view of covering problems, Preprint 1994.

5. H. Brönnimann and M. Goodrich. Almost optimal set covers in finite VC-dimension, *Proc 10th Annual Symp. Computational Geometry*, 293-301, 1994.

6. S. Chopra, I. Gilboa and T. Sastry. Source sink flows with capacity installation in batches, preprint, 1995.

7. G. Cornuejols, M. Fisher and G. Nemhauser. Location of Bank Accounts to optimize float: An analytic study of exact and approximate algorithms, *Management Science*, 23, 789-810, 1977.

8. V. Chvátal. A greedy heuristic for the set-covering problem. *Mathematics of Operations Research*, 4, 233-235, 1979.

9. W. Fernandez de la Vega and G. Lueker. Bin packing solved within $1+\epsilon$ in linear time, *Combinatorica*, 1, 349-355, 1981.

10. M.X. Goemans and D. Williamson. A new 3/4 approximation algorithm for MAX SAT, *Proc. 3rd IPCO Conference*, 313-321, 1993.

11. M.X. Goemans and D. Williamson. .878 approximation algorithms for MAX-CUT and MAX 2SAT, *Proc. 26th Annual ACM STOC*, 422-431,1994.

12. M. Grötschel and L. Lovász. Combinatorial Optimization: A survey, DIMACS Technical Report 93-29, 1993.

13. P. Hansen and B. Simeone. Unimodular functions, *Discrete Applied Mathematics*, 14, 269-281, 1986.

14. F. Harary. On the notion of balance of a signed graph, *Michig. Math. Journal*, 2, 143-146, 1953.

15. D. Hochbaum. Approximation algorithms for Set Covering and Vertex Cover Problems, *SIAM Journal on Computing*, 11, 555-556, 1982.

16. R. Kohli, R. Khrishnamurti and P. Mirchandani. The minimum satisfiability problem, *SIAM Journal of Discrete Mathematics*, 7(2), 275-283, 1994.

17. R. Loulou. Minimal cut cover of a graph with an application to the testing of electronic boards, *Operations Research Letters*, 12, 301-305, 1992.

18. T. Magnanti and P. Mirchandani. Shortest paths, single origin-destination network design, and associated polyhedra, *Networks*, 23, 103-121, 1993.

19. M. Marathe and S. Ravi. On approximation algorithms for the minimum satisfiability problems, Preprint, 1995.

20. R. Motwani and J. Naor. On exact and approximate cut covers of graphs, Preprint, 1993.

21. R. Motwani and P. Raghavan. Randomized Algorithms, Cambridge University Press, 1995.

22. G. Nemhauser and L. Wolsey. *Integer and Combinatorial Optimization*, Wiley, New York, 1988.

23. P. Raghavan and C. Thompson. Randomized rounding : a technique for provably good algorithms and algorithmic proofs, *Combinatorica*, 7, 365-374, 1987.

24. D. Shmoys and E. Tardos. An approximation algorithm for the generalized assignment problem, *Math Programming*, 62, 461-474, 1993.

25. A. Srinivasan. Improved Approximations of Packing and Covering Problems, *Proceedings of the 28th Annual Symposium on Theory of Computing*, 268-276, 1995.

26. M. Yannakakis. On the approximation of maximum satisfiability, *Proc. 3rd ACM-SIAM Symp. on Discrete Algorithms*, 1-9, 1992.

Coloring Bipartite Hypergraphs

Hui Chen and Alan Frieze

Carnegie Mellon University, Pittsburgh PA 15213, USA

Abstract. It is NP-Hard to find a proper 2-coloring of a given 2-colorable (*bipartite*) hypergraph H. We consider algorithms that will color such a hypergraph using *few* colors in polynomial time. The results of the paper can be summarized as follows: Let n denote the number of vertices of H and m the number of edges. (i) For bipartite hypergraphs of dimension k there is a polynomial time algorithm which produces a proper coloring using $\min\{O(n^{1-1/k}), O((m/n)^{\frac{1}{k-1}})\}$ colors. (ii) For 3-uniform bipartite hypergraphs, the bound is reduced to $\tilde{O}(n^{2/9})$. (iii) For a class of dense 3-uniform bipartite hypergraphs, we have a randomized algorithm which can color optimally. (iv) For a model of random bipartite hypergraphs with edge probability $p \geq dn^{-2}$, $d > 0$ a sufficiently large constant, we can almost surely find a proper 2-coloring.

1 Introduction

A hypergraph $H = (V, E)$ has vertex set V, $|V| = n$, edge set E, $|E| = m$ and each edge $e \in E$ is simply a subset of V. Its dimension $\dim(H)$ is the size of the largest edge in E. A set $S \subseteq V$ is said to be *independent* if it contains no edge of H. A *proper k-coloring* of the vertex set V is a partition of V into k independent sets. A hypergraph *bipartite* if it admits a proper 2-coloring.

Lovász [9] showed that it is in general NP-Hard to determine whether or not a hypergraph H is bipartite. (There are important special cases where such a coloring can be found, see for example Beck [4], Alon and Spencer [3] or McDiarmid [10]). In such circumstances it is of some interest to see if one can find a proper coloring of a hypergraph H which is known to be bipartite, but for which no proper 2-coloring is given. This problem is very similar in flavour to that of finding a good coloring of a given 3-colorable graph. In this paper we modify recent ideas for tackling this latter problem and apply them in the context of bipartite hypergraphs. The results of this paper can be summarised as follows:

- We first consider an algorithm based on ideas of Wigderson [12] and show how to color a bipartite hypergraph of dimension k with $\min\{O(n^{1-1/k}), O((m/n)^{\frac{1}{k-1}})\}$ colors.
- We then modify the techniques of Karger, Motwani and Sudan [7] to derive a smaller $\tilde{O}(n^{2/9})$[1] upperbound on the number of colors needed to color a

---

[1] $\tilde{O}$ notation suppresses a factor $(\log n)^d$ for some positive constant d.

3-uniform bipartite hypergraph in polynomial time.

- We next consider *dense* hypergraphs and show, using similar ideas to those of Edwards [5], that dense 3-uniform bipartite hypergraphs can be 2-colored in polynomial time.
- We then consider the case where H is chosen randomly from some natural distribution. We use a spectral method introduced by Alon and Kahale [2] to show that **whp**[2] we can 2-color a random bipartite hypergraph with edge density $p \geq dn^{-2}$.

2 Approximate Coloring for General Bipartite Hypergraphs

In this section we consider an algorithm for coloring bipartite hypergraphs which is a development of Wigderson's graph coloring algorithm. We assume that the edges of H are pairwise incomparable, if $e \supseteq e'$ then delete edge e.

For $U \subset V$, U let $N(U) = \{x \in V : \{x\} \cup U \in E\}$. Note that $N(U) \neq \emptyset$ implies that U is independent. Let $E_U = \{U \cup \{x\} : x \in N(U)\}$ and $H \star U = (V, (E \setminus E_U) \cup \{U\})$. We begin with the following simple lemma.

Lemma 1. *Suppose $U \subseteq V$ and $N(U)$ does not contain an edge. Then*

(a) $H \star U$ *is bipartite.*

(b) *An s-coloring of $H \star U$ is also an s-coloring of H.*

Proof   (a) Let A, B be a partition of V into independent sets. If for example, $U \subseteq A$ then $N(U) \subseteq B$ and so is independent.

(b) Fix some s-coloring $\mathcal{K}$ of $H \star U$. The edges of H which are not in $H \star U$ all contain U. Since $\mathcal{K}$ properly colors U it properly colors all such edges. $\square$

We can now describe the algorithm

Phase 1

begin

   while there exists $U \subseteq V$ with $N(U)$ dependent;

         $H \leftarrow H \star U.$

end

This can easily be carried out in time polynomial in m, n.

At the end of Phase 1, H satisfies

$$U \subseteq V, N(U) \neq \emptyset \text{ implies } N(U) \text{ is independent.} \tag{1}$$

[2] with high probability i.e. probability $1 - o(1)$

If $N(U)$ is independent then its elements can be colored with a single color. For $S \subseteq V$ let $E^S = \{e \in E : e \cap S = \emptyset\}$ and $H \setminus S = (V \setminus S, E^S)$.

Phase 2

begin

> **while** there exists $U \subseteq V$ with $|N(U)| \geq n^{1/k}$;
>
> > Assign a new color to the elements in $N(U)$;
> >
> > $H \leftarrow H \setminus N(U)$.

end

Phase 2 is also executable in polynomial time and requires no more than $n^{1-1/k}$ colors.

By the end of Phase 2 the maximum possible number of edges in H has been reduced.

Lemma 2. *At the end of Phase 2, H contains at most $n^{t-1+1/k}$ edges of size t, $2 \leq t \leq k$.*

Proof   Consider the set of pairs (U, x) where $|U| = t - 1$ and $x \in N(U)$. Clearly there are at most $\binom{n}{t-1} n^{1/k}$ such pairs. On the other hand, each edge of size t gives rise to t distinct pairs. $\square$

We complete the coloring randomly.

Phase 3

begin

> Randomly color each vertex with one of $r = \lceil 4n^{1-1/k} \rceil$ colors;
>
> $E' \leftarrow \{e \in E : e \text{ is not properly colored}\}$;
>
> $S \leftarrow \bigcup_{e \in E'} e$;
>
> $H \leftarrow H \setminus (V \setminus S)$

end

Lemma 3. *Let S be as in Phase 3. Then*

$$\Pr(|S| \geq n/2) \leq 1/2.$$

Proof   Let m_t denote the number of edges of size t which remain in H and let $p_t = r^{-(t-1)}$ denote the probability that an edge of size t is not properly colored. Then

$$\mathbf{E}(|S|) \leq \sum_{t=3}^{k} t m_t p_t$$

$$\leq \sum_{t=3}^{k} tn^{t-1+1/k}4^{-t}n^{-(t-1)(k-1)/k}$$

$$\leq \sum_{t=3}^{k} t4^{-t}n^{t/k}$$

$$\leq n/7,$$

for n large. The result follows from the Markov inequality. $\qquad\square$

We repeat Phase 3, with the same initial set of edges, until $|S| \leq n/2$. The expected number of repetitions is less than 2. When this happens we have succeeded in reducing the number of vertices in H by a factor of 2 and we have used at most $5n^{1-1/k}$ colors.

We can repeat Phases 1 – 3, using new colors and halving the number of vertices at each iteration until the number of vertices remaining is less than $n^{1-2/k}$, in which case we give each vertex a unique color. The total number of colors used is then at most

$$5 \sum_{i=1}^{\infty} \left(\frac{n}{2^i}\right)^{1-1/k} + n^{1-2/k} \leq 10n^{1-1/k}.$$

It can also be seen that if the hypergraph does not have many edges, i.e. when $m \leq n^{k-1-1/k}$, the simple random coloring idea will give a better bound (use $r = \lceil 2(m/n)^{\frac{1}{k-1}} \rceil$). To summarise

Theorem 4. *If H is a bipartite hypergraph of dimension k which has m edges then there is a polynomial time algorithm which properly colors the vertices of H in*
$$\min\{O(n^{1-1/k}), O((m/n)^{\frac{1}{k-1}})\} \text{ colors.}$$

3 3-Uniform Bipartite Hypergraphs

We now consider coloring a bipartite hypergraph H in which each edge has size 3. The approach here is very similar to that of Karger, Motwani and Sudan [7]. Consequently we will be somewhat brief in our exposition. As H is bipartite there is a partition A, B of V such that each edge of H meets both A and B. Putting $y_i = +1$ for $i \in A$ and $y_i = -1$ for $i \in B$ we see that

$$\{i, j, k\} \in E \text{ implies } y_iy_j + y_iy_k + y_jy_k \leq -1. \tag{2}$$

Arguing as in [7] we consider the semi-definite program

SDP

> Minimise α
> Subject to
> $$\mathbf{v}^{(i)} \cdot \mathbf{v}^{(j)} + \mathbf{v}^{(i)} \cdot \mathbf{v}^{(k)} + \mathbf{v}^{(j)} \cdot \mathbf{v}^{(k)} \leq \alpha, \qquad \forall\{i, j, k\} \in E$$
> $$\mathbf{v}^{(i)} \qquad\qquad\qquad\qquad\qquad\quad \in S_{n-1}, i \in [n],$$

where $S_{n-1} = \{v \in R^n : |v| = 1\}$.

We see from (2) that SDP has an optimal solution with $\alpha \leq -1$. We can compute a solution with $\alpha \leq -1 + \epsilon$ in time polynomial n and $\log 1/\epsilon$ – see for example Alizadeh [1]. Thus we can take ϵ as zero on the understanding that the errors introduced are swamped by other errors in the approximation.

So assume that we have computed $v^{(1)}, v^{(2)}, \ldots, v^{(n)} \in S_{n-1}$ such that

$$v^{(i)} \cdot v^{(j)} + v^{(i)} \cdot v^{(k)} + v^{(j)} \cdot v^{(k)} \leq -1, \quad \forall \{i, j, k\} \in E \tag{3}$$

We now choose t (t defined later) random vectors $x^{(1)}, x^{(2)}, \ldots, x^{(t)}$ such that the components $x_j^{(i)}$ are independent standard normal $N(0,1)$ random variables. We say that $x^{(r)}$ captures $v^{(i)}$ if

$$x^{(r)} \cdot v^{(i)} \geq x^{(s)} \cdot v^{(i)}, \quad \forall s \neq r.$$

Let
$$S_r = \{i \in V : x^{(r)} \text{ captures } v^{(i)}\}, \quad 1 \leq r \leq t.$$

With probability one, $S_1, S_2, \ldots, S_t$ is a partition of V and so defines a coloring $\mathcal{K}$. Some edges of H may not be properly colored. Let m' be the number of such edges. We say that $\mathcal{K}$ is a semi-coloring if $m' \leq n/4$. In which case, the number of vertices in edges which are monochromatic is at most $3n/4$. Hence we can easily find a set S of at least $n/4$ vertices such that if $e \cap S \neq \emptyset$ then e is properly colored. We remove S, along with the associated colors and apply our algorithm to $H \setminus S$. This yields an $O(t)$ proper coloring of H (The number of colors needed in each next round will be geometrically decreasing).

We show later that for $\Delta = m/n$ and $t = O(\Delta^{1/8}(\log \Delta)^{9/8})$ then

$$\Pr(m' \geq n/4) \leq 1/2. \tag{4}$$

The initial hypergraph could have Δ as large as $n^2/6$ and this would lead to the use of $\tilde{O}(n^{1/4})$ colors. But by applying Phases 1 and 2 of the previous section, with $n^{1/k}$ in A of Phase 2 replaced by $n^{7/9}/\log n$, we can reduce Δ to $n^{16/9} \log n$ at the expense of using at most $n^{2/9} \log n$ colors. This (modulo the proof of (4)) leads to

Theorem 5. *If H is a bipartite hypergraph of dimension 3 then there is a polynomial time algorithm which properly colors the vertices of H in $\tilde{O}(n^{2/9})$ colors.*

Proof of (4)

We need a simple geometric fact:

Lemma 6. *Let $v^{(1)}, v^{(2)}, v^{(3)}, v \in S_{n-1}$ and suppose*

$$v^{(1)} \cdot v^{(2)} + v^{(1)} \cdot v^{(3)} + v^{(2)} \cdot v^{(3)} \leq -1.$$

Then

$$\min_{i=1,2,3} v^{(i)} \cdot v \leq 1/3.$$

Proof   Observe first that

$$(\mathbf{v}^{(1)} + \mathbf{v}^{(2)} + \mathbf{v}^{(3)})^2 = (|\mathbf{v}^{(1)}|^2 + |\mathbf{v}^{(2)}|^2 + |\mathbf{v}^{(3)}|^2)$$
$$+2(\mathbf{v}^{(1)} \cdot \mathbf{v}^{(2)} + \mathbf{v}^{(1)} \cdot \mathbf{v}^{(3)} + \mathbf{v}^{(2)} \cdot \mathbf{v}^{(3)})$$
$$\leq 1.$$

So $|\mathbf{v}^{(1)} + \mathbf{v}^{(2)} + \mathbf{v}^{(3)}| \leq 1$. Hence

$$\mathbf{v}^{(1)} \cdot \mathbf{v} + \mathbf{v}^{(2)} \cdot \mathbf{v} + \mathbf{v}^{(3)} \cdot \mathbf{v} \leq 1,$$

and the lemma follows.   □

Let now $P(n, t)$ denote the maximum over vectors $\mathbf{v}^{(i)}, i = 1, 2, 3$ of the probability that the three vectors are captured by the same vector from $\mathbf{x}^{(j)}, 1 \leq j \leq t$. We consider the probability that they are all captured by $\mathbf{x}^{(1)}$ and multiply this by t. By Lemma 6 we can assume without loss of generality that the angle between $\mathbf{x}^{(1)}$ and $\mathbf{v}^{(1)}$ is at least $\theta = \arccos(1/3)$. Put $\zeta = 1/\log \Delta, p = \zeta/\pi$ and $q = 1/\cos(\theta - \zeta)^2$. Consider the 2-dimensional subspace L generated by $\mathbf{v}^{(1)}$ and $\mathbf{x}^{(1)}$ and let R denote the wedge of this plane within an angle ϵ from $\mathbf{v}^{(1)}$. If $\mathbf{x}^{(1)}$ captures $\mathbf{v}^{(1)}$ then, as observed in [7] (proof of Theorem 7.7), the projection of $\mathbf{x}^{(1)}$ onto the nearer of the two lines bounding R exceeds the length of any $\mathbf{x}^{(j)}$ which lies in R. The probability of this event is shown in [7] to be $O((pt)^{-q})$. Thus

$$P(n, t) = O(t(pt)^{-q})$$
$$= O(t(\Delta \log \Delta)^{-\frac{8}{3}+O(\zeta)})$$
$$= O(1/\Delta). \tag{5}$$

By choosing the constant in the definition of t sufficiently large, the hidden constant in (5) can be made less than $1/8$. In which case, the expected number of improperly colored edges will be less than $n/8$. Equation (4) follows from the Markov inequality.   □

4   Dense Hypergraphs

Suppose that H is k-uniform. For $X \subset V, |X| = k - 1$, let $N(X) = \{v \in V : v \cup X \in E\}$. Let $\alpha > 0$ be fixed. We say that H is α-dense if $|N(X)| > \alpha n$ for all $X \subset V$.

Theorem 7. *If H is bipartite, 3-uniform and α-dense then H can be 2-colored in $n^{O(1/\alpha)}$ time.*

Proof

Let $A \cup B$ be a partition of V into 2 independent sets. Clearly $|A|, |B|$
Choose S, $|S| = 3\alpha^{-1} \log n$ randomly from V. Then

$$\Pr(\exists x, y \in V : N(x,y) \cap S = \emptyset) \leq n^2 \left(1 - \frac{|S|}{n}\right)^{\alpha n}$$

$$\leq n^{-1}.$$

So we can assume that $N(x,y) \cap S \neq \emptyset$ for all $x, y \in V$. By consider.
$2^{|S|} = n^{O(1/\alpha)}$ possibilities we can guess $S_A = S \cap A$ and $S_B = S \cap B$.
see Edwards [5], we construct an instance of 2-SAT with variables $x_v, v \in$
and clauses $\mathcal{C}$. $x_v = 1$ will stand for $v \in A$ and the clauses will be d
$C_{u,v}$, $u, v \in V \setminus S$ where

$$C_{u,v} = \begin{cases} \{\bar{x}_u, \bar{x}_v\} : & N(u,v) \cap S_A \neq \emptyset, N(u,v) \cap S_B = \emptyset, \\ \{x_u, x_v\} : & N(u,v) \cap S_A = \emptyset, N(u,v) \cap S_B \neq \emptyset, \\ \text{NO CLAUSE} : \text{otherwise} \end{cases}$$

Now $\mathcal{C}$ is satisfiable by $x_v = 1$ for $v \in A \setminus S$ and $x_v = F$ for $v \in B \setminus S$
if $A' = \{v : x_v = 1\}$ is the 2-SAT solution we construct then $|A' \setminus A| \leq$
if $v_1, v_2 \in A' \setminus A \subseteq B \setminus S$ then the clause $C_{v_1, v_2} = \{\bar{x}_{v_1}, \bar{x}_{v_2}\}$ is not sa
Similarly, $|B' \setminus B| \leq 1$ and a simple brute force final check can correct any
in $O(n^2)$ time.

The proof of correctness of the algorithm does not seem to generalize to
dimensions. Though there seems to be no intrinsic reason why coloring be
harder when $k \geq 4$, a completely different method seems to be necessary.

5 Random Hypergraphs

We first describe our model of a random bipartite hypergraph. Let $W_1 = \{1, 2,$
and $W_2 = \{n+1, n+2, \ldots, 2n\}$. There are $N = 2n\binom{n}{2}$ triples contai
$W_1 \cup W_2$ which contain at least one element from both of W_1 and W_2. We
ate $H = H_{2n,3,p} = H(V, \mathcal{E})$ by independently including each possible tripl
probability p. This is a natural analogue of the standard model of a r:
3-colorable graph.

We will show that there exists a constant $d_0 > 0$ such that if $p \geq d_0 n^{-2}$ the
$H_{2n,3,p}$ can be properly 2-colored in polynomial time whp (without kno
of the partition W_1, W_2). We only consider the case $p = dn^{-2}$, $d \geq d_0$ co
Things get easier if $d \to \infty$. The method used is an adaptation of the s
method of Alon and Kahale [2].

5.1 The reduction

To apply the methodology of [2] we need a graph. So let $G = (V, E)$

triple into a triangle in G and merge multiple edges into one. We now proceed more or less as in [2].

1. Construct $G' = (V, E')$ by deleting all edges incident with vertices of degree at least $5d$ in G.
2. Compute the eigenvector v corresponding to the most negative eigenvalue of the adjacency matrix A' of G'.
3. Let $X = \{i \in V : v'_i \geq 0\}$ and $Y = V \setminus X$.
4. Use X, Y as the start of an iterative process to 2-color H.

5.2 Eigenvalues of G'

The intuition behind the approach is as follows: if each vertex of H lies in its expected number of triples with 1 or 2 members of W_1 ($d/2$ and d for a vertex in W_1) then $\mathbf{f} = (-1, -1, \ldots, -1, +1, +1, \ldots, +1)$ (-1 for $i \in W_1$ and $+1$ for $i \in W_2$) is an eigenvector of the adjacency matrix A of G, with eigenvalue $-d$. Since this is approximately true **whp**, there should be an eigenvector close to $\mathbf{f}$ which will give us a good idea of W_1 and W_2.

Let $\lambda_1 \geq \lambda_2 \geq \cdots \geq \lambda_{2n}$ be the eigenvalues of A', and $e_1, e_2, \ldots, e_{2n}$ be corresponding eigenvectors which form an orthonormal basis of R^{2n}.

Lemma 8. *The following are true* **whp**:

(i) $\lambda_1 \geq (1 - 2^{-\Omega(d)})3d$,

(ii) $\lambda_{2n} \leq -(1 - 2^{-\Omega(d)})d$,

(iii) $|\lambda_i| = O(\sqrt{d})$ *for all* $2 \leq i \leq 2n - 1$.

Proof The proof is very similar to that of Proposition 2.1 of [2] which is based on ideas of Kahn and Szemerèdi [8]. We give only a bare outline. We use the fact (see for example [11]) that

$$\lambda_i = \min_L \max_{\substack{s \in L \\ s \neq 0}} \frac{x^T A' x}{x^T x}, \tag{7}$$

where L ranges over all subspaces of R^{2n} of dimension $2n - i + 1$.

The matrices A, A' partition naturally into 4 blocks arising from the partition of V into W_1, W_2. The off-diagonal elements of $A_{i,i}$, $i = 1, 2$ (corresponding to edges of G edges within the same W_i) are 0/1 where the 1's occur independently with probability

$$p_1 = 1 - (1 - p)^n$$
$$= \frac{d}{n} + O\left(\frac{d^2}{n^2}\right)$$

The off-diagonal elements of $A_{1,2}$ (corresponding to edges of G joining W_1 and W_2) do not occur independently, but this can be sidestepped. Consider $A_{1,2}$. For each of the N triples contained in V there are 2 edges of G which have one end in W_1 and one end in W_2. Randomly color one edge red and the other blue. Let $A_{1,2} = A_{1,2,R} + A_{1,2,B}$ where, for example, $A_{1,2,R}$ is the adjacency matrix of the bipartite grpah $G_R = (W_1, W_2, E_R)$ defined by the red edges. We claim that the edges of G_R occur independently with probability

$$p_2 = (1 - (1-p)^{2(n-1)})/2$$
$$= \frac{d}{n} + O\left(\frac{d^2}{n^2}\right)$$

Independence comes from the fact that the occurrence or non-occurrence of distinct red edges depends on the occurrence or non-occurrence of disjoint sets of triples.

We are now in good shape to appply the ideas of [2].

(i) Observe first that simple calculations show that **whp**

$$|E| = (1 - o(1))3d/2 \tag{8}$$
$$|E \setminus E'| \leq ne^{-\Omega(d)}. \tag{9}$$

Now apply (7) with $L = R^{2n}$ and $x = g = (1, 1, 1, \ldots, 1)$.

(ii) Apply (7) with $L = F = \{\lambda f : \lambda \in R\}$.

(iii) Let S be the set of all unit vectors $x \in R^{2n}$ such that $\sum_{x \in W_j} x_v = 0$, $j = 1, 2$. We first need to show that **whp** and uniformly over $x \in S$,

$$|x^T A' x| = O(\sqrt{d}). \tag{10}$$

It is enough to separately bound the contributions of each of the $A_{i,j}$ to $x^T A' x$ by $O(\sqrt{d})$. Further split the contribution of $A_{1,2}$ into that from $A_{1,2,R}$ and $A_{1,2,B}$. Similarly for $A_{2,1}$. We thus have to bound the contribution of 6 $n \times n$ 0/1 matrices with off-diagonal entries occurring independently with probability $(d + o(1))/n$. This is precisely what is done in [2] Lemma 2.4 and the preceding discussion. Thus, we can consider (10) to be proved.

The next two equations are proved in a similar manner to Lemma 2.8 of [2]. It helps to use the above decomposition into 6 matrices, to avoid problems of independence.

$$|(A' + dI)f| = O(|f|\sqrt{d}) \qquad \textbf{whp} \qquad (11)$$
$$|(A' - 3dI)g| = O(|g|\sqrt{d}) \qquad \textbf{whp}$$

The proof of (iii) can now be completed. To show $\lambda_2 = O(\sqrt{d})$ we take L in (7) to be the set $\{x \in R^n : x^T g = 0\}$. Write $x \in L$ as $\alpha f + s$ where $s \in S$. Then,

$$x^T A' x = \alpha^2 f^T A' f + 2\alpha s^T A' f + s^T A' s$$

$$= \alpha^2 f^T A' f + 2\alpha s^T (A' + dI) f + s^T A' s$$
$$\leq -\alpha^2 (1 - e^{-\Omega(d)}) d |f|^2 + 2\alpha |f| |s| O(\sqrt{d}) + |s|^2 O(\sqrt{d})$$
$$\leq |x|^2 O(\sqrt{d}).$$

To show $|\lambda_{2n-1}| = O(\sqrt{d})$ let L be any 2-dimensional subspace of R^{2n}. L contains x such that $x^T f = 0$. Write $x = \alpha g + s$ where $s \in S$. Then

$$x^T A' x = \alpha^2 g^T A' g + 2\alpha s^T (A' - 3d) g + s^T A' s$$
$$\geq \alpha^2 (1 - e^{-\Omega(d)}) d |g|^2 - 2\alpha |g| |s| O(\sqrt{d}) - |s|^2 O(\sqrt{d})$$
$$\geq -|x|^2 O(\sqrt{d}).$$

$\square$

Let $v_1, v_2, \ldots, v_{2n}$ be an orthonormal set of eigenvectors. If $f = \sum_{i=1}^{2n} c_i v_i$ then **whp**

$$\left| \sum_{i=1}^{2n-1} c_i v_i \right| = O(n/d). \tag{12}$$

To prove (12) we use

$$|(A' + dI) f|^2 = \sum_{i=1}^{2n} c_i^2 (\lambda_i + d)^2$$

$$= \Omega(d^2) \sum_{i=1}^{2n} c_i^2,$$

from Lemma 8. Applying (11) we get $\sum_{i=1}^{2n-1} c_i^2 = O(n/d)$ which is (12). Let $v_{2n} = (\xi_1, \xi_2, \ldots, \xi_{2n})$ and $U_1 = \{i : \xi_i \geq 0\}$ and $U_2 = \{i : \xi_i < 0\}$. It follows from (12) that we can assume without loss of generality that

$$|W_j \setminus U_j| = O(n/d) \text{ for } j = 1, 2. \tag{13}$$

5.3 Perfecting the coloring

We can therefore assume that the choice of U_1, U_2 as a coloring leaves all but at most $n/1000$ vertices properly colored. We proceed as in [2] to perfect the coloring.

We first list some properties that that G will have **whp**: $\gamma > 0$ is some small absolute constant.

P1 All but $n(1 - e^{-\gamma d})$ members of W_i have between $.99d$ and $1.01d$ neighbours in W_{3-i} and between $.49d$ and $.51d$ neighbours in W_i for $i = 1, 2$.

P2 For all $A, B \subseteq V$ with $ne^{-\gamma d} = |A| \geq |B|/2$ the number of edges joining A and B is at most $.001d|A|$.

P3 There are no two subsets $U, W \subseteq V$ such that $|U| \leq 0.001n$ and $|W| = |U|/2$, and every vertex of W has at least $d/5$ neighbors in U.

The proofs of these assertions are straightforward and are omitted.

An Iterative Procedure for $i = 0, 1, \ldots, \lceil \log n \rceil$ **do**

begin

    Simultaneously, for all $v \in V$, re-color v with the minority color of its

    neighbours in the previous round.

end

Analysis

Let H_i be the set of vertices with at most $1.01d$ G-neighbors in W_{3-i} and $0.51d$ G-neighbours in W_i. Let $H = H_0 = H_1 \cup H_2$. Then, while possible, delete from H_i, $i = 1, 2$ a vertex h with at most $0.99d$ H-neighbors in W_{3-i} and $0.51d$ H-neighbours in W_i.

Lemma 9. *H has at least $2n(1 - e^{-\Omega(d)})$ vertices* **whp.**

Proof   Property **P1** shows that H_0 is large. Let the vertices removed from H be $h_1, h_2, \ldots, h_m$. Let $m_0 = ne^{-\gamma d}$. If $m \geq m_0$ then there are at least $.002m_0d$ edges joining $A = \{h_1, h_2, \ldots, h_{m_0}\}$ and $A \cup (V \setminus H_0)$, contradicting **P2**. $\square$

Lemma 10. *At the end of the iterative procedure H is properly colored,* **whp.**

Proof   Let U_i be the set of wrongly colored vertices in H at the start of iteration i. If $v \in U_i$ then by the minority recoloring rule, at least $d/5$ of its neighbors are in U_{i-1}. Since **whp** $|U_1| = e^{-\Omega(d)}n$ we can apply **P3** repeatedly to show that $|U_i| \leq |U_{i-1}|/2$. $\square$

Brute force re-coloring In this phase, we simply uncolor any vertex in the set $V \setminus H$ to ensure that all the colored vertices are correctly colored. We then use exhaustive search to re-color each component of the graph Γ induced by $V \setminus H$. This generally takes polynomial time as

Proposition 11. **Whp** *the largest connected component of Γ has at most $\lceil \log_2 n \rceil$ vertices.*

Proof   We sketch the proof which is similar to Proposition 3.9 of [2].

Let T be a fixed tree on $\log_2 n$ vertices of V. Let $E(T)$ and $V(T)$ denote the the edge set and vertex set of T. Let I be the subset of $V(T)$ all of whose vertices having degree at most 4 in T. So $|I| \geq |V(T)|/2$. Build H' in the following way:

i. Let H' be the set of vertices with at most $1.01d - 4$ G-neighbors in W_{3-i} and $0.51d$ G-neighbors in W_i, $i = 1, 2$.

ii. Delete from H' all vertices of $V(T) - I$.

iii. Repeatedly delete from H' all vertices having at most $0.99d$ H'-neighbors in W_{3-i} and $0.51d$ H'-neighbors in W_i.

The following two claims from [2] are also true in our case. For a set of triples F we let F_e denote the set of edges they induce in the graph G.

CLAIM Let F be any subset of $\mathcal{E}$. Let $H(F \cup T)$ be the value of H in case $E = F_e \cup T$, and $H'(F)$ be the value of H' in the case $E = F_e$. Then $H'(F) \subseteq H(F \cup T)$.

Proof   We first show that the initial value of $H'(F)$, obtained after step (i) and (ii), is a subset of $H(F \cup T)$. Let v be any vertex that does not belong to the initial value of $H(F \cup T)$. Then v has more than $1.01d$ neighbors in the opposite color class of $(V, F \cup T)$ or more than $.51d$ neighbors into own color class. Therefore if:

Case 1. $v \in V(T) - I$. Then $v \notin H'(F)$ as it will be deleted at step (ii).

Case 2. $v \notin V(T) - I$. Then v is incident with at most 4 edges of T , thus it either has at least $1.01d - 4$ neightbors in the opposite color class in (V, F) or has at least $.51d$ neighbors in its own color class.

In either case, v does not belong to the initial value of $H'(F)$. By a similar argument to that for Lemma 3.8 from [2](notice that the assumption of a tripartition is not significant here) any vertex which is deleted in the process of constructing H will be deleted in the process of constructing H' as well and this completes the proof.

□

CLAIM

Pr[T is a subgraph of $G \wedge V(T) \cap H = \emptyset$] $\leq$ **Pr**[T is a subgraph of G]**Pr**[$I \cap H' = \emptyset$].

Proof   It is sufficient to show that

$$\mathbf{Pr}[I \cap H = \emptyset \mid T \text{ is a subgraph of } G] \leq \mathbf{Pr}[I \cap H' = \emptyset].$$

Let $\hat{T}$ be the set of triples that contain an edge of T. For a set of triples Z let $Z' = Z \setminus \hat{T}$ and $Z'' = Z \setminus Z'$. By the previous claim, we have

$$\mathbf{Pr}[I \cap H' = \emptyset] = \sum_{F: I \cap H'(F) = \emptyset} \mathbf{Pr}[\mathcal{E} = F]$$

$$\geq \sum_{F: I \cap H(F \cup T) = \emptyset} \mathbf{Pr}[\mathcal{E} = F]$$

$$= \sum_{F: I \cap H(F \cup T) = \emptyset} \mathbf{Pr}[F' = \mathcal{E}']\mathbf{Pr}[F'' = \mathcal{E}'']$$

$$= \sum_{F': I \cap H(F' \cup T) = \emptyset, F' \cap \hat{T} = \emptyset} \mathbf{Pr}[\mathcal{E}' = F']$$

$$= \sum_{F':I\cap H(F'\cup T)=\emptyset, F'\cap \hat{T}=\emptyset} \mathbf{Pr}[\mathcal{E}' = F' \mid \text{T is a subgraph of G}]$$

$$= \mathbf{Pr}[I \cap H = \emptyset \mid \text{T is a subgraph of G}]$$

where F ranges over all the subset of triples , and F' ranges over those that do not contain an edge in T while F''' ranges over those that contain at least an edge in T. $\square$

Assume that $d \leq \alpha \log\log n$ for some fixed constant α, as otherwise almost surely $H = V$. One can show that **whp** H' misses at most $2^{-\Omega(d)}n$ vertices in each color classes. Let Φ be the events that there are at most $\log n$ pair of vertices u, v such that the number of triples containing u, v is at least 2. CLAIM

$$\mathbf{Pr}[\overline{\Phi}] = O(\frac{\log\log n}{\log n}) = o(1).$$

The proof is a simple first moment calculation. We omit it here. Now we can delete all pairs of vertices that have at least 2 triples containing them and their neighbors, since there are only $O(\log n)$ of them, a simple brute force coloring will find the correct one.

The intuition now is that since the conditioning between edges is small, we can assume that a large portion (nearly half) of the edges in T can be treated as unconditioned. Given edge $e = \{u, v\}$, let

$$N(e) = \{\{w, u\}, \{w, v\} : \{w, u, v\} \in \mathcal{E}\}.$$

By the previous claim,**whp** after deleting $O(\log n)$ vertices, $\forall e \in E(T)$, $|N(e) \cap E(T)| \leq 1$.

Since the choice of H' is independent of I the probability that there exists some T of size at least $\log_2 n$ is

$$\mathbf{Pr}[\text{T is a subgraph of G}|\Phi]$$

$$\leq \frac{\mathbf{Pr}[e_1 \in E(G)]}{\mathbf{Pr}[\Phi]}\mathbf{Pr}[E(T)\backslash N(e_1) \subseteq E(G)|\Phi]$$

$$\leq C(d/n^2)^{(|V(T)|-1)/2} \qquad \text{using an inductive argument}$$

$$\leq C'(d/n^2)^{\log_2 n/2}$$

for some large constants C, C' and

$$\mathbf{Pr}[I \cap H' = \emptyset] \leq \frac{\binom{n-|H'|}{|I|}}{\binom{n}{|I|}}$$

$$\leq 2^{-\Omega(d|I|)}$$

$$\leq 2^{-\Omega(d\log_2 n)}$$

Since the total number of possible connected trees of this size or more is at most

,

$$\binom{2n}{\log_2 n}(\log_2 n)^{\log_2 n - 2}.$$

The multiplication of the above terms is of $O(n^{-\Omega(d)})$, which is the probability that the algorithm will fail in the third phase. We have thus completed the proof of Proposition 11. $\qquad\square$

Acknowledgement We thank Avrim Blum for reminding us of the problem of 2-coloring hypergraphs.

References

1. F.Alizadeh, *Combinatorial Optimizations with Semi-definite matrices*, 2nd Conference of Integer Programming and Combinatorial Optimizations (1992). pp385-405.
2. N.Alon, N.Kahale , *A spectral technique for coloring random 3-colorable graphs*, DIMACS TR-94-35.
3. N.Alon, J.Spencer, *The Probabilistic Method*, John Wiley & Sons (1992).
4. J.Beck, *An algorithmic approach to Lovász Local Lemma I*, Random Structures & Algorithms (1991). pp343-365.
5. K.Edwards, *The complexity of coloring problems on dense graphs*, Theoretical Computer Science 43 (1986) 337-343.
6. M.Goemans, D.Williamson, *.878-Approximation Algorithms for MAX CUT and MAX 2SAT*. Proceedings of the 26th ACM Symposium on Theory of Computing (1994).
7. D.Karger, R.Motwani, M.Sudan, *Approximation Graph Coloring by Semidefinite Programming*. 35th Foundations of Computer Science (1995). pp2-13.
8. J.Kahn, Szemerédi,J.Friedman, *On the second eigenvalue in random regular graphs*, Proceedings of the 21st ACM STOC (1989). pp587-598.
9. L.Lovász, *Covering and coloring of hypergraphs*, Proceding of the 4th Sourtheastern Conference on Combinatorics, Graph Theory and Computing. Utilitas Mathematica Publishing. Winnipeg (1973). pp3-12.
10. C.J.H McDiarmid, *A random recoloring method for graph and hypergraph*, Combinatorial Probability and Computing 2 (1993). pp363-365.
11. G.Strang, *Linear algebra and its applications*, Hardcourt Brace Jovanovich Publishing (1988) .
12. A.Widgerson, *Improving the performance gurantee of approximate graph coloring*, Journal of ACM 30 (1983). pp729-735.

Improved Randomized Approximation Algorithms for Lot-Sizing Problems

Chung-Piaw Teo[1]   Dimitris Bertsimas[1]

Sloan School of Management and Operations Research Center, MIT

1   Introduction

We consider in this paper multi-product, lot-sizing problems that arise in manufacturing and inventory systems. We describe the problem in a manufactruring setting. There is a set N of products. For each product $j \in N$ there is a set π_j (called predecessors of product j) of products consumed in producing product j. We define the product network G to be a directed network with node set N and arc set $A = \{(i, j) : i \in \pi_j\}$. In other words, the network G corresponds to the flow of materials in the system and contains no circuit.

External demand d_i for product i is assumed to be constant in time. Clearly in order to satisfy the demand orders should be placed for the products dynamically in time. If an order is placed for product i, an ordering cost K_i is incurred. Moreover, an incremental echelon holding cost h_i is incurred per unit time the item spends in inventory. The production rate is assumed to be infinite. The objective is to schedule orders for each of the products over an infinite horizon so as to minimize long-run average cost.

As the optimal dynamic policy can be very complicated, the research community (see for instance Roundy [18, 19], Jackson, Maxwell and Muckstadt [10], Muckstadt and Roundy [14]) has focused on stationary and nested policies defined as follows: Orders are placed periodically in time at equal intervals, for each of the products in the system (stationary policies). If product j precedes product i, then an order is placed for product j only when an order is placed for product i at the same time (nested policies). Therefore, under a stationary and nested policy the objective is to decide the period T_i that an order is placed. The reason stationary and nested policies are attractive is that they are easy to implement. Muckstadt and Roundy [14] discuss in detail the rationale of using order intervals T_i as variables.

The problem of designing an optimal stationary and nested policy can then be formulated (see [18]) as the following nonlinear integer programming problem.

$$(P_{NS}) \quad = \min \ G(T) \equiv \sum_{i \in N} \left(\frac{K_i}{T_i} + H_i T_i \right)$$

$$\frac{T_i}{T_j} \in \{1, 2, 3 \ldots\} \text{ if } (i, j) \in A,$$

$$T_i = k_i T_L \text{ for each } i, \ k_i \in \mathcal{Z}_+$$

Period T_L is called the base period and it can be constant or allowed to vary, depending on the model. The coefficient H_i is given by $H_i = (h_i - \sum_{j \in \pi_i} h_j) D_i$ and D_i represents the aggregate demand, which is calculated recursively starting from products with $s_i = \emptyset$ by $D_i = d_i + \sum_{k \in s_i} D_k$ (s_i is the set of successors of product i).

We consider the following convex relaxation of the problem:

$$(P_R) \quad Z_R = \min \sum_{j \in N} \left(\frac{K_i}{T_i} + H_i T_i \right)$$
$$T_i \geq T_j \text{ if } (i,j) \in A,$$
$$T_i \geq T_L \text{ for each } i.$$

Notice that the constraints $T_i \geq T_j$ model the condition that policies are nested.

As the objective function is convex, the relaxation (P_R) can be solved in polynomial time using interior point algorithms or the algorithm by Hochbaum and Shanktikumar [9]. For systems with special structure the runnning time can be improved substantially. For instance, if G is a tree, Jackson and Roundy [11] show that the relaxed problem can be solved in $O(n \log n)$ time, where $n = |N|$. When G corresponds to a star graph, Queyranne [15], and also Lu Lu and Posner [12] showed that the relaxed problem can be solved in $O(n)$ time, using a linear time median finding algorithm.

Regarding approximation algorithms, Roundy ([18, 19]), and Maxwell and Muckstadt [13] showed how to round an optimal solution of the relaxed problem (P_R) to a feasible solution for (P_{NS}). The policies constructed are called power-of-two policies, where each T_i is of the form $2^{p_i} T_L$, where p_i is integer. Let Z_H be the value of the heuristic used. They obtained the following bounds:

1. If T_L is not fixed, but subject to optimization, then

$$\frac{Z_H}{Z_R} \leq \frac{1}{\sqrt{2} \log 2} \approx 1.02.$$

2. If T_L is fixed, then

$$\frac{Z_H}{Z_R} \leq \left(\sqrt{2} + \frac{1}{\sqrt{2}} \right) \approx 1.06.$$

The technique used is deterministic rounding and convex duality. The technique utilizes the properties of the optimal relaxed solution. In both cases the bounds are tight. These results are often referred in the literature as 98% and 94% effective lot-sizing policies respectively.

These results have been extensively studied and extended to other versions of lot-sizing problems: finite production rates (Atkins, Queyranne and Sun [1]), individual capacity bounds of the form $2^{l_i} T_L \leq T_i \leq 2^{u_i} T_L$ and more general cost structures (Zheng [21]), and backlog (Atkins et al. [1]). All these extensions use determinisitc rounding to generate power-of-two policies with the same 94% and 98% bounds.

In this paper, we propose a new approach to these lot-sizing problems that uses randomized rounding. This design technique has been used extensively by the discrete optimization community. It was first introduced by Raghavan and Thompson [16], and was used subsequently for a variety of other combinatorial problems. See for instance Goemans and Williamson [7, 8], Bertsimas and Vohra [3], Bertsimas et al. [2]. Our contributions in this paper are as follows:

1. We propose new 94% and 98% randomized rounding algorithms for Problem (P_{NS}) under both the fixed and the variable base period models. Our proof is simple and unlike the original deterministic rounding does not depend on the structure of the optimal solution. Roundy's 98% algorithm can be obtained by derandomizing our algorithm. However, derandomizing the 94% algorithm leads to a different deterministic algorithm. The randomized rounding method is interesting in its own right as it introduces dependencies in the rounding process and generates random variables with distributions with nonlinear density functions.

2. We study a generalization of the fixed based model by allowing the base period T_L to vary over a finite set of choices $\{2^{k/p} T_L : k \text{ integer}\}$, with p, T_L fixed. We propose a randomized rounding algorithm that produces a power-of two policy with bound $\dfrac{2^{\frac{1}{p}} + 1}{2\sqrt{2} p \left(2^{\frac{1}{p}} - 1 \right)}$, where p denote the number of points T_L is allowed to vary. For $p = 1$ and $p = \infty$, the bound reduces to 1.06 and 1.02 respectively. For the one warehouse, multi-retailer problem (OWMR), Lu Lu and Posner [12] have also obtained a similar bound for a class of integer-ratio policies.

3. For a general production distribution network under nested policies, we propose new convex relaxations and randomized rounding algorithms that use $T_i = 2^{p_i} T_L$ or $3 \cdot 2^{p_i} T_L$. This improves the bound for the fixed base period case from 1.06 to 1.043 and for the special case of Problem (OWMR) to 1.031.

4. Our techniques generalize to several other extensions considered in the literature (capacitated versions, submodular cost functions and multiple resource constrained problems)

2 Randomized rounding and lot-sizing problems

In this section, we introduce the key randomized rounding ideas used in this paper.

2.1 A new 94% approximation algorithm

In this section we consider the case of fixed base period T_L. We consider the following rounding scheme:

> **Algorithm A:**
> Let $T = (T_1, \ldots, T_n)$ be a feasible solution to relaxation (P_R), and $T_i = 2^{p_i} z_i T_L$, where $1 \leq z_i \leq 2$. Generate a point Y in the interval $[1, 2]$, with probability distribution $F(y) = \frac{y^2 - 1}{1 + y^2/2}$. If $z_i < Y$, then $T_i^o = 2^{p_i} T_L$, else $T_i^o = 2^{p_i + 1} T_L$.

The above rounding scheme always generates a feasible solution $(T_1^o, T_2^o, \ldots, T_n^o)$ to problem (P_{NS}). We only need to check that the precedence constraints $T_i \leq T_j$ are preserved. If $p_j > p_i$, then $T_j^o \geq T_i^o$. If $p_j = p_i$, then since $T_j \geq T_i$, we must have $z_j \geq z_i$. Hence $z_i \geq y$ only if $z_j \geq y$.

Theorem 1. *Given any feasible solution* $(T_1, \ldots, T_n)$ *to Problem* (P_R) *with cost* $G(T)$, *Algorithm A returns a power-of-two policy (with fixed base* T_L) *with an expected cost of not more than* $1.06\, G(T)$.

Proof: It is easy to see that

$$E(T_i^o) = 2^{p_i} T_L (1 - F(z_i)) + 2^{p_i+1} T_L F(z_i) = T_i \frac{1 + F(z_i)}{z_i}$$

$$= T_i \frac{3z_i}{z_i^2 + 2} \leq \frac{\sqrt{2} + 1/\sqrt{2}}{2} \, T_i \approx 1.06 \, T_i.$$

Similarly,

$$E\left(\frac{1}{T_i^o}\right) = \frac{1}{2^{p_i} T_L}(1 - F(z_i)) + \frac{1}{2^{p_i+1} T_L} F(z_i) = \frac{1}{T_i}\left(1 - \frac{F(z_i)}{2}\right) z_i$$

$$= \frac{1}{T_i} \frac{3z_i}{z_i^2 + 2} \leq \frac{\sqrt{2} + 1/\sqrt{2}}{2} \frac{1}{T_i}.$$

The bound follows since the maximum value of the function $\frac{3z_i}{z_i^2 + 2}$ is at most $3\sqrt{2}/4$. □

Note that the distribution function $F(y)$ is chosen so that $(1 + F(y))/y = y(1 - F(y)/2) = 3y/(y^2 + 2)$. The maximum is attained at the point $y = \sqrt{2}$ with a value of $\frac{3\sqrt{2}}{4} \approx 1.06$. Furthermore, using the optimal solution to (P_R) as input to the rounding process, we obtain a 94% approximation algorithm to the original lot-sizing problem.

De-randomization. The above randomized algorithm can be made deterministic: Sort the z_i's in non-decreasing order, say $z_1 \leq z_2 \leq \ldots \leq z_n$. For all y in $[z_i, z_{i+1})$, the randomized algorithm returns the same solution. Hence, there are at most $n + 1$ distinct solutions obtained. Thus the best solution can be obtained in $O(n \log n)$ time, which is the time needed for the sorting operation.

2.2 The 98% approximation algorithm revisited

The same insensitivity result can also be improved to a 98% guarantee, if one allows the base period T_L to vary, i.e., T_L is a variable in (P_R). In fact, Roundy's 98% algorithm [18, 19] already has this feature. We recast Roundy's algorithm into the following randomized rounding algorithm:

Algorithm B:
Let $T = (T_1, \ldots, T_n, T_L)$ be a feasible solution to (P_R), with $T_L > 0$. Let $T_i = 2^{p_i} T_L z_i$, where $1 \leq z_i \leq 2$. Generate a point Y in the interval $[1, 2]$, with probability distribution $F(y) = \frac{\log y}{\log 2}$. If $Y > z_i$, then $T_i^o = 2^{p_i} \frac{Y}{\sqrt{2}}$ else $T_i^o = 2^{p_i+1} \frac{Y}{\sqrt{2}}$. Let $T_L^o = \frac{Y}{\sqrt{2T_L}}$.

The rounded solution T_i^o is chosen to ensure that it lies in the interval $[\frac{T_i}{\sqrt{2}}, \sqrt{2}T_i]$. Furthermore, it is clear that $(T_1^o, T_2^o, \ldots, T_n^o, T_L^o)$ is a feasible solution to (P_{NS}).

Theorem 2. *Given any feasible solution $(T_1, \ldots, T_n, T_L)$ to Problem (P_R) with cost $G(T)$, Algorithm B returns a power-of-two policy $(T_1^o, T_2^o, \ldots, T_n^o, T_L^o)$ with expected cost at most $\frac{G(T)}{\sqrt{2}\log(2)} \approx 1.02\, G(T)$.*

Proof: Without loss of generality, we may assume $T_L = 1$. Then

$$E(T_i^o) = \frac{\int_1^{z_i} 2^{p_i+1} dy + \int_{z_i}^2 2^{p_i} dy}{\sqrt{2}\log 2}$$

$$= \frac{2^{p_i}[2(z_i - 1) + (2 - z_i)]}{\sqrt{2}\log 2} = \frac{T_i}{\log 2\sqrt{2}}.$$

Similarly,

$$E(1/T_i^o) = \frac{\sqrt{2}\int_1^{z_i} 2^{-p_i-1}(1/y^2)dy + \sqrt{2}\int_{z_i}^2 2^{-p_i}(1/y^2)dy}{\log 2}$$

$$= \frac{\sqrt{2}2^{-p_i}(1/2 - \frac{1}{2z_i} - 1/2 + \frac{1}{z_i})}{\log 2} = \frac{1}{T_i \log 2\sqrt{2}},$$

and the theorem follows. □

De-randomization. Suppose $z_1 \le z_2 \le \ldots \le z_n$. For y in $[z_i, z_{i+1})$, suppose the algorithm returns a policy with cost $A/y + By$, then for all other y' in the same interval, the algorithm returns a policy with cost $A/y' + By'$. By choosing a y' in the interval that maximizes this term, and doing the same for each interval partitioned by the z_i's, we obtained an $O(n \log n)$ deterministic algorithm, which is exactly Roundy's rounding procedure.

The argument used above can easily be adapted to analyze more general costs in the objective function. For instance, we have the following:

Theorem 3. *Under Algorithm B,*

$$E\{(T_i^o)^2\}/T_i^2 = T_i^2 E\{(\frac{1}{T_i^o})^2\} = \frac{3}{4\log(2)} \approx 1.082 \; ;$$

$$\frac{1}{\sqrt{2}\log(2)} \le E(\frac{1}{T_i^o T_j^o})T_i T_j \le \frac{3}{4\log(2)} \; ;$$

$$\frac{1}{\sqrt{2}\log(2)} \le E(T_i^o T_j^o)\frac{1}{T_i T_j} \le \frac{3}{4\log(2)} \; ;$$

$$E(\frac{T_i^o}{T_j^o}) \le 1.06\frac{T_i}{T_j}.$$

The above inequalities imply new bounds (91.8%) if there are $T_i^2, \frac{1}{T_i^2}, T_i T_j$ or $\frac{T_i}{T_j}$ terms in the objective function.

2.3 Unification of the 94% and 98% bounds

The 94% and 98% performance bounds assume that the base period is fixed and optimally selected respectively. The 94% bound is attained by a power-of-two policy, where every order interval is a fixed multiple of a preselected base period. The 98% approximation algorithm, however, cannot ensure that the base planning period belongs to a preselected set. In this section, we propose a technique to bridge the gap between the performance of these two algorithms, by giving progressively more flexibility to the choice of base periods. We assume that the allowed base periods are in the set $S = \{2^{\frac{j}{p}} : j \text{ integer}\}$.

Consider the following randomized rounding algorithm.

Algorithm C:

Let $T_i = 2^{p_i} z_i$, where $1 \leq z_i < 2$. Let Y be a random number generated in the interval $[2^{-\frac{1}{2p}}, 2^{\frac{1}{2p}})$, with distribution function $F(y) = \frac{2^{1/p} y^2 - 1}{(2^{1/p} - 1)(1 + y^2)}$. Construct a power-of-two policy as follows:

Select the base period $T_L = 2^{j/p}$ with probability $\frac{1}{p}$.

$$T_i^o = \begin{cases} 2^{p_i + 1} 2^{\frac{i}{p}} & \text{if } z_i > \sqrt{2}\, 2^{\frac{i}{p}} Y \\ 2^{p_i - 1} 2^{\frac{i}{p}} & \text{if } z_i < \frac{2^{\frac{i}{p}} Y}{\sqrt{2}} \\ 2^{p_i} 2^{\frac{i}{p}} & \text{otherwise.} \end{cases}$$

Theorem 4. *Given any feasible lot-sizing policy* $(T_1, \ldots, T_n)$ *in* (P_R) *with cost* $G(T)$, *Algorithm C returns a power-of-two policy* T^o *with expected cost at most*

$$\left(\frac{2^{\frac{1}{p}} + 1}{2\sqrt{2} p (2^{\frac{1}{p}} - 1)} \right) G(T).$$

Note that for $p = 1$ and $p = \infty$, we obtain the 94% and 98% bounds respectively. For $p = 2$, the bound already improves to 97%. This observation implies that for the fixed base period model, the 94% bound might be improved considerably by considering only two distinct base periods, both integral multiples of T_L. In the next section, we use this observation to derive an improved approximation algorithm.

3 An improved approximation algorithm for the fixed base period model

In this section, we propose an improved approximation algorithm for the general problem (P_{NS}) under the fixed base period model. The improvement over the 94% bound comes from having a tighter representation of the objective function over the discrete points $\{T_L, 2T_L, 3T_L\}$ in the interval $[T_L, 3T_L]$. We consider an improved relaxation of the original problem:

$$(P_R') \quad \min \sum_{i=1}^{n} f_i'(T_i)$$

subject to $T_i \geq T_j$ if $(i, j) \in A$, $T_i \geq 0$,

where $f'(\cdot)$, which is depicted in Figure 1, represents the piecewise linearization of $f_i(T_i) = K_i/T_i + H_i T_i$ over the points $\{T_L, 2T_L, 3T_L\}$, i.e.,

$$f_i'(T_i) = \begin{cases} (T_i - T_L) f_i(2T_L) + (2T_L - T_i) f_i(T_L) & \text{if } T_L \leq T_i \leq 2T_L, \\ (T_i - 2T_L) f_i(3T_L) + (3T_L - T_i) f_i(2T_L) & \text{if } 2T_L \leq T_i \leq 3T_L. \end{cases}$$

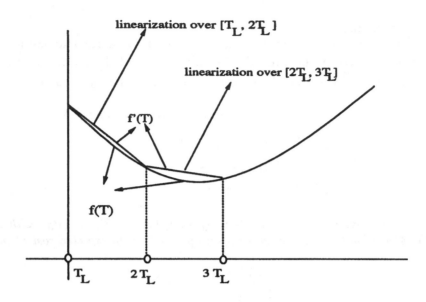

Fig. 1. Piecewise linearization of the objective function over the points $T_L, 2T_L, 3T_L$.

We introduce the following notation. Let p, q be nonnegative numbers, such that $p + q = 1$. Let

$$a(p) = 2\sqrt{\frac{p + 3q/2}{p + 2q/3}}, \quad b(p) = 2\sqrt{\frac{2p + 3q/2}{p/2 + 2q/3}},$$

and

$$F(p, z) = \frac{\frac{1}{4}(p + \frac{2}{3}q)z^2 - \frac{3}{2}q - p}{p(1 + \frac{1}{8}z^2)},$$

$$F'(p, z) = \frac{\frac{1}{9}(q + \frac{3}{4}p)z^2 - \frac{4}{3}p - q}{q(1 + \frac{1}{18}z^2)}.$$

Note that $F(p, a(p)) = F'(p, b(p)) = 0$, and $F(p, b(p)) = F'(p, 2a(p)) = 1$. Furthermore, F and F' are nondecreasing in z and are valid distribution functions. Suppose further that T^* is an optimal solution to (P_R'). Note that for $T_i^* \leq 3T_L$, we may assume that $T_i^* \in \{T_L, 2T_L, 3T_L\}$. This follows from the following lemma:

Lemma 5. *There exists an optimal solution T^* with the property that $T_i^* \in \{T_L, 2T_L, 3T_L\}$ if $T_i^* \leq 3T_L$.*

Consider the following rounding algorithm:

Algorithm D: Let $p = 0.7, q = 0.3$. Let $a = a(0.7)$ and $b = b(0.3)$. Note that $a < b < 2a$. Select Policy 1 below with probability p, and Policy 2 with probability q.

Policy 1: Let $T_i^* = 2^{p_i} T_L z_i$, where z_i is in the interval $[1, 2)$. Let Y be a random number generated in the interval $[a(p), b(p)]$ with distribution function $F(p, y)$. Let

$$T_i^1 = \begin{cases} 2^{p_i} T_L & \text{if } 2z_i < Y, \\ 2^{p_i+1} T_L & \text{if } 2z_i \geq Y. \end{cases}$$

Policy 2: Let $T_i^* = 3 \cdot 2^{p_i} T_L z_i'$, where z_i' is in the interval $[1, 2)$. Let Y' be a random number generated in the interval $[b(p), 2a(p)]$ with distribution function $F'(p, y)$. For $T_i \geq 3T_L$, let

$$T_i^2 = \begin{cases} 3 \cdot 2^{p_i} T_L & \text{if } 3z_i' < Y', \\ 3 \cdot 2^{p_i+1} T_L & \text{if } 3z_i' \geq Y'. \end{cases}$$

For all items i with $T_i = 2T_L$, we round them (simultaneously) to $3T_L$ with probability $\frac{9}{14}$ and to T_L with probability $\frac{5}{14}$. Note that in this way, for $T_i = 2T_L$,

$$\frac{E(T_i^2)}{2T_L} = \frac{8}{7} = 2T_L E(\frac{1}{T_i^2}).$$

Finally, if $T_i^* = T_L$, $T_i^2 = T_L$.

Let T denote the vector of ordering intervals under the selected policy.

Theorem 6. *The expected cost of the policy T produced by algorithm D is at most 1.043 times the value of the continuous relaxation (P_R').*

Proof: Without loss of generality, we assume $T_L = 1$. If $T_i^* = 2$, then

$$\frac{E(T_i)}{2T_L} = 2T_L E(\frac{1}{T_i}) = p + \frac{8}{7}q = 1.0428...$$

Thus we only need to consider the case when T_i^* greater than 3. Suppose T_i^* lies in (1) $[2^{k_i}a, 2^{k_i}b]$ or (2) $(2^{k_i}b, 2^{k_i+1}a]$. In case (1), Policy 2 always rounds T_i to $3 \cdot 2^{k_i}$, whereas in case (2), Policy 1 always rounds T_i to 2^{k_i+2}.
Case (1) : T_i^* lies in $[2^{p_i}a, 2^{p_i}b]$, i.e., $T_i^* = 2^{p_i}w_i$, where $w_i \in [a, b]$. Then

$$E(T_i) = pE(T_i^1) + qE(T_i^2) = T_i^* (\frac{3q}{w_i} + p\frac{2(1 + F(p, w_i))}{w_i}),$$

and

$$E(\frac{1}{T_i}) = pE(\frac{1}{T_i^1}) + qE(\frac{1}{T_i^2}) = \frac{1}{T_i^*}(q\frac{w_i}{3} + p(1 - \frac{F(p, w_i)}{2})w_i/2).$$

We have chosen $F(p, \cdot)$ such that

$$q\frac{3}{w_i} + p\frac{2(1 + F(p, w_i))}{w_i} = q\frac{w_i}{3} + p(1 - \frac{F(p, w_i)}{2})w_i/2.$$

With this choice of $F(p, \cdot)$, and $p = 0.7, q = 0.3$, we can optimize the bound over the range of w_i to obtain

$$\frac{E(T_i)}{T_i^*} = T_i^* E(\frac{1}{T_i}) \le 1.043.$$

Case (2) : T_i^* lies in $(2^{p_i}b, 2^{p_i+1}a]$, i.e. $T_i^* = 2^{p_i}w_i$ where $w_i \in (b, 2a]$. Then

$$E(pT_i^1 + qT_i^2) = T_i^*(p\frac{4}{3w_i} + q\frac{3(1 + F'(p, w_i))}{w_i}),$$

and

$$E(p\frac{1}{T_i^1} + q\frac{1}{T_i^2}) = \frac{1}{T_i^*}(p\frac{3w_i}{4} + q(1 - \frac{F'(p, w_i)}{2})w_i/3).$$

We have chosen $F'(p, \cdot)$ such that

$$p\frac{4}{3w_i} + q\frac{3(1 + F'(p, w_i))}{w_i} = p\frac{3w_i}{4} + q(1 - \frac{F'(p, w_i)}{2})w_i/3.$$

With this choice of $F'(p, \cdot)$, again we have

$$\frac{E(pT_i^1 + qT_i^2)}{T_i^*} = T_i^* E(p\frac{1}{T_i^1} + q\frac{1}{T_i^2}) \le 1.043.$$

Hence the result follows. □

We next show that if $T_i^* \ge \sqrt{6}T_L$ for all i we can improve the approximation guarantee. This result will be useful in the next section. We consider the following modified rounding algorithm:

Algorithm E: Let $p = 0.5, q = 0.5$. Select Policy 1 with probability p and Policy 2 otherwise:
Policy 1: The same as in Algorithm D.
Policy 2: For $T_i^* \ge 3T_L$, the same as in Algorithm D. For T_i^* in $[\sqrt{6}T_L, 3T_L]$, we round T_i to $3T_L$.

The following result follows from a similar analysis to Theorem 6.

Theorem 7. *If $T_i^* \ge \sqrt{6}T_L$ for all i, then the expected cost of the policy T obtained from Algorithm E is at most 1.031 times the optimal value of the continuous relaxation (P_R).*

4 An improved approximation algorithm for the (OWMR) problem

In this section we improve the guarantee of 1.043 to 1.031 for the problem of a single warehouse supplying and distributing items to a group of n retailers. For distribution systems Roundy [18] has showed that the optimal nested policy can be arbitrarily bad compared to the optimal stationary policy. Under the assumption that the retailers place their order only when their inventory level is zero, he showed that there is an optimal stationary policy which satisfies the integer ratio property, i.e., the ratio of the ordering interval T_i for retailer i and the ordering interval T_0 of the warehouse is either an integer or 1 over an integer. He has also constructed similar 94% and 98% approximation algorithms for problem $(OWMR)$, with fixed and variable base period respectively.

The problem can be modelled as follows (see [18]):

$$(P_{OWMR}) \quad \min \ C(T) = \sum_{i=0}^{n}(K_i/T_i) + \sum_{i=1}^{n}(g_i \max(T_0, T_i) + H_i T_i)$$

$$\text{subject to } \frac{T_i}{T_0} \in \{k_i, \frac{1}{k_i} : k_i \text{ integer}\},$$

$$\frac{T_i}{T_L} \text{ integer for all } i = 0, \ldots, n,$$

where $g_i = \frac{1}{2}h_0 d_i$, and $H_i = \frac{1}{2}(h_i - h_0)d_i$. We consider the following relaxation:

$$(P_{ROWMR}) \quad \min \ \sum_{i=0}^{n}(K_i/T_i) + \sum_{i=1}^{n}(g_i \max(T_0, T_i) + H_i T_i)$$

$$\text{subject to } T_i \geq T_L \text{ for all } i = 0, \ldots, n.$$

The constraint $T_i \geq T_L$ is a relaxation of the condition that each T_i is an integral multiple of T_L. Let T_i^*, $i = 0, 1, \ldots, n$ be a solution of the relaxation (P_{ROWMR}).

In this section, we improve on the approximation bound for the fixed base period model, by using six stronger relaxations. These relaxations correspond to the requirement that either $T_0^* \geq 6T_L$ or $T_0^* = kT_L$ for k in $\{1, 2, 3, 4, 5\}$.

We first consider the relaxation

$$(P_6) \quad Z_6 = \min\{\sum_{i=1}^{n}(f_i'(T_i) + g_i \max(T_0, T_i)) + K_0/T_0 : T_0 \geq 6T_L, T_i \geq T_L\},$$

where $f_i'(T_i) = f_i(T_i) = K_i/T_i + H_i T_i$ if $T_i \geq 3T_L$, and $f_i'(T_i)$ is the piecewise linearization of $f_i(T_i)$ over the points $\{T_L, 2T_L, 3T_L\}$. Note that this relaxation provides a lower bound to the optimal value of (P_{OWMR}). Z_6 can be computed in $O(n)$ time by using a linear time median finding algorithm, as suggested in

Queyranne [15] or Lu Lu and Posner [12]. Let T_i^* be the optimal solution of relaxation (P_6). Following Lemma 5 we may assume that $T_i^* \in \{T_L, 2T_L, 3T_L\}$ if $T_i^* \leq 3T_L$. We apply Algorithm E that leaves those T_i^* with values T_L or $2T_L$ unchanged.

Lemma 8. *Algorithm E applied to an optimal solution to relaxation (P_6) produces an integer ratio policy with cost within 1.031 of Z_6.*

Proof: Policies 1 and 2 of Algorithm E round those T_i^* with values greater than or equal to $3T_L$ to a power-of-two policy of the type $2^{p_i}T_L$ or $2^{p_i}(3T_L)$. Those T_i^* with values T_L or $2T_L$ are left unchanged. The expected gap between T_i^* and the rounded value T_i again satisfies

$$\frac{E(T_i)}{T_i^*} \leq 1.031, \quad \frac{E(1/T_i)}{1/T_i^*} \leq 1.031.$$

Note that in addition, because of the dependence in the rounding process,

$$E[\max(T_i, T_0)] = \max(E[T_i], E[T_0])] \leq 1.031 \max(T_i^*, T_0^*).$$

Note that since $T_0^* \geq 6T_L$, T_0^* is rounded to a multiple of $4T_L$ (under Policy 1) or multiples of $6T_L$ (under Policy 2). Therefore, the policy constructed need not satisfy the condition $T_0 \geq 6T_L$, since Policy 1 might round T_0^* down to $4T_L$. However, the policy obtained is an integer ratio policy. $\qquad \square$

We next consider the case that $T_0^* = kT_L$, $k \in \{1, 2, 3, 4, 5\}$. Let $f_i^k(T_i)$ denote a partial piecewise linearization of $f_i(T_i)$ in the interval $[T_L, 3kT_L]$, over the points $T_L, kT_L, 2kT_L, 3kT_L$. Particularly for $k = 4$, in addition to $T_L, 4T_L, 8T_L, 12T_L$ we include the point $2T_L$ in the linearization. For $k \in \{1, 2, 3, 4, 5\}$ we consider the following five relaxations, in which we fix the value of T_0 to be kT_L and consider the linearization $f_i^k(T_i)$ instead of $f_i(T_i)$:

$$(P_k) \quad Z_k = \min\{C^k(T) = K_0/(kT_L) + \sum_{i=1}^{n}(g_i \max(kT_L, T_i) + f_i^k(T_i)) : T_i \geq T_L\}.$$

Note that each relaxation can be solved in $O(n)$. Moreover,

Lemma 9. *There exists an optimal solution T^k to Z_k with the property that if $T_i^k \leq 3kT_L$, then*

$$T_i^k \in \{T_L, kT_L, 2kT_L, 3kT_L\} \text{ for } k = 1, 2, 3, 5$$

$$T_i^k \in \{T_L, 2T_L, 4T_L, 8T_L, 12T_L\} \text{ for } k = 4.$$

We next show that Algorithm E applied to the optimal solution of relaxation (P_k) produces an integer ratio policy within 1.031 of Z_k.

Lemma 10. *For $k = 1, \ldots, 5$ Algorithm E applied to an optimal solution of relaxation (P_k) that satisfies Lemma 9 produces an integer ratio policy with cost within 1.031 of Z_k.*

Combining Lemmas 8 and 10 we obtain

Theorem 11. *For the one-warehouse-multi-retailer problem with fixed base period, there is an $O(n)$ time 96.9% approximation algorithm.*

5 Extensions

Since our prior analysis did not utilize any structure of the optimal solution, our proof techniques cover several extensions of the basic models almost effortlessly. Our techniques produce randomized rounding algorithms for the following problems considered in the literature:

1. Capacitated lot-sizing problems, in which we add constraints $2^{l_i}T_L \le T_i \le 2^{u_i}T_L$ for each i. Since the Algorithms A and B preserve these properties, Theorems 1 and 2 apply also for this capacitated version of the problem, giving rise to 94% and 98% power-of-two policies respectively. The same result was also derived in Federgruen and Zheng [6] by extending Roundy's approach to the capacitated version.

2. Submodular ordering costs introduced in Federgruen et al. [5] and Zheng [21]:

$$(P_{SUB}) \quad Z = \min_T \max_k \sum_{j \in N} (\frac{k_i}{T_i} + H_i T_i)$$

$$T_i \le T_j \text{ if } (i, j) \in A,$$
$$T_i \ge T_L \text{ for each } i.$$
$$k \in \mathcal{P},$$

where

$$\mathcal{P} = \{k : \sum_{j \in S} k_j \le K(S), \sum_{j \in N} k_j = K(N), k_j \ge 0\},$$

and $K(S)$ submodular. Algorithms A and B can be used to round the fractional optimal solution in (P_{SUB}) to 94% and 98% optimal power-of-two solutions. Furthermore, if $T_i^* \ge \sqrt{6}T_L$ for all i, then the fixed base period bound can be improved further to 96.9%, using Theorem 7.

3. Resource constrained lot-sizing problems considered in Roundy [20], in which we add to (P_{NS}) constraints of the type

$$\sum_j a_{ij}/T_j \le A_i, \quad i = 1, \ldots, m.$$

He showed that there is a power-of-two policy for the variable base period case with cost at most 1.44 times the optimal solution. We can generalize this result to the lot-sizing problems with submodular joint cost function. Consider the following algorithm:

Algorithm F:
Let (k^*, T^*) be an optimal solution to (P_{SUB}) with the resource constraints added. Use $T_j = \sqrt{2}T_j^*$ in Algorithm B to obtain a power-of-two policy T^o.

First note that T_j^o lies in the interval $[T_j/\sqrt{2}, T_j\sqrt{2}]$ and hence $T_j^o \geq T_j^*$. Therefore, T_j^o satisfies the resource constraints.

Theorem 12. *Let T^* be an optimal solution to the resource constrained version of (P_{SUB}). Using Algorithm F on T^*, we obtain a power-of-two policy with cost at most 1.44 times of the optimal.*

Proof: Since scaling by $\sqrt{2}$ does not affect the ordering of T_j^*, the solution k^* is also a maximum solution to $G(T^o)$. Therefore, the result follows directly from the following observation:

$$E(T_j^o) \leq \frac{1}{\sqrt{2}\log(2)}T_j = \frac{1}{\sqrt{2}\log(2)}\sqrt{2}T_j^* \approx 1.44T_j^*$$

and

$$E(\frac{1}{T_j^o}) \leq \frac{1}{\sqrt{2}\log(2)T_j} \leq \frac{1}{T_j^*}.$$

$\square$

References

1. D. Atkins, M. Queyranne and D. Sun. Lot sizing policies for finite production rate assembly systems, *Operations Research*, 40, 126-141, 1992.

2. D. Bertsimas, C. Teo and R. Vohra. Nonlinear relaxations and improved randomized approximation algorithms for multicut problems, *Proc. 4th IPCO Conference*, 29-39, 1995.

3. D. Bertsimas and R. Vohra. Linear programming relaxations, approximation algorithms and randomization : a unified view of covering problems, Preprint 1994.

4. G. Dobson. The Economic Lot-Scheduling Problem: Achieving Feasibility using Time-Varying Lot Sizes, *Operations Research*, 35, 764-771, 1987.

5. A. Federgruen, M. Queyranne and Y.S. Zheng. Simple power-of-two policies are close to optimal in a general class of production/distribution system with general joint setup costs, *Mathematics of Operations Research*, 17, 4, 1992.

6. A. Federgruen and Y.S. Zheng. Optimal power-of-two replenishment strategies in capacitated general production/distribution networks, *Management Science*, 39, 6, 710-727, 1993.

7. M.X. Goemans and David Williamson. A new 3/4 approximation algorithm for MAX SAT, *Proc. 3rd IPCO Conference*, 313-321, 1993.

8. M.X. Goemans and David Williamson. .878 approximation algorithms for MAX-CUT and MAX 2SAT, *Proc. 26th Annual ACM STOC*, 422-431, 1994.

9. D. Hochbaum and G. Shanthikumar. Convex separable optimization is not much harder than linear optimization, *Journal of ACM*, 37, 843-861, 1990.

10. P. Jackson, W. Maxwell and J. Muckstadt. The joint replenishment problem with power-of-two restriction. *AIIE Trans.*, 17, 25-32, 1985.

11. P. Jackson and R. Roundy. Minimizing separable convex objective on arbitrarily directed trees of variable upperbound constraints, *Mathematics of Operations Research*, 16, 504-533, 1991.

12. Lu Lu and M. Posner. Approximation procedures for the one-warehouse multi-retailer system. *management Science*, 40, 1305-1316, 1994.

13. W.L. Maxwell and J.A. Muckstadt. Establishing consistent and realistic reorder intervals in Production-distribution systems, *Operations Research*, 33, 1316-1341, 1985.

14. J.A. Muckstadt and R.O. Roundy. Analysis of Multisatage Production Systems, in: S.C. Graves, A.H.G. Rinnooy Kan and P.H. Zipkin (ed.) *Logistics of Production and Inventory*, North Holland, 59-131, 1993.

15. M. Queyranne. Finding 94%-effective policies in linear time for some production/inventory systems, Unpublished manuscript, 1987.

16. P. Raghavan and C. Thompson. Randomized rounding : a technique for provably good algorithms and algorihmic proofs, *Combinatorica* 7, 365-374, 1987.

17. M. Rosenblatt and M. Kaspi. A dynamic programming algorithm for joint replenishment under general order cost functions, *Management Science*, 31, 369-373, 1985.

18. R.O. Roundy. 98% Effective integer-ratio lot-sizing for one warehouse multi-retailer systems, *Management Science*, 31(11),1416-1430, 1985.

19. R.O. Roundy. A 98% Effective lot-sizing rule for a multi-product, multi-stage production inventory system, *Mathematics of Operations Research*, 11, 699-727, 1986.

20. R.O. Roundy. Rounding off to powers of two in continuous relaxations of capacitated lot sizing problems, *Management Science*, 35, 1433-1442, 1989.

21. Y.S. Zheng. Replenishment strategies for production/distribution networks with general joint setup costs, Ph.D. Thesis, Columbia University, New York, 1987.

Minimizing Total Completion Time in a Two-Machine Flowshop: Analysis of Special Cases

Han Hoogeveen[1] and Tsuyoshi Kawaguchi[2]

[1] Department of Mathematics and Computing Science
Eindhoven University of Technology
P.O. Box 513, 5600 MB Eindhoven, The Netherlands
email address: slam@bs.win.tue.nl
[2] Department of Computer Science and Intelligent Systems
Oita University
Oita 870-11, Japan

Abstract. We consider the problem of minimizing total completion time in a two-machine flowshop. We present a heuristic with worst-case bound $2\beta/(\alpha+\beta)$, where α and β denote the minimum and maximum processing time of all operations. Furthermore, we analyze four special cases: equal processing times on the first machine, equal processing times on the second machine, processing a job on the first machine takes time no more than its processing on the second machine, and processing a job on the first machine takes time no less than its processing on the second machine. We prove that the first special case is $\mathcal{NP}$-hard in the strong sense and present an $O(n \log n)$ approximation algorithm for it with worst-case bound 4/3; we show that the other three cases are solvable in polynomial time.

1980 Mathematics Subject Classification (Revision 1991): 90B35.
Keywords and Phrases: Flowshop, total completion time, $\mathcal{NP}$-hardness, heuristics, worst-case analysis, special cases, polynomial algorithms.

1 Introduction

We consider the two-machine flowshop problem, which is described as follows. There are two machines, M_1 and M_2, that are continuously available from time zero onwards for processing a set of n independent jobs $\mathcal{J} = \{J_1, \ldots, J_n\}$. Each machine can handle no more than one job at a time. Each job consists of a *chain* of two operations, that is, the second operation cannot be started before the first one has been completed. The first (second) operation of each job J_j has to be executed by machine M_1 (M_2) during a given uninterrupted time A_j (B_j). Without loss of generality, we assume A_j and B_j ($j = 1, \ldots, n$) to be integral. Note that all jobs go through the machines in the same order. A *schedule* σ specifies a completion time $C_{ij}(\sigma)$ for the ith operation ($i = 1, 2$) of each J_j ($j = 1, \ldots, n$) such that the above conditions are met. The completion time $C_j(\sigma)$ of job J_j is then equal to $C_{2j}(\sigma)$. We omit the argument σ if there is no confusion possible as to the schedule to which we are referring. Our objective is to minimize *total completion time*, that is, $\sum_{j=1}^{n} C_j$. Following the three-field notation scheme introduced by Graham, Lawler, Lenstra, and Rinnooy Kan (1979), we denote this problem as $F2||\sum C_j$.

This problem was first studied by Ignall and Schrage (1965); they present a branch-and-bound algorithm for this problem based on two lower bounds. Conway, Maxwell, and Miller (1967) show that for this problem it suffices to optimize over all permutation schedules with no idle time on the first machine; a *permutation schedule* is a schedule in which both machines process the jobs in the same order. Gonzalez and Sahni (1978) provide an approximation algorithm for the m-machine problem and show that it has *worst-case performance ratio m*, that is, this algorithm is guaranteed to produce a schedule with cost no more than m times the cost of an optimal schedule. Garey, Johnson, and Sethi (1976) provide justification of these approximation algorithms by establishing $\mathcal{NP}$-hardness in the strong sense of the problem. Van de Velde (1990) applies Lagrangian relaxation to the problem; he proves that his lower bound generalizes and dominates the two bounds by Ignall and Schrage. Hoogeveen and Van de Velde (1995) show that this bound can be strengthened by addressing the corresponding slack variable problem.

In Section 2, we prove that for $m = 2$ the worst-case performance ratio of the approximation algorithm by Gonzalez and Sahni is equal to $2\beta/(\alpha + \beta)$, where α and β denote the minimal and maximal processing time of all $2n$ operations; this is a slight improvement over the bound of 2 established by Gonzalez and Sahni. We also analyze four special cases:

- Equal processing times of all jobs on M_1, that is, $A_1 = \cdots = A_n$. In Section 3, we show that this problem is $\mathcal{NP}$-hard in the strong sense, and we present an approximation algorithm for it that is guaranteed to deliver a solution with cost at most 4/3 times the cost of an optimal schedule.
- Equal processing times of all jobs on M_2, that is, $B_1 = \cdots = B_n$. In Section 4, we show that this problem is solvable in $O(n \log n)$ time.
- The processing of any J_j on M_1 takes time no less than its processing on M_2, that is, $A_j \geq B_j$ for $j = 1, \ldots, n$. In Section 4, we show that this problem is solvable in $O(n \log n)$ time.
- The processing of any J_j on M_1 takes time no more than its processing on M_2, that is, $A_j \leq B_j$ for $j = 1, \ldots, n$. In Section 4, we show that this problem is solvable in $O(n^3)$ time.

2 Worst case behavior for the general case

In this section, we analyze the worst-case behavior of the approximation algorithm presented by Gonzalez and Sahni (1978) for the m-machine flowshop problem and show that in case of two machines its worst-case performance ratio is equal to $2\beta/(\alpha + \beta)$, where α and β denote the minimal and maximal processing time of all $2n$ operations.

The approximation algorithm proceeds by reindexing the jobs in order of nondecreasing $(A_j + B_j)$, settling ties arbitrarily, and by subsequently scheduling the jobs in that order on M_1 and M_2 such that unnecessary machine idle time is avoided. This leads to the set of completion times

$$C_{11} = A_1, \quad C_{21} = A_1 + B_1, \text{ and}$$

$$C_{1j} = C_{1,j-1} + A_j, \quad C_{2j} = \max\{C_{2,j-1}, C_{1j}\} + B_j \text{ for } j = 2, \ldots, n.$$

Note that this approximation algorithm runs in $O(n \log n)$ time and that the resulting schedule is a permutation schedule with no idle time on M_1 between the execution of the jobs.

In order to analyze the worst-case behavior of the algorithm, we need two preliminary results, which are stated in Lemmas 1 and 2.

Lemma 1. *Let σ^* be an optimal schedule. Then*

$$\sum_{j=1}^{n} C_j(\sigma^*) \geq \left[\sum_{j=1}^{n}((n-j+1)(A_j + B_j)) + \sum_{j=1}^{n} B_j + n \min_{1 \leq j \leq n}\{A_j\} \right] / 2.$$

Proof. Consider any schedule σ. Let $J_{[j]}$ denote the job that occupies the jth position in σ; $C_{[j]}$, $A_{[j]}$, and $B_{[j]}$ are defined accordingly. We derive the lower bound stated in the lemma by combining two lower bounds. The first of these stems from $C_{2,[j]} = \max\{C_{2,[j-1]}, C_{1[j]}\} + B_{[j]} \geq C_{2,[j-1]} + B_{[j]}$, which implies that $C_{[j]} \geq A_{[1]} + \sum_{k=1}^{j} B_{[k]}$, for $j = 1, \ldots, n$; hence,

$$\sum_{j=1}^{n} C_{[j]} \geq n A_{[1]} + \sum_{j=1}^{n} \sum_{k=1}^{j} B_{[k]} = n A_{[1]} + \sum_{j=1}^{n}(n-j+1)B_{[j]}.$$

The second lower bound comes from $C_{2[j]} = \max\{C_{2,[j-1]}, C_{1[j]}\} + B_{[j]} \geq C_{1[j]} + B_{[j]}$, which implies that $C_{[j]} \geq \sum_{k=1}^{j} A_{[k]} + B_{[j]}$, for $j = 1, \ldots, n$; hence,

$$\sum_{j=1}^{n} C_{[j]} \geq \sum_{j=1}^{n} \sum_{k=1}^{j} A_{[k]} + \sum_{j=1}^{n} B_{[j]} = \sum_{j=1}^{n}((n-j+1)A_{[j]}) + \sum_{j=1}^{n} B_{[j]}.$$

Adding up these bounds yields

$$2 \sum_{j=1}^{n} C_j(\sigma) \geq \sum_{j=1}^{n}((n-j+1)(A_{[j]} + B_{[j]})) + n A_{[1]} + \sum_{j=1}^{n} B_{[j]} \geq$$

$$\sum_{j=1}^{n}((n-j+1)(A_j + B_j)) + n \min_{1 \leq j \leq n}\{A_j\} + \sum_{j=1}^{n} B_j,$$

as the term $\sum_{j=1}^{n}(n-j+1)(A_{[j]} + B_{[j]})$ is minimized by matching the largest $(A_{[j]} + B_{[j]})$ value to the smallest multiplication factor, and so on (recall that we have reindexed the jobs according to nondecreasing $(A_j + B_j)$). The lemma follows immediately. □

Lemma 2. $|B_k - A_{k+1}| \leq (A_{k+1} + B_{k+1})(\beta - \alpha)/(\beta + \alpha)$, for $k = 1, \ldots, n-1$.

Proof. First, consider the case that $|B_k - A_{k+1}| = B_k - A_{k+1}$; suppose to the contrary that $B_k - A_{k+1} > (\beta - \alpha)(A_{k+1} + B_{k+1})/(\beta + \alpha)$ for some k. Hence, $B_k - A_{k+1} > (\beta - \alpha)(A_k + B_k)/(\beta + \alpha)$, as $A_{k+1} + B_{k+1} \geq A_k + B_k$; this inequality can be rewritten as $2\alpha B_k > (\beta - \alpha)A_k + (\beta + \alpha)A_{k+1}$. Since, $(\beta - \alpha)A_k + (\beta + \alpha)A_{k+1} \geq (\beta - \alpha)\alpha + (\beta + \alpha)\alpha = 2\beta\alpha$, we obtain the contradiction that $B_k > \beta$. The case that $|B_k - A_{k+1}| = A_{k+1} - B_k$ can be dealt with similarly. $\square$

Theorem 3. *The approximation algorithm determines a schedule with cost no more than $2\beta/(\alpha + \beta)$ times the cost of an optimal schedule, and this bound can be approximated arbitrarily closely.*

Proof. First, we prove that the heuristic never delivers a schedule with cost more than $2\beta/(\alpha + \beta)$ times the cost of an optimal solution. We have $C_1 = A_1 + B_1$ and

$$C_j = B_j + \max\{C_{1j}, C_{2,j-1}\} \leq B_j + \max\{C_{1,j-1} + A_j, (C_{2,j-2} + \max\{A_j, B_{j-1}\})\}$$

$$\leq B_j + \max\{C_{1,j-1}, C_{2,j-2}\} + \max\{A_j, B_{j-1}\} \leq A_1 + \sum_{k=1}^{j-1} \max\{B_k, A_{k+1}\} + B_j,$$

for $j \geq 2$. Using the equality $2\max\{B_k, A_{k+1}\} = B_k + A_{k+1} + |B_k - A_{k+1}|$, we get, after rearranging the terms, that

$$2C_j \leq \sum_{k=1}^{j} A_k + \sum_{k=1}^{j} B_k + A_1 + B_j + \sum_{k=1}^{j-1} |B_k - A_{k+1}| \quad \forall j = 1, \ldots, n,$$

from which, after regrouping, follows that

$$2\sum_{j=1}^{n} C_j \leq (W + nA_1 + \sum_{j=1}^{n} B_j + \sum_{j=1}^{n-1} \sum_{k=1}^{j} |B_k - A_{k+1}|),$$

where $W = \sum_{j=1}^{n}(n - j + 1)(A_j + B_j)$. Using Lemma 2, we obtain

$$2\sum_{j=1}^{n} C_j \leq (W + nA_1 + \sum_{j=1}^{n} B_j) + (W - n(A_1 + B_1))(\beta - \alpha)/(\beta + \alpha).$$

Rewriting the right-hand side, we obtain that

$$2\sum_{j=1}^{n} C_j \leq W(2\beta/(\beta + \alpha)) + \sum_{j=1}^{n} B_j + nA_1(2\alpha/(\beta + \alpha)).$$

Since $\alpha A_1 \leq \alpha\beta \leq \beta \min_{1 \leq j \leq n} A_j$, we obtain the inequality

$$2\sum_{j=1}^{n} C_j \leq (2\beta/(\beta + \alpha))(W + \sum_{j=1}^{n} B_j + n \min_{1 \leq j \leq n} A_j),$$

and the desired worst-case bound then follows immediately from Lemma 1.

Second, we show that the bound $2\beta/(\alpha+\beta)$ is tight in the limit. Consider the following instance. There are $2m$ jobs $J_1, \ldots, J_{2m}$ with processing times $A_j = \beta$, $B_j = \alpha$ for $j = 1, \ldots, m$ and $A_j = \alpha$ and $B_j = \beta$ for $j = m+1, \ldots, 2m$. Since $A_j + B_j$ is equal for all jobs, our heuristic may schedule the jobs in any order. Suppose we execute the jobs in the order $(1, 2, \ldots, 2m)$; the resulting schedule has cost equal to $2\beta m^2 + (\beta + 2\alpha)m$. An optimal schedule has the jobs executed in the order $(m+1, 1, m+2, 2, \ldots, 2m, m)$; it has cost $(\beta + \alpha)m^2 + (\beta + 2\alpha)m$. The ratio goes to $2\beta/(\alpha + \beta)$ for $m \to \infty$. $\qquad\square$

3 Equal processing times on the first machine

In this section, we present an approximation algorithm with worst-case bound $4/3$ for $F2|A_j = A|\sum C_j$. First, we justify our using an approximation algorithm by proving that the problem is $\mathcal{NP}$-hard in the strong sense.

3.1 Strong $\mathcal{NP}$-hardness proof for $F2|A_j = A|\sum C_j$

Theorem 4. *The problem of deciding whether for a given instance of the problem $F2|A_j = A|\sum C_j$ there exists a schedule with cost no more than a given threshold value y is $\mathcal{NP}$-complete in the strong sense.*

Proof. Our proof is based upon a reduction from the problem NUMERICAL MATCHING WITH TARGET SUMS or, in short, TARGET SUM, which is known to be $\mathcal{NP}$-hard in the strong sense (Garey and Johnson, 1979).

TARGET SUM
Given two multisets $\mathcal{F} = \{f_1, \ldots, f_q\}$ and $\mathcal{G} = \{g_1, \ldots, g_q\}$ of positive integers and a target vector $(h_1, \ldots, h_q)$, where $\sum_{j=1}^{q}(f_j + g_j) = \sum_{j=1}^{q} h_j$, is there a partition of the set $\mathcal{F} \cup \mathcal{G}$ into q disjoint sets $T_1, \ldots, T_q$, each containing exactly one element from each of $\mathcal{F}$ and $\mathcal{G}$, such that the sum of the numbers in T_j equals h_j, for $j = 1, \ldots, q$?

Consider any instance of TARGET SUM. Since an equivalent problem is obtained by adding a constant to each g_j and h_j, we may assume without loss of generality that $g_i > 2\sum_{j=1}^{q} f_j$, for $i = 1, \ldots, q$. We construct the following instance of $F2|A_j = A|\sum C_j$. Let A be some positive integer with $A > \max_{1 \le j \le q} h_j$. Each f_j $(j = 1, \ldots, q)$ corresponds to an F-job with $A_{F_j} = A$ and $B_{F_j} = A + f_j$. Each g_j $(j = 1, \ldots, q)$ corresponds to a G-job with $A_{G_j} = A$ and $B_{G_j} = A + g_j$. Each h_j $(j = 1, \ldots, q)$ corresponds to an H-job with $A_{H_j} = A$ and $B_{H_j} = A - h_j$. Furthermore, there are three E-jobs, E_1, E_2, and E_3, with processing time A on the first and $A + M$ on the second machine, with $M = 4\sum_{j=1}^{q} h_j$. The threshold is equal to $y = (9q^2 + 27q + 18)A/2 + 6M + 2\sum_{j=1}^{q} f_j + \sum_{j=1}^{q} g_j$.

Instead of analyzing this problem, we rather analyze $F2|A_j = A|\sum Q_j$ with the additional constraint that there is no machine idle time on M_1 between the execution of the jobs; Q_j denotes the delay time, that is, the time that elapses between its completion on M_1 and its start on M_2. For each job J_j, we have $C_{2j} = C_{1j} + B_j + Q_j$, which implies that $\sum C_j = \sum C_{1j} + \sum B_j + \sum Q_j$. Since $\sum C_{1j}$ is constant for each schedule with no idle time on M_1, there exists a schedule for $F2|A_j = A|\sum C_j$ with cost no more than y if and only if there exists a schedule for $F2|A_j = A|\sum Q_j$ with cost no more than $\bar{y} = 2\sum_{j=1}^{q} f_j + \sum_{j=1}^{q} g_j + 3M$.

Suppose that TARGET SUM is answered affirmatively. Consider the permutation schedule $F_{[1]}, G_{[1]}, H_1, \ldots, F_{[q]}, G_{[q]}, H_q, E_1, E_2, E_3$ without any avoidable idle time on M_1, where $F_{[j]}$ and $G_{[j]}$ are the jobs corresponding to the elements in T_j $(j = 1, \ldots, q)$; this schedule is depicted on the next page in Figure 1. A straightforward computation shows that this schedule has cost no more than $\bar{y}$.

Conversely, suppose that there exists a schedule σ for $F2|A_j = A|\sum Q_j$ with cost no more than $\bar{y}$. Without loss of generality, we assume that E_1, E_2, and E_3 are processed in this order in σ. First, we will show that the E-jobs are processed last in σ.

Suppose to the contrary that the E-jobs are not last in σ. Consider any job $J_{[j]}$ $(j = 2, \ldots, 3q + 3)$. We have

$$Q_{[j]} = C_{[2j]} - B_{[j]} - C_{[1j]} \geq C_{[2,j-1]} - (C_{[1,j-1]} + A) = Q_{[j-1]} + B_{[j-1]} - A;$$

hence, the jobs immediately scheduled after E_1, E_2, and E_3 have delay time greater than or equal to M, $2M - \sum_{j=1}^{q} h_j$, and $3M - \sum_{j=1}^{q} h_j$. Therefore, if E_3 has a successor, then

$$\sum Q_j(\sigma) \geq 6M - 2\sum_{j=1}^{q} h_j = 3M + 10\sum_{j=1}^{q} h_j > 3M + 2\sum_{j=1}^{q} f_j + \sum_{j=1}^{q} g_j = \bar{y},$$

which is a contradiction. Similarly, if E_2 has a successor other than E_3, then $\sum Q_j(\sigma) \geq 5M - 2\sum_{j=1}^{q} h_j = 3M + 6\sum_{j=1}^{q} h_j > \bar{y}$, and if E_1 has a successor other than E_2 and E_3, then $\sum Q_j(\sigma) \geq 4M - 2\sum_{j=1}^{q} h_j = 3M + 2\sum_{j=1}^{q} h_j > \bar{y}$. These contradictions show that the E-jobs are last in σ. Hence, we have that the total delay time of E_2 and E_3 amounts to at least $3M$.

Moreover, any immediate successor J_j of job G_i $(i = 1, \ldots, q)$ faces an unavoidable idle time of size $Q_j \geq g_i$, and the immediate successor J_k of J_j also has $Q_k \geq g_i$, unless J_j is an H-job. Since the E-jobs are last in σ, any other job has at least three successors. Suppose that some G_i is not followed by an H-job in σ. Then the total delay time of the immediate successors of all G-jobs amounts to no less than $\sum_{j=1}^{q} g_j$, and the second successor of G_i has delay time at least equal to $M + g_i$ if it is E_2 and to g_i, otherwise. Since the delay time of E_3 is greater than or equal to $2M$, we have that the combined delay time of the immediate successors of the G-jobs, of the E-jobs, and of the second successor of G_i, where we avoid double counting, amounts to no less than $3M + \sum_{j=1}^{q} g_j + g_i > \bar{y}$, which is a contradiction. Hence, each G-job has an H-job as its immediate successor, which implies that no F-job has an H-job as its immediate successor. Therefore, the delay time of each immediate and each second successor of any job F_j is no less than f_j.

As a result, we have that each f_j occurs at least twice in $\sum Q_j(\sigma)$, both in the delay time of the immediate successor and the second successor of F_j,

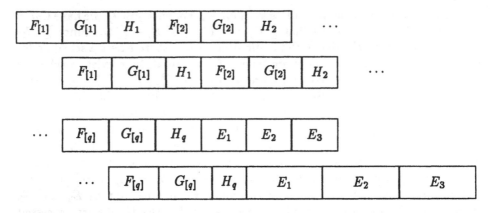

Fig. 1. OPTIMAL SCHEDULE FOR A 'YES'-INSTANCE

that each g_j occurs at least once in $\sum Q_j(\sigma)$, and that $\sum Q_j(\sigma)$ contains a term $3M$. So $\bar{y}$ is completely accounted for, and we are not allowed to have any additional delay time. This implies that if any F_j has a second successor that is not an H-job, then f_i will be accounted for at least three times in $\sum Q_j(\sigma)$, and accordingly $\sum Q_j(\sigma)$ will exceed $\bar{y}$. Hence, σ must have the following form: it starts with q triples in which an F-job is scheduled first, a G-job second, and an H-job third, followed by the E-jobs. It follows straightforwardly that each triple should have total processing time equal to $3A$ on the second machine in order to let $\sum Q_j(\sigma)$ not exceed $\bar{y}$. A solution that answers TARGET SUM affirmatively is readily derived from σ.

Since the reduction is clearly achieved in polynomial time and the decision variant of $F2|A_j = A|\sum Q_j$ is a member of $\mathcal{NP}$, we have that $F2|A_j = A|\sum Q_j$ is $\mathcal{NP}$-hard in the strong sense. Since $F2|A_j = A|\sum Q_j$ and $F2|A_j = A|\sum C_j$ are identical but for a constant, it follows immediately that $F2|A_j = A|\sum C_j$ is $\mathcal{NP}$-hard in the strong sense. $\qquad\square$

3.2 Worst-case behavior for the special case $A_j = A$

We now turn to the worst-case analysis. Because of lack of space, we do not provide any proofs in this subsection. For more details, we refer to the original paper. We start by giving an example to show that 4/3 is the best worst-case bound we can hope for for this heuristic.

Lemma 5. *The worst-case performance ratio of Gonzalez and Sahni's heuristic for $F2|A_j = A|\sum C_j$ is at least equal to 4/3.* $\qquad\square$

Given this lower bound on the worst-case performance ratio, we proceed our analysis by characterizing the properties of an instance, $\mathcal{J}$, that attains the worst-case bound; once we have fully characterized $\mathcal{J}$, we can compute the worst-case bound.

Without loss of generality, we assume $A = 1$. Note that the processing times on the second machine then do not have to be integral any more.

Lemma 6. *Every job $J_j \in \mathcal{J}$ has either $B_j = 0$ or $B_j > 1$.* $\quad\quad\quad\quad\quad\square$

From now on, we shall refer to the jobs with $B_j = 0$ as the *short* jobs. The other jobs are referred to as the *long* jobs. Let m denote the number of short jobs and r denote the number of long jobs in a given instance. Given an arbitrary permutation schedule σ, let $J_{[j]}$ denote the jth $(j = 1,\ldots,r)$ long job in σ.

Lemma 7. *Any permutation schedule σ can be transformed into a permutation schedule $\bar{\sigma}$ with $\sum C_j(\bar{\sigma}) \leq \sum C_j(\sigma)$ that satisfies the following conditions:*

$$-1 < C_{2,[j]} - C_{1,[j+1]} < 1 \text{ for all } j = 1,\ldots,q-1, \text{ and} \tag{1}$$

$$C_{2,[j]} = C_{1,[j+1]} \text{ for all } j = q,\ldots,r-1, \tag{2}$$

where $J_{[q]}$ is the last long job in $\bar{\sigma}$ that is succeeded by a short job. $\quad\square$

To simplify the analysis, we show that the ratio between the cost of the heuristic solution and a lower bound on the optimal cost never exceeds $4/3$; this result together with Lemma 5 shows that the heuristic has worst-case ratio $4/3$. The bound on the optimal cost is derived by allowing preemption of the short jobs, by adjusting the objective function, and by allowing an instance of our original problem to contain one 'dustbin' job.

The objective function is adjusted in the following way. Consider any permutation schedule σ that satisfies 1 and 2. Then we have that

$$\sum C_j(\sigma) = N_0(N_0 + 1)/2 + \sum_{j=1}^{q}(N_j + 1)C_{2,[j]} + \sum_{j=q+1}^{r} C_{2,[j]},$$

where N_j $(j = 1,\ldots,q)$ is defined as the number of short jobs processed between $A_{[j]}$ and $A_{[j+1]}$, and N_0 as the number of short jobs processed before $A_{[1]}$.

We adjust the objective function to

$$MZ(\sigma) = n_0(n_0 + 1)/2 + \sum_{j=1}^{q}(n_j + 1)C_{2,[j]} + \sum_{j=q+1}^{r} C_{2,[j]},$$

where n_j $(j = 1,\ldots,q)$ is defined as the total amount of work of short jobs executed between $A_{[j]}$ and $A_{[j+1]}$, that is, the amount of time that M_1 is busy processing short jobs between $A_{[j]}$ and $A_{[j+1]}$, and n_0 as the total amount of work of short jobs executed before $A_{[1]}$.

Note that $\sum C_j = MZ$ for any permutation schedule that satisfies 1 and 2. We now allow preemption of the short jobs.

Lemma 8. *If preemption of the short jobs is allowed, then without increasing the total completion time, any permutation schedule satisfying 1 and 2 can be transformed into a schedule in which*

$$C_{2,[j]} = C_{1,[j+1]} \text{ for all } j = 1, \ldots, r-1.$$

$\square$

For any instance $\mathcal{I}$, we try to minimize MZ, which minimum is denoted by $MZ(\mathcal{I})^*$, and compare it to the cost of the schedule obtained through the heuristic, which is equal to $\sum C_j(\mathcal{I})$. We will show that for any instance $\mathcal{I}$ the ratio $\sum C_j(\mathcal{I})/MZ(\mathcal{I})^*$ is at most 4/3. Since the minimal MZ value is a lower bound on the total completion time, our result proves that for each 'perturbed' instance of the original problem the heuristic produces a schedule with total completion time no more than 4/3 times the minimum total completion time. The combination with the 4/3 example of Lemma 5 then establishes the worst-case bound. We continue our description of the instance $\mathcal{J}$, but we now assume that it maximizes the ratio $\sum C_j/MZ^*$.

Lemma 9. *Any schedule for instance $\mathcal{J}$ that is optimal with respect to MZ starts with a long job and contains no idle time on M_2 once the first long job has started.*

$\square$

Lemma 10. *There exists an optimal schedule for $\mathcal{J}$ in which the long jobs are executed in order of nondecreasing B_j.*

$\square$

Let B_j $(j = 1, \ldots, r)$ simultaneously denote the second operation of the jth smallest long job and the processing time of this operation. Lemma's 8 to 10 imply that, if $\sum C_j(\mathcal{J})/MZ(\mathcal{J})^* > 4/3$, then MZ^* is given by

$$MZ^* = \sum_{j=1}^{q}(n_j + 1)C_{2,[j]} + \sum_{j=q+1}^{r} C_{2,[j]},$$

where $C_{2,[j]} = 1 + \sum_{i=1}^{j} B_i$ for all $j = 1, \ldots, r$, $n_j = B_j - 1$ for $j = 1, \ldots, q-1$, and $n_q \leq B_q - 1$.

Lemma 11. *For the instance $\mathcal{J}$, we may assume that $B_1 = \cdots = B_q = X/q - 1$, where X denotes the total length of the first operations of the short jobs and the dustbin job.*

$\square$

Lemma 12. *For the instance $\mathcal{J}$, we may assume that $B_{q+1} = \cdots = B_r = B_1$.*

$\square$

Lemma 13. *For any instance $\mathcal{J}$ that meets the descriptions specified in the previous lemmas, $\sum C_j(\mathcal{J})/MZ^*(\mathcal{J}) \leq 4/3$.*

$\square$

Theorem 14. *The heuristic has worst-case bound equal to 4/3 and this bound can be approximated arbitrarily closely.*

$\square$

4 Solvable cases

In this section, we present polynomial algorithms for three special cases. These are that the processing times of all job on M_2 are equal, that is, $B_1 = \cdots = B_n$, that each job J_j needs at least as much time for processing on M_1 as it does on M_2, that is, $A_j \geq B_j$ for $j = 1, \ldots, n$, and that the processing of any J_j on M_1 takes time no more than its processing on M_2, that is, $A_j \leq B_j$ for $j = 1, \ldots, n$.

Until further notice, we assume that the jobs are indexed in order of nondecreasing A_j, where ties are settled arbitrarily.

Theorem 15. *The special case of $F2||\sum C_j$ in which $B_1 = \cdots = B_n = B$ is solved by the permutation schedule without any avoidable idle time in which the jobs are processed in order of nondecreasing A_j.*

Proof. Let σ denote the corresponding schedule; due to the numbering, J_j occupies the jth position in σ. Let π denote an arbitrary schedule, let $J_{[j]}$ $(j = 1, \ldots, n)$ denote the job that occupies the jth position in π.

We prove the theorem by showing that $C_j(\sigma) \leq C_{[j]}(\pi)$ for $j = 1, \ldots, n$. The proof is based on induction. For $j = 1$, we have that $C_1(\sigma) = A_1 + B \leq A_{[1]} + B = C_{[1]}(\pi)$. Suppose that $C_j(\sigma) \leq C_{[j]}(\pi)$ for $j = 1, \ldots, k$. We have that $C_{k+1}(\sigma) = \max\{C_k(\sigma), \sum_{j=1}^{k+1} A_j\} + B$ and $C_{[k+1]}(\pi) = \max\{C_{[k]}(\pi), \sum_{j=1}^{k+1} A_{[j]}\} + B$. Since $\sum_{j=1}^{k+1} A_j \leq \sum_{j=1}^{k+1} A_{[j]}$ and $C_k(\sigma) \leq C_{[k]}(\pi)$, we have that $C_{k+1}(\sigma) \leq C_{[k+1]}(\pi)$. $\square$

Theorem 16. *The special case of $F2||\sum C_j$ in which $A_j \geq B_j$ for $j = 1, \ldots, n$ is solved by the permutation schedule without any avoidable idle time in which the jobs are processed in order of nondecreasing A_j.*

Proof. Let σ denote the corresponding schedule. Since $A_{j+1} \geq A_j \geq B_j$, we have that $C_j = \sum_{i=1}^{j} A_i + B_j$ for $j = 1, \ldots, n$. Hence, $\sum_{j=1}^{n} C_j(\sigma) = \sum_{j=1}^{n}(n - j + 1)A_j + \sum_{j=1}^{n} B_j$, which is known to be a lower bound on the minimal total completion time. $\square$

We now address the special case $A_j \leq B_j$ for $j = 1, \ldots, n$; we denote this problem as $F2|A_j \leq B_j|\sum C_j$.

Consider an arbitrary schedule σ; let $J_{[j]}$ $(j = 1, \ldots, n)$ denote the job that occupies the jth position in σ, and let $I_{[j]}$ denote the total amount of idle time on M_2 before $J_{[j]}$. We have that $C_{[j]}(\sigma) = \sum_{i=1}^{j} B_{[i]} + I_{[j]}$. If we use $W_{[j]}$ to denote $\sum_{i=1}^{j} B_{[i]}$, then we obtain $\sum C_j = \sum(W_j + I_j)$.

Since $\sum C_j = \sum(W_j + I_j)$, we have that any optimal schedule for $F2|A_j \leq B_j|\sum C_j$ corresponds to a *Pareto optimal* point for $(\sum W_j, I_{[1]}, \ldots, I_{[n]})$, that is, there exists no other point $(\sum \bar{W}_j, \bar{I}_{[1]}, \ldots, \bar{I}_{[n]})$ such that each component is no more than the corresponding component of $(\sum W_j, I_{[1]}, \ldots, I_{[n]})$. Hence, we can solve $F2|A_j \leq B_j|\sum C_j$ by identifying all Pareto optimal points for $(\sum W_j, I_{[1]}, \ldots, I_{[n]})$.

Unfortunately, the number of Pareto optimal points is not polynomially bounded, and therefore enumerating all Pareto optimal points is no sensible approach. We generate only a part of the set of Pareto optimal points through a dynamic programming approach, and we prove that the Pareto optimal point that corresponds to an optimal solution of $F2|A_j \leq B_j| \sum C_j$ is an element of this set.

We start by identifying the Pareto optimal point for which $I_{[1]}, \ldots, I_{[n]}$ is minimal. In order to find this point, we have to solve $F2|A_j \leq B_j, I_{[j]} \leq K_j| \sum W_j$, where the constraint $I_{[j]} \leq K_j$ indicates that the amount of idle time on M_2 before the start of the job on the jth position should not exceed a given upper bound K_j, for $j = 1, \ldots, n$. Since $I_{[j]} \leq I_{[j+1]}$, we assume that $K_j \leq K_{j+1}$, for $j = 1, \ldots, n - 1$.

We reindex the jobs according to nondecreasing B_j, where ties are settled according to nondecreasing A_j. A job J_k is available for processing at position j if $\sum_{h=1}^{j-1} A_{[h]} + A_k \leq W_{[j-1]} + K_j$. Let σ denote the feasible schedule for the $F2|A_j \leq B_j, I_{[j]} \leq K_j| \sum W_j$ problem that is obtained by the SMALLEST INDEX rule: *choose from the set of jobs that are available for processing at time* $W_{[j-1]} + K_j$ *the job with smallest index, for* $j = 1, \ldots, n$, *where* $W_{[0]} = 0$.

Before showing that σ is optimal for $F2|A_j \leq B_j, I_{[j]} \leq K_j| \sum W_j$, we first derive some useful properties concerning the structure of σ.

Lemma 17. *Consider an arbitrary job J_i in σ. There are no two jobs J_k and J_l in σ with $i < k$ and $i < l$ that are scheduled before J_i in σ.*

Proof. Without loss of generality, let J_k precede J_l in σ. Due to the way the jobs are chosen in σ, we are done if we show that J_i is available for processing when J_k is completed. Suppose that J_k occupies position j in σ, then we know that J_k starts no later than time $W_{[j-1]} + K_j$, which implies that $\sum_{h=1}^{j-1} A_{[h]} + A_k \leq W_{[j-1]} + K_j$. Since $k > i$, we have that $A_i \leq B_i \leq B_k$, and since $K_j \leq K_{j+1}$, it follows that $\sum_{h=1}^{j} A_{[h]} + A_i = \sum_{h=1}^{j-1} + A_k + A_i \leq W_{[j-1]} + K_j + A_i \leq W_{[j-1]} + K_{j+1} + B_k = W_{[j]} + K_{j+1}$. □

Corollary 18. *Let $\mathcal{U} = \{J_{u_1}, \ldots, J_{u_b}\}$ contain the jobs that are not preceded by a higher indexed job, where $u_1 < \ldots < u_b$. Then σ can be partitioned in b blocks $Z_1, \ldots, Z_b$ that are processed in this order in which block Z_{q+1} consists of the jobs $J_{u_{q+1}}, J_{u_q+1}, \ldots, J_{u_{q+1}-1}$, which are processed in this order in σ without any idle time in between.* □

Theorem 19. *The schedule σ that is obtained through the SMALLEST INDEX rule solves $F2|A_j \leq B_j, I_{[j]} \leq K_j| \sum W_j$.*

Proof. Suppose to the contrary that σ is not optimal; let π denote an optimal schedule. Compare σ and π, starting at the first position; suppose that the first difference occurs at position j. Let J_k and J_l be the jobs that occupy the jth position in σ and π. Due to the way σ is constructed, we must have $k < l$. Suppose that $B_k < B_l$. Consider the part of π that follows after J_l. Let $\mathcal{V}$ denote the jobs with index smaller than l that are scheduled after J_l in π. An easy interchange computation shows that the schedule $\bar{\pi}$ that is obtained by scheduling all jobs in $\mathcal{V}$ in order of nondecreasing index immediately after J_l and before all other jobs while preserving the order of all other jobs is also feasible and has cost no more than the cost of π. If we swap J_k and J_l in $\bar{\pi}$, however, than we obtain a feasible schedule with cost smaller than the cost of $\bar{\pi}$, contradicting the optimality of π. Hence, we must have $B_k = B_l$, and it follows immediately that we can swap J_k and J_l in π without increasing the cost and without creating an infeasibility.

Proceeding in this way yields that σ is an optimal schedule. $\Box$

Theorem 20. *Let σ be obtained by the* SMALLEST INDEX *rule for $F2|A_j \leq B_j, I_{[j]} \leq K_j| \sum W_j$. Then the point $(\sum W_j(\sigma), I_{[1]}(\sigma), \ldots, I_{[n]}(\sigma))$ is Pareto optimal.*

Proof. Since σ is optimal for $F2|A_j \leq B_j, I_{[j]} \leq I_{[j]}(\sigma)| \sum W_j$, we have to show that there is no other point with the same $\sum W_j$ value that dominates the point corresponding to σ. This is easily shown by an interchange argument like the one employed in the proof of Theorem 19; it is due to the way the ties are settled when the jobs are indexed. $\Box$

Theorem 21. *Let K and $\bar{K}$ denote two nondecreasing vectors, with $K_j \leq \bar{K}_j$ $(j = 1, \ldots, n)$. Let σ and $\bar{\sigma}$ be the schedules obtained through the* SMALLEST IN-DEX *rule for the problems $F2|A_j \leq B_j, I_{[j]} \leq K_j| \sum W_j$ and $F2|A_j \leq B_j, I_{[j]} \leq \bar{K}_j| \sum W_j$. Let Z be any block in σ. The jobs that precede Z in σ precede any job from Z in $\bar{\sigma}$.*

Proof. This follows immediately from the observation that all jobs that precede Z in σ have index smaller than the index of any job in Z. Hence, these jobs will be preferred by the SMALLEST INDEX rule to any job from Z; there is always one of these jobs available, since $K_j \leq \bar{K}_j$, for $j = 1, \ldots, n$. $\Box$

We have now arrived at the point of how to find the Pareto optimal point that corresponds to an optimal solution of $F2|A_j \leq B_j| \sum C_j$.

Given a Pareto optimal schedule σ_1 with minimal idle time vector $I_{[1]}, \ldots, I_{[n]}$, which is easily determined, it follows immediately from Theorem 21 that in order to determine another Pareto optimal point we have to increase the idle time vector such that in some block another job can be scheduled first in that block. The amount of idle time by which the idle time vector has to be increased is readily computed.

We will now describe a dynamic programming procedure to derive the Pareto optimal point that corresponds to an optimal solution for $F2|A_j \leq B_j| \sum C_j$. Although we do not determine it by the SMALLEST INDEX rule, we make extensive use of the properties that any schedule obtained by this rule possesses. The dynamic programming procedure is based on the optimality principle that any optimal partial schedule for the jobs $J_k, \ldots, J_n$ with initial idle time $I_{[k-1]}$ should constitute an optimal partial schedule for the jobs $J_l, \ldots, J_n$ with initial idle time $I_{[l-1]}$, for $l = k + 1, \ldots, n$. The unknown factor is the initial idle time $I_{[k-1]}$. We circumvent this problem by determining for each $k = 2, \ldots, n$ a set of optimal partial schedules for the jobs $J_k, \ldots, J_n$ containing the partial schedule that is optimal in case of initial idle time equal to $I_{[k-1]}$. After having added J_1, we can determine an optimal schedule, since $I_{[0]} = 0$.

We show that this approach leads to a polynomial algorithm by proving that when we add J_k there are at most $n + 1 - k$ partial schedules for the jobs $J_k, \ldots, J_n$ that are not dominated for some value of $I_{[k-1]}$ and that we can identify these schedules and the interval of $I_{[k-1]}$ on which they are not dominated in polynomial time.

Given σ_1, determine the set Z that contains all blocks that will occur when the idle time is increased. The number of different blocks amounts to no more than $(2n-1)$, as is easily shown by induction; the n blocks that contain only one job are among these $(2n - 1)$ blocks. Index the blocks according to the position that the first and last job occupy: a block Z_{km} occupies the positions $k, \ldots, m$. Due to Lemma 17, this block contains the jobs $J_k, \ldots, J_m$. Note that the blocks that contain a job J_k are due to a strict *hierarchy*, since each block except the ones σ_1 was partitioned into has been obtained by splitting a higher ranked block. We denote the idle time that has to be inserted to obtain block Z_{km} by $I(Z_{km})$. Each Pareto optimal schedule can be partitioned into a set of blocks that forms a subset of Z.

Suppose that we know the extension that we should choose to cover the jobsets $\{J_l, \ldots, J_n\}$ $(l = n, \ldots, k + 1)$ when we are given the value $I_{[l-1]}$. Now we describe how to determine for which value of $I_{[k-1]}$ we should choose which extension to cover the jobs $J_k, \ldots, J_n$.

Let Z_k denote the blockset containing all blocks $Z_{km} \in Z$. Given the value $I_{[k-1]}$, we can for each block Z_{km} determine $I_{[m]}$; hence, we can immediately determine the optimal extension. A straightforward comparison argument shows that for each $I_{[k-1]}$ the optimal partial schedule for the jobs $J_k, \ldots, J_n$ is found by starting with some Z_{km} followed by its optimal extension: we denote this partial schedule by $\sigma_{km}(I_{[k-1]})$. The total completion time of the jobs in $\sigma_{km}(I_{[k-1]})$, which we from now on will denote by $f(\sigma_{km}(I_{[k-1]}))$, is a stepwise linear function of $I_{[k-1]}$. We want to determine the interval of $I_{[k-1]}$ values on which a certain $\sigma_{km}(I_{[k-1]})$ is optimal.

Lemma 22. *Let I_1 and I_2 denote two values of $I_{[k-1]}$ with $I_1 < I_2$; let $\sigma_{km_1}(I_1)$ and $\sigma_{km_2}(I_2)$ denote the optimal partial schedules for I_1 and I_2, settling ties in favor of the schedule with smallest initial block. Then we have that $m_1 \geq m_2$, that is, the first block in σ_1 contains at least as many jobs as the first block in σ_2.*

Proof. Suppose to the contrary that $m_2 > m_1$. First consider the case that $I_2 \geq I(Z_{km_1})$; this implies that partitioning Z_{km_2} into Z_{km_1} and some other blocks does not increase I_j for any J_j in $\sigma_{km_2}(I_2)$, from which we get the contradiction that $\sigma_{km_1}(I_2)$ is at least as good and hence preferred to $\sigma_{km_2}(I_2)$. Therefore, we must have $I_2 < I(Z_{km_1})$. This implies that all jobs J_j in $\sigma_{km_1}(I_1)$ and $\sigma_{km_1}(I_2)$ have $I_j \geq I(Z_{km_1}) > I_2 > I_1$. Hence, $f(\sigma_{km_1}(I_1)) = f(\sigma_{km_1}(I_2))$. As $f(\sigma_{km_2}(I_1)) \geq f(\sigma_{km_1}(I_1))$ and $f(\sigma_{km_1}(I_2)) > f(\sigma_{km_2}(I_2))$, we obtain the inequality $f(\sigma_{km_2}(I_1)) > f(\sigma_{km_2}(I_2))$, which implies that the increase of the initial idle time has reduced the completion times. This is clearly a contradiction. □

Hence, in our search to find the optimal partial schedules for a given $I_{[k-1]}$ value, we can restrict ourselves to compare the schedules in the order imposed by the block hierarchy.

Suppose that we compare the costs of two partial schedules that start with the same block Z_{km}. As we know the $I_{[m]}$ value that corresponds to the breakpoint in the dominance relation for the two extensions, we can immediately tell what $I_{[k-1]}$ value will correspond to the breakpoint in the dominance relation for the two partial schedules.

If we compare the costs of two partial schedules that start with blocks Z_{km_1} and Z_{km_2}, where $m_1 > m_2$ and Z_{km_1} and Z_{km_2} are consecutive in the hierarchy, then we know that the schedule that starts with Z_{km_2} dominates the one that starts with Z_{km_1} if $I_{[k-1]} \geq I(Z_{km_2})$. Hence, we know that the breakpoint between these two partial schedules occurs at some $I_{[k-1]}$ value that is smaller than $I(Z_{km_2})$; for these values of $I_{[k-1]}$, the optimal extension for the partial schedule starting with Z_{km_2} is the optimal partial schedule for the jobs $J_{m_2+1}, \ldots, J_n$ with $I_{[m_2]}$ value equal to $I(Z_{km_2})$. The breakpoint is found by a straightforward comparison of the schedule that starts with Z_{km_2} and the schedules that start with Z_{km_1} that are nondominated on some interval of $I_{[k-1]}$.

The last point we have to bother about is the number of partial schedules for the jobs $J_k, \ldots, J_n$ that are nondominated on some interval of $I_{[k-1]}$.

Theorem 23. *Consider any block Z_{km} from Z. The number of different orders in which the jobs $J_k, \ldots, J_m$ appear in the set of nondominated partial schedules for $J_j, \ldots, J_n$, for any $j \leq k$, without being preceded by a higher indexed job, amounts to no more than $m + 1 - k$.*

Proof. The proof proceeds by induction on $l = m - k$. The case $l = 0$ is trivial. Suppose that it holds for any $l \leq q - 1$. We have to show that it also holds for $l = q$.

First, note that our algorithm does not alter any extension that covers $J_j, \ldots, J_n$ in which the jobs $J_k, \ldots, J_m$ are scheduled without any higher indexed predecessor, although it may remove some of these extensions when we add a job. Hence, we are done if we prove it for $j = k$.

Suppose that in the first partitioning the block Z_{km} splits into h disjoint blocks, which we denote by $R_1, \ldots, R_h$. Each R_i $(i = 1, \ldots, h)$ contains a set of consecutively indexed jobs, its cardinality, which is denoted by $card(R_i)$, is less than q; due to the induction hypothesis, we know that the jobs in R_i appear in at most $card(R_i)$ different orders in the set of nondominated partial schedules for $J_k, \ldots, J_n$, which includes the order in which the jobs occur in R_i. The monotonicity imposed by Lemma 22 implies that the switch from one order for the jobs in R_i to another occurs only once. Hence, each change of the order the jobs $J_k, \ldots, J_m$ occur in the set of nondominated partial schedules for $J_k, \ldots, J_n$

corresponds to a switch from one order to another in some R_i. The number of switches in each R_i amounts to at most $card(R_i) - 1$, implying that the total number of switches for all R_i amounts to no more than $\sum card(R_i) - h = q - h$, since the R_i's are disjoint. The number of orders is one higher than the number of switches, and the order in Z_{km} has not been counted yet. Hence, the number of orders in which the jobs $J_k, \ldots, J_m$ appear in the set of nondominated partial schedules without being preceded by any higher indexed job amounts to no more than $q - h + 2 \leq q$. $\qquad \square$

Corollary 24. *The number of partial schedules for the jobs $J_k, \ldots, J_n$ that are nondominated on some interval of $I_{[k-1]}$ amounts to no more than $n + 1 - k$.* $\square$

The analysis described above leads to the following algorithm for $F2|A_j \leq B_j| \sum C_j$ that runs in $O(n^3)$ time.

ALGORITHM SOLVED

Step 1. Determine σ_1 and $I(Z_{nn})$; $k \leftarrow n$.
Step 2. Given the set of possibly optimal extensions that cover $J_k, \ldots, J_n$ and their dominance intervals, determine the set of possibly optimal extensions for $J_{k-1}, \ldots, J_n$.
Step 3. $k \leftarrow k - 1$; if $k > 1$, then go to Step 2.
Step 4. Select from among the set of possibly optimal extensions the one whose dominance interval contains 0.

Acknowledgement. Part of the research was conducted when the first author visited Princeton University, which visit was made possible by a grant from Netherlands Organization for Scientific Research (NWO).

References

1. R.W. CONWAY, W.L. MAXWELL, AND L.W. MILLER (1967). *Theory of Scheduling*, Addison-Wesley, Reading, Massachusetts.
2. M.R. GAREY, D.S. JOHNSON, AND R. SETHI (1976). The complexity of flowshop and jobshop scheduling. *Mathematics of Operations Research 13*, 330-348.
3. M.R. GAREY AND D.S. JOHNSON (1979). *Computers and Intractability: A Guide to the Theory of NP-Completeness*, Freeman, San Francisco.
4. R.L. GRAHAM, E.L. LAWLER, J.K. LENSTRA, AND A.H.G. RINNOOY KAN (1979). Optimization and approximation in deterministic sequencing and scheduling: a survey. *Annals of Discrete Mathematics 5*, 287-326.
5. T. GONZALEZ AND S. SAHNI (1978). Flowshop and jobshop schedules: Complexity and approximation. *Operations Research 26*, 36-52.
6. J.A. HOOGEVEEN AND S.L. VAN DE VELDE (1995). Stronger Lagrangian bounds by use of slack variables: applications to machine scheduling problems. *Mathematical Programming 70*, 173-190.
7. E. IGNALL AND L. SCHRAGE (1965). Application of the branch and bound technique for some flow-shop scheduling problems. *Operations Research 13*, 400-412.
8. S.L. VAN DE VELDE (1990). Minimizing the sum of the job completion times in the two-machine flow shop by Lagrangian relaxation. *Annals of Operations Research 26*, 257-268.

A New Approach to Computing Optimal Schedules for the Job-Shop Scheduling Problem

Paul Martin * and David B. Shmoys **

School of Operations Research and Industrial Engineering, Cornell University, Ithaca, NY 14853.

Abstract. From a computational point of view, the job-shop scheduling problem is one of the most notoriously intractable $\mathcal{NP}$-hard optimization problems. In spite of a great deal of substantive research, there are instances of even quite modest size for which it is beyond our current understanding to solve to optimality. We propose several new lower bounding procedures for this problem, and show how to incorporate them into a branch-and-bound procedure. Unlike almost all of the work done on this problem in the past thirty years, our enumerative procedure is not based on the disjunctive graph formulation, but is rather a time-oriented branching scheme. We show that our approach can solve most of the standard benchmark instances, and obtains the best known lower bounds on each.

1 Introduction

In the *job-shop scheduling problem* we are given a set of n jobs, $\mathcal{J}$, a set of m machines, $\mathcal{M}$, and a set of operations, $\mathcal{O}$. Each job consists of a chain of operations, let $\mathcal{O}_j$ be the chain of operations required by job j, whereas $\mathcal{O}^i$ is the set of operations that must be processed on machine i. Each operation k requires a specified amount of processing, p_k, on a specified machine $M(k)$. A schedule is an assignment of starting times to the operations such that each operation runs without interruption, no operation starts before the completion time of its predecessor in its job and such that each machine processes at most one operation at any moment in time. The goal is to find the schedule which minimizes the *makespan*, the time at which the last operation completes.

In this paper we focus on the special case in which each job has exactly one operation on each machine. Therefore, $|\mathcal{O}_j| = m$ for each $j \in \mathcal{J}$. The main ideas of this paper are easily extended to the general case. The *feasibility problem* is, given a time T, to determine whether there exists a schedule that completes by time T. Throughout this paper we consider the feasibility problem and use a bisection search to reduce the optimization problem to the feasibility problem.

The job-shop scheduling problem is $\mathcal{NP}$-hard [12] and has proven to be very difficult even for relatively small instances. An instance with 10 jobs and 10

---

* Research supported in part by the NEC Research Institute and in part by NSF grant CCR-9307391.

** Research supported in part by NSF grant CCR-9307391.

machines posed in a 1963 book by Muth & Thompson [10] remained unsolved until Carlier & Pinson [7] finally solved it in 1986, and there is a 20 job and 10 machine instance of Lawrence [15] that remains unsolved despite a great amount of effort that has been devoted to improving optimization codes for this problem.

The most effective optimization techniques to date have been branch-and-bound algorithms based on the disjunctive graph model for the job-shop scheduling problem. The decision variables in this formulation indicate the order in which operations are processed on each machine. Once these variables are set, the time at which each operation starts processing can be easily computed. Algorithms of this type have been proposed by Carlier & Pinson [7], Brucker, Jurisch & Sievers [6] and Applegate & Cook [2].

Although solving job-shop scheduling problems to optimality is difficult, recently there has been progress developing heuristics that find good schedules. Adams, Balas & Zawack [1] proposed the *shifting bottleneck procedure*, which uses a primitive form of iterated local search to produce substantially better schedules than were previously computed. Currently, all of the best known algorithms use iterated local search. The current champions are a *taboo search* algorithm by Nowicki & Smutnicki [16] and a technique called *reiterated guided local search* by Balas & Vazacopoulos [4]. Vaessens, Aarts & Lenstra [21] provide an excellent survey of recent developments in this area.

There has been less progress in finding good lower bounds for the job-shop scheduling problem. The *job bound* is the maximum total processing required for the operations on a single job. Similarly, the maximum amount of processing on any one machine gives the *machine bound*. We can improve upon the machine bound by considering the *one-machine subproblem* obtained by relaxing the capacity constraint on all machines except one. This problem is the one-machine scheduling problem with heads and tails for each operations: $1 \mid r_j \mid L_{\max}$ in the notation of [13]. Here each operation $k \in \mathcal{O}^i$ has a head r_k, equal to the total processing requirement of all operations which precede k in its job, a tail q_k, the analogous sum for operations which follow k, and a body p_k, the amount of processing required by k. Each operation k may not begin processing before r_k and must be processed for p_k time units; if C_k denotes the time at which k completes processing, the *length* of the schedule is $\max_{k \in \mathcal{O}^i} \{C_k + q_k\}$. The *one-machine bound* of Bratley, Florian, & Robillard [5] is the the maximum over all machines $i \in \mathcal{M}$ of the bound for the one-machine subproblem for i. This subproblem is $\mathcal{NP}$-hard, although in practice it is not difficult to solve. The *preemptive one-machine subproblem* is the relaxation of the one-machine subproblem obtained by relaxing the constraint that each operation must be processed without interruption. Jackson [14] gives an $O(n \log n)$ algorithm for computing the optimal schedule for this subproblem. Carlier & Pinson [8] give an algorithm that strengthens the preemptive one-machine bound but maintains polynomial computability. Their algorithm updates the heads and tails of operations to account for the fact that we are looking for a non-preemptive schedule. In Section 3 we show that their updates can be seen as adjusting each operation's head (or tail) to the earliest time that the operation can start such that it processes continuously and such that there is a preemptive schedule for the remaining operations on that machine.

Fisher, Lageweg, Lenstra & Rinnooy Kan [11] examine surrogate duality while Balas [3] and Applegate & Cook [2] attempt to find bounds by examining the polyhedral structure of the problem. Neither approach has proven to be practical so far. The strongest computational results of Applegate & Cook [2] use only combinatorial bounds in their branch-and-bound algorithm.

In this paper we present new lower bounding techniques based on a time-oriented approach to job-shop scheduling. Recently there has been significant success applying time-oriented approaches to other scheduling problems [19] [22].

The first approach we propose is based upon the *packing formulation*, a time-indexed integer programming formulation of the job-shop scheduling problem. Here we attempt to pack schedules for each job so that they do not overlap on any machine. The bound we obtain is based on the *fractional packing formulation* which is the linear relaxation of the packing formulation. This approach was motivated by the packing approximation algorithm of Plotkin, Shmoys & Tardos [17] along with an application to scheduling given by Stein [20]. This bound gives stronger bounds than those obtained by other linear programming relaxations but appears to be too time consuming.

To improve on the packing bound we attempt to restrict the time in which operations are allowed to start processing. We call this interval a *processing window*. The Carlier-Pinson head and tail modification algorithm can be used in an iterated fashion to reduce each operation's processing window. In addition to reducing processing windows for use with the fractional packing formulation, the iterated Carlier-Pinson can itself be a bound since if the feasible processing window is empty then we have solved the feasibility problem: no feasible schedule of length T exists. Iterated Carlier-Pinson gives results that are approximately the same as those obtained from the fractional packing formulation and it obtains these bounds in only a fraction of the time. Combining the two techniques provides only a slightly improved bound.

In order to reduce the size of the processing windows we introduce a technique called shaving. Here we suppose that an operation starts at one end of its processing window and then try to determine whether we can prove that there is no feasible schedule satisfying this assumption. If we can prove this, then we can conclude that the operation cannot start at that end of its processing window and we can therefore reduce its window. This technique depends on our ability to determine that there is no schedule after restricting the processing window of one operation. When we use the preemptive one-machine relaxation to determine that there is no schedule we call it *one-machine shaving*. We prove that this is equivalent to iterated Carlier-Pinson. When we use the non-preemptive one-machine relaxation we call it *exact one-machine shaving* and we show that this is only a slightly better bound then iterated Carlier-Pinson but significantly more costly to compute. If we use iterated Carlier-Pinson to show that there is no schedule we call it *C-P shaving*. This bound is significantly better than iterated Carlier-Pinson but requires more time to compute. Finally if we use C-P shaving to show that there is no schedule, then we call it *double shaving*. This bound is excellent but takes a substantial amount of time to compute, much too long to be practical in most branch-and-bound schemes.

Traditional job-shop branch-and-bound schemes are not particularly well-suited to a time-oriented lower bound. We give two new schemes designed for use with a processing window approach. The first is a simplistic approach which starts at time zero and either assigns an available operation to start immediately or delays it. This approach works well only if through the use of one of the shaving techniques we have been able to greatly reduce the processing windows. The second technique works by examining a machine over a time period where we know that there is very little idle time. We can then branch on which operation goes first in this time period. This technique is very effective on a wide range of problems. It is particularly effective when the one-machine bound is tight, a situation in which many other algorithms seem to perform poorly.

Using these techniques we were able to improve upon the best known lower bound for several of the standard benchmark problems, and in some cases we establish optimality. We have tested all of our algorithms on the standard benchmark problems. We have used all the problems for which Vaessens, Aarts & Lenstra [21] give recent local search results as well as three larger problems from Adams, Balas & Zawack [1]. All computations were performed on a 90MHz Pentium computer. Table 1 gives the bounds that we were able to compute along with the amount of time (in seconds) that it took to compute them.

One important advantage of our time-oriented approach is that we can trivially extend it to incorporate additional constraints, such as release dates and deadlines for jobs, as well as internal release dates and deadlines for operations. In fact, one would expect that adding these constraints would improve the performance of our algorithms.

2 A Packing Formulation

In spite of a great deal of effort, the disjunctive integer programming formulation of the job-shop problem appears to be of little assistance in solving instances of even moderate size; furthermore, its natural linear programming relaxation has been shown to give very poor lower bounds for the problem. We propose a new linear programming formulation which gives better bounds.

A *job-schedule* is an assignment of starting times to the operations of one job such that no operation starts before its predecessor completes and such that all operations of that job complete by time T. The job-shop feasibility problem can then be restated: does there exist a set of job-schedules, one for each job, such that no two job-schedules require the same machine at the same time? This gives rise to the following *packing formulation*, where we attempt to pack job-schedules so as to minimize the maximum number of jobs concurrently assigned to some machine.

For each job $j \in \mathcal{J}$, let $\mathcal{F}_j$ denote the set of all job-schedules, and let $x_{j\sigma}$ be a 0-1 variable that indicates if j is scheduled according to job-schedule $\sigma \in \mathcal{F}_j$.

Let $\gamma(\sigma, i, t)$ indicate whether job-schedule σ requires machine i at time t and let λ be the maximum number of jobs concurrently on any machine. Then the integer programming formulation is to minimize λ subject to

$$\sum_{\sigma \in \mathcal{F}_j} x_{j\sigma} = 1, \quad \text{for all } j \in \mathcal{J}, \tag{1}$$

$$\sum_{j \in \mathcal{J}} \sum_{\sigma \in \mathcal{F}_j} \gamma(\sigma, i, t) x_{j\sigma} \leq \lambda, \quad \text{for all } i \in \mathcal{M}, t = 1, \ldots, T, \tag{2}$$

$$x_{j\sigma} \in \{0, 1\}, \text{for all } j \in \mathcal{J}, \sigma \in \mathcal{F}_j. \tag{3}$$

There is a feasible job-shop schedule which completes by T if and only if the optimal value of the above integer programming problem is 1.

A natural linear programming relaxation can be obtained by replacing (3) by $x_{j\sigma} \geq 0$, for all $j \in \mathcal{J}$, $\sigma \in \mathcal{F}_j$; let λ^* denote its optimal value. If we show that $\lambda^* > 1$, then we have proved that there is no schedule with makespan at most T: $T + 1$ is a valid lower bound for this instance.

One natural way to show that $\lambda^* > 1$ is to exhibit a feasible dual solution of value greater than 1. We will give the dual linear program in a compact form, which is easy to derive. The dual variables are $y_{i,t}$, $i \in \mathcal{M}$, $t = 1, \ldots, T$, where each can be viewed as the cost of scheduling on machine i at time t. Thus, the cost of $\sigma \in \mathcal{F}_j$ is $C_\sigma(y) = \sum_{i \in \mathcal{M}} \sum_{t=1}^T \gamma(\sigma, i, t) y_{i,t}$. If we let $\mathcal{S} = \{y \geq 0 : \sum_{i \in \mathcal{M}} \sum_{t=1}^T y_{i,t} = 1\}$, then the dual is $\max_{y \in \mathcal{S}} \sum_{j \in \mathcal{J}} \min_{\sigma \in \mathcal{F}_j} C_\sigma(y)$.

Solving the packing relaxation The packing formulation and its linear relaxation contain an exponential number of variables; fortunately, fractional packing problems can be approximately solved by algorithms that are insensitive to the number of variables. Plotkin, Shmoys, & Tardos [17] provide a framework for obtaining solutions within a specified accuracy for such problems.

We view the relaxation as the fractional packing problem where we are attempting to pack job-schedules subject to a restriction on the amount of work on each machine at each time unit. Let P^j be the set consisting of all convex combinations of the job-schedules for job j and let $P = P^1 \times \ldots \times P^n$. Then our relaxation can be rewritten as $\min\{\lambda \mid x \in P, x \text{ satisfies (2)}\}$. To apply the algorithm of [17], we need to be able to efficiently evaluate the dual objective for a given y, and to efficiently optimize over P.

To evaluate the dual, we need to find the minimum-cost job-schedule for each job $j \in \mathcal{J}$. This can be done by a simple dynamic programming algorithm. We look at each operation $k \in \mathcal{O}_j$ in processing order and for each possible completion time t we calculate $c_k(t)$, the minimum cost of k and all operations preceding k in $\mathcal{O}_j$, such that k completes by t. Then if l is the last operation of job j, $c_l(T)$ gives the minimum cost job-schedule for j with respect to y. We can then use this dynamic program to find an optimal solution in P.

The fractional packing algorithm works as follows. Start with some $x \in P$ and an error parameter $\epsilon > 0$. Calculate λ for x; if $\lambda \leq 1$, then we are done. Otherwise, compute $y_{i,t}$ as an exponential function of the load on machine i at time t, scaled appropriately. Then apply the dynamic programming algorithm to evaluate y and to find the point $\tilde{x} \in P$ of minimum cost with respect to y. If the dual value exceeds 1 or is within a $(1 + \epsilon)$ factor of the primal value then stop. Update x by taking a convex combination of x and $\tilde{x}$ and start over. If the dual value is greater than 1 when we terminate, then we can conclude that $T + 1$ is a valid lower bound. Otherwise we can not determine a lower bound. The running time of the algorithm is proportional to T and ϵ^{-2}.

Computational results The times that we report in Table 1 include only the time that our packing algorithm uses to show that the bound is valid: it does not include any time to search for the correct value of T. We cut off our algorithm after 20,000 iterations so there is no guarantee that the bounds we have given are the best possible.

For a given T, we are trying to determine whether $\lambda^* > 1$. When λ^* is very close to 1, this can be difficult; the number of iterations required to determine whether $\lambda^* > 1$ increases dramatically as λ^* approaches 1.

The bounds found by the packing algorithm are significantly better than other bounds in the literature based on LP-relaxations, such as the cutting plane approach of Applegate & Cook [2]. The algorithm however is still to slow to be of practical significance.

3   Processing Windows

When considering an instance of the feasibility problem with target makespan T, we can view each operation as constrained to be started within a given time interval. For example, consider an operation $k \in \mathcal{O}^i$, and let r_k and q_k denote, respectively, its head and tail in the corresponding one-machine instance of $1|r_j|L_{\max}$; that is, r_k is the total processing requirement of operations preceding k in job j, and q_k is the analogous sum for operations following k in j. Then the job bound implies that the starting time of operation k must occur in the interval $[r_k, T - q_k - p_k]$. In the packing formulation, we consider only job-schedules in which each operation starts in this window. We can strengthen this formulation by computing restricted *processing windows*; that is, for each operation k, specify a smaller interval $[u_k, v_k]$ within which the operation must begin processing.

In this section, we describe several approaches to compute restricted processing windows, by relying on bounds stronger than the job bound. In particular, we rely on lower bounds given by the one-machine subproblem. In describing our algorithms, we focus only on increasing the start of the processing window, since updating the end of the processing window can be done analogously, due to the symmetry of the problem.

One-machine shaving We shall first consider the one-machine relaxation in order to compute improved processing windows.

Focus on machine i and suppose that we are given a proposed processing window $[u_k, v_k]$ for each operation $k \in \mathcal{O}^i$: we wish to decide if machine i can be scheduled so that each operation starts within its designated window. It is equivalent to decide if there is a feasible schedule of the one-machine subproblem in which k has head u_k, body p_k, and tail $T - p_k - v_k$. Hence, we can rephrase the problem of computing a reduced processing window to that of adjusting heads and tails, which was an idea first introduced by Carlier & Pinson [7].

We can exploit the fact that the problem $1|r_j, pmtn|L_{\max}$ can be solved in polynomial time, and in fact, admits a nice min-max theorem. For any $S \subseteq \mathcal{O}^i$, let $p(S) = \sum_{k \in S} p_k$, $r(S) = \min_{k \in S} r_k$, and $q(S) = \min_{k \in S} q_k$. Since the first job in S can start no earlier than $r(S)$ and the tail of the last job of S to complete processing is at least $q(S)$, the length of the schedule is at least $r(S) + p(S) + q(S)$. Remarkably, Carlier & Pinson [7] have shown that minimum length of a preemptive schedule is equal to $\max_{S \subseteq \mathcal{O}^i} \{r(S) + p(S) + q(S)\}$.

Consider an operation $k \in \mathcal{O}^i$ with current processing window $[u_k, v_k]$; we shall maintain the invariant that if there is feasible schedule for the job-shop instance (of length T), then there is a schedule in which each operation starts within its current window. We shall compute $\bar{q}_k$, the maximum tail for operation k for which there still is a feasible preemptive schedule for the one-machine subproblem (where the other data are unchanged). Operation k is processed without interruption in any job-shop schedule, therefore its window may be updated to $[T - \bar{q}_k - p_k, v_k]$.

While one could merely apply a bisection search to find $\bar{q}_k$, for each $k \in \mathcal{O}^i$, we can instead apply the following more efficient procedure, **One-Machine Shave**, which is based on the min-max theorem above. For each ordered pair of operations $(b, c) \in \mathcal{O}^i$, first compute $P_{b,c} = \sum_{k \in S} p_k$, where $S = \{k \in \mathcal{O}^i \mid r_k \geq r_b \text{ and } q_k \geq q_c\}$, provided $S \neq \emptyset$. If $r_b + P_{b,c} + q_c > T$, then there does not exists a feasible job-shop schedule of length T. For each operation $k \in \mathcal{O}^i$ and each ordered pair (b, c) such that (i) $q_k < q_c$, (ii) $r_k \geq r_b$, and (iii) $r_b + P_{b,c} + p_k + q_c > T$, update $u_k \leftarrow \max\{u_k, r_b + P_{b,c}\}$

Lemma 1. One-Machine Shave *correctly updates the processing window of each operation on the machine in* $O(n^3)$ *time.*

Once we have updated the starting point of each interval, we must also update the ending point. We can then repeat these updates until no further changes can be made for this machine.

Carlier-Pinson updates Carlier & Pinson [8] gave the first procedure to update heads and tails for the one-machine problem. An *ascendent set* (k, S) consists of an operation k and a set of operations S, $k \notin S$ such that:

$$r(S) + p(S) + p_k + q(S) > T, \tag{4}$$
$$r_k + p(S) + p_k + q(S) > T. \tag{5}$$

Ascendent sets are important because condition (4) implies that in any feasible schedule, k is processed first or last among $S \cup \{k\}$, and condition (5) implies that k cannot be processed first. Therefore, if there is a feasible schedule, k must be processed last. Once we find an ascendent set (k, S), we can update

$u_k \leftarrow \max\{r_k, c(S)\}$, where $c(S) = \max\{r(T) + p(T)|T \subseteq S\}$ is a lower bound on the time at which the operations in S can complete processing.

Carlier and Pinson [8] give an $O(n^2)$ algorithm based on Jackson's algorithm to solve $1|r_j, pmtn|L_{\max}$ that finds ascendent sets and updates all of the heads or all of the tails for a one-machine problem. We can use their algorithm directly on the one-machine relaxation to improve the processing windows. Recently, Carlier & Pinson [9] have given an $O(n \log n)$ implementation of this algorithm.

Once again, we need to apply this procedure alternately to heads and tails until no further updates are possible. Surprisingly, the two algorithms, One-Machine Shave and Carlier-Pinson Updates yield the same result.

Theorem 2. *For any one-machine subproblem, applying* Carlier-Pinson Updates *until no more updates can be found is equivalent to applying* One-Machine Shave *until no updates are found.*

Another $O(n^2)$ algorithm for performing Carlier-Pinson updates We shall give another algorithm to compute Carlier-Pinson updates, by combining ideas from the previous two algorithms, which we will see in §4 is faster when solving a sequence of closely related problems. We can perform all updates for the ascendent sets in $O(n^2)$ time by repeatedly applying an $O(n)$ algorithm that performs updates for all ascendent sets (l, S) with a fixed value of $q(S)$.

Suppose that we wish to find all ascendent sets (l, S) in which $q(S) = q_k$, $k \in S$. We first construct a nested list of candidates for the set S. Set $S^0 = \{l|r_l \geq r_k, q_l \geq q_k\}$. Find the operation $l \in \mathcal{O}^i - S^0$ with maximum head length r_l. If $q_l \geq q_k$, then create a new set, $S^1 = S^0 \cup \{l\}$. Continue in this manner, iteratively choosing the operation with the next largest head, and creating a new set whenever its tail is at least q_k. Let $S^0 \subset \cdots \subset S^b$ be the sets constructed. The construction ensures that $r(S^0) \geq r(S^1) \geq \cdots \geq r(S^b)$. Finally, as we construct each S^i, also compute $c(S^i)$, $i = 0, \ldots, b$.

Let L denote the set of operations $l \in \mathcal{O}^i$ for which $q_l < q_k$. Let l be the operation in L for which p_l is maximum. If (l, S^b) does not satisfy condition (4), then each (l', S^b), $l' \in L$, also does not satisfy it, by the choice of l. So S^b can be eliminated as a candidate set. If (l, S^b) does not satisfy condition (5), then each (l, S^i), $i = 0, \ldots, b$ also does not satisfy it, and so l can be removed from L. If both ascendent set conditions are satisfied, then (l, S^b) is an ascendent set. We can then increase the head of l to $c(S^b)$. We can remove l from L, since S^b has the latest completion time of all candidate sets. We can continue this process, removing either one operation or one set until one list is empty. We have now found all updates for ascendent sets (l, S) with $q(S) = q_k$.

Once again, we must repeatedly apply this algorithm until no further updates are made.

Iterated one-machine window reduction When we update the processing window $[u_k, v_k]$ for an operation $k \in \mathcal{O}_j$, this can refine the windows of other operations of job j. For example, if we increase the start of k's interval from u_k to $\tilde{u}_k$, then its successor in j cannot start until $\tilde{u}_k + p_k$. Hence, updating windows on machine i can cause further updates on all other machines. This approach was used by Applegate & Cook [2], Brucker, Sievers, & Jurisch [6], and by Carlier & Pinson [8] as well, although this is apparent only from their computational results.

We shall refer to the procedure of iteratively updating the processing windows for one machine and propagating the implications of these updates through each job, until no further updates can be made, as an *iterated one-machine windows reduction algorithm*.

Exact one-machine shaving We can also base the update of the processing window for an operation $k \in \mathcal{O}^i$ on finding the maximum tail $\bar{q}_k$ such that there exists a feasible non-preemptive schedule for this one-machine subproblem. Once again, after computing $\bar{q}_k$, we can update the start of k's interval to $\max\{u_k, T - p_k - \bar{q}_k\}$. To compute $\bar{q}_k$ requires solving several instances of $1|r_j|L_{\max}$; while this is an "easy" $\mathcal{NP}$-hard problem, it is still fairly time-consuming to compute $\bar{q}_k$. This technique is interesting because it gives the maximum window reduction obtainable by considering only the one-machine relaxation. This technique does only slightly better than the techniques based on the preemptive model and requires much more time to compute.

Computational results By simply reducing some windows until they are empty, iterated Carlier-Pinson provided lower bounds that are approximately the same as those found by the packing algorithm and it obtained those bounds in less than 1/4 of a second for each problem tested. This time includes the time for a search to determine the correct value of T. So this algorithm is several orders of magnitude faster than the packing algorithm.

Exact one-machine shaving gave better bounds for several of the smaller problem that we examined but it gave virtually no improvement on the larger problems and is significantly slower. The running times include a bisection search between a lower bound provided by iterated Carlier Pinson and an upper bound provided by Applegate & Cook's implementation of the shifting bottleneck procedure. The time to compute the initial bounds is not included in the running times. The exact one-machine shaving results were obtained using the one-machine scheduling code of Applegate & Cook and no effort was made to optimize performance for our application.

4 Shaving

We shall show how to strengthen the ideas from One-Machine Shave to further reduce the size of the processing windows. For some operation $k \in \mathcal{O}$ and its operation window $[u_k, v_k]$ we attempt to reduce the window by showing that the operation could not have started at one end of the interval. If we assume that the operation starts at one end of the interval and show that this implies that there is no feasible schedule, then we can conclude that the operation cannot start then. Thus we can shave a portion of the window by applying this idea. Carlier & Pinson [9] have independently proposed this idea; a preliminary version of our results for this was announced in [18].

We shall describe our algorithm only in terms of shaving the start of the windows; of course these ideas apply to the end as well. Let the window for operation k be $[u_k, v_k]$ and fix w_k such that $u_k \leq w_k < v_k$. Consider the problem where $[u_v, v_k]$ is replaced by $[u_k, w_k]$ and all other windows are unchanged. If we can show that there is no feasible schedule for the modified data then if there is a feasible schedule for the original instance, k must start in the interval $[w_k + 1, v_k]$.

This approach relies on the fact that we are able to show quickly that no feasible schedule exists in which the given operation is constrained to start within a small window. The algorithms described in §3 are ideally suited for this. In particular, in our procedure C-P Shave, we use Iterated Carlier-Pinson to determine the maximum value w_k such that $[u_k, w_k]$ is infeasible. Once again, updates for one operation's window may lead to further updates for other operations. Hence we repeatedly apply this until no further updates are possible.

The algorithm C-P Shave can also be viewed as an algorithm to verify that there exists a schedule consistent with a particular window, and hence we can apply this idea recursively. However, if we apply more than one level of recursion then the algorithm becomes impractical, so we consider only Double Shave, which simply calls C-P Shave.

Implementation issues We wish to find the maximum amount to shave off a given window. Since frequently we cannot shave off anything, we initially attempt to shave just one unit, and repeatedly double this trial length until the shaving procedure fails. We then perform a normal bisection search to find the optimal amount to shave.

We can specify exact starting times for some operations while ensuring that feasibility is maintained. For example, if $k \in \mathcal{O}^i$ is the first operation of a job and for each other $l \in \mathcal{O}^i$, $u_l \geq u_k + p_k$ then we can schedule k to start at u_k. We preprocess the instance to find all of these operations.

Each time we attempt to shave a portion of a window we compute a bound for an instance that differs only slightly from the original one. If we have already computed a bound for the original instance, it may be possible to exploit this fact. An advantage of our new $O(n^2)$ algorithm implementation to compute iterated Carlier-Pinson updates is that after updating one head we can compute the implied updates for all of the tails in essentially linear time.

Computational results For each algorithm we report a lower bound T and two running times: the first is the time to show that $T - 1$ is infeasible, whereas the second also includes the bisection search for T. For C-P Shave, we use Iterated Carlier-Pinson as our initial lower bound and an upper bound provided by Applegate & Cook's implementation of the shifting bottleneck procedure. For Double Shave, we use C-P Shave as our initial lower bound and the best known upper bound, primarily due to Balas & Vazacopoulos [4].

C-P Shave provides a significant improvement over Iterated Carlier Pinson. The computation time for C-P Shave ranged from 3 seconds for the fastest problem to 160 seconds for the slowest; the more difficult problems are generally in the one to two minute range.

Double Shave is an excellent bound: in all but three of the instances it is tight. The three problems where Double Shave is not tight are currently unsolved. However we report better bounds in §5 using a branch-and-bound procedure. Double Shave is quite slow: running times range from 10 minutes for some of the easier problems to several days for the hardest. In addition to finding tight bounds for some problems, Double Shave was able to reduce the processing windows to such an extent that it is straightforward to construct a schedule.

It is worth noting the sensitivity to scaling of the various algorithms. The running time of the packing algorithm is proportional to T whereas the running time of iterated Carlier-Pinson is independent of T. The running times of the various Shave algorithms are all proportional to $\log T$, since they involve bisection searches to find the amount to shave.

5 Time-Oriented Branching

Almost all branch-and-bound algorithms for the job shop scheduling problem have been based on the disjunctive graph formulation. We give two branching rules that more directly exploit our processing windows formulation.

Branching on the next available operation We first consider a simple strategy: focus on the first available operation that has not yet been assigned a starting time: this operation either starts immediately or it is delayed. This gives us two branches. If an operation is delayed, we know that it must start at the completion time of some operation on the same machine (since we can restrict our attention to active schedules): in this case we can increase the start of the processing window to the earliest time that an unscheduled operation on that machine can finish.

While the above branching scheme is technically correct, it is not as efficient as possible. Since we do not restrict our search to active schedules, we can find the same schedule in both branches. When we update the head of an operation, we do not require that it immediately follow another operation on the same machine. Hence, it is possible that we delay an operation $k \in \mathcal{O}^i$, but still do not schedule another operation on machine i before we schedule k. The schedule obtained might also fail to be active in that there might be idle time immediately prior to some operation that could be used to process another operation which is currently being processed later.

To solve the first problem, when we delay an operation $k \in \mathcal{O}^i$ we mark it delayed and then require that it be scheduled to start at the completion of some other operation $l \in \mathcal{O}^i$. (To ensure this, we further delay k until it is true.) The second problem requires a bit more care, but we can nonetheless modify this approach to ensure that we generate only active schedules.

Branching on tight set and nearly tight sets A *tight set* is a set $S \subset \mathcal{O}^i$ such that

$$r(S) + P(S) + q(S) = T. \tag{6}$$

An α-*tight set* is a set $S \subseteq \mathcal{O}^i$ such that

$$r(S) + P(S) + q(S) \geq T - \alpha. \tag{7}$$

A tight set $S \subset \mathcal{O}^i$ is significant because we know that machine i must process S without interruption from $r(S)$ until $T - q(S)$. Thus we first attempt to compute the order in which the operations of S are processed. To do this, we start at $r(S)$ and branch depending on which operation starts at time $r(S)$.

Unfortunately, for many instances, tight sets are hard to find, and so we relax the concept and consider α-tight sets. For any α-tight set S, we know that some operation starts in the interval $[r(S), r(S) + \alpha]$. If α is sufficiently small, each permutation of operations of the α-tight set will result in a set of disjoint processing windows. Thus, a natural extension of the branch-and-bound algorithm for tight sets can be used for α-tight sets.

We branch depending on which operation in S is processed first. If operation k is first, we can reduce the size of its window to be at most α and update the window size of the other operations to reflect the fact that they cannot start until k is completed. This creates an α-tight set for the remainder of the operations.

An interesting question is "Are the branches mutually exclusive?" The answer to that is "yes", provided α is not too large relative to the length of the short operations of the tight set.

Lemma 3. *If S is an α-tight set, and k and l are the two shortest operations of $\mathcal{O}^i$ then the branches are mutually exclusive if $\alpha < p_k + p_l$*

Computational Results The results of our branch-and-bound algorithms can be found in Table 1; when we did not obtain a result we report NR in the table.

Our first implementation ran Double Shave to reduce the windows and then branched on the next available operation to obtain a schedule. We used iterated Carlier-Pinson as the lower bounding technique for the branch-and-bound. This approach was successful in solving 11 of the 16 problems. We separate the time to find the optimal schedule into two parts: the time to reduce the windows using Double Shave and the time to find the schedule using branch-and-bound. In all cases where we found a schedule, the branch-and-bound portion took less than a second while the windows reduction phase usually took several thousand seconds. This approach was ineffective for instances where the critical path contained a long chain of operations on the same machine since, in this case, Double Shave performed badly.

Next we did not perform the windows reduction phase but instead used the stronger lower bound, C-P Shave. Using this technique we were able to solve fewer instances. But the amount of time required to solve some of the instances was greatly reduced. This technique seems to be appropriate only for relatively easy instances.

We also implemented α-tight set branching using C-P Shave as our lower bounding technique. We branched on the machine i with the highest one-machine bound and chose S to be the tightest subset of $\mathcal{O}^i$ subject to $|S| \geq n/2$. We indicate the running time to find the optimal schedule given the optimal T, the time to prove that the schedule is optimal and then the time to find the optimal schedule by starting T at the C-P Shave bound and increasing it by one until we find an optimal schedule. Using this technique we are able to solve 11 of the sixteen instances and obtain the best lower bounds known for three others. We were able to solve one other instance, LA27, when we used iterated Carlier-Pinson as our lower bound. The one instance that this technique has so far shown itself ineffective in solving is LA38: this might be due to the large gap between the one-machine bound and the optimal schedule length: consequently it is very difficult to find nearly tight sets.

There seem to be at least two very different classes of difficult job shop scheduling instances. In the first class, the one-machine bound is tight or nearly tight, but it is still very difficult to find an optimal schedule. LA27 is an example of this class: here the one-machine bound is tight but this problem was just recently solved when a new local search technique was finally able to reduce the upper bound to the one-machine bound. In the second class, the one-machine bound is very poor and it seems to be difficult to obtain good lower bounds. LA38 is an example of this sort; in this case the upper bound was established before the lower bound.

It seems likely that techniques that work for one type of problem might not work well for the other. The α-tight set branching seems to work well on the first type of problems. LA37 is a problem in this class. It seems to be difficult for local search techniques: only the guided local search techniques of Balas and Vazacopoulos are able to find an optimal schedule, but using α-tight set branching we are able to solve this problem in under two minutes. As we discussed above, this technique does not work as well on LA38, an instance at the other extreme.

Overall, our time-oriented branch-and-bound algorithms gave good results for the instances that we tested. We were able to solve to optimality four standard benchmark problems which were unsolved when we began our effort, including ABZ7 which to our knowledge we were the first to solve. For the other three unsolved problems that we examined, we were able to improve upon the best known lower bounds. In all cases, the performance of our algorithm is at least competitive with disjunctive graph based approaches and in many cases our algorithm performs significantly better.

It seems likely that further progress can be made by combing the purely combinatorial approaches with our time-oriented approach. From our insight into the dual nature of hard instances, we believe that it is likely such a "multiple-pronged attack" might be most effective.

References

1. J. Adams, E. Balas, and D. Zawack. The shifting bottleneck procedure for job shop scheduling. *Management Sci.*, 34:391–401, 1988.
2. D. Applegate and W. Cook. A computational study of the job-shop scheduling problem. *ORSA J. Comput.*, 3:149–156, 1991.

3. E. Balas. On the facial structure of scheduling polyhedra. *Math. Programming Stud.*, 24:179–218, 1985.

4. E. Balas and A. Vazacopoulos. *Guided Local Search with Shifting Bottleneck for Job Shop Scheduling*. Management Science Research Report #MSRR-609, Graduate School of Industrial Administration, Carnegie Mellon University, Pittsburgh, PA, 1994.

5. P. Bratley, M. Florian, and P. Robillard. On sequencing with earliest starts and due dates with application to computing bounds for the $(n/m/G/F_{max})$ problem. *Naval Res. Logist. Quart.*, 20:57–67, 1973.

6. P. Brucker, B. Jurisch, and B. Sievers. A branch and bound algorithm for the job-shop scheduling problem. *Discrete Appl. Math.*, 49:107–127, 1994.

7. J. Carlier and E. Pinson. An algorithm for solving the job-shop problem. *Management Sci.*, 35:164–176, 1989.

8. J. Carlier and E. Pinson. A practical use of Jackson's preemptive schedule for solving the job-shop problem. *Ann. Oper. Res.*, 26:269–287, 1990.

9. J. Carlier and E. Pinson. Adjustments of heads and tails for the job-shop problem. *European J. Oper. Res.*, 78:146–161, 1994.

10. H. Fisher and G.L. Thompson. Probabilistic learning combinations of local job-shop scheduling rules. In J.F. Muth and G.L. Thompson, editors, *Industrial Scheduling*, pages 225–251. Prentice Hall, Englewood Cliffs, NJ, 1963.

11. M.L. Fisher, B.J. Lageweg, J.K. Lenstra, and A.H.G. Rinnooy Kan. Surrogate duality relaxation for job shop scheduling. *Discrete Appl. Math.*, 5:65–75, 1983.

12. M.R. Garey, D.S. Johnson, and R. Sethi. The complexity of flowshop and jobshop scheduling. *Math. Oper. Res.*, 1:117–129, 1976.

13. R.L. Graham, E.L. Lawler, J.K. Lenstra, and A.H.G. Rinnooy Kan. Optimization and approximation in deterministic sequencing and scheduling. *Ann. Discrete Math.*, 5:287–326, 1979.

14. J.R. Jackson. An extension of Johnson's results on job lot scheduling. *Naval Res. Logist. Quart.*, 3:201–203, 1956.

15. S. Lawrence. *Resource Constrained Project Scheduling: an Experimental Investigation of Heuristic Scheduling Techniques (Supplement)*. Graduate School of Industrial Administration, Carnegie Mellon University, Pittsburgh, PA, 1984.

16. E. Nowicki and C. Smutnicki. A fast taboo search algoritm for the job shop problem. *Management Sci.* To appear.

17. S. A. Plotkin, D. B. Shmoys, and É. Tardos. Fast approximation algorithms for fractional packing and covering problems. *Math. Oper. Res.*, 20:257–301, 1995.

18. D.B. Shmoys. Solving scheduling problems via linear programming. Talk at *ORSA/TIMS*, Boston, MA, May 1994.

19. J.P. Sousa and L.A. Wolsey. A time-indexed formulation of non-preemptive single-machine scheduling problems. *Math. Programming*, 54:353–367, 1992.

20. C. Stein. *Approximation Algorithms for Multicommodity Flow and Shop Scheduling Problems*. PhD thesis, MIT/LCS/TR-550, Laboratory for Computer Science, MIT, Cambridge, MA, 1992.

21. R.J.M. Vaessens, E.H.L. Aarts, and J.K. Lenstra. Job shop scheduling by local search. *Math. Programming B*. To appear.

22. J.M. Van den Akker. *LP-based solution methods for single-machine scheduling problems*. PhD thesis, Eindhoven University of Technology, Eindhoven, The Netherlands, 1994.

	MT10		LA21		LA24		LA25		LA27		LA29		LA37		LA38		LA40		ABZ7	
size (mxn)	10x10		10x15		10x15		10x15		10x20		10x20		15x15		15x15		15x15		15x20	
optimal value or (LB,UB)	930		1046[a]		935		977		1235[b]		(1130,1157)		1397		1196[c]		1222		(655,665)	
	bnd	time	bnd	time	bnd	time	bnd	time	bnd	time	bnd	time	bnd	time	bnd	time	bnd	time	bnd	time
job bound	655		717		704		723		686		723		986		943		955		410	
one-machine bound	808		995		881		894		1235		1114		1355		1077		1170		650	
packing bound	859	431	1014	4536	886	4536	920	481	1235	8082	1116	4307	1382	15946	1110	2053	1176	2406	651	2204
iterated Cartier-Pinson	855	0.1	1033	0.1	889	0.1	919	0.1	1235	0.1	1119	0.1	1397	0.1	1106	0.1	1190	0.2	651	0.1
exact one-machine shave	868	0.7	1033	4.2	898	8.7	919	3.5	1235	12.1	1119	14.2	1397	4.3	1108	4.9	1192	14.9	651	19.0
C-P shave	919	13	1033	0	918	20	958	12	1235	0	1119	0	1397	0	1154	22	1210	32	651	0
with search		45		35		92		57		32		89		0		130		110		113
double shave	930	180	1046	10585	935	3132	977	2176	1235	0	1140	0	1397	0	1196	49337	1222	5000	656	15343
with search		530		34815		18325		12629		0		0		0		132564		45951		65835
double shave then branch on next available job (LB: iterated Cartier-Pinson)																				
double shave time	483		18953		2978		4756		NR		NR		NR		98954		6500		93562	
branch-and-bound time	0.1		0.1		0.1		0.1								0.1		0.1		0.1	
branch on α-tight set (LB: C-P shave)																				
LB found	930	166	1046	637	935	696	977	120	1235	0	1141		1397		1222		1222	468	656	
find schedule time	15		583		332		256		20562[d]		NR		124		NR		507		NR	
total (with search)	1100		3588		4227		5258						124				1407			

Table 1: Computational Results

[a]Prior to the start of our work, the lower bound was 1040
[b]Prior to the start of our work, the upper bound was 1236
[c]Prior to the start of our work, the lower bound was 1184
[d]Using Iterated Cartier-Pinson as lower bound

Optimal On-Line Algorithms for Single-Machine Scheduling

J.A. Hoogeveen and A.P.A. Vestjens

Department of Mathematics and Computing Science,
Eindhoven University of Technology,
P.O.Box 513, 5600 MB, Eindhoven, The Netherlands

Abstract. We consider single-machine on-line scheduling problems where jobs arrive over time. A set of independent jobs has to be scheduled on the machine, where preemption is not allowed and the number of jobs is unknown in advance. Each job becomes available at its release date, which is not known in advance, and its characteristics, e.g., processing requirement, become known at its arrival. We deal with two problems: minimizing total completion time and minimizing the maximum time by which all jobs have been delivered. For both problems we propose and analyze an on-line algorithm based on the following idea: As soon as the machine becomes available for processing, choose an available job with highest priority, and schedule it if its processing requirement is not too large. Otherwise, postpone the start of this job for a while. We prove that our algorithms have performance bound 2 and $(\sqrt{5}+1)/2$, respectively, and we show that for both problems there cannot exist an on-line algorithm with a better performance guarantee.

Keywords: on-line algorithms, single-machine scheduling, worst-case analysis.

1 Introduction

Until a few years ago, one of the basic assumptions made in deterministic scheduling was that all of the information needed to define the problem instance was known in advance. This assumption is usually not valid in practice, however. Abandoning it has led to the rapidly emerging field of on-line scheduling. Two on-line models have been proposed. The first one assumes that there are no release dates and that the jobs arrive in a list. The on-line algorithm has to schedule the first job in this list before it sees the next job in the list (e.g., see Graham (1966) and Chen, Van Vliet & Woeginger (1994)). The second model assumes that jobs arrive over time. Next to the presence of release dates, the main difference between the models is that in the second model jobs do not have to be scheduled immediately upon arrival. At each time that the machine is idle, the algorithm decides which one of the available jobs is scheduled, if any. In this paper we consider two single-machine on-line scheduling problems with release dates.

We deal with the single-machine scheduling problems of minimizing total completion time and maximum time by which all jobs have been delivered, respectively. In the latter problem, after their processing on the machine, the jobs need to be delivered, which takes a certain *delivery time*. The corresponding off-line problems are both strongly NP-hard, but the preemptive versions can be solved in polynomial time through an on-line algorithm (e.g., see Lawler, Lenstra, Rinnooy Kan & Shmoys (1993)). Well known on-line algorithms for the problems are the SPT-rule and LDT-rule: choose from among the available jobs the one with the shortest processing time and largest delivery time, respectively. If all release dates are equal, then the problems are solved by these algorithms. For the case that the release dates are not equal, Mao, Kincaid & Rifkin (1995) prove that SPT has a performance guarantee of n, where n is the number of jobs, and Kise, Ibaraki & Mine (1979) prove that LDT has a performance guarantee of 2. The question is of course: can we do better from a worst-case point of view?

Throughout the paper we use J_j to denote job j, and r_j, p_j, and q_j to denote the release date, processing requirement, and delivery time of J_j, respectively. We denote by $S_j(\sigma)$, $C_j(\sigma)$, and $L_j(\sigma)$, the starting time, completion time, and the time by which J_j is delivered in schedule σ. We use σ to denote the schedule produced by the heuristic and π to denote an optimal schedule.

This paper is organized as follows. In Section 2 we consider the problem of minimizing total completion time on a single-machine. We prove that any on-line algorithm for this problem has a worst-case ratio of at least 2, and we present an algorithm that achieves this bound. Independent of this work both Phillips, Stein & Wein (1995) and Stougie (1995) developed algorithms with equal performance guarantee; the lower bound of 2 was achieved by Stougie as well. We present both algorithms and compare them to our algorithm. In Section 3 we consider the problem of minimizing the time by which all jobs have been delivered. We show that any on-line algorithm has a worst-case ratio of at least $(\sqrt{5} + 1)/2$. Moreover, we present an algorithm that achieves this bound.

2 Total completion time

In this section, we present an on-line 2-approximation algorithm for the single-machine scheduling problem of minimizing total completion time and show that no on-line algorithm can do better from a worst-case point of view. At the end of this section, we compare the algorithms of Phillips et al. and Stougie to our algorithm.

We first show that 2 is a lower bound on the worst-case ratio of any on-line algorithm; this result follows from an example. For a given schedule σ, we use $C(\sigma)$ as a short notation for the total completion time of σ, i.e., $C(\sigma) = \sum_j C_j(\sigma)$.

Theorem 1. *Any on-line algorithm has a worst-case ratio of at least 2.*

Proof. We show this result by describing a set of instances for which no on-line algorithm can guarantee an outcome strictly less than twice the optimum. Consider the following situation. The first job arrives at time 0 and has processing requirement p. The on-line algorithm decides to schedule the job at time S. Depending on S, either no jobs arrive anymore or $n-1$ jobs with processing requirement 0 arrive at time $S+1$. In the first case we get a ratio of $C(\sigma)/C(\pi) = (S+p)/p$, whereas in the second case we get a ratio of $C(\sigma)/C(\pi) \geq n(S+p)/(n(S+1)+p)$. Hence,

$$\frac{C(\sigma)}{C(\pi)} \geq \max\left\{\frac{S+p}{p}, \frac{n(S+p)}{n(S+1)+p}\right\}.$$

The algorithm may choose S so as to minimize this expression. Some simple algebra shows that the best choice for S is

$$S = \frac{n-1}{n}p - 1.$$

This implies a worst-case ratio of

$$\frac{C(\sigma)}{C(\pi)} \geq 2 - \frac{1}{n} - \frac{1}{p}.$$

If we let both n and p tend to infinity, then we get the desired ratio of 2.   □

We can use the example of Theorem 1 to show that any on-line algorithm that schedules a job as soon as the machine is available will have an unbounded worst-case ratio. If an algorithm wants to guarantee a better performance bound, then it needs a waiting strategy. For example, if an available job has a large processing requirement compared to the optimal solution of the currently available instance, the algorithm should wait for extra information. To incorporate this, we slightly modify the SPT-rule and call the new rule the delayed SPT-rule (D-SPT).

ALGORITHM D-SPT
If the machine is idle and a job is available at time t, determine an unscheduled job with smallest processing requirement, say J_i. If there is a choice, take the job with the smallest release date. If $p_i \leq t$, then schedule J_i; otherwise, wait until time p_i or until a new job arrives, whichever happens first.

As we have already seen, the worst-case bound of any algorithm is at least equal to 2. If the performance guarantee exceeds 2, then there exists an instance, which we call *counterexample*, for which the algorithm produces a schedule with value more than twice the optimal value. We show that our algorithm has a performance bound exactly equal to 2, by showing that there does not exist such a counterexample. Thereto, we first derive some characteristics of a *smallest counterexample*, i.e., a counterexample consisting of a minimum number of jobs. Let $\mathcal{I}$ be such a smallest counterexample, and let σ be the schedule created by D-SPT for this instance.

Observation 2. The schedule σ consists of a single block: it possibly starts with idle time after which all jobs are executed contiguously.

Proof. Suppose that σ contains idle time between the execution of the jobs. The jobs scheduled before this idle period do not influence the scheduling decision concerning the jobs scheduled after this idle period, and vice versa. Therefore, the instance can be split into two independent smaller instances. For at least one of these partial instances D-SPT creates a schedule with value more than twice the optimal value, which contradicts the assumption that we considered an instance with a minimum number of jobs. $\quad\Box$

From now on, we assume that the jobs are numbered according to their position in the schedule σ. We partition σ into *subblocks*, such that within every subblock the jobs are ordered according to the SPT-rule, and that the last job of a subblock is larger than the first job of the succeeding subblock if it exists. We denote these subblocks by $B_1, \ldots, B_k$; subblock B_{i+1} consists of the jobs $J_{b(i)+1}, \ldots, J_{b(i+1)}$, where the indices $b(i)$ are determined recursively as $b(i) = \min\{j > b(i-1) \mid p_j > p_{j+1}\}$. The number of subblocks, k, in which the schedule is partitioned, follows from the recursion scheme.

For ease of exposition, we define a dummy job J_0 with $p_0 = S_1(\sigma)$, which will not be scheduled. Although J_0 will not be scheduled, we define $S_0(\sigma) = p_0$. Let $m(i)$ be the index of the job that has the largest processing requirement in the first i blocks, i.e., $p_{m(i)} = \max_{0 \le j \le b(i)} p_j$. We define a *pseudo-schedule* ψ for the schedule σ as follows. The order of the jobs in ψ is the same as in σ, but the first job in B_{i+1} starts at time $S_{b(i)+1}(\sigma) - p_{m(i)}$. Furthermore, all jobs in a block are scheduled contiguously. It is easy to verify that ψ is not a real schedule, since some jobs start before their release date and some jobs overlap. Note that ψ contains no idle time. Let ϕ be an optimal preemptive schedule for $\mathcal{I}$.

Lemma 3. *For all $J_j \in \mathcal{I}$, we have that $C_j(\sigma) - C_j(\psi) \le C_j(\phi)$.*

Proof. Consider an arbitrary job, say J_j, and suppose that $J_j \in B_{i+1}$. For this job $C_j(\sigma) - C_j(\psi) = p_{m(i)}$. If $p_j < p_{m(i)}$, then $r_j > S_{m(i)}(\sigma) \ge p_{m(i)}$, because D-SPT always schedules the smallest available job first and never starts a job before a time smaller than its own processing time. Therefore, either $p_j \ge p_{m(i)}$ or $r_j > p_{m(i)}$, which implies that $C_j(\phi) \ge r_j + p_j \ge p_{m(i)}$. Hence, $C_j(\sigma) - C_j(\psi) \le C_j(\phi)$. $\quad\Box$

Lemma 4. $C(\psi) \le C(\phi)$.

Proof. Let I denote the job set corresponding to the smallest counterexample. Using this instance and the pseudo-schedule ψ for this instance we create a new instance $\mathcal{I}'$. The instance $\mathcal{I}'$ consists of all jobs in I. The processing requirements of the jobs remain the same, but the release dates r_j' are set equal to $\min\{r_j, S_j(\psi)\}$.

Let ϕ' be the optimal preemptive schedule for the instance $\mathcal{I}'$. Determine the first job in ϕ' that starts earlier in ϕ' than in ψ; suppose this job, say J_j, belongs to B_{i+1} in σ. If $p_j \ge p_{m(i)}$, then all jobs scheduled before J_j in ψ have a higher priority, i.e., either they have a smaller processing requirement or they have equal processing requirement and a smaller release date. This implies that in the preemptive schedule these jobs also have a higher priority and hence will be scheduled before J_j, which contradicts the fact that $S_j(\phi') < S_j(\psi)$. If $p_j < p_{m(i)}$, then all jobs that are executed in the interval $[r_j + p_{m(i)}, S_j(\sigma)]$ in σ have a higher priority than J_j. Hence, all jobs executed in the interval $[r_j, S_j(\psi)]$ in ψ have a higher priority than J_j; let V denote the set containing all these jobs. Since J_j is the first job in ϕ' with $S_j(\phi') < S_j(\psi)$, there is no room to start one of the jobs in V before time r_j. Hence, one of the jobs in V must be postponed in ϕ' to enable J_j to start before time $S_j(\psi)$, which is inconsistent with the way ϕ' has been constructed. Therefore, no job starts earlier in ϕ' than in ψ, which implies $C_j(\phi') \geq C_j(\psi)$ for all $j = 1, \ldots, n$.

As the release dates in $\mathcal{I}'$ are smaller than or equal to the release dates in $\mathcal{I}$, we have that $C(\phi) \geq C(\phi')$. Together this implies that $C(\psi) \leq C(\phi') \leq C(\phi)$.
□

Theorem 5. $C(\sigma) \leq 2C(\phi)$.

Proof. Combining Lemmas 3 and 4 we obtain that $C(\sigma) \leq C(\phi) + C(\psi) \leq 2C(\phi)$.
□

Corollary 6. *The on-line algorithm* D-SPT *has performance bound 2.*

The algorithm ONE-MACHINE (1-M) developed by Phillips et al. uses the preemptive schedule. The algorithm maintains a list of jobs that have been completed in the preemptive schedule. As soon as a job has been finished in the preemptive schedule it will be appended to the end of this list. As soon as the machine becomes idle, the first job in this list will be assigned to the machine.

The algorithm developed by Stougie modifies the release dates of the jobs, before they are presented to the on-line algorithm. The release date of each job is increased by its own processing requirement. For this new instance the algorithm uses the SPT-rule. Since the algorithm first shifts the release dates and then uses SPT, we call it the shifted SPT-rule (S-SPT).

All three algorithms 1-M, S-SPT, and D-SPT create schedules with cost no more than twice the value of the optimal preemptive schedule. It is not to difficult to see that D-SPT does not create more idle time than S-SPT, which again does not create more idle time than 1-M. Hence, we might expect that on average D-SPT performs slightly better than the other two algorithms. There exist instances, however, for which one algorithm performs twice as well as the other ones. Table 1, which is displayed on the next page, shows these instances. The values of $C(\sigma)$ are the limiting values for $\varepsilon \downarrow 0$, and σ denotes only the order in which the jobs are scheduled.

3 Maximum delivery time

In this section, we present an on-line α-approximation algorithm for the single-machine scheduling problem of minimizing the time by which all jobs have been

Table 1. Instances to compare the worst-case behavior of D-SPT, S-SPT, and 1-M.

Instance					Algorithm	σ	$C(\sigma)$
j	1	2	$\cdots$	n	D-SPT	$(2,\ldots,n,1)$	n+1
r_j	0	1	$\cdots$	1	S-SPT	$(1,\ldots,n)$	2n
p_j	1	ε	$\cdots$	ε	1-M	$(1,\ldots,n)$	2n
j	1	2	$\cdots$	n	D-SPT	$(1,\ldots,n)$	2n
r_j	ε	$1+\varepsilon/2$	$\cdots$	$1+\varepsilon/2$	S-SPT	$(2,\ldots,n,1)$	n+1
p_j	1	$\varepsilon/2$	$\cdots$	$\varepsilon/2$	1-M	$(1,\ldots,n)$	2n
j	1	2	3	$\cdots$	n+1	D-SPT	$(1,\ldots,n+1)$	2n
r_j	0	0	$1+\varepsilon/2$	$\cdots$	$1+\varepsilon/2$	S-SPT	$(1,\ldots,n+1)$	2n
p_j	ε	1	0	$\cdots$	0	1-M	$(1,3,\ldots,n+1,2)$	n+1

delivered, where $\alpha = (\sqrt{5}+1)/2$, and show that no on-line algorithm can do better from a worst-case point of view. We start with the latter. Again, we prove the lower bound on the worst-case ratio by means of an example for which any on-line algorithm will have at least the required ratio. Let $L_{\max}(\pi)$ denote the minimum time by which all jobs can be delivered, and let $L_{\max}(\sigma)$ denote the time by which all jobs are delivered in schedule σ, where σ is the schedule obtained through some on-line algorithm.

Theorem 7. *Any on-line algorithm has a worst-case ratio of at least α.*

Proof. Consider the following situation. The first job arrives at time 0 and has processing requirement $p_1 = p$ and delivery time $q_1 = 0$. The on-line algorithm decides to schedule the job at time S. Depending on S, either no jobs arrive any more or one job with processing requirement $p_2 = 1$ and delivery time $q_2 = p$ arrives at time $r_2 = S+1$. In the first case we get a ratio of $L_{\max}(\sigma)/L_{\max}(\pi) = (S+p)/p$; in the second case we get a ratio of $L_{\max}(\sigma)/L_{\max}(\pi) \geq (S+2p+1)/(S+p+2)$. Hence,

$$\frac{L_{\max}(\sigma)}{L_{\max}(\pi)} \geq \max\left\{ \frac{S+p}{p}, \frac{S+2p+1}{S+p+2} \right\}.$$

The algorithm may choose S so as to minimize this expression. Some simple algebra shows that the best choice for S is

$$S = \frac{p}{2}\left(\sqrt{5 - \frac{4}{p^2}} - 1\right) - 1.$$

This implies a worst-case ratio of

$$\frac{L_{\max}(\sigma)}{L_{\max}(\pi)} \geq \frac{1}{2}(\sqrt{5 - \frac{4}{p^2}} + 1) - \frac{1}{p}.$$

If we let p tend to infinity, then we get the desired ratio of α. $\qquad\square$

We can use the example of Theorem 7 to show that any on-line algorithm that schedules a job as soon as the machine is available will have a worst-case ratio of at least 2. Note that a simple algorithm like LDT already achieves this bound. Again, if an algorithm wants to guarantee a better performance bound, then it needs a waiting strategy. Therefore, we modify the LDT-rule and call the new rule the delayed LDT-rule (D-LDT). The basic idea behind the algorithm is that, if no jobs with a large processing requirement are available, then we should schedule the job with the largest delivery time; otherwise, we should decide whether to schedule the large job, the job with the largest delivery time, or no job at all.

Throughout this section, we use the following notation:

- $p(S)$ denotes the total processing time of all jobs in S;
- $J(t)$ is the set containing all jobs that arrived at or before time t;
- $U(t)$ is the set containing all jobs in $J(t)$ that have not been started at time t;
- t_1 denotes the start time of the last idle time period before time t; if there is no idle time, then define $t_1 = 0$.
- We call a job J_j *big* if $p_j > (\alpha - 1)p((J(t) \setminus J(t_1)) \cup U(t_1))$.
  Note that $(J(t) \setminus J(t_1)) \cup U(t_1)$ contains all jobs that arrived at or before time t and that were not completed at time t_1;
- $J_i(t)$ denotes the job with the largest processing time in $U(t)$.
- $J_m(t)$ denotes the job with the largest delivery time in $U(t)$.

ALGORITHM D-LDT
Wait until the machine is idle and a job is available. Suppose this happens at time t. If there is no big job available, then schedule $J_m(t)$. Otherwise, do the following.

- If $J_i(t)$ is the only available job, then wait until a new job arrives or until time $r_i + (\alpha - 1)p_i$, whichever happens first.
- Otherwise,
  if $t + p(U(t)) > r_i + \alpha p_i$, schedule $J_m(t)$ if $q_m > (\alpha - 1)p_i$ and $J_i(t)$, otherwise; else, if $J_m(t) \neq J_i(t)$, schedule $J_m(t)$, else schedule the job with the second largest delivery time.

Again we work with a smallest counterexample, where smallest refers to the number of jobs. Let $\mathcal{I}$ be such a smallest counterexample, and let σ be the schedule created by D-LDT for $\mathcal{I}$. We suppose that J_l denotes the first completed job in σ that assumes the value $L_{\max}(\sigma)$.

Observation 8. The schedule σ consists of a single block: it possibly starts with idle time after which all jobs are executed contiguously.

Proof. Suppose to the contrary that σ does not have this form. We will show that then either we can find a counterexample that consists of a smaller number of jobs, or that this alleged counterexample is not a counterexample at all.

Suppose that σ consists of more than one block. Suppose that block B is a block that contains a job J_l with $L_l(\sigma) = L_{\max}(\sigma)$; consider any block that precedes B. Since the algorithm bases its choices on the set $(J(t) \setminus J(t_1)) \cup U(t_1)$, the existence of the jobs that are completed before the start of block B does not influence the start time of B and the order in which the jobs are executed. Therefore, we can remove all jobs that are completed before the start of block B without changing the value $L_{\max}(\sigma)$ and without increasing $L_{\max}(\pi)$. Similarly, we can remove all jobs from $\mathcal{I}$ that are released after the start of J_l in σ. Therefore, we may assume that our counterexample consists of the jobs from block B and the jobs that are available at the start of J_l in σ but that are scheduled in another block. Since the algorithm always starts a job if more than one job is available and the machine is empty, we know that there is at most one job that is available at time $S_l(\sigma)$ and does not belong to B; moreover, we know that this job, which we denote by J_i, must be marked big. Let $S(B)$ and $C(B)$ denote the start time of the first job and the completion time of the last job in B. Since J_i is big, $(\alpha-1)p_i > p(B)$. Let J_1 be the first available job, which may be equal to J_i. Due to the operation of the algorithm, $S(B) = \min\{r_1+(\alpha-1)p_1, r_2\}$, where r_2 denotes the release date of the second available job. Since J_l is a job in B, we know that $L_{\max}(\sigma) = L_l(\sigma) = C_l(\sigma) + q_l \le C(B) + q_l$. If J_1 is the first job in π, then $L_{\max}(\pi) \ge L_l(\pi) \ge r_1 + p_1 + p_l + q_l > S(B) + q_l$, from which we derive that $L_{\max}(\sigma) - L_{\max}(\pi) < C(B) - S(B) = p(B) < (\alpha-1)p_i \le (\alpha-1)L_{\max}(\pi)$, which disproves the validity of our counterexample. If J_1 is not the first job in π, then the first job in π cannot start before time $r_2 \ge S(B)$, which implies that $L_{\max}(\pi) \ge L_l(\pi) \ge S(B) + p_l + q_l$, and we again have that $L_{\max}(\sigma) - L_{\max}(\pi) \le C(B) - S(B)$, from which we deduce that $\mathcal{I}$ does not correspond to a counterexample. □

From now on, we let J_0 denote the job that arrives first in $\mathcal{I}$. Note that without loss of generality we may assume that $r_0 = 0$.

Observation 9. For all $J_j \in I \setminus \{J_0\}$, we have that $p_j \le (\alpha-1)p(\mathcal{I})$.

Proof. Suppose to the contrary that there does exist a job J_1 with $r_1 \ge r_0$ that has $p_1 > (\alpha-1)p(\mathcal{I})$, i.e., $\alpha p_1 > p(\mathcal{I})$. Then at time r_1 there are at least two jobs available, which implies that the algorithm starts a job if it had not done so already. On basis of Observation 8, we may conclude that there is no idle time in the remainder of the schedule. But since J_1 is marked as big by the algorithm, this can only be the case if the other jobs are able to keep the machine busy from time r_1 to time $r_1 + (\alpha-1)p_1$. In that case, however, $(\alpha-1)p_1 \le p(\mathcal{I}) - p_1 < \alpha p_1 - p_1 = (\alpha-1)p_1$, which is a contradiction. □

We let J_k denote the last job in σ before J_l with a delivery time smaller than q_l, and we let $G(l)$ denote the set containing J_l and all jobs between J_k and J_l in σ. Note that all jobs in $G(l)$ have delivery time greater than or equal to q_l.

Observation 10. $p_k > (\alpha - 1)p(\mathcal{I})$.

Proof. If J_k does not exist, then

$$L_{\max}(\pi) \geq \sum_{j \in G(l)} p_j + q_l.$$

Since the first job in the block starts at time $(\alpha - 1)p_0$ at the latest,

$$L_{\max}(\sigma) = C_l(\sigma) + q_l \leq (\alpha-1)p_0 + \sum_{j \in G(l)} p_j + q_l \leq (\alpha-1)p_0 + L_{\max}(\pi) \leq \alpha L_{\max}(\pi),$$

which contradicts the fact that we consider a counterexample. Therefore, we assume from now on that such a job J_k exists. There are two possibilities for the algorithm to select J_k and not one of the jobs from $G(l)$:

(1) All jobs in $G(l)$ have a release date larger than $S_k(\sigma)$.
(2) There is one job from $G(l)$ available, which we denote by J_1, that is marked as big and cannot be started yet. Note that, since J_1 cannot be started yet, we must have that $S_k(\sigma) + p_k \leq r_1 + (\alpha - 1)p_1$.

For case (1), we have that

$$L_{\max}(\pi) \geq \min_{j \in G(l)} r_j + \sum_{j \in G(l)} p_j + q_l > S_k(\sigma) + \sum_{j \in G(l)} p_j + q_l,$$

and since $L_{\max}(\sigma) = C_l(\sigma) + q_l = S_k(\sigma) + p_k + \sum_{j \in G(l)} p_j + q_l$, we deduce that

$$L_{\max}(\sigma) - L_{\max}(\pi) < p_k \leq (\alpha - 1)p(\mathcal{I}) \leq (\alpha - 1)L_{\max}(\pi).$$

Concerning case (2), we have that

$$L_{\max}(\pi) \geq \min_{j \in G(l)} r_j + \sum_{j \in G(l)} p_j + q_l = r_1 + \sum_{j \in G(l)} p_j + q_l,$$

from which we deduce that

$$L_{\max}(\sigma) - L_{\max}(\pi) < S_k(\sigma) + p_k - r_1 \leq r_1 + (\alpha - 1)p_1 - r_1 \leq (\alpha - 1)L_{\max}(\pi).$$

Since neither of both cases corresponds to a counterexample, we conclude that J_k must be big. $\square$

Corollary 11. $J_k = J_0$.

For our analysis in Theorem 13, we need the following lemma.

Lemma 12. *Either J_0 is the first job in π, or $C_{\max}(\pi) \geq C_{\max}(\sigma) \geq \alpha p_0$.*

Proof. Let J_1 be the first job other than J_0 that becomes available. As there are two jobs available at time r_1, the algorithm starts one of the jobs if the machine is still idle. Therefore, the first job in σ starts no later than the first job in π, and since there is no idle time in σ, we have $C_{\max}(\sigma) \leq C_{\max}(\pi)$. It is easily checked that $C_{\max}(\sigma) \geq \alpha p_0$. □

Theorem 13. *The on-line algorithm D-LDT has performance bound α.*

Proof. Suppose to the contrary that there exists an instance for which the algorithm finds a schedule σ with $L_{\max}(\sigma) > \alpha L_{\max}(\pi)$. Obviously, then there exists a counterexample $\mathcal{I}$ with a minimum number of jobs. On basis of Observations 8 through 10, we may assume that the first job available in $\mathcal{I}$, which is defined to be J_0, has $p_0 > (\alpha - 1)p(\mathcal{I})$. Note that, due to Corollary 11, J_0 is the last job before J_l in σ with a delivery time smaller than q_l. J_0 starts no later than at time $(\alpha - 1)p_0$ unless some job with delivery time greater than $(\alpha - 1)p_0$ is available. Let $G(h)$ denote the set of jobs that were selected instead of J_0 when J_0 was eligible for being scheduled; $G(h)$ may be empty. Let $S_h(\sigma)$ denote the start time of the first job in this set if available; $S_h(\sigma) \leq (\alpha - 1)p_0$. Note that, if $S_0(\sigma) > (\alpha - 1)p_0$, then $G(h) \neq \emptyset$.

The proof proceeds by a case-by-case analysis. There are two reasons possible for starting J_0 at time $S_0(\sigma)$ instead of a job from $G(l)$. The first one is that simply none of the jobs in $G(l)$ were available, i.e., $r_j > S_0(\sigma)$ for all $J_l \in G(l)$. The second one is that the available jobs in $G(l)$ all have a delivery time at most equal to $(\alpha - 1)p_0$. We cover both cases by distinguishing between

(1) $r_j > S_0(\sigma)$ for all $J_j \in G(l)$, and
(2) $q_j \leq (\alpha - 1)p_0$ for some $J_j \in G(l)$.

Case 1. Since none of the jobs in $G(l)$ is available at time $S_0(\sigma)$,

$$L_{\max}(\pi) > S_0(\sigma) + \sum_{j \in G(l)} p_j + q_l, \text{ and}$$
$$L_{\max}(\sigma) = S_0(\sigma) + p_0 + \sum_{j \in G(l)} p_j + q_l.$$

Hence, $L_{\max}(\sigma) - L_{\max}(\pi) < p_0$. If J_0 is not the first job in π, then according to Lemma 12 $L_{\max}(\pi) \geq C_{\max}(\pi) \geq \alpha p_0$, which implies that $L_{\max}(\sigma) - L_{\max}(\pi) < (\alpha - 1)L_{\max}(\pi)$. Therefore, we assume that J_0 is the first job in π. Then

$$L_{\max}(\pi) \geq p_0 + \sum_{j \in G(l)} p_j + q_l,$$

and hence, $L_{\max}(\sigma) - L_{\max}(\pi) \leq S_0(\sigma)$. Now, either $S_0(\sigma) \leq (\alpha - 1)p_0$, which disqualifies the counterexample, or $G(h) \neq \emptyset$. Note that all jobs in $G(h)$ have a delivery time greater than $(\alpha-1)p_0$. Since J_0 is the first job in π, $L_{\max}(\pi) > \alpha p_0$, and we do not have a counterexample.

Case 2. Since all jobs in $G(l)$ have a delivery time that is at least as large as q_l, we have that $q_l \leq (\alpha - 1)p_0$. If J_0 is not the first job in π, then according to Lemma 12, $C_{\max}(\sigma) \leq C_{\max}(\pi)$, and we get

$$L_{\max}(\sigma) = C_l(\sigma) + q_l \leq C_{\max}(\sigma) + q_l \leq C_{\max}(\pi) + q_l \leq L_{\max}(\pi) + q_l \leq L_{\max}(\pi) + (\alpha - 1)p_0 \leq \alpha L_{\max}(\pi).$$

Therefore, we assume that J_0 is the first job in π. Since all jobs in $G(h)$ have a delivery time greater than $(\alpha - 1)p_0$, J_l is the job with the smallest delivery time in $G(h) \cup G(l)$. Combining all this yields

$$L_{\max}(\pi) \geq p_0 + \sum_{j \in G(h) \cup G(l)} p_j + q_l, \text{ and}$$
$$L_{\max}(\sigma) = S_h(\sigma) + p_0 + \sum_{j \in G(h) \cup G(l)} p_j + q_l,$$

which implies that $L_{\max}(\sigma) - L_{\max}(\pi) \leq S_h(\sigma)$, and we are done since $S_h(\sigma) \leq (\alpha - 1)p_0$.

Since we have checked all possibilities, we conclude that there is no counterexample to Theorem 13. $\qquad\qquad\qquad\qquad\qquad\qquad\qquad\qquad\qquad\qquad\Box$

References

CHEN, B., A. VAN VLIET, AND G.J. WOEGINGER [1994], New lower and upper bounds for on-line scheduling, *Operations Research Letters* **16**, 221–230.

GRAHAM, R.L. [1966], Bounds for certain multiprocessing anomalies, *Bell System Technical Journal* **45**, 1563–1581.

KISE, H., T. IBARAKI, AND H. MINE [1979], Performance analysis of six approximation algorithms for the one-machine maximum lateness scheduling problem with ready times, *Journal of the Operations Research Society of Japan* **22**, 205–224.

LAWLER, E. L., J.K. LENSTRA, A.H.G. RINNOOY KAN, AND D.B. SHMOYS [1993], Sequencing and scheduling: Algorithms and complexity, in: S.C. Graves, A.H.G. Rinnooy Kan, and P.H. Zipkin (eds.), *Logistics of Production and Inventory*, Handbooks in OR & MS 4, Elsevier Science Publishers B.V., Amsterdam, Chapter 9, 445–522, ISBN 0-444-87472-0.

MAO, W., R.K. KINCAID, AND A. RIFKIN [1995], On-line algorithms for a single machine scheduling problem, in: S.G. Nash and A. Sofer (eds.), *The impact of emerging technologies on computer science and operations research*, Kluwer Academic Press, Chapter 8, 157–173.

PHILLIPS, C., C. STEIN, AND J. WEIN [1995], Scheduling jobs that arrive over time, *Proceedings of the Fourth Workshop on Algorithms and Data Structures*, Lecture Notes in Computer Science 955, Springer.

STOUGIE, L. [1995], personal communication.

The Strongest Facets of the Acyclic Subgraph Polytope Are Unknown

Michel X. Goemans[1*] and Leslie A. Hall[2**]

[1] Dept. of Mathematics, Room 2-382, M.I.T., Cambridge, MA 02139.[***]
[2] Dept. of Math. Sciences, The Johns Hopkins University, Baltimore, MD 21218.[†]

Abstract. We consider the acyclic subgraph polytope and define the notion of strength of a relaxation as the maximum improvement obtained by using this relaxation instead of the most trivial relaxation of the problem. We show that the strength of a relaxation is the maximum of the strengths of the relaxations obtained by simply adding to the trivial relaxation each valid inequality separately. We also derive from the probabilistic method that the maximum strength of any inequality is 2. We then consider all (or almost all) the known valid inequalities for the polytope and compute their strength. The surprising observation is that their strength is at most slightly more than 3/2, implying that the strongest inequalities are yet unknown. We then consider a pseudo-random construction due to Alon and Spencer based on quadratic residues to obtain new facet-defining inequalities for the polytope. These are also facet-defining for the linear ordering polytope.

1 Introduction

Given weights w_a on the arcs of a complete directed graph (or digraph) $D = (V, A)$, the *acyclic subgraph problem* is that of determining a set of arcs of maximum total weight that define an acyclic subgraph. The complement of an acyclic subgraph is called a *feedback arc set*. For general graphs, the acyclic subgraph problem is NP-hard, even for graphs with unit weights and with total indegree and outdegree of every vertex no more than three [GJ79], although the problem is polynomially solvable for planar graphs as was shown by Lucchesi and Younger [LY78]. For any number n of vertices, the acyclic subgraph polytope P_{AC}^n is defined as the convex hull of incidence vectors of acyclic subgraphs of the complete digraph on n vertices; for simplicity, we will omit the superscript n. The acyclic subgraph polytope P_{AC} was extensively studied by Grötschel, Jünger and Reinelt [GJR85a, Ju85, Re85]. At the present time, many classes of facet-defining valid inequalities are known (in addition to the references just cited,

---

[*] Research supported in part by NSF contract 9302476-CCR, a Sloan fellowship, and ARPA Contract N00014-95-1-1246.
[**] Research supported in part by NSF RIA award DMI-9496153.
[***] Email: goemans@math.mit.edu.
[†] Email: lah@jhu.edu.

see, e.g., [Gi90, Ko95, LL94, Su92]; for a survey of many of these inequalities, see Fishburn [Fi92]).

Goemans [Go95] introduced a notion for evaluating the *strength* of a linear programming relaxation for a combinatorial problem, relative to a weaker relaxation of that problem. He applied these results to compute the relative strength of classes of facet-defining inequalities for the traveling salesman problem. Motivated by that definition, we set out to determine the strengths of the known classes of facet-defining inequalities for the maximum acyclic subgraph problem, using as our relaxation for comparison the completely trivial relaxation of P_{AC} given by

$$P = \{x : x_{ij} + x_{ji} \leq 1 \text{ for all } i, j \in V\}.$$

In general, we were aware that certain probabilistic results imply that the strength of P_{AC} itself relative to P must be close to 2 (specifically, $2 - o(1)$[5]), something that follows from the fact that, for large random tournaments, the maximum acyclic subgraph has at most about half the total number of arcs (with high probability). Thus we were quite surprised to discover, in computing the strengths of the known inequalities, that in every case except one, their strength was at most $3/2$ (see Table 1). In the last case, the strength was still no more than $55/36$. This "gap" implies, in particular, that if we were to choose a large random graph and optimize a unit-weight function over the polytope consisting of *all* known valid inequalities, the relaxed solution value would (with high probability) be off by at least 30 percent from the true optimum!

Inequality Type		Reference	General Strength	Max Strength	
k-dicycle	$(k \geq 3)$	[GJR85a, Ju85]	$k/(k-1)$	$3/2$	
k-fence	$(k \geq 3)$	[GJR85a, Ju85]	$k^2/(k^2-k+1)$	$9/7$	$= 1.2857...$
augmented k-fence $(k \geq 3)$		[LL94, Mc90]	$(3k^2-4k)/(2k^2-3k+1)$	$55/36$	$= 1.5277...$
r-reinforced k-fence $(k \geq 3, 1 \leq r \leq k-2)$		[LL94, Su92]	$\dfrac{k^2+(r-1)k}{k^2-k+(r^2+r)/2}$	$\dfrac{1+\sqrt{3}}{2}$	$= 1.3660...$
k-wheel	$(k \geq 3)$	[Ju85]	$10k/(7k-1)$	$3/2$	
Z_k	$(k \geq 4)$	[Re85]	$(4k+3)/(3k+2)$	$19/14$	$= 1.3571...$
diagonal		[Gi90]		$< 3/2$	
α-critical fence		[Ko95]	$(|V|+2|E|)/(\alpha(G)+2|E|)$	$9/7$	$= 1.2857...$
node-disjoint k-Möbius ladder $(k \geq 3)$		[GJR85a, Ju85]	$\leq 6k/(5k-1)$	$9/7$	

Table 1. Strength of various classes of valid inequalities for P_{AC}.

---

[5] $o(1)$ means that this term is nonnegative and tends to 0 as the number of vertices n tends to ∞.

Armed with these results, we started hunting for new facet-defining inequalities with strength closer to 2. The probabilistic proof that the strength of P_{AC} is $2 - o(1)$ shows that an inequality based on a uniformly selected random tournament has strength $2 - o(1)$ with high probability. Many results in random graph theory of this flavor, however, are highly "existential" in nature and indicate no way of explicitly constructing a graph with the desired (highly probable) property. However, in some cases, explicit constructions (often based on number-theoretic arguments) are known which exhibit almost the same properties as their random counterparts. For the maximum acyclic subgraph problem, Alon and Spencer [AS92] explicitly construct such *pseudo-random* tournaments with an upper bound on the size of the maximum acyclic subgraph asymptotically close to one-half. These tournaments are known as *Paley* tournaments.

Asymptotically, these tournaments induce valid inequalities which have strength arbitrarily close to 2, even though there is no guarantee that they define facets of P_{AC}. We have considered Paley tournaments on several small numbers of vertices, and have discovered that the associated Paley inequality on 11 vertices already has strength larger than that of all known valid inequalities, and also defines a facet for P_{AC}. The same results hold for 19 vertices.

The polytope P_{AC} is a cousin of the *linear ordering* polytope P_{LO} (see [GJR85b] and references therein). A linear ordering is a permutation π on the vertex set $\{1, 2, \cdots, n\}$ and will be denoted by $\langle \pi(1), \pi(2), \cdots, \pi(n) \rangle$. Any linear ordering induces an acyclic subgraph $\{(\pi(i), \pi(j)) : i < j\}$. The linear ordering polytope P_{LO} is the convex hull of the maximal acyclic subgraphs of complete graphs induced by linear orderings. Clearly, P_{LO} is a face of P_{AC} [GJR85b]. More precisely, $P_{LO} = \{x \in P_{AC} : x_{ij} + x_{ji} = 1\}$. Moreover, for any nonnegative weight function, any optimal solution over P_{LO} is also an optimal solution over P_{AC}. For technical reasons, our results about the strengths of inequalities will apply only to the acyclic subgraph polytope; nonetheless, our new facet-defining inequalities turn out to be facet-defining for the linear ordering polytope, as well.

The extended abstract is structured as follows. In the next section, we generalize the results on the strength of relaxations derived in [Go95] to polytopes of anti-blocking type, such as P_{AC}. In Section 3, we establish that the strength of P_{AC} is $2 - o(1)$, and we compute the strength of almost all known inequalities for P_{AC} in Section 4. In Section 5, we present the Paley inequalities and give a general condition under which they are facet-defining. Finally, in Section 6 we consider a different notion of strength by defining it relative to a stronger LP relaxation of P_{AC}.

2 Strength

A notion of *strength* of a relaxation was introduced by Goemans [Go95] for polyhedra of blocking type. The strength of a relaxation is one measure of how well a relaxation approximates a polyhedron in comparison to another weaker relaxation. In this section, we derive the equivalent results for polytopes of anti-blocking type, i.e., for polyhedra $P \subseteq \mathbb{R}^n_+$ such that $y \in P$ and $0 \le x \le y$

imply that $x \in P$. The exposition is adapted from [Go95]. We state the main result (Theorem 2) in the case of polytopes restricted to the nonnegative orthant, although the result holds for more general anti-blocking polyhedra as well.

If P and Q are polyhedra in $\mathbb{R}^n$ then we say that P is a *relaxation* of Q or Q is a *strengthening* of P if $P \supseteq Q$. For polytopes P and Q of anti-blocking type, we say that P is an α-relaxation of Q ($\alpha \in \mathbb{R}, \alpha \geq 1$) or Q is an α-strengthening of P if $Q \supseteq P/\alpha = \{x/\alpha : x \in P\}$, i.e., Q is a relaxation of P/α. Any α-relaxation is also a β-relaxation for any $\beta \geq \alpha$. Also, let $t(P, Q)$ denote the minimum value of α such that P is an α-relaxation of Q. Notice that $t(P, Q) \geq 1$, $t(P, Q) = 1$ if and only if $P = Q$, and that $t(P, Q)$ could be infinite.

The following lemma follows trivially from the separating hyperplane theorem.

Lemma 1. *Let P be a relaxation of a polytope Q, P and Q being of anti-blocking type. Then P is an α-relaxation of Q if and only if, for any nonnegative vector $w \in \mathbb{R}^n$,*

$$Max\{wx : x \in Q\} \geq \frac{1}{\alpha} Max\{wx : x \in P\}.$$

As a corollary, $t(P, Q)$ is equal to

$$t(P, Q) = \text{Sup}_{w \in \mathbb{R}^n_+} \frac{Max\{wx : x \in P\}}{Max\{wx : x \in Q\}}, \tag{1}$$

where, by convention, $\frac{0}{0} = 1$.

The following result gives an alternative characterization of $t(P, Q)$ when a description of Q in terms of linear inequalities is known. The proof is similar to that of Theorem 2 in [Go95].

Theorem 2. *Let P and Q be polytopes of anti-blocking type, and let P be a relaxation of Q. Assume $Q = \{x : a_i x \leq b_i \text{ for } i = 1, \ldots, m, x \geq 0\}$, $a_i, b_i \geq 0$ for $i = 1, \ldots, m$. Then*

$$t(P, Q) = Max_i \frac{d_i}{b_i},$$

where $d_i = Max\{a_i x : x \in P\}$.

Proof. From (1), it is clear that

$$t(P, Q) \geq Max_i \frac{Max\{a_i x : x \in P\}}{Max\{a_i x : x \in Q\}} \geq Max_i \frac{d_i}{b_i}.$$

We therefore need to prove the reverse inequality.

Let w be any nonnegative weight function. By strong duality, we know that

$$
\begin{array}{ll}
\text{Max } wx & = \text{Min } b^T y \\
\text{s.t. } Ax \leq b & \text{s.t. } A^T y \geq w^T \\
\quad x \geq 0 & \quad y \geq 0,
\end{array}
$$

where ^{T} denotes the transpose. Let y^* be the optimal dual solution of the above program. Then

$$\begin{array}{c} \text{Max } wx \\ s.t. \ x \in P \end{array} \leq \begin{array}{c} \text{Max } (y^*)^T Ax \\ s.t \ x \in P \end{array} \leq \sum_i \left\{ \begin{array}{c} \text{Max } a_i x \\ s.t. \ x \in P \end{array} \right\} y_i^* = \sum_i d_i y_i^*$$

(the middle inequality follows from the fact that $x \geq 0$ and $(y^*)^T A \geq w^T$). Hence,

$$\frac{\text{Max}\{wx : x \in P\}}{\text{Max}\{wx : x \in Q\}} \leq \frac{\sum_i d_i y_i^*}{\sum_i b_i y_i^*} = \sum_i \left(\frac{b_i y_i^*}{\sum_j b_j y_j^*} \right) \frac{d_i}{b_i}.$$

Since $d_i \geq 0$ (since $a_i \geq 0$) and $y_i^* \geq 0$, the latter quantity can be interpreted as a convex combination of $\frac{d_i}{b_i}$ and is therefore less than or equal to

$$\text{Max}_i \frac{d_i}{b_i}.$$

The result is proved by taking the supremum over all nonnegative weight functions w.

Theorem 2 can be rephrased as follows. To compute $t(P, Q)$, one only needs to consider the cases in which a single inequality of Q is added to P. This motivates the following definition. The *strength* of an inequality $ax \leq b$ with respect to a polytope P is defined as

$$\frac{\text{Max}\{ax : x \in P\}}{b}.$$

Theorem 2 implies that $t(P, Q)$ is equal to the maximum strength with respect to P of a facet-defining inequality for Q.

3 Strength of the Acyclic Subgraph Polytope

The acyclic subgraph polytope P_{AC} is of anti-blocking type, and therefore the results of the previous section apply.

Theorem 3. *The strength $t(P, P_{AC})$ of the acyclic subgraph polytope P_{AC} on n vertices is $2 - O(1/\sqrt{n})$.*

Proof. We first show that the strength of P_{AC} is at most 2. This is a well-known result. Given any nonnegative weight function w, consider the acyclic subgraphs induced by two opposite linear orderings, i.e. $\langle 1, 2, \cdots, n \rangle$ and $\langle n, n-1, \cdots, 2, 1 \rangle$. Clearly, every arc is in one of these linear orderings and, therefore, the maximum weight of the acyclic subgraphs induced by these 2 orderings is at least $\frac{1}{2} \sum_{i,j} w_{ij}$. Thus, using (1), $t(P, P_{AC})$ is at most

$$t(P, P_{AC}) \leq \text{Sup}_{w \in \mathbf{R}_+^n} \frac{\sum_{i<j} \max(w_{ij}, w_{ji})}{\frac{1}{2} \sum_{i<j} (w_{ij} + w_{ji})} \leq 2.$$

To establish a lower bound on $t(P, P_{AC})$, we use a probabilistic result. A *tournament* is a directed graph $D = (V, A)$ in which for every (i, j) exactly one of (i, j) or (j, i) belongs to A. We can associate a weight function to every tournament: $w_{ij} = 1$ if $(i, j) \in A$ and 0 otherwise. The strength $t(P, P_{AC})$ given in (1) must be at least the maximum ratio obtained by considering only the weight functions associated with tournaments. Consider now a random tournament chosen uniformly among all tournaments. Erdös and Moon [EM65] have shown that the size $f(n)$ of the maximum acyclic subgraph in a random tournament is $f(n) = \frac{1}{2}\binom{n}{2} + O(n^{3/2}\sqrt{\ln n})$ with high probability and this was refined by Spencer [Sp71] (and de la Vega [Ve83]) to $f(n) = \frac{1}{2}\binom{n}{2} + \Theta(n^{3/2})$ with high probability. As a result, a random tournament gives a ratio of $\frac{\binom{n}{2}}{f(n)} = 2 - O(1/\sqrt{n})$ with high probability, and the worst tournament must give a ratio at least as high.

4   Strength of Known Valid Inequalities

We begin by characterizing certain properties of valid inequalities for P_{AC}; these facts can be found in Jünger [Ju85]. Since P_{AC} is of anti-blocking type, all its facet-defining valid inequalities, except the nonnegativity constraints, are of the form $ax \leq b$ with $a \geq 0$ and $b > 0$. Moreover, any facet-defining valid inequality $ax \leq b$, except the nonnegativity constraints and the inequalities $x_{ij} + x_{ji} \leq 1$, must satisfy $\min(a_{ij}, a_{ji}) = 0$ for all (i, j). We can therefore restrict our attention to such *support reduced* inequalities [BP91].

For the acyclic subgraph problem and the trivial relaxation P given in Section 1, it is very easy to compute the strength of any support reduced inequality $ax \leq b$ with $a, b \geq 0$. Indeed, when optimizing over P, the problem decomposes over all pairs of indices and, as a result, the strength of $ax \leq b$ is given by

$$\frac{\sum_{i<j} \max(a_{ij}, a_{ji})}{b} = \frac{\sum_{i,j} a_{ij}}{b}. \tag{2}$$

In particular, if the inequality is a *rank* inequality, i.e. an inequality of the form $x(A) \leq b$ where $x(A) = \sum_{(i,j)\in A} x_{ij}$, as are most known facet-defining valid inequalities for P_{AC}, then the strength is simply $|A|/b$.

We can use these formulas to compute the strength of most classes of known facet-defining valid inequalities for P_{AC}. However, because of space limitations, we do not include the description of the known inequalities. We refer the reader to Fishburn [Fi92] for a survey. We summarize our findings in Table 1. All the inequalities listed in the table are rank inequalities except the augmented k-fence, the r-reinforced k-fence, and the diagonal inequalities. The column "Max Strength" indicates the maximum value over all possible values of the parameters of an inequality in the class considered. As mentioned before, the strength of any of these inequalities appears to be at most $3/2$, except for the augmented k-fences, whose strength is maximized for $k = 5$ at $55/36$.

We would like to comment on three of the entries. The node-disjoint k-Möbius ladder inequalities are rank inequalities $x(A) \leq b$ with $b = |A| - \frac{k+1}{2}$. The value of the strength indicated in the table follows from the fact that $|A|$ can be seen to be at least $3k$ when the cycles in the definition of the Möbius ladder are node-disjoint. For the diagonal inequalities, we have not computed their exact strength, but we use the following result.

Lemma 4. *Let $ax \leq b$ be a support reduced valid inequality for P_{AC} and assume that $ax \leq b$ is implied (over P_{LO}) by the valid inequalities $c_i x \leq d_i$ for P_{AC} and the equalities $x_{ij} + x_{ji} = 1$ for all i, j valid only over P_{LO}. Then the strength of $ax \leq b$ is at most the maximum strength of the inequalities $c_i x \leq d_i$. Moreover, the strength is strictly less than the maximum strength if the equalities $x_{ij} + x_{ji} = 1$ are needed.*

Proof. Let l_i/d_i be the strength of inequality $c_i x \leq d_i$, where l_i denotes the sum of the entries of c_i. Since $ax \leq b$ is implied by $c_i x \leq d_i$ and the equalities $x_{ij} + x_{ji} = 1$, there must exist $\lambda_i \geq 0$ and μ_{ij} such that $b = \sum_i \lambda_i d_i - \sum_{ij} \mu_{ij}$ and $a = \sum_i \lambda_i c_i - \sum_{ij} \mu_{ij} e_{ij}$, where e_{ij} is the incidence vector of $x_{ij} + x_{ji}$. Since $ax \leq b$ is support reduced, all μ_{ij}'s must be nonnegative.

The strength of $ax \leq b$ is therefore

$$\frac{\sum_i \lambda_i l_i - 2\sum_{ij} \mu_{ij}}{\sum_i \lambda_i d_i - \sum_{ij} \mu_{ij}} \leq \frac{\sum_i \lambda_i l_i}{\sum_i \lambda_i d_i} \leq \max_i \frac{l_i}{d_i},$$

the first inequality following from the fact that the strength of any inequality is less than 2. Moreover, the first inequality is strict if any $\mu_{ij} > 0$.

As an illustration, the lemma shows that the strength of the k-dicycle inequalities ($k > 3$) is less than the strength of the 3-dicycle inequalities since the k-dicycle inequalities are implied by the 3-dicycle inequalities over P_{LO} (but not over P_{AC}). For the diagonal inequalities, Leung and Lee [LL94] show that they are implied by the 3-dicycle inequalities, the r-reinforced k-fence inequalities and the equalities $x_{ij} + x_{ji} = 1$. This in conjunction with the lemma shows that the strength of the diagonal inequalities is less than $3/2$.

The α-critical fence inequalities [Ko95] are generalizations of k-fence inequalities in which the structure of the inequality's supporting digraph is related to a connected undirected graph $G = (V, E)$ with $|V| \geq 3$. Koppen proves that the associated inequality is a facet if and only if the graph is α-critical, i.e., if any edge is removed from G then the independence number of G, $\alpha(G)$, increases. For this inequality, $\sum_{i,j} a_{ij}$ is $|V| + 2|E|$, while $b = \alpha(G) + 2|E|$, implying that their strength is $\frac{|V|+2|E|}{\alpha(G)+2|E|}$. We show next that the strength of these inequalities is bounded by $9/7$.

Lemma 5. *For any connected, α-critical graph $G = (V, E)$ with $|V| \geq 3$,*

$$\frac{|V| + 2|E|}{\alpha(G) + 2|E|} \leq \frac{9}{7},$$

and this bound is attained if G is a triangle.

Proof. For simplicity, let $v = |V|$, $e = |E|$, and $\alpha = \alpha(G)$. Since G is α-critical and $|V| \geq 3$, it cannot be a tree [LP86, Th. 12.1.8.], and therefore $e \geq v$.

From Turán's theorem [Tu41], we have that $\alpha \geq \frac{v}{d+1}$ where d is the average degree. Thus $\alpha \geq \frac{v}{1+2e/v}$. Letting $x = e/v \geq 1$, the bound on the strength becomes

$$\frac{v+2e}{\alpha+2e} \leq \frac{v+2e}{\frac{v^2}{2e+v}+2e} = \frac{(2e+v)^2}{v^2+4e^2+2ev} = \frac{1+4x+4x^2}{1+2x+4x^2} \leq \frac{9}{7},$$

the value of $9/7$ being attained at $x = 1$. When $x = 1$, G must be a cycle; but, among all cycles, Turán's theorem is tight only for the triangle.

Given a valid inequality for P_{AC}, several operations are known for deriving other valid inequalities [Ju85]. For example, both the operations of node-splitting and arc-subdivision take an inequality $ax \leq b$ and transform it into another inequality $cx \leq d$ with the property that $d - b = \sum c_{ij} - \sum a_{ij} \geq 0$. Such a transformation results in an inequality of lesser strength since both the numerator and denominator of $\sum a_{ij}/b$ increase by the same amount.

Finally, we note that the class of Möbius ladder inequalities introduced by Grötschel, Jünger, and Reinelt [Ju85, GJR85a] is much more general than what we have described here. It is quite possible that there exist general Möbius inequalities whose strength is greater than $3/2$. However, the description of these general inequalities gives no systematic way of recognizing classes for which the strength might be greater. (Indeed, our new inequalities, described in the next section, could in fact be Möbius ladders; we comment further about this possibility at the end of the next section.) Also, we have omitted the web inequalities [Ju85, GJR85a] because they are defined relative to general Möbius ladders. When they are defined relative to node-disjoint Möbius ladders, their strength is at most $(6k + 10)/(5k + 9) < 6/5$.

5   Paley Inequalities

Motivated by the results of Section 3 and 4, we introduce a class of valid inequalities based on a construction of Alon and Spencer [AS92] and we give conditions under which these inequalities define a facet of the linear ordering and acyclic subgraph polytopes.

5.1   Number-Theoretic Preliminaries

We collect here some basic number theoretic results which will be useful for the definition and properties of Paley inequalities. The results can be found in Hardy and Wright [HW79], for example.

The letter p will always denote a prime number. Given p, any x not congruent to 0 modulo p has a unique inverse modulo p, denoted by x^{-1}, i.e. $xx^{-1} \equiv 1$ (mod p). We can thus define a division modulo p by $x/y = xy^{-1}$.

A given $a \not\equiv 0 \pmod{p}$ is called a *quadratic residue* of p if the congruence $x^2 \equiv a \pmod{p}$ has a solution x; otherwise a is called a *quadratic non-residue*. There are exactly $(p-1)/2$ quadratic residues of an odd prime p. If we let $\chi(a)$ be 1 if a is a quadratic residue, 0 if a is 0, and -1 if a is a quadratic non-residue, then it is known that $\chi(a) \equiv a^{(p-1)/2} \pmod{p}$ [HW79, Th. 83]. Since $\chi(ab) = \chi(a)\chi(b)$, the product of two quadratic residues (or two quadratic non-residues) is a quadratic residue, and the product of a quadratic residue and a quadratic non-residue is a quadratic non-residue. For primes of the form $4k+3$, exactly one of a or $-a$ is a quadratic residue [HW79, Th. 82]. For any odd prime p, there exist quadratic residues a, b and c such that $a + b + c \equiv 0 \pmod{p}$ [HW79, Th. 87].

The number of positive integers not greater than and prime to m is denoted by $\phi(m)$. The function ϕ is called *Euler's function*. For a prime p, $\phi(p) = p - 1$. In general, if m is expressed in its standard form $m = p_1^{a_1} p_2^{a_2} \cdots p_l^{a_l}$ where the p_i's are distinct primes then $\phi(m) = m \left(1 - \frac{1}{p_1}\right) \left(1 - \frac{1}{p_2}\right) \cdots \left(1 - \frac{1}{p_l}\right)$ [HW79, Th. 62]. The *order* of an element $a \pmod{m}$ is the smallest positive value d such that $a^d \equiv 1 \pmod{m}$. If a is prime to m then the order of $a \pmod{m}$ divides $\phi(m)$ [HW79, Th. 88]. The integer a is called a *primitive root* of m if its order is equal to $\phi(m)$. It is known that every prime p has exactly $\phi(p-1)$ primitive roots [HW79, Th. 110]. We will refer to the square of a primitive root as a *squared primitive root*. Observe that any squared primitive root is a quadratic residue with order exactly $(p-1)/2$. If a is a primitive root of a prime p then $\chi(a) \equiv a^{(p-1)/2} \equiv -1 \pmod{p}$, i.e. a is a quadratic non-residue. Therefore, if p is of the form $4k+3$, then $-a$ is a quadratic residue and cannot be a primitive root. This implies that for a prime p of the form $4k+3$ the squares of the primitive roots are distinct, and thus there are $\phi(p-1) = \phi((p-1)/2)$ squared primitive roots. In particular, if $(p-1)/2$ is prime, all quadratic residues, except the residue 1, are squared primitive roots.

5.2 Paley tournaments

A *Paley tournament* can be constructed as follows. Take a prime p of the form $4k+3$. Let the vertex set V be the residues modulo p or $\{0, 1, \cdots, p-1\}$, and let the arc set A be $\{(i,j) : \chi(j-i) = 1\}$. The fact that $\chi(a) = -\chi(-a)$ implies that (V, A) is a tournament. Let $s(\pi)$ denote the number of arcs in the acyclic subgraph defined by a linear ordering π on V, and let $l(p)$ be the maximum of $s(\pi)$ over all π. Alon and Spencer [AS92] show that $l(p) \leq \frac{1}{2}\binom{p}{2} + O(p^{3/2} \log p)$. Given a Paley tournament (V, A), we can associate the inequality

$$\sum_{(i,j) \in A} x_{ij} \leq l(p), \tag{3}$$

which is valid for both P_{AC} and P_{LO}. We refer to this inequality as a *Paley inequality*. The term Paley comes from R.E.A.C. Paley [Pa33] who introduced a very closely related construction for a Hadamard matrix of size $p + 1$.

5.3 Facets

In this section, we give conditions under which the Paley inequality defines a facet of either P_{LO} or P_{AC}.

We first consider the case $p = 7$. The quadratic residues modulo 7 are 1, 2 and 4. We claim that, for $p = 7$, the Paley inequality is implied by the 3-dicycle inequalities. Indeed, if we consider the seven dicycle inequalities corresponding to vertices i, $i - 1$, and $i - 4$, $i = 0, \ldots, 6$, where all differences are modulo 7, then we cover each arc of the Paley tournament exactly once, and so we have $\sum_{(i,j) \in A} x_{ij} \leq 14$. This value is attainable, e.g., by the permutation $\langle 0, 1, 2, 3, 4, 5, 6 \rangle$.

For higher values of p, we need the following definition. Given a linear ordering π, we say that π *contains the loose triple* (u, v, w), where u, v, w are (not necessarily all distinct) quadratic residues summing to 0 modulo p, if there is an index i such that $\pi(i+1) - \pi(i) = u$, $\pi(i+2) - \pi(i+1) = v$ and $\pi(i) - \pi(i+2) = w$.

Theorem 6. *For a prime $p \geq 11$ of the form $4k + 3$, the inequality (3) defines a facet of P_{LO} if there exists an optimal linear ordering π (i.e. for which $s(\pi) = l(p)$) which contains a loose triple (u, v, w) where one of $v/u, w/v$ or u/w is a squared primitive root.*

Proof. We start by observing that $v/u, w/v$ and u/w cannot all be equal. If they were all equal say to x then $x^3 \equiv 1 \pmod{p}$, which contradicts the fact that x must be a squared primitive root (unless $p = 7$ which is ruled out by assumption).

We will consider several operations on linear orderings of Paley tournaments which preserve optimality. Given π and a quadratic residue a, we let $a\pi$ denote the permutation defined by $(a\pi)(i) = \pi(i) \cdot a \pmod{p}$. Similarly, for any residue a (not necessarily quadratic), we let $\pi + a$ denote the permutation defined by $(\pi + a)(i) = \pi(i) + a \pmod{p}$. Observe that, in both cases, $s(a\pi) = s(\pi)$ and $s(\pi + a) = s(\pi)$. The effect of these operations on a loose triple (u, v, w) of π is the following: (u, v, w) is still a loose triple of $\pi + a$, and (au, av, aw) is a loose triple of $a\pi$. Also, if π has a loose triple, say given by the index i, then we can obtain two other permutations $(\rho_i \circ \pi)$ and $(\rho_i^2 \circ \pi)$ by applying (once or twice) the permutation ρ_i that only rotates elements $\pi(i)$, $\pi(i + 1)$ and $\pi(i+2)$. The important observation is that, by definition of a loose triple, we have $s(\pi) = s(\rho_i \circ \pi) = s(\rho_i^2 \circ \pi)$. Moreover, by applying these rotations, we can assume without loss of generality that the loose triple (u, v, w) in some optimal ordering π satisfies: (i) v/u is a squared primitive root, and (ii) $w/v \not\equiv v/u$. Furthermore, by considering $a\pi$, we can assume that $u = 1$, and then by considering $\pi + a$, we can assume that $\pi(i) = 0$. We therefore assume the existence of an optimal linear ordering π with a loose triple $(1, k, kl)$ at index i where k is a squared primitive root, $k \neq l$ and $\pi(i) = 0$.

To prove that (3) defines a facet of P_{LO}, we consider the set $\mathcal{O} = \{\pi : s(\pi) = l(p)\}$ of optimal orderings and show that if the equality $c^T x = b$ is satisfied for all incidence vectors x^π corresponding to permutations π in $\mathcal{O}$ then $c^T x = b$

is implied by the equality version of (3) and the equalities $x_{ij} + x_{ji} = 1$ for all i, j. Because of the latter equalities, we can assume that the only non-zero coefficients of c correspond to arcs in the Paley tournament, and, as a result, we only need to show that such restricted c's are multiples of the equality version of (3). For simplicity of notation, we let $c_{u,a}$ denote the coefficient of the arc $(u, u + a)$ where a is a quadratic residue. The subscripts of c should always be taken modulo p.

Comparing $c^T x^{\pi_1}$ and $c^T x^{\pi_2}$ where $\pi_1 = \rho_i \circ \pi$ and $\pi_2 = \rho_i^2 \circ \pi$, we derive that $c_{0,1} = c_{1,k}$. If we now consider $\pi' = a\pi + b$ where a is a quadratic residue, the same argument shows that $c_{b,a} = c_{a+b,ak}$. In particular, taking $(a, b) = (k, 1)$, we get that $c_{0,1} = c_{1,k} = c_{k+1,k^2}$. Repeating this process, we derive that, for any positive integer j, $c_{0,1} = c_{d(k,j),k^j}$ where $d(k, j) = k^{j-1} + k^{j-2} + \cdots + 1$. Since k was assumed to be a squared primitive root, as j runs from 0 to $(p-1)/2 - 1$, k^j runs over all quadratic residues modulo p.

If we compare now $c^T x^\pi$ with $c^T x^{\pi_2}$, we derive that $c_{0,1} = c_{1+k,kl}$. But we already know that $c_{0,1} = c_{d(k,j),kl}$ where j is such that k^j is congruent to the quadratic residue kl modulo p. We claim that $k + 1 \not\equiv d(k, j) \pmod{p}$. Assuming the claim, we are almost done. We have just proved that $c_{b_1,kl} = c_{b_2,kl}$ for two distinct values b_1, b_2. By repeatedly adding $b_2 - b_1$ to the permutations, we derive that $c_{0,kl} = c_{b,kl}$ for any b. By multiplying by all quadratic residues, we get that $c_{0,a} = c_{b,a}$ for all b and all quadratic residues a. Now using the fact derived in the previous paragraph that, for any quadratic residue, $c_{0,1} = c_{m,a}$ for some m, we get that all coefficients of $c^T x$ are identical, proving the result.

We still need to prove the claim that $k + 1 \not\equiv d(k, j) \pmod{p}$. To avoid having to deal with the cases $j = 0$ or $j = 1$ separately, we add $(p-1)/2$ to j in such cases (remember that $k^{(p-1)/2} \equiv 1 \pmod{p}$), and therefore we assume without loss of generality that $2 \le j \le (p-1)/2 + 1$. We need to show that $d(k, j) - k - 1 \not\equiv 0 \pmod{p}$. We observe that

$$(k-1)(d(k,j) - k - 1) \equiv (k-1)(k^{j-1} + k^{j-2} + \cdots k^2) \pmod{p}$$
$$\equiv k^j - k^2 \pmod{p}$$
$$\equiv k^2(k^{j-2} - 1) \pmod{p}$$

Since k is a squared primitive root, we have that $k \ne 1$ and $k^{j-2} \not\equiv 1 \pmod{p}$ unless $j = 2$. However, our assumption that $k \ne l$ implies that $j \ne 2$. Therefore, $d(k, j) - k - 1 \not\equiv 0 \pmod{p}$, proving the claim.

Since the inequality (3) is support reduced, if it defines a facet of P_{LO} then it also defines a facet of P_{AC}. (See Boyd and Pulleyblank [BP91] and Leung and Lee [LL94] for the details of why generally such an implication holds.)

Corollary 7. *Under the same conditions as in Theorem 6, inequality (3) defines a facet of P_{AC}.*

For $p = 11$, 19, and 23, we have been able to compute the correct right-hand side value $l(p)$ using CPLEX branch-and-bound code; see Table 2. In the case

$p = 11$, the following permutation, which yields a solution containing 35 arcs, satisfies the conditions of the proof:

$$\langle 1, 6, 7, 10, 4, 8, 2, 0, 5, 9, 3 \rangle.$$

Notice that the first three elements induce a loose triple $(5, 1, 5)$. Since 5 is a squared primitive root, the theorem can be applied. For $p = 19$, we have the following permutation containing 107 arcs:

$$\langle 8, 17, 18, 15, 5, 3, 12, 10, 2, 0, 9, 7, 16, 14, 6, 4, 13, 11, 1 \rangle.$$

The first three elements induce the loose triple $(9, 1, 9)$. Since 9 is a squared primitive root for 19, we again have the necessary conditions that imply the constraint is facet-defining.

Since the Paley inequalities can be trivially lifted to larger instances, we therefore have the following corollary to Theorem 6 and Corollary 7.

Corollary 8. *The Paley inequalities on 11 and on 19 vertices are facet-defining for P_{AC}^n and P_{LO}^n, for all $n \geq 11$ and 19, respectively.*

In the case of $p = 23$, two rather striking things occurred. First, the only permutations we have found that lie on the facet have no loose triples; moreover, the solutions we have found are all isomorphic to the permutation $\langle 0, 1, 2, \cdots, 21, 22 \rangle$. Second, the strength 253/161 of this inequality is *identical* to that of $p = 11$, 55/35. For both of these reasons we conjecture that the Paley inequality in this case is not facet-defining.

For $p = 31$ and $p = 43$, we have been unable to ascertain whether the best feasible solutions generated (with 285 and 543 edges, respectively) are optimal. In the case of $p = 31$, if the true right-hand side is 285, then we are able to show that the inequality is facet-defining. Although the proof of this fact does not follow directly from the previous theorem, it uses a similar technique of combining "loose tuples" (in this case, tuples larger than triples) in order to equate coefficients.

We do not know whether the Paley inequalities are special cases of Möbius ladders. However, we do know that if they are, then, in the case of $p = 11$, the Möbius ladder would have to have 39 cycles, while the $p = 19$ case would require 127 cycles! Finally, we note that the Paley inequalities are but one type of tournament with pseudo-random properties; a larger class that subsumes the Paley inequalities can be extracted from Hadamard matrices. Our experiments with additional inequalities generated from small instances of Hadamard matrices have thus far yielded inequalities that are neither strong nor provably facet-defining.

6 Dicycle strengths

One drawback to the definition of the strength of an inequality is that strength is defined relative to a given relaxation. An inequality which is strong relative

p	$l(p)$	# Edges	Strength
$p = 11$	35	55	1.57142857...
$p = 19$	107	171	1.59813084...
$p = 23$	161	253	1.57142857...
$p = 31$	≥ 285	465	$\leq 1.64893617...$
$p = 43$	≥ 543	903	$\leq 1.66298342...$

Table 2. Examples of Paley inequalities for specific values of p.

to one relaxation might actually be much weaker relative to another relaxation. In this section, we consider a stronger and classical relaxation of P_{AC} as our "benchmark". Some, but not all, of the results corroborate the observations from the previous sections.

Let

$$P_d = \{x : x_{ij} + x_{ji} \leq 1 \text{ for all } i, j \in V \text{ and}$$
$$\sum_{(i,j) \in C} x_{ij} \leq |C| - 1 \text{ for all dicycles } C\},$$

namely, P_d is the relaxation obtained by adding all dicycle inequalities to P. We can apply the results of Section 2 to evaluate the strength of any relaxation relative to P_d instead of relative to P. To avoid any confusion, let the *dicycle strength* of any relaxation Q of P_{AC} with $P_d \supseteq Q$ be defined as $t(P_d, Q)$. Because of Theorem 2, we know that the dicycle strength of any relaxation Q is equal to the maximum dicycle strength of (the relaxation obtained by adding to P_d) any inequality defining Q. The computations of dicycle strengths are slightly more tedious than the computations of strengths; we have nevertheless been able to compute the dicycle strengths of most of the inequalities considered before. Table 3 summarizes these results.

By the asymptotic properties of Paley tournaments, we know that the dicycle strength of the Paley inequalities approaches $4/3$ as p tends to infinity. However, the convergence is very slow, and in fact the dicycle strength for $p = 11$ or $p = 23$ is only $22/21 = 1.0476...$ while the dicycle strength for $p = 19$ is $114/107 = 1.0654...$. Thus the dicycle strengths of the Paley inequalities for $p = 11, 19$, and 23 are actually smaller than the dicycle strength of (for example) the 4-fence, which is $14/13$; these results are in contrast to those concerning the (regular) strengths of these inequalities. For $p = 31$, the dicycle strength would be $62/57 = 1.0877...$ if in fact $l(31) = 285$.

The dicycle strength of P_{AC} itself must be at least $4/3$ (asymptotically, as the number of vertices grows to infinity) since the Paley inequalities achieve this bound. However, we do not know if the dicycle strength of P_{AC} is $4/3$ or whether it is larger (possibly as large as 2). On the one hand, if the dicycle strength of P_{AC} is more than $4/3$, then there must exist inequalities which are stronger

Inequality Type	Dicycle Strength	Max Value	
k-fence	$\dfrac{k^2 - k/2}{k^2 - k + 1}$	$14/13$	$= 1.0769...$
augmented k-fence	$\dfrac{2k^2 - 5k/2}{2k^2 - 3k + 1}$	$21/20$	$= 1.05$
r-reinforced k-fence	$\dfrac{k(k-1) + rk/2}{k(k-1) + r(r+1)/2}$	$1/2 + \sqrt{6}/4$	$= 1.1123...$
k-wheel	$7k/(7k - 1)$	$21/20$	$= 1.05$
Z_k	$(6k + 5)/(6k + 4)$	$29/28$	$= 1.0357...$
α-critical fence	$\dfrac{|V|/2 + 2|E|}{\alpha(G) + 2|E|}$	$\leq \dfrac{12 + 7\sqrt{3}}{12 + 6\sqrt{3}}$	$= 1.0773...$
Paley	$\dfrac{p(p - 1)}{3l(p)}$	$4/3$	$= 1.3333...$

Table 3. Dicycle strength of various classes of valid inequalities for P_{AC}.

than the Paley inequalities in the dicycle-strength sense. On the other hand, if the dicycle strength of P_{AC} were bounded away from 2, say at most $c < 2$, this fact would be extremely interesting from an approximation point-of-view since it would imply that the value obtained by optimizing over P_d (and this can be done in polynomial time) is within a ratio of c of the value of the maximum acyclic subgraph. The problem of finding an approximation algorithm with a performance ratio better than $2 - o(1)$ for the directed acyclic subgraph problem has been a long-standing open problem, at least in part due to the poor upper bounds used in proving the performance guarantees.

Acknowledgments The authors would like to thank Michael Jünger and Gerhard Reinelt for providing us with pointers to the literature, and for their help in generating feasible solutions to the Paley tournaments on 31 and 43 vertices. We also thank Andreas Schulz for helpful discussions, and Sebastián Ceria and Gérard Cornuéjols for pointing out the importance of considering stronger LP relaxations for strength comparisons.

References

[AS92] N. Alon and J.H. Spencer, *The Probabilistic Method*, John Wiley & Sons, 1992.

[BP91] S.C. Boyd and W.R. Pulleyblank, "Facet Generating Techniques", Technical Report 91-31, Department of Computer Science, University of Ottawa, Ottawa, 1991.

[Ve83] W.F. de la Vega, "On the Maximum Cardinality of a Consistent Set of Arcs in a Random Tournament", *Journal of Combinatorial Theory*, Series B **35**, 328–332 (1983).

[EM65] P. Erdös and J.W. Moon, "On Sets of Consistent Arcs in a Tournament", *Canad. Math. Bull.* **8**, 269–271 (1965).

[Fi92] P.C. Fishburn, "Induced Binary Probabilities and the Linear Ordering Polytope: A Status Report", *Math. Social Sci.* **23**, 67–80 (1992).

[GJ79] M.R. Garey and D.S. Johnson, *Computers and Intractability: A Guide to the Theory of NP-completeness*, W.H. Freeman and Company, 1979.

[Gi90] I. Gilboa, "A Necessary but Insufficient Condition for the Stochastic Binary Choice Problem", *J. Math. Psych.* **34**, 371–392 (1990).

[Go95] M.X. Goemans, "Worst-case Comparison of Valid Inequalities for the TSP", *Mathematical Programming* **69**, 335–349 (1995).

[GJR85a] M. Grötschel, M. Jünger and G. Reinelt, "On the Acyclic Subgraph Polytope", *Mathematical Programming* **33**, 28–42 (1985).

[GJR85b] M. Grötschel, M. Jünger and G. Reinelt, "Facets of the Linear Ordering Polytope", *Mathematical Programming* **33**, 43–60 (1985).

[HW79] G.H. Hardy and E.M. Wright, *An Introduction to the Theory of Numbers*, Oxford Science Publications, 1979.

[Ju85] M. Jünger, *Polyhedral Combinatorics and the Acyclic Subdigraph Problem.* Volume 7, *Research and Exposition in Mathematics*, ed. K. Hofmann and R. Wille. Heldermann Verlag, Berlin (1985).

[Ko95] M. Koppen, "Random Utility Representation of Binary Choice Probabilities: Critical Graphs Yielding Critical Necessary Conditions", *J. of Mathematical Psychology* **39**, 21–39 (1995).

[LL94] J. Leung and J. Lee, "More Facets from Fences for Linear Ordering and Acyclic Subgraph Polytopes", *Discrete Applied Math.* **50**, 185–200 (1994).

[LP86] L. Lovász and M.D. Plummer, "Matching Theory", North-Holland, 1986.

[LY78] C.L. Lucchesi and D.H. Younger, "A Minimax Theorem for Directed Graphs", *J. of the London Mathematical Soc.* **17**, 369–374 (1978).

[Mc90] A. McLennon, "Binary Stochastic Choice", in: J.S. Chipman, D. McFaddon, and M.K. Richter, eds., *Preferences, Uncertainty, and Optimality*, Westview Press, Boulder, 187–202 (1990).

[Pa33] R.E.A.C. Paley, "On Orthogonal Matrices", *Journal of Mathematics and Physics*, **12**, 311–320 (1933).

[Re85] G. Reinelt, *The Linear Ordering Problem: Algorithms and Applications.* Volume 8, *Research and Exposition in Mathematics*, ed. K. Hofmann and R. Wille. Heldermann Verlag, Berlin (1985).

[Sp71] J. Spencer, "Optimal Ranking of Tournaments", *Networks* **1**, 135–138 (1971).

[Su92] R. Suck, "Geometric and combinatorial properties of the polytope of binary choice probabilities", *Mathematical Social Sciences* **23**, 81–102 (1992).

[Tu41] P. Turán, "On an Extremal Problem in Graph Theory" (in Hungarian), *Mat. Fiz. Lapok* **48**, 436–452 (1941). English translation in: P. Erdös, ed., *Collected Papers of Paul Turán*, Akadémiai Kiadó, Budapest, Vol. 1, Chap. 24 (1990).

Transitive Packing *

Rudolf Müller[1] and Andreas S. Schulz[2]

[1] Humboldt–Universität zu Berlin, Institut für Wirtschaftsinformatik,
D–10178 Berlin, Germany, rmueller@wiwi.hu–berlin.de
[2] Technische Universität Berlin, Fachbereich Mathematik,
D–10623 Berlin, Germany, schulz@math.tu–berlin.de

Abstract. This paper is intended to give a concise understanding of the facial structure of previously separately investigated polyhedra. We introduce the notion of transitive packing and the transitive packing polytope and give cutting plane proofs for huge classes of valid inequalities of this polytope. We introduce generalized cycle, generalized clique, generalized antihole, generalized antiweb, generalized web, and odd partition inequalities. These classes subsume several known classes of valid inequalities for several of the special cases but also give many new inequalities for several others. For some of the classes we also prove a nontrivial lower bound for their Chvátal rank. Finally, we relate the concept of transitive packing to generalized (set) packing and covering as well as to balanced and ideal matrices.

1 Introduction

Various types of packing problems and related polyhedra play a central role in combinatorial optimization. Due to both a large variety of practical applications and interesting structural properties they have found considerable attention in the literature, see, e. g., [BP76, Pad79] for an overview. One of the classic examples is the *node packing problem* in graphs and the associated *node packing polytope*. If we denote by A the edge–node incidence matrix of a given graph, it can be formulated as $\max\{cx : Ax \leq 1, x_u \in \{0,1\}\}$ where c is an arbitrary vector, and 1 denotes (here and henceforth) the all–one vector of compatible dimension. The node packing polytope is defined as the convex hull of feasible solutions and has been studied, among others, in [Pad73, NT74, Tro75, GLS88].

The node packing problem can be extended to hypergraphs where it reads $\max\{cx : Ax \leq p_A - 1, x_u \in \{0,1\}\}$ where A is now an arbitrary 0/1 matrix (the edge–node incidence matrix of the hypergraph), and the i–th component of the vector p_A gives the number of positive entries in row i of the matrix A. The undominated rows of A can be interpreted as the incidence vectors of the circuits of an *independence system*. Hence the above problem can be seen as the problem of finding an independent set of maximum weight. The convex hull of incidence vectors of independent sets is known as the *independence system polytope*. Much

---

* The authors acknowledge support by the Deutsche Forschungsgemeinschaft under the grants SFB 373 and We 1265/2–1.

work has been done to find classes of valid inequalities for special cases of the independence system polytope. Among them are, just to name a few, the *acyclic subdigraph polytope* [GJR85b, Jün85], the *bipartite subgraph polytope* [BGM85], and the *planar subgraph polytope* [JM93]. Only recently the thorough study of the facial structure of the independence system polytope in general has begun, see [EJR87, Lau89].

In Sect. 2, we introduce an extension of the node packing problem in hypergraphs, called *transitive packing*, by taking a kind of transitive elements into account. The problems we consider can be described as

$$
\begin{aligned}
\max \quad & c\,x \\
\text{s. t.} \quad & A\,x \le p_A - \mathbb{1} \\
& x_u \in \{0,1\}
\end{aligned}
\tag{1}
$$

where A is now an arbitrary $0/\pm 1$ matrix and the i–th component of the vector p_A gives the number of positive entries in row i of matrix A. Many combinatorial optimization problems can be modeled as *transitive packing problems*. We do not (and cannot) list all problems that fit with this novel framework but name a few of them we are going to revisit later. Indeed, besides those that can be interpreted as finding an independent set of maximum weight there are the *clique partitioning problem* [GW89, GW90, ORS95], the *transitive acyclic subdigraph problem* [Mül93], the *interval graph completion problem* [MS95, Sch95], and the *relatively transitive subgraph problem* [KL89, SB].

One of our main purposes is to derive broad classes of valid inequalities for the *transitive packing polytope*, the convex hull of feasible solutions to (1). In Sect. 3, we present *generalized cycle, generalized clique, generalized antihole, generalized antiweb, generalized web,* and *odd partition inequalities* which are valid for the transitive packing polytope. These classes explain and classify many known inequalities for polytopes that fit with this general framework. Thereby we emphasize the relations between and the common structure of (inequalities for) different, formerly independently studied polyhedra, and we provide new insights as well as new inequalities for some of the special polytopes that arise from certain hypergraphs and choices of transitive elements. We show how the knowledge of structural properties of the transitive packing polytope makes possible to derive results for the specialized problems.

We derive most of the inequalities for the transitive packing polytope by *integer rounding*. This provides *cutting plane proofs* for many of the known inequalities for special polytopes that do not seem to be observed before. It may also be seen as a guide for using certain patterns of the (initial) constraint matrix A to obtain new inequalities in a systematic way. The latter is of particular importance for solving general $0/1$ integer programs. Moreover, the derivation of the inequalities may be seen as a guide–line to generalize each valid inequality for the node packing polytope whose cutting plane proof is known.

Section 4 is concerned with an interesting subclass of the transitive packing polytopes formed by those whose corresponding hypergraph is actually a graph.

In Sect. 5, we discuss the separation problem associated with the classes of inequalities introduced before. Finally, in Sect. 6 we describe the strong relation between *set covering* and independence system polytopes, point out its extension to *generalized set covering* and transitive packing polytopes, translate our results into this context, and discuss the relation of our work to $0/\pm 1$ matrices that are *balanced* or *ideal*.

Due to space limitations we have to omit all proofs in this paper. A complete version can be obtained from the authors [MS96] (see also [Sch95, Chap. 4]).

2 The Transitive Packing Polytope

A *hypergraph* is an ordered pair $(N, \mathcal{H})$ where N is a finite ground set, the set of *nodes*, and $\mathcal{H}$ is a collection of distinct subsets of N, the set of *(hyper)edges*. We only deal with hypergraphs without loops, i. e., we always assume that $|H| \geq 2$ for all $H \in \mathcal{H}$. Here, we are interested in hypergraphs with additional node subsets associated with each edge.

Definition 1. Let $(N, \mathcal{H})$ be a hypergraph, and let $tr : \mathcal{H} \to 2^N$ be a mapping from the set of edges to the powerset of N with the property that $tr(H) \subseteq N \setminus H$. We call the ordered triple $(N, \mathcal{H}, tr)$ an *extended hypergraph*, and $tr(H)$ the set of *transitive elements* associated with the edge H.

We are interested in packing nodes of an extended hypergraph whereby the restrictions imposed by the edges may be compensated by picking transitive elements.

Definition 2. Let $(N, \mathcal{H}, tr)$ be an extended hypergraph. A subset S of the nodes is a *transitive packing* (in $(N, \mathcal{H}, tr)$) if, for every $H \in \mathcal{H}$ such that $H \subseteq S$, there exists a node $u \in S \cap tr(H)$.

In other words, a transitive packing S is a set of nodes that contains an edge only if S contains at least one node from the set of transitive elements associated with that edge. Given in addition to $(N, \mathcal{H}, tr)$ a weight function $c : N \to \mathbb{Q}$, the *(maximum weight) transitive packing problem* consists of finding a transitive packing $S \subseteq N$ of maximum weight $c(S)$. As indicated in the introduction, the transitive packing problem is equivalent to the integer linear programming problem

$$
\begin{array}{llll}
\max & c\,x & & \\
\text{s. t.} & x(H) - x(tr(H)) & \leq & |H| - 1 \quad \text{for all } H \in \mathcal{H} \quad (2) \\
& x & \leq & \mathbb{1} \quad (3) \\
& x & \geq & 0 \quad (4) \\
& x & \in & \mathbb{Z}^N \quad (5)
\end{array}
$$

Note that the constraint matrix of the inequalities (2) is the edge–node incidence matrix of the hypergraph $(N, \mathcal{H})$, with additional -1's for the transitive elements of the edge represented by the particular row. We call the inequalities (2) *transitivity constraints*. In the following, we study the transitive packing polytope

$$P_{\text{TP}}(N, \mathcal{H}, tr) := \text{conv}\{\chi^S \in \mathbb{R}^N : S \text{ transitive packing in } (N, \mathcal{H}, tr)\} \ .$$

So $P_{\text{TP}}(N, \mathcal{H}, tr)$ is equal to the integer hull of the feasible solutions to (2) – (4). At this point, it seems to be reasonable to introduce a few examples that serve to improve the accessibility of the discussions to follow. Of course, if $tr(H) = \emptyset$ and $|H| = 2$ for all edges $H \in \mathcal{H}$, a transitive packing reduces to a node packing in the graph $(N, \mathcal{H})$. But to motivate hypergraphs and transitive elements we show now that the acyclic subdigraph polytope as well as the clique partitioning polytope can be obtained by special choices of the hypergraph and the transitive elements. Other examples will be discussed later.

The *acyclic subdigraph polytope* is the convex hull of incidence vectors of acyclic arc subsets of a given digraph $D = (V, A)$. It was mainly studied by Grötschel, Jünger, and Reinelt, see [GJR85b, GJR85a, Jün85]. If we choose the arc set A of the digraph D as the node set of the hypergraph, if we declare the directed cycles in D as the edges of this hypergraph, and if we let $tr(H) = \emptyset$ for all $H \in \mathcal{H}$, it appears as a special transitive packing polytope.

Given an undirected graph $G = (V, E)$, a set $F \subseteq E$ of edges is called a *clique partitioning* of G if there is a partition of V into nonempty, disjoint sets $W_1, W_2, \ldots, W_k$ such that the subgraph induced by each W_i is a clique and such that $F = \bigcup_{i=1}^k \{\{u, v\} : u, v \in W_i\}$. The clique partitioning polytope is the convex hull of the incidence vectors of all clique partitionings in G. It was introduced by Grötschel and Wakabayashi [GW89, GW90] and has very recently been considered again by Oosten, Rutten, and Spieksma [ORS95]. To show that it is an instance of the transitive packing polytope we consider the *line graph* of G. Indeed, we take as the set N of nodes the edges of G, and two nodes are adjacent (form a hyperedge) if and only if the associated edges are incident in the original graph G. The transitive element that we attach to a pair of incident edges $\{u, v\}$, $\{v, w\}$ in G is the edge $\{u, w\}$, if it exists.

Before starting the study of the transitive packing polytope we shall discuss an algorithmic aspect. How is $(N, \mathcal{H}, tr)$ given? Having in mind problems like the acyclic subdigraph problem, it does not seem to be satisfactory to assume that it is given as a list of hyperedges and its transitive elements. It rather seems to be reasonable to assume that the linear programming problem arising from (2) – (4) is solvable in polynomial time. In particular, this guarantees that the decision version of the transitive packing problem belongs to the class NP. Since the node packing problem on graphs is NP–hard the same holds for the transitive packing problem.

Proposition 3. *Let $(N, \mathcal{H}, tr)$ be an extended hypergraph.*

(a) *The transitive packing polytope $P_{\text{TP}}(N, \mathcal{H}, tr)$ is full dimensional.*
(b) *The nonnegativity constraint $x_u \geq 0$ defines a facet of $P_{\text{TP}}(N, \mathcal{H}, tr)$ for each node $u \in N$.*

(c) *If $tr(H)$ is the empty set for all edges $H \in \mathcal{H}$ such that $|H| = 2$, then an inequality $x_u \leq 1$ with $u \in N$ defines a facet of $P_{TP}(N, \mathcal{H}, tr)$ if and only if $|H| \geq 3$ for all edges $H \in \mathcal{H}$ that contain u.*

There are other conditions which suffice such that an upper bound constraint is facet defining. Observe that a transitivity constraint $x(H') - x(tr(H')) \leq |H'| - 1$ is dominated by $x(H) - x(tr(H)) \leq |H| - 1$ if $H \subseteq H'$ and $tr(H) \subseteq tr(H')$.

3 Valid Inequalities

Let P be a rational polyhedron, for instance the initial relaxation of $P_{TP}(N, \mathcal{H}, tr)$ defined by (2) – (4). One way to produce a characterization of the integer hull P_I of P by means of linear inequalities is integer rounding. For a thorough discussion of this topic and its applications to integer programming and combinatorial optimization we refer the reader to the textbooks of Nemhauser and Wolsey [NW88, Chap. II.1] and Schrijver [Sch86, Chap. 23].

If we set $P' := \{x \in P : ax \leq \beta$ whenever $a \in \mathbb{Z}^N, \beta \in \mathbb{Z}$, and $\max\{ax : x \in P\} < \beta + 1\}$, then P' can be seen as obtained from P by one step of rounding. If we define $P^{(0)} := P$ and, recursively, $P^{(t+1)} := (P^{(t)})'$ for all nonnegative integers t, then $P_I \subseteq P^{(t)}$ for all nonnegative integers t. Schrijver [Sch80] showed that P' is again a polyhedron and that there is a nonnegative integer t such that $P^{(t)} = P_I$. The *rank* of P is the smallest t such that $P^{(t)} = P_I$. Let $ax \leq \beta$ be a valid inequality for P_I. Its *depth* relative to P is the smallest d such that $ax \leq \beta$ is valid for $P^{(d)}$. Let $Ax \leq b$ be a system of linear inequalities, and let $cx \leq \delta$ be an inequality. Moreover, let $c_1 x \leq \delta_1, c_2 x \leq \delta_2, \ldots, c_m x \leq \delta_m$ be a sequence of linear inequalities such that each vector c_i, $i = 1, \ldots, m$, is integral, $c_m = c$, $\delta_m = \delta$, and, for $i = 1, \ldots, m$, the inequality $c_i x \leq \delta_i'$ is a nonnegative linear combination of the inequalities $Ax \leq b, c_1 x \leq \delta_1, \ldots, c_{i-1} x \leq \delta_{i-1}$ for some δ_i' with $\lfloor \delta_i' \rfloor \leq \delta_i$. Such a sequence is called a *cutting plane proof* of $cx \leq \delta$ from $Ax \leq b$, and m is the *length* of this proof. Every integer solution of $Ax \leq b$ satisfies $cx \leq \delta$. Let $P = \{x : Ax \leq b\}$. Since $P^{(t)} = P_I$ for some t the converse is true as soon as P_I is nonempty. That is, every inequality $cx \leq \delta$ with c integral and valid for P_I has a cutting plane proof from $Ax \leq b$. Clearly, the length of a cutting plane proof of a valid inequality for P_I is at least its depth.

The idea of deriving cutting planes by rounding based on exploitation of problem structure can in particular be used to obtain valid inequalities for the transitive packing polytope. Thereby we also show that many inequalities valid for the polytopes that arise from $P_{TP}(N, \mathcal{H}, tr)$ by certain choices of $(N, \mathcal{H}, tr)$ have nice and insightful cutting plane proofs from the initial relaxation (2) – (4).

Subclasses of the classes of valid inequalities which we introduce have been presented earlier for the independence system polytope: generalized cycle, generalized clique, and generalized antihole inequalities by Euler, Jünger, and Reinelt [EJR87] and generalized antiweb inequalities by Laurent [Lau89]. It will turn out that our inequalities are more general even if we restrict ourselves to the independence system polytope. Nevertheless, in order to keep the terminology simple we will give them the same names and point out the restrictions that lead to the known inequalities, respectively.

3.1 Generalized Cycle Inequalities

We first use cycles of the hypergraph $(N, \mathcal{H})$ to obtain a class of valid inequalities for the transitive packing polytope each of which has a cutting plane proof from (2) – (4) of length 1.

Definition 4. Let $(N, \mathcal{H})$ be a hypergraph, and let q, s, and r be positive integers such that $q \geq 2$ and $1 \leq r \leq q - 1$. For convenience, we set $k := sq + r$. Let $N_1, \ldots, N_k$ be a sequence of pairwise disjoint nonempty subsets of N. For $i = 1, \ldots, k$, let $H_i \in \mathcal{H}$ be an edge such that $\bigcup_{j=i}^{i+q-1} N_j \subseteq H_i$. (Indices greater than k are taken modulo $k + 1$ and shifted by $+1$.) Let $C := \bigcup_{i=1}^{k} H_i$, and $m(u) := |\{i \in \{1, \ldots, k\} : u \in H_i\}|$. We assume that $m(u) \leq q$ for all nodes $u \in C$. Then, we call the hypergraph $(C, \{H_i : i = 1, 2, \ldots, k\})$ a *generalized* (k, q)–*cycle* (contained in $(N, \mathcal{H})$).

Let $(C, \{H_i : i = 1, 2, \ldots, k\})$ be a generalized (k, q)–cycle in $(N, \mathcal{H}, tr)$ and assume that the set $tr(C) := \bigcup_{i=1}^{k} tr(H_i)$ does not interact with C itself, i. e., $tr(H_i) \cap C = \emptyset$ for $i = 1, \ldots, k$. For a node $u \in N \setminus C$ we define $n(u) := |\{i \in \{1, \ldots, k\} : u \in tr(H_i)\}|$. Furthermore, we let $\lceil \alpha \rceil_q$ be the smallest integer that is bigger than or equal to the scalar α as well as divisible by q.

Theorem 5. *Let $(N, \mathcal{H}, tr)$ be an extended hypergraph, and let, for $k > q$, $k \not\equiv 0 \bmod q$, $(C, \{H_i : i = 1, 2, \ldots, k\})$ be a generalized (k, q)–cycle in $(N, \mathcal{H})$ such that $tr(H_i) \cap C = \emptyset$ for $i = 1, \ldots, k$. Then, the generalized (k, q)–cycle inequality*

$$\sum_{u \in C} x_u - \sum_{u \in tr(C)} \frac{\lceil n(u) \rceil_q}{q} x_u \leq |C| - \left\lceil \frac{k}{q} \right\rceil$$

is valid for the transitive packing polytope $P_{\mathrm{TP}}(N, \mathcal{H}, tr)$.

We now relate this first class of inequalities for the transitive packing polytope $P_{\mathrm{TP}}(N, \mathcal{H}, tr)$ to the three selected examples. For the node packing polytope we obtain exactly the *odd cycle inequalities* introduced by Padberg [Pad73]. This is true because all edges of the (hyper)graph have size 2 and hence all sets N_i have to be singletons.

The class of *Möbius ladder inequalities* [GJR85b] is a very prominent class of facet defining inequalities for the acyclic subdigraph polytope. It can be shown that a large subclass is contained in the class of generalized (k, q)–cycle inequalities for the acyclic subdigraph polytope.

In the case of the clique partitioning polytope, we are obviously restricted to generalized $(k, 2)$–cycles as the underlying hypergraph is a graph. This class contains two known classes of inequalities that are facet defining if G is a complete graph. The first class is formed by the *2–chorded odd cycle inequalities*, introduced by Grötschel and Wakabayashi [GW90]. Also structures that are not cycles in G lead to generalized $(k, 2)$–cycle inequalities, for instance, the *odd wheel inequalities* of Chopra and Rao [CR93].

Euler, Jünger, and Reinelt [EJR87] introduced generalized cycle inequalities for the independence system polytope. Our generalized cycles, restricted to independence systems, extend those, since they assumed that the nodes of $C \setminus \bigcup_{i=1}^{k} N_i$ are arranged in a certain sequence corresponding to that of the sets N_i.

Finally, we introduce a class of inequalities supported by generalized cycles, too, which are in general weaker than the generalized cycle inequalities. It arises from the class of generalized cycle inequalities when we do not take care of repetitions of transitive elements. We call this class of valid inequalities *weak generalized cycle inequalities*. For ease of referencing, we state this as a lemma.

Lemma 6. *Let $(N, \mathcal{H}, tr)$ be an extended hypergraph, and let, for $k > q$, $k \not\equiv 0 \bmod q$, the hypergraph $(C, \{H_i : i = 1, 2, \ldots, k\})$ be a generalized (k, q)–cycle in $(N, \mathcal{H})$ such that $tr(H_i) \cap C = \emptyset$ for $i = 1, \ldots, k$. Then, the weak generalized (k, q)–cycle inequality*

$$\sum_{u \in C} x_u - \sum_{u \in tr(C)} n(u) \, x_u \leq |C| - \left\lceil \frac{k}{q} \right\rceil \tag{6}$$

is valid for the transitive packing polytope $P_{\mathrm{TP}}(N, \mathcal{H}, tr)$.

Clearly, in case $n(u) \leq 1$ for all nodes $u \in N$ a generalized (k, q)–cycle inequality and its weak version coincide.

3.2 Generalized Clique Inequalities

A second well–known class of valid inequalities for the node packing polytope are *clique inequalities*, see, e. g., [Pad73]. We now describe how the clique inequalities can be extended to the transitive packing polytope.

Definition 7. Let $(N, \mathcal{H})$ be a hypergraph, and let $N_1, \ldots, N_k$, for integers $k \geq q \geq 2$, be a collection of mutually disjoint nonempty subsets of the node set N. For each q–element subset $\{i_1, \ldots, i_q\} \subseteq \{1, \ldots, k\}$ of indices we let $H_{i_1, \ldots, i_q} \in \mathcal{H}$ be an edge such that $\bigcup_{j=1}^{q} N_{i_j} \subseteq H_{i_1, \ldots, i_q}$. We assume that the edges in any collection of intersecting edges all have one common index. Let $C := \bigcup_{1 \leq i_1 < i_2 < \cdots < i_q \leq k} H_{i_1, \ldots, i_q}$. Then, we call the hypergraph

$$(C, \{H_{i_1, \ldots, i_q} : 1 \leq i_1 < i_2 < \cdots < i_q \leq k\})$$

a *generalized (k, q)–clique* (contained in $(N, \mathcal{H})$).

As for generalized cycles, we assume in the following that C and its set $tr(C)$ of transitive elements are disjoint. We denote by $mtr(C)$ the multiset that arises from the union of the transitive elements $tr(H_{i_1, \ldots, i_q})$. In other words, the multiplicity of a node $u \in mtr(C)$ is precisely the number of edges $H_{i_1, \ldots, i_q}$ of which u is a transitive element.

Theorem 8. *Let $(N, \mathcal{H}, tr)$ be an extended hypergraph, and let, for $k \geq q \geq 2$, the hypergraph $(C, \{H_{i_1, \ldots, i_q} : 1 \leq i_1 < i_2 < \cdots < i_q \leq k\})$ be a generalized (k, q)-clique in $(N, \mathcal{H})$ such that $tr(H_{i_1, \ldots, i_q}) \cap C = \emptyset$ for $1 \leq i_1 < i_2 < \cdots < i_q \leq k$. Then, the generalized (k, q)-clique inequality*

$$x(C) - x(mtr(C)) \leq |C| - k + q - 1 \tag{7}$$

is valid for $P_{\mathrm{TP}}(N, \mathcal{H}, tr)$.

Again, if we consider the case of independence systems the definition of generalized cliques given above is slightly more general than the one of Euler, Jünger, and Reinelt [EJR87]. They assumed that a node $u \in C \setminus \bigcup_{i=1}^{k} N_i$ cannot be contained in more than $\binom{k-1}{q-1} - 1$ edges (with common index) of the generalized (k, q)-clique. They showed that the corresponding generalized clique inequalities are facet inducing for the independence system with ground set C and circuits $H_{i_1, \ldots, i_q}$ only. Euler, Jünger, and Reinelt also observed that in case of the acyclic subdigraph polytope the *simple k-fence inequalities* [GJR85b] are contained in the class of generalized clique inequalities. We can even show that the facet defining k-fence inequalities (not necessarily simple) are contained in the class of generalized $(k, 2)$-clique inequalities. Whereas the class of generalized (k, q)-clique inequalities for the acyclic subdigraph polytope is much richer than the class of k-fence inequalities, the class of generalized $(k, 2)$-clique inequalities turns out to be precisely the class of $(1, k)$-2-*partition inequalities* for the clique partitioning polytope of a graph $G = (V, E)$ (again, $q > 2$ is not possible). The latter inequalities are due to Grötschel and Wakabayashi [GW90] and are facet defining if G is complete.

In case $q = 2$, the length of the cutting plane proof for (7) is at most $\lceil \log(k - 1) \rceil$. This bound is almost best possible.

Theorem 9. *Let $(C, \{H_{ij} : 1 \leq i < j \leq k\})$ be a generalized $(k, 2)$-clique of the extended hypergraph $(N, \mathcal{H}, tr)$. Assume that $N_i = \left(\bigcap_{i < j \leq k} H_{ij} \right) \cap \left(\bigcap_{1 \leq j < i} H_{ji} \right)$, for $i = 1, 2, \ldots, k$, and that each edge $H \in \mathcal{H}$ such that $H \subseteq C$ satisfies $N_i \cup N_j \subseteq H$ for some $i, j \in \{1, 2, \ldots, k\}$, $i \neq j$. Then the depth of the generalized $(k, 2)$-clique inequality (7) relative to (2) – (4) is at least $\log k - 1$.*

Theorem 9 shows that the depth of the generalized $(k, 2)$-clique inequalities tends to infinity with k. It was proved before for the special instances formed by the clique inequalities of the node packing polytope [Chv73] and by the simple k-fence inequalities of the acyclic subdigraph polytope [CCH89]. The assumption of Theorem 9 is also satisfied by the k-fence inequalities of the acyclic subdigraph polytope. Moreover, Theorem 9 also applies to the $(1, k)$-2-partition inequalities of the clique partitioning polytope.

3.3 Generalized Antiweb Inequalities

For reasons of brevity, we omit the development of *generalized antihole inequalities* here. Anyhow, as for the node packing polytope they form a subclass of *generalized antiweb inequalities*.

The main idea in the derivation of the generalized antiweb inequalities is to combine generalized clique inequalities in a manner oriented on the cutting plane proof of generalized cycle inequalities.

Definition 10. Let $(N, \mathcal{H})$ be a hypergraph, and let k, s, and q be integers such that $k \geq s \geq q \geq 2$. Let $N_1, N_2, \ldots, N_k$ be a sequence of mutually disjoint nonempty subsets of the node set N. For each $\ell \in \{1, 2, \ldots, k\}$ and each q-element set of indices $\{i_1, i_2, \ldots, i_q\} \subseteq \{\ell, \ell+1, \ldots, \ell+s-1\}$ (where indices are taken modulo $k+1$ and shifted by $+1$) we let $H^\ell_{i_1, i_2, \ldots, i_q} \in \mathcal{H}$ be an edge such that $\bigcup_{j=1}^q N_{i_j} \subseteq H^\ell_{i_1, i_2, \ldots, i_q}$. In addition, we assume, for each $\ell \in \{1, 2, \ldots, k\}$, that the edges in any collection of intersecting edges of type $H^\ell_{i_1, i_2, \ldots, i_q}$ all have one common (sub)index. For each ℓ we denote by W^ℓ the union of the associated edges, $W^\ell := \bigcup_{\ell \leq i_1 < i_2 < \cdots < i_q \leq \ell+s-1} H^\ell_{i_1, i_2, \ldots, i_q}$. Moreover, we let W denote the union of all these edges, $W := \bigcup_{\ell=1}^k W^\ell$. Again, for $u \in W$ we let $\tilde{m}(u)$ be the multiplicity of u with respect to its occurrence in W^ℓ, $\ell = 1, 2, \ldots, k$, i. e., $\tilde{m}(u) := |\{\ell \in \{1, 2, \ldots, k\} : u \in W^\ell\}|$. If $\tilde{m}(u) \leq s$ for all $u \in W$, then we call the hypergraph

$$(W, \{H^\ell_{i_1, i_2, \ldots, i_q} : \ell \leq i_1 < i_2 < \cdots < i_q \leq \ell+s-1 \text{ for some } \ell \in \{1, 2, \ldots, k\}\})$$

a *generalized (k, s, q)-antiweb* (contained in $(N, \mathcal{H})$).

Theorem 11. *Let $(N, \mathcal{H}, tr)$ be an extended hypergraph, and let the hypergraph $(W, \{H^\ell_{i_1, i_2, \ldots, i_q} : \ell \leq i_1 < i_2 < \cdots < i_q \leq \ell+s-1 \text{ for some } \ell \in \{1, 2, \ldots, k\}\})$ be a generalized (k, s, q)-antiweb in $(N, \mathcal{H})$ such that $tr(W) \cap W = \emptyset$. Then, the generalized (k, s, q)-antiweb inequality*

$$\sum_{u \in W} x_u - \sum_{u \in tr(W)} \frac{\lceil \tilde{n}(u) \rceil_s}{s} x_u \leq \left\lfloor \frac{s|W| - k(s-q+1)}{s} \right\rfloor \tag{8}$$

is valid for $P_{\text{TP}}(N, \mathcal{H}, tr)$. It has a cutting plane proof from (2) – (4) of length at most $\lceil \log(s-1) \rceil + 1$.

It follows from their construction that generalized (k, s, q)-antiweb inequalities subsume all the former classes of inequalities for the transitive packing polytope $P_{\text{TP}}(N, \mathcal{H}, tr)$. In fact, if $q = s$ and if s does not divide k, we obtain the class of generalized (k, q)-cycle inequalities; if $s = k$, the class of generalized antiweb inequalities contains the class of generalized (k, q)-clique inequalities; if $k = qs + 1$, we have the class of generalized (s, q)-antihole inequalities.

It is interesting that, although Laurent [Lau89] extended antiwebs to the independence system polytope, our inequalities restricted to this setting are still far more general. She only used one-element sets N_i and edges that are precisely the union of q of these. We are also able to extend the web inequalities of the node packing polytope [Tro75]. For brevity, we omit the development of the *generalized web inequalities* here.

3.4 Odd Partition Inequalities

In this section, we introduce another new class of inequalities for the transitive packing polytope. It is an extension of a class of inequalities recently proposed by Caprara and Fischetti [CF95] for the acyclic subdigraph polytope.

Let $H_1, \ldots, H_k$ be a collection of distinct edges of $\mathcal{H}$, and let $m(u) := |\{i \in \{1, \ldots, k\} : u \in H_i\}|$ and $n(u) := |\{i \in \{1, \ldots, k\} : u \in tr(H_i)\}|$. We denote the difference of these two numbers by $d(u)$, $d(u) := m(u) - n(u)$. Let W be the union of all the nodes involved, $W := \bigcup_{i=1}^{k}(H_i \cup tr(H_i))$, and let W^{odd} be the set of those nodes that occur either in an odd number of edges H_i or in an odd number of transitive sets $tr(H_i)$, but not both, $W^{odd} := \{u \in W : d(u) \text{ odd}\}$. Furthermore, let (W_1^{odd}, W_2^{odd}) be a partition of W^{odd} such that $\sum_{i=1}^{k} |H_i| + |W_1^{odd}| - k$ is odd. ($W_1^{odd} = \emptyset$ or $W_2^{odd} = \emptyset$ is possible.)

The following inequality is valid for the transitive packing polytope,

$$\sum_{u \in W \setminus W^{odd}} \frac{d(u)}{2} x_u + \sum_{u \in W_1^{odd}} \frac{d(u)+1}{2} x_u +$$

$$\sum_{u \in W_2^{odd}} \frac{d(u)-1}{2} x_u \leq \frac{\sum_{i=1}^{k} |H_i| + |W_1^{odd}| - k - 1}{2}. \tag{9}$$

We call inequalities of this type *odd partition inequalities*. We are able to point out some special cases in which inequality (9) is dominated by other inequalities as well as some other cases in which it has depth 1 relative to (2) – (4), and is therefore interesting.

As mentioned before, Caprara and Fischetti [CF95] introduced the odd partition inequalities for the acyclic subdigraph polytope in order to show that a subclass of the Möbius ladder inequalities can be derived from the initial relaxation by a cutting plane proof of length 1 where all coefficients used are either 0 or $\frac{1}{2}$. Indeed, if $(C, \{H_i : i = 1, 2, \ldots, k\})$ is a generalized $(k, 2)$–cycle we obtain the associated generalized $(k, 2)$–cycle inequality as an odd partition inequality by setting $W_1^{odd} := \{u \in C : m(u) \text{ odd}\}$ and $W_2^{odd} := \{u \in tr(C) : n(u) \text{ odd}\}$. The subclass of Möbius ladder inequalities where each triple of participating dicycles has empty intersection is contained in the class of generalized $(k, 2)$–cycle inequalities for the acyclic subdigraph polytope. This implies Caprara and Fischetti's observation.

4 Transitive Packing in Graphs

An important subproblem of the transitive packing problem is formed by the instances where the given hypergraph is actually a graph. This section is devoted to discuss the polytopes associated with these instances in more detail. To avoid confusions we still use the notation $(N, \mathcal{H}, tr)$ but assume throughout this section that $|H| = 2$ for all $H \in \mathcal{H}$ and call the triple $(N, \mathcal{H}, tr)$ an *extended graph*. The transitive packing polytope is then given as $P_{TP}(N, \mathcal{H}, tr) = \text{conv}\{x \in$

$\{0,1\}^N : x_u + x_v - \sum_{w \in tr(\{u,v\})} x_w \leq 1$ for $\{u,v\} \in \mathcal{H}\}$. Recall that both the node packing polytope and the clique partitioning polytope are of this flavor. It is known that for the node packing polytope all facet defining inequalities with right–hand side 1 are clique inequalities, see [Pad73]. This remains true for the transitive packing polytope of many extended graphs. The assumptions made in the following theorem are satisfied, for instance, by the clique partitioning problem.

Theorem 12. *Let $(N, \mathcal{H}, tr)$ be an extended graph such that for every clique C in $(N, \mathcal{H})$ the following condition is satisfied: each node $u \in tr(C)$ is associated with exactly one edge $\{v, w\}$ induced by C and satisfies either $\{u, v\}, \{u, w\} \notin \mathcal{H}$, or $\{u, v\} \notin \mathcal{H}, \{u, w\} \in \mathcal{H}$, and $v \in tr(\{u, w\})$, or $\{u, w\} \notin \mathcal{H}, \{u, v\} \in \mathcal{H}$, and $w \in tr(\{u, v\})$, or $\{u, v\}, \{u, w\} \in \mathcal{H}$, and $v \in tr(\{u, w\})$ and $w \in tr(\{u, v\})$. Then, any facet defining inequality $c\,x \leq 1$ (with c integral) of the transitive packing polytope $P_{TP}(N, \mathcal{H}, tr)$ is either of the form $x_u \leq 1$ or is a generalized $(k, 2)$–clique inequality.*

A clique and its transitive elements form the support of another class of valid inequalities where the nodes of the cliques have coefficients greater than one.

Theorem 13. *Let $(N, \mathcal{H}, tr)$ be an extended graph, and let C be the node set of a generalized $(k, 2)$–clique in $(N, \mathcal{H})$ such that $tr(C) \cap C = \emptyset$. Moreover, let $t \geq 1$ be an integer. Then, the t–reinforced generalized $(k, 2)$–clique inequality*

$$t\,x(C) - x(mtr(C)) \leq \frac{t(t+1)}{2} \tag{10}$$

is valid for the transitive packing polytope $P_{TP}(N, \mathcal{H}, tr)$.

The proof of Theorem 13 implies a range on t in order to ensure that the intersection of the transitive packing polytope and the hyperplane defined by a t–reinforced generalized $(k, 2)$–clique inequality is nonempty. The bound on t can even be strengthened if we assume that the t–reinforced generalized $(k, 2)$–clique inequality is facet defining.

Lemma 14. *Let $(N, \mathcal{H}, tr)$ be an extended graph, and let C be the node set of a generalized $(k, 2)$–clique in $(N, \mathcal{H})$ such that $tr(C) \cap C = \emptyset$. Let $t \geq 1$ be an integer. If the t–reinforced generalized $(k, 2)$–clique inequality (10) induces a facet of the transitive packing polytope $P_{TP}(N, \mathcal{H}, tr)$, then $t \leq |C| - 2$.*

One might ask whether there exist transitive packing polytopes of extended graphs such that the t–reinforced generalized $(k, 2)$–clique inequalities are facet defining. This is indeed the case. Oosten, Rutten, and Spieksma [ORS95] subsequently showed that the t–reinforced generalized $(k, 2)$–clique inequalities define facets of the clique partitioning polytope of a complete graph, for $t \leq k - 2$ of course.

5 Separation

After introducing several classes of valid inequalities for the transitive packing polytope, the natural question arises whether we can use these inequalities efficiently in cutting plane algorithms for attacking the transitive packing problem. We concentrate on generalized cycle, generalized clique, and odd partition inequalities. In fact, there is not much hope for clique inequalities. Grötschel, Lovász, and Schrijver [GLS81] showed that the optimization problem for the clique–constrained node packing polytope is NP–hard. Moreover, Müller [Mül93] proved that it is NP–hard to decide whether any simple k–fence inequality of the acyclic subdigraph polytope is violated by a given point that already satisfies the nonnegativity constraints, the upper bound constraints, and the dicycle inequalities. The separation problem for the last three classes of inequalities is polynomially solvable. The following theorem captures in particular all transitive packing problems in graphs. So, it covers for instance the 2–chorded odd cycle inequalities and the odd wheel inequalities for the clique partitioning polytope.

Theorem 15. *Let $(N, \mathcal{H}, tr)$ be an extended hypergraph, and let $P_{\mathrm{TP}}(N, \mathcal{H}, tr)$ be the associated transitive packing polytope. The separation problem for a class of valid inequalities for $P_{\mathrm{TP}}(N, \mathcal{H}, tr)$ that contains all weak generalized $(k, 2)$–cycle inequalities supported by cycles $(C, \{H_i : i = 1, 2, \ldots, k\})$ such that $|H_i| = 2$, $i = 1, 2, \ldots, k$ can be solved in polynomial time.*

Theorem 16. *Let $(N, \mathcal{H}, tr)$ be an extended hypergraph, and let $P_{\mathrm{TP}}(N, \mathcal{H}, tr)$ be the associated transitive packing polytope. The separation problem for a class of valid inequalities for $P_{\mathrm{TP}}(N, \mathcal{H}, tr)$ that contains certain odd partition inequalities can be solved in polynomial time.*

6 Additional Notes and References

Besides the node packing polytope, the acyclic subdigraph polytope, and the clique partitioning polytope there are several other polytopes that can be interpreted as transitive packing polytopes; for instance, the transitive acyclic subdigraph polytope [Mül93], the interval order polytope [MS95, Sch95] and the relatively transitive subgraph polytope [KL89, SB]. Due to space limits we cannot discuss these and other related polytopes here. We are not only able to explain several of the known valid inequalities but also to present several new inequalities by identifying generalized cycles, cliques, antiholes, antiwebs, and webs in the respective setting. Notice that the inequalities presented above remain valid when we allow for hypergraphs with loops. Then, we cover, for instance, the *cut polytope* [BM86] as well. Moreover, our inequalities are also useful for *plant location* and *frequency assignment* polytopes as well as in *propositional logic* and *constraint logic programming*.

It is well–known (see [Edm62]) that every *set packing problem* $\max\{c\,x : A\,x \leq \mathbb{1}, x_u \in \{0, 1\}\}$ where A is a 0/1 matrix can be transformed into an equivalent node packing problem on the *intersection graph* of A. Hence transitive packing covers set packing as well since it subsumes node packing. However, from the transitive packing polytope we cannot derive *generalized set packing polytopes* as special instances, neither in the sense of Laurent [Lau89], nor in the sense of Conforti and Cornuéjols [CC92]. Laurent called $\text{conv}\{x \in \{0,1\}^n : Ax \leq b\}$ the generalized set packing polytope where A is a 0/1 matrix and b a vector with nonzero integer components. Given a $0/\pm 1$ matrix A and the vector n_A whose components count the number of negative entries in the corresponding row of A, Conforti and Cornuéjols defined $\{x : Ax \leq 1 - n_A, 0 \leq x \leq 1\}$ as a generalized set packing polytope. (We would rather call it *fractional* generalized set packing polytope and save the name generalized set packing polytope for its integer hull.) On the other hand, as already pointed out, the transitive packing polytope of an extended hypergraph with no transitive elements reduces to an independence system polytope. There is a close relation between independence system polytopes and set covering polytopes (see, e. g., [Lau89, NS89]). A *set covering polytope* is of the form $\text{conv}\{y \in \{0,1\}^n : Ay \geq 1\}$ where A is a 0/1 matrix. The points y in the set covering polytope and the points x in the independence system polytope of the circuit system defined by the undominated rows of A are related by the affine transformation $x = 1 - y$. Consequently, set covering polytopes and independence system polytopes are equivalent – modulo the above transformation. An implication is that any result stated for the independence system polytope can be translated to the set covering polytope, and vice versa. Thus the work of Balas and Ng [BN89a, BN89b], of Cornuéjols and Sassano [CS89], Nobili and Sassano [NS89], as well as Sassano [Sas89] on the set covering polytope can be seen as a contribution to the knowledge of the independence system polytope. This implies especially that our extension of the class of antiweb inequalities for the independence system polytope extends the known rose inequalities [Sas89] for the set covering polytope, too.

If we apply the complementing of variables to the transitive packing polytope $P_{\text{TP}}(N, \mathcal{H}, tr) = \text{conv}\{x \in \{0,1\}^N : Ax \leq p_A - 1\}$ where the $0/\pm 1$ matrix A is the extended edge–node incidence matrix of the extended hypergraph $(N, \mathcal{H}, tr)$, it turns out to be equivalent (modulo this affine transformation) to the polytope $Q(A) := \text{conv}\{x \in \{0,1\}^N : Ax \geq 1 - n_A\}$. The linear relaxation of the polytope $Q(A)$ has been recently introduced by Conforti and Cornuéjols [CC92] in the context of balanced $0/\pm 1$ matrices as the *generalized set covering polytope*. We call it *fractional* generalized set covering polytope here and rather $Q(A)$ the generalized set covering polytope. Conforti and Cornuéjols [CC92] as well as Nobili and Sassano [NS95] try to characterize when the fractional generalized set covering polytope is integral, i. e., when it coincides with the generalized set covering polytope. Our work can be seen as a first contribution to the study of the generalized set covering polytope when it is properly contained in the corresponding fractional one. Conforti and Cornuéjols showed that a $0/\pm 1$ matrix A is balanced if and only if the fractional generalized set covering (or packing) polytope is integral, for each submatrix of A. An extension of the concept of balanced $0/\pm 1$ matrices are ideal matrices. A $0/\pm 1$ matrix A is *ideal* if its fractional generalized set covering polytope is integral, or, equivalently, if its

443

transitive packing polytope is integral. It would be very interesting for problems
that can be interpreted as transitive packing problems, to characterize when the
extended edge–node incidence matrices of their associated extended hypergraphs
are ideal.

References

[BGM85] F. Barahona, M. Grötschel, and A. R. Mahjoub. Facets of the bipartite sub-
         graph polytope. *Mathematics of Operations Research*, 10:340 – 358, 1985.
[BM86]   F. Barahona and A. R. Mahjoub. On the cut polytope. *Mathematical Pro-
         gramming*, 36:157 – 173, 1986.
[BN89a] E. Balas and S. M. Ng. On the set covering polytope: I. All the facets with
         coefficients in {0, 1, 2}. *Mathematical Programming*, 43:57 – 69, 1989.
[BN89b] E. Balas and S. M. Ng. On the set covering polytope: II. Lifting the facets
         with coefficients in {0, 1, 2}. *Mathematical Programming*, 45:1 – 20, 1989.
[BP76]   E. Balas and M. W. Padberg. Set partitioning: A survey. *SIAM Review*,
         18:710 – 760, 1976.
[CC92]   M. Conforti and G. Cornuéjols. Balanced 0, ±1 matrices, bicoloring and total
         dual integrality. Preprint, Carnegie Mellon University, Pittsburgh, USA,
         1992.
[CCH89] V. Chvátal, W. Cook, and M. Hartmann. On cutting–plane proofs in com-
         binatorial optimization. *Linear Algebra and its Applications*, 114/115:455 –
         499, 1989.
[CF95]   A. Caprara and M. Fischetti. {0, ½}–Chvátal–Gomory cuts. Technical Re-
         port, DEIS, University of Bologna, Bologna, Italy, 1993, revised 1995.
[Chv73] V. Chvátal. Edmonds polytopes and a hierarchy of combinatorial problems.
         Discrete Mathematics, 4:305 – 337, 1973.
[CR93]   S. Chopra and M. R. Rao. The partition problem. *Mathematical Program-
         ming*, 59:87 – 115, 1993.
[CS89]   G. Cornuéjols and A. Sassano. On the 0, 1 facets of the set covering polytope.
         Mathematical Programming, 43:45 – 55, 1989.
[Edm62] J. Edmonds. Covers and packings in a family of sets. *Bulletin of the Amer-
         ican Mathematical Society*, 68:494 – 499, 1962.
[EJR87] R. Euler, M. Jünger, and G. Reinelt. Generalizations of cliques, odd cycles
         and anticycles and their relation to independence system polyhedra. *Math-
         ematics of Operations Research*, 12:451 – 462, 1987.
[GJR85a] M. Grötschel, M. Jünger, and G. Reinelt. Acyclic subdigraphs and linear
         orderings: Polytopes, facets, and cutting plane algorithms. In I. Rival, ed-
         itor, *Graphs and Order*, pages 217 – 266. D. Reidel Publishing Company,
         Dordrecht, 1985.
[GJR85b] M. Grötschel, M. Jünger, and G. Reinelt. On the acyclic subgraph polytope.
         Mathematical Programming, 33:28 – 42, 1985.
[GLS81] M. Grötschel, L. Lovász, and A. Schrijver. The ellipsoid method and its
         consequences in combinatorial optimization. *Combinatorica*, 1:169 – 197,
         1981. (Corrigendum: 4 (1984), 291 – 295).
[GLS88] M. Grötschel, L. Lovász, and A. Schrijver. *Geometric Algorithms and
         Combinatorial Optimization*, volume 2 of *Algorithms and Combinatorics*.
         Springer, Berlin, 1988.
[GW89]   M. Grötschel and Y. Wakabayashi. A cutting plane algorithm for a clustering
         problem. *Mathematical Progamming*, 45:59 – 96, 1989.
[GW90]   M. Grötschel and Y. Wakabayashi. Facets of the clique partitioning poly-
         tope. *Mathematical Programming*, 47:367 – 388, 1990.

[JM93]   M. Jünger and P. Mutzel. Solving the maximum weight planar subgraph. In
         G. Rinaldi and L. A. Wolsey, editors, *Integer Programming and Combinato-*
         rial Optimization, pages 479 – 492, 1993. Proceedings of the 3rd International
         IPCO Conference.

[Jün85] M. Jünger. *Polyhedral Combinatorics and the Acyclic Subdigraph Problem*,
         volume 7 of *Research and Expositions in Mathematics*. Heldermann Verlag
         Berlin, 1985.

[KL89]   B. Korte and L. Lovász. Polyhedral results for antimatroids. In G. S. Bloom,
         R. L. Graham, and J. Malkevitch, editors, *Combinatorial Mathematics*, pages
         283 – 295. Academy of Sciences, New York, 1989. Proceedings of the Third
         International Conference.

[Lau89] M. Laurent. A generalization of antiwebs to independence systems and their
         canonical facets. *Mathematical Programming*, 45:97 – 108, 1989.

[MS95]   R. Müller and A. S. Schulz. The interval order polytope of a digraph. In
         E. Balas and J. Clausen, editors, *Integer Programming and Combinatorial*
         Optimization, number 920 in Lecture Notes in Computer Science, pages 50
         – 64. Springer, Berlin, 1995. Proceedings of the 4th International IPCO
         Conference.

[MS96]   R. Müller and A. S. Schulz. Transitive packing. Preprint, Department of
         Mathematics, Technical University of Berlin, Berlin, Germany, 1996.

[Mül93] R. Müller. On the transitive acyclic subdigraph polytope. In G. Rinaldi and
         L. A. Wolsey, editors, *Integer Programming and Combinatorial Optimization*,
         pages 463 – 477, 1993. Proceedings of the 3rd International IPCO Confer-
         ence.

[NS89]   P. Nobili and A. Sassano. Facets and lifting procedures for the set covering
         polytope. *Mathematical Programming*, 45:111 – 137, 1989.

[NS95]   P. Nobili and A. Sassano. $(0, \pm 1)$ ideal matrices. In E. Balas and J. Clausen,
         editors, *Integer Programming and Combinatorial Optimization*, number 920
         in Lecture Notes in Computer Science, pages 344 – 359. Springer, Berlin,
         1995. Proceedings of the 4th International IPCO Conference.

[NT74]   G. L. Nemhauser and L. E. Trotter Jr. Properties of vertex packing and in-
         dependence system polyhedra. *Mathematical Programming*, 6:48 – 61, 1974.

[NW88]   G. L. Nemhauser and L. A. Wolsey. *Integer and Combinatorial Optimization*.
         John Wiley & Sons, New York, 1988.

[ORS95] M. Oosten, J. H. G. C. Rutten, and F. C. R. Spieksma. The clique partition-
         ing polytope: Facets. Department of Mathematics, University of Limburg,
         Maastricht, The Netherlands, 1995.

[Pad73] M. W. Padberg. On the facial structure of set packing polyhedra. *Mathe-*
         matical Programming, 5:199 – 215, 1973.

[Pad79] M. W. Padberg. Covering, packing and knapsack problems. *Annals of Dis-*
         crete Mathematics, 4:265 – 287, 1979.

[Sas89] A. Sassano. On the facial structure of the set covering polytope. *Mathemat-*
         ical Programming, 44:181 – 202, 1989.

[SB]   D. F. Shallcross and R. G. Bland. On the polyhedral structure of relatively
         transitive subgraphs. Technical report, Cornell University, Ithaca, NY.

[Sch80] A. Schrijver. On cutting planes. In M. Deza and I. G. Rosenberg, editors,
         Combinatorics '79, Part II, volume 9 of *Annals of Discrete Mathematics*,
         pages 291 – 296. North–Holland, Amsterdam, 1980.

[Sch86] A. Schrijver. *Theory of Linear and Integer Programming*. John Wiley &
         Sons, Chichester, 1986.

[Sch95] A. S. Schulz. *Polytopes and Scheduling*. PhD thesis, Technical University of
         Berlin, Berlin, Germany, 1995.

[Tro75] L. E. Trotter Jr. A class of facet producing graphs for vertex packing poly-
         hedra. *Discrete Mathematics*, 12:373 – 388, 1975.

A Polyhedral Approach to the Feedback Vertex Set Problem

Meinrad Funke and Gerhard Reinelt

Institut für Angewandte Mathematik, Universität Heidelberg
Im Neuenheimer Feld 294, D-69120 Heidelberg
meinrad.funke@iwr.uni-heidelberg.de

Abstract. Feedback problems consist of removing a minimal number of arcs or nodes of a directed or undirected graph in order to make it acyclic. In this paper we consider a special variant, namely the problem of finding a maximum weight node induced acyclic subdigraph. We discuss valid and facet defining inequalities for the associated polytope and present computational results with a branch–and–cut algorithm.

1 Introduction

The feedback vertex set problem for an undirected or directed graph consists of finding a set W of nodes with minimum cardinality and the property that W contains at least one node of every directed or undirected cycle of the graph. In other words, all cycles have to be broken by deleting as few nodes as possible. Variations are possible if node weights and/or arc weights are given.

One application of the feedback vertex set problem is the removal of deadlocks as explained in [12]. Deadlocks can arise in many contexts. For example, consider an operating system where different processes request resources, which they need to use exclusively, before they can release the resources they have already allocated. For modelling these resource requirements we use a digraph which has a node for each process and arcs (u, v) whenever process u requests a resource already allocated to process v. If this digraph contains a dicycle we have a deadlock situation since all processes on this dicycle are waiting for each other. No process can allocate new resources and make previously allocated resources available. In order to break such deadlocks we must eliminate in every dicycle at least one process by deleting it or setting it to a waiting state without allocated resources. It is clear that we want to delete as few nodes as possible. Further applications in circuit design and numerics are reported in [4], [9], and in [13].

The feedback vertex set problem is $\mathcal{NP}$-complete ([7]). There exist several approximative algorithms ([8], [11], [12]), polynomial exact algorithms for special cases ([4], [14], [15]), and exact algorithms with exponential time complexity ([13]). In this paper we discuss the possible variants of feedback problems and present a branch–and–cut approach for one of them.

Definitions for different feedback problems are given in section 2 and the corresponding polyhedra are introduced in section 3. Section 4 deals with the integer programming formulation of one specific problem. Sections 5 and 6 discuss facets and their separation. Primal heuristics are addressed in section 7 and computational results are presented in 8.

2 Definitions

We first define several feedback problems and give the equivalent formulations as acyclic subdigraph problems. We consider only the case of directed graphs. All definitions carry over to undirected graphs in the obvious way.

2.1 Feedback Problems

A *feedback arc set* for a digraph $D = (V, A)$ is a set $B \subseteq A$ of arcs such that $(V, A \setminus B)$ is acyclic, i.e. contains no dicycle. The *feedback arc set problem* consists of determining a feedback arc set of minimum cardinality.

A *feedback vertex set* is a set $W \subseteq V$ of nodes such that $(V \setminus W, A(V \setminus W))$, i.e. the subdigraph of D induced by $V \setminus W$, is acyclic. The *feedback vertex set problem* consists of determining a feedback vertex set of minimum cardinality.

If arcs or nodes have associated weights c_a for an arc a or c_v for a node v, we can define the *weighted feedback arc set problem* or the *weighted feedback vertex set problem* where the task is to find a feedback arc set or feedback vertex set with minimum sum of weights of arcs or nodes.

A subdigraph $D' = (W, B)$ is *node induced* if $B = A(W)$ holds. The *node induced feedback arc set problem* consists of finding a minimum set B of arcs such that there exists a set $W \subseteq V$ with $A(W) = A \setminus B$ and $(W, A \setminus B)$ is acyclic.

An arc is deleted if at least one of its end nodes is deleted. We have therefore in general various possibilities to delete arc sets by deleting different node sets. For example, arcs (v_1, v_2), (v_2, v_3), (v_3, v_4), and (v_4, v_5) can be eliminated by deleting the node sets $\{v_2, v_3, v_4\}$ or $\{v_2, v_4\}$. Hence, it is clear that in order to model the node induced feedback arc set problem as a linear program we need arc variables even though the dicycles of D must be broken by eliminating nodes.

2.2 Acyclic Subdigraph Problems

The *acyclic subdigraph problem* for a digraph $D = (V, A)$ consists of finding an acyclic subset $B \subseteq A$ of maximum cardinality. The *node induced acyclic subdigraph problem* consists of finding an acyclic node induced subdigraph of maximum cardinality.

Variants are possible if node induced acyclic subdigraphs are required. Depending on whether the number of arcs, the number of nodes, or both are to be maximized we use the terms *node induced acyclic subdigraph problem with arc variables*, *with node variables*, or *with arc and node variables* to make the differences clear.

These problems are equivalent to feedback problems. For example suppose a digraph $D = (V, A)$ is given. If W is a feedback vertex set of minimum cardinality then $V \setminus W$ is the node set of a node induced acyclic subdigraph with maximum number of nodes.

The acyclic subdigraph problem has been extensively studied in [2] and in [5]. Here we will concentrate on the node induced acyclic subdigraph problem with arc variables.

3 Polyhedra

In the following, we suppose that all digraphs do not contain parallel arcs or loops. For a given ground set A and $B \subseteq A$ the *incidence vector* $\chi^B \in \{0,1\}^A$ of B is defined by setting $\chi_a^B = 1$ if $a \in B$ and $\chi_a^B = 0$ if $a \notin B$.

Definition 1. Let $D = (V, A)$ be a digraph. Then we define the following polytopes.

a) $P_{NAC}^V(D) := \operatorname{conv}\{\chi^W \mid (W, A(W))$ is a node induced acyclic subdigraph$\}$.

b) $P_{NAC}^A(D) := \operatorname{conv}\{\chi^B \mid (V(B), B)$ is a node induced acyclic subdigraph$\}$.

c) $P_{NAC}^{V \cup A}(D) := \operatorname{conv}\{\chi^{W \cup B} \mid (W, B)$ is a node induced acyclic subdigraph$\}$.

For an arc $a = (u, v)$ we define the *inverted arc* $\overleftarrow{a} := (v, u)$ and for a given arc set A the set of *inverted arcs* $\overleftarrow{A} := \{(v, u) \mid (u, v) \in A\}$. The dimensions of the polytopes are now easily obtained.

Theorem 2. Let $D = (V, A)$ be a digraph. Then we have the following dimensions.

a) $\dim P_{NAC}^V(D) = |V|$.

b) $\dim P_{NAC}^A(D) = |A| - |A \cap \overleftarrow{A}|$.

c) $\dim P_{NAC}^{V \cup A}(D) = |V| + |A| - |A \cap \overleftarrow{A}|$.

Proof. This proposition is clear, since $\chi^{\{v\}}$ is always feasible for $P_{NAC}^V(D)$ and $P_{NAC}^{V \cup A}(D)$ and $\chi^{\{a\}}$ or $\chi^{\{a,v,u\}}$ for $a = (u, v)$ is feasible for $P_{NAC}^A(D)$ or $P_{NAC}^{V \cup A}(D)$ if and only if $\overleftarrow{a} \notin A$. □

In the following we will focus on the polytope $P_{NAC}^A(D)$.

4 Integer Programming Formulation

For obtaining an integer programming formulation we have to characterize incidence vectors of node induced acyclic arc sets.

Theorem 3. Let $D = (V, A)$ be a digraph. A 0-1 vector x is the incidence vector of a node induced acyclic arc set if and only if x satisfies the following conditions.

$$x(C) \leq |C| - 1 \quad \text{for all dicycles } C \subseteq A . \tag{1}$$

$$x_e - x_f + x_g \leq 1 \quad \text{for all } e, f, g \in A \text{ such that } V(f) \subseteq V(\{e, g\}) . \tag{2}$$

□

Inequalities (1) model the fact, that we must delete at least one arc in every dicycle. Inequalities (2) guarantee that the subdigraph is node induced. We denote these inequalities *node inducing inequalities*. Therefore, the set of feasible 0-1 solutions of (1)-(2) is the intersection of the set of acyclic subdigraphs of D and the set of node induced subdigraphs.

It will turn out that, in general, this formulation can be improved considerably. E.g. observe that $x_a = x_{\overleftarrow{a}} = 0$ must hold for any pair of anti-parallel arcs a and $\overleftarrow{a}$ in A since they form a dicycle of length two. Furthermore, it is obvious that $x(C) \leq |C| - 1$ can be strengthened to $x(C) \leq |C| - 2$ because deleting only one arc from C would violate a node inducing inequality in (2). If $f, \overleftarrow{f} \in A$, the upper bound $x_e \leq 1$ is implied by the node inducing inequality $x_e - x_f + x_g \leq 1$ and $x_f = 0$. For an undirected cycle of length three, two node inducing inequalities $x_{uv} - x_{vw} + x_{uw} \leq 1$ and $x_{uv} + x_{vw} - x_{uw} \leq 1$ imply the upper bound inequality $x_{uv} \leq 1$. An upper bound inequality $x_{uv} \leq 1$ is also implied by the dicycle inequality $x_{uv} + x_{vw} + x_{wu} \leq 1$ for the dicycle $\{(u,v), (v,w), (w,u)\}$.

5 Polyhedral Structure of $P_{NAC}^A(D)$

For describing the polyhedral structure of $P_{NAC}^A(D)$ we need further inequalities in addition to (1)-(2) of section 4. Furthermore, we have to examine which of the inequalities (1)-(2) are redundant in the sense that they do not define facets of $P_{NAC}^A(D)$.

To reduce tedious technical details, we will from now on assume that all digraphs contain no pair of antiparallel arcs. In this case $P_{NAC}^A(D)$ is full dimensional.

We can use the following transformation to get rid of antiparallel arcs. Suppose that we are given a digraph $D = (V, A)$ with arc weights c_a, $a \in A$, and that A contains the pair $(u,v), (v,u)$ of antiparallel arcs. We define $D' = (V', A')$ as follows.

$$V' := V \cup \{x, y\}, \quad A' := (A \setminus \{(v, u)\}) \cup \{(v, x), (x, y), (y, u)\} .$$

For every arc $a \in A \cap A'$ different from (u, v) we set $c_a' := c_a$ and

$$c_{uv}' = c_{uv} + c_{vu}, \quad c_{vx}' = 0, \quad c_{xy}' = \min\left\{ \sum_{a \in \delta(u)} c_a, \sum_{a \in \delta(v)} c_a \right\} + 1, \quad c_{yu}' = 0 .$$

Clearly, every node induced acyclic arc set of D is a node induced acyclic arc set of D' having the same weight. Consider now a node induced acyclic arc set of D' containing (u, v). Because of the dicycle $\{(u,v), (v,x), (x,y), (y,u)\}$, this set cannot contain (x, y). We obtain a node induced acyclic arc set with larger weight by including (x, y) and either deleting u and all of its incident arcs or v and all of its incident arcs. The choice between u and v depends on whether the arcs incident with u or incident with v have smaller sum of weights. Therefore, a maximum weight node induced acyclic subdigraph of D' does not contain (u, v) and induces a node induced acyclic subdigraph of D with equal and maximum weight.

5.1 Trivial Inequalities

We first examine the trivial inequalities $0 \leq x_a \leq 1$, for every $a \in A$.

Theorem 4. *For every arc* $a \in A$, *the inequality* $x_a \geq 0$ *is facet defining for* $P_{NAC}^A(D)$.

Proof. Let $a \in A$. For every $b \in A \setminus \{a\}$, $\{b\}$ is a node induced acyclic arc set and $\chi^{\{b\}}$ satisfies $x_a = 0$. These incidence vectors together with the zero vector give a set of $|A|$ affinely independent vectors satisfying $x_a = 0$. □

A *triangle* in D is a set of three arcs induced by three nodes, i.e. an undirected cycle of length three if the directions of the arcs are neglected. In the following we use the notation $[u, v]$ for arcs (u, v) or (v, u) if the direction of an arc does not matter. In such a case we do not differentiate between x_{uv} and x_{vu}. Due to our assumption at most one of (u, v) and (v, u) exists.

Theorem 5. *For every arc* $a = (u, v) \in A$, *the inequality* $x_a \leq 1$ *is facet defining for* $P_{NAC}^A(D)$ *if and only if the following conditions are satisfied.*
  (i)   a is not contained in a triangle.
  (ii) Neither u nor v are nodes of a dicycle of length three.
  (iii) a is not contained in a dicycle of length four.

Proof. Let $a = (u, v)$ be contained in a triangle $\{[u, v], [v, w], [w, u]\}$. Summing up the node inducing inequalities $x_{uv} - x_{vw} + x_{wu} \leq 1$ and $x_{uv} + x_{vw} - x_{wu} \leq 1$ implies $2x_{uv} \leq 2$ and therefore $x_{uv} \leq 1$ is not facet defining.

Let u be a node of a dicycle of length three and let (w, z) be the arc of this dicycle not incident with u. Then $x_{uv} = 1$ implies $x_{wz} = 0$ and therefore $x_{uv} \leq 1$ cannot define a facet of $P_{NAC}^A(D)$.

Consider a dicycle $C = \{(u, v), (v, w), (w, z), (z, u)\}$ of length four. The inequality $x_{uv} + x_{wz} \leq 1$ is valid for $P_{NAC}^A(D)$ because if both (u, v) and (w, z) are present in a subset of arcs then also (v, w) and (z, u) have to be present and the subset contains a dicycle. This valid inequality $x_{uv} + x_{wz} \leq 1$ implies $x_a \leq 1$.

Now assume that conditions (i)-(iii) are satisfied. The incidence vector $\chi^{\{a\}}$ satisfies $x_a \leq 1$ with equality. Let $b \in A \setminus \{a\}$. If a and b have a common node then $\chi^{\{a,b\}}$ is the incidence vector of a node induced acyclic arc set due to condition (i). Now let $b = (w, z)$ and $\{u, v\} \cap \{w, z\} = \emptyset$. Because of (ii) and (iii), the arc set $B = A(\{u, v, w, z\})$ induced by u, v, w, and z is acyclic and χ^B satisfies $x_a = 1$. We have constructed $|A|$ feasible incidence vectors satisfying $x_a \leq 1$ with equality. If we denote by $A_1 \subseteq A \setminus \{a\}$ the set of arcs having one common endnode with a and by A_2 the set of arcs having no common endnode with a, after a suitable reordering the matrix of these incidence vectors looks as follows.

$$
\begin{array}{ccc}
a & A_1 & A_2 \\
\end{array}
$$

$$
\begin{pmatrix}
1 & 0\cdots0 & 0\cdots0 \\
\vdots & \begin{smallmatrix} 1 & & 0 \\ & \ddots & \\ 0 & & 1 \end{smallmatrix} & 0 \\
\vdots & * & \begin{smallmatrix} 1 & & 0 \\ & \ddots & \\ 0 & & 1 \end{smallmatrix}
\end{pmatrix}
$$

This matrix clearly has full rank proving that $x_a \leq 1$ is facet defining. □

In the proof of this theorem we constructed a special matrix of $|A|$ incidence vectors to show the facet defining property in three steps. First, an incidence vector equal to zero for all arcs $b \neq a$ was given, followed by appropriate incidence vectors for every arc b incident with an endnode of a. Finally, all arcs having no endnode in common with a were considered. We will use this technique for further inequalities with the only difference that instead of a single arc a the set of arcs induced by the endnodes of the corresponding support set (arcs with non zero coefficients) has to be considered.

5.2 Node Inducing Inequalities

Theorem 6. *Let $D = (V, A)$ and $\{e, f, g\} \subseteq A$ such that $e = [u, v]$, $f = [v, x]$, $g = [x, y]$ and all four nodes u, v, x, y are distinct. Then the inequality $x_e - x_f + x_g \leq 1$ is facet defining for $P_{NAC}^A(D)$ if and only if the following conditions are satisfied.*

  (i)   *The undirected subgraph induced by $\{u, v, x, y\}$ is not isomorphic to K_4.*
  (ii) *$A(\{u, v, x, y\})$ is acyclic.*
  (iii) *$A(\{w, u, v\})$ and $A(\{w, x, y\})$ are acyclic for every node $w \in V \backslash \{u, v, x, y\}$.*
  (iv) *$A(\{w, z, u, v\})$ or $A(\{w, z, x, y\})$ is acyclic for any arc $(w, z) \in A$ with $w, z \in V \setminus \{u, v, x, y\}$.*

Proof. Let $a_1 = [u, x]$, $a_2 = [u, y]$, $a_3 = [v, y]$. The possible incidence vectors (restricted to e, f, g, a_1, a_2, and a_3) that satisfy $x_e - x_f + x_g = 1$ are represented by the matrix

$$
\begin{array}{cccccc}
e & f & g & a_1 & a_2 & a_3
\end{array}
$$
$$
\begin{pmatrix}
1 & 0 & 0 & 0 & 0 & 0 \\
0 & 0 & 1 & 0 & 0 & 0 \\
1 & 1 & 1 & 1 & 1 & 1 \\
1 & 0 & 0 & 0 & 1 & 1 \\
0 & 0 & 1 & 1 & 1 & 0
\end{pmatrix} .
$$

If one of the arcs a_1, a_2, or a_3 is not present in A, then the corresponding column has to be deleted. Only if not all arcs a_1, a_2, and a_3 are present and if no dicycle is contained in the subdigraph induced by $\{u, v, x, y\}$, this matrix has full column rank. Therefore, conditions (i) and (ii) are necessary for $x_e - x_f + x_g \leq 1$ to be facet defining.

Suppose $A(\{u, v, w\})$ contains a dicycle for a node w. In this case there exists no feasible incidence vector satisfying $x_e - x_f + x_g = 1$ and $x_b = 1$ for $b = [v, w]$. Hence, $x_e - x_f + x_g = 1$ implies $x_e - x_f + x_g - x_b = 1$ and $x_e - x_f + x_g \leq 1$ is not facet defining.

Suppose now that for an arc $a = [w, z]$ both $A(\{w, z, u, v\})$ and $A(\{w, z, x, y\})$ are not acyclic. In this case $x_a = 0$ must hold for every feasible incidence vector satisfying $x_e - x_f + x_g = 1$ and $x_e - x_f + x_g \leq 1$ cannot define a facet.

Now assume that conditions (i)-(iv) are satisfied. Since (i) and (ii) hold, we have a submatrix M of the matrix above with the rank $|A \cap \{e, f, g, a_1, a_2, a_3\}|$.

Consider a node $w \notin \{u, v, x, y\}$ and arcs $b_1 = [w, u]$, $b_2 = [w, v]$, $b_3 = [w, x]$, and $b_4 = [w, y]$. Restricted to $e, f, g, b_1, \ldots, b_4$ we obtain the matrix of incidence vectors shown below. Note that the corresponding arc sets are acyclic due to condition (iii) and can be linearly transformed such that the submatrix for $b_1, \ldots, b_4$ is the identity matrix.

$$
\begin{array}{ccccccc}
e & f & g & b_1 & b_2 & b_3 & b_4
\end{array}
$$
$$
\begin{pmatrix}
1 & 0 & 0 & 1 & 1 & 0 & 0 \\
1 & 0 & 0 & 1 & 1 & 0 & 1 \\
0 & 0 & 1 & 0 & 0 & 1 & 1 \\
0 & 0 & 1 & 1 & 0 & 1 & 1
\end{pmatrix} .
$$

Finally, an arc $a = [w, z]$ having no endnodes in $\{u, v, x, y\}$ has to be considered. W.l.o.g. we can assume that $B = A(\{w, z, u, v\})$ does not contain a dicycle due to condition (iv). χ^B satisfies $x_a = 1$ and $x_e - x_f + x_g = 1$.

We have now $|A|$ fesible incidence vectors satisfying $x_e - x_f + x_g = 1$. Denoting $A_1 = A(\{u, v, x, y\})$, $A_2 = \{[w, z] \mid w \notin \{u, v, x, y\}$ and $z \in \{u, v, x, y\}\}$, $A_3 = \{a \mid a$ has no endnode in $\{u, v, x, y\}\}$, the reordered and transformed matrix of these vectors looks as follows.

$$
\begin{array}{ccc}
A_1 & A_2 & A_3
\end{array}
$$
$$
\begin{pmatrix}
M & 0 & 0 \\
* & \begin{smallmatrix} 1 & & 0 \\ & \ddots & \\ * & & 1 \end{smallmatrix} & 0 \\
* & * & \begin{smallmatrix} 1 & & 0 \\ & \ddots & \\ * & & 1 \end{smallmatrix}
\end{pmatrix}
$$

This matrix has linear rank $|A|$ and proves that $x_e - x_f + x_g \leq 1$ defines a facet. □

Theorem 7. Let $D = (V, A)$ and $\{e, f, g\} \subseteq A$ be a triangle such that $e = [u, v]$, $f = [v, x]$, $g = [x, u]$. Then the inequality $x_e - x_f + x_g \leq 1$ is facet defining for $P_{NAC}^A(D)$ if and only if the following conditions are satisfied.

(i) $A(\{w, u, v\})$ and $A(\{w, x, u\})$ are acyclic for every node $w \in V \setminus \{u, v, x\}$.

(ii) If the undirected graph induced by $A(\{w, u, v, x\})$ is isomorphic to K_4 for a node $w \notin \{u, v, x\}$ then $A(\{w, v, x\})$ must be acyclic.

(iii) $A(\{w, z, u, v\})$ or $A(\{w, z, x, u\})$ is acyclic for any arc $(w, z) \in A$ with $w, z \in V \setminus \{u, v, x, u\}$.

(iv) $\{e, f, g\}$ is not a dicycle.

Proof. The proof proceeds along the same line as the previous one for theorem 6. We only outline the differences.

Since $\{e, f, g\}$ is a triangle and acyclic we have the initial matrix

$$
\begin{array}{ccc}
 & e & f & g
\end{array}
$$
$$
M = \begin{pmatrix}
1 & 0 & 0 \\
0 & 0 & 1 \\
1 & 1 & 1
\end{pmatrix} .
$$

Consider now an arbitrary node $w \in V \setminus \{u, v, x\}$. For $b_1 = [w, u]$, $b_2 = [w, v]$, $b_3 = [w, x]$, we obtain the following matrix of incidence vectors satisfying $x_e - x_f + x_g = 1$ restricted to e, f, g, $b_1, \ldots, b_3$.

$$
\begin{array}{cccccc}
e & f & g & b_1 & b_2 & b_3 \\
\end{array}
$$
$$
\begin{pmatrix}
1 & 0 & 0 & 1 & 1 & 0 \\
0 & 0 & 1 & 0 & 1 & 1 \\
1 & 1 & 1 & 1 & 1 & 1
\end{pmatrix}
$$

If $A(\{w, v, x\})$ is not acyclic then the undirected graph induced by $A(\{w, u, v, x\})$ is not isomorphic to K_4 due to (ii). In this case, we only need the first two rows. If the undirected subgraph induced by $A(\{w, u, v, x\})$ is isomorphic to K_4, then (i), (ii), and (iv) guarantee that $A(\{w, u, v, x\})$ is acyclic. In both cases, the rank of the corresponding submatrix is equal to $|\{b_1, b_2, b_3\} \cap A|$.

Condition (ii) is necessary. Otherwise the matrix above could only have two rows and having the full column rank if $b_1, \ldots, b_3$ are present.

The necessity of conditions (i) and (iii) is obtained as in the proof of the previous theorem. If $\{e, f, g\}$ is not acyclic then $x_e - x_f + x_g \leq 1$ is implied by the dicycle inequality $x_e + x_f + x_g \leq 1$ and is not facet defining. $\qquad\square$

5.3 Cycle Inequalities

Clearly, for a dicycle $C \subseteq A$ the associated inequality $x(C) \leq |C| - 2$ is valid for $P_{NAC}^A(D)$. The analysis whether it is facet defining is complicated by the fact that we also have to take into account that only node induced subgraphs are feasible.

Theorem 8. *For every dicycle $C \subseteq A$ the inequality $(C) \leq |C| - 2$ defines a facet of $P_{NAC}^A(D)$ if and only if the following conditions are satisfied.*

- *(i) C has no chord.*
- *(ii) C has odd length.*
- *(iii) For a node $v \notin V(C)$ define $A_1 := \{(v, w) \in A \mid w \in V(C)\}$ and $A_2 := \{(w, v) \in A \mid w \in V(C)\}$. Then, for every $v \notin V(C)$ one of the following conditions is true.*
  - *(a) $A_1 = \emptyset$ or $A_2 = \emptyset$.*
  - *(b) $|A_1| = |A_2| = 1$.*
  - *(c) $|A_1| + |A_2| = 3$ and $A(V(C) \cup \{v\})$ contains no dicycle of length three different from C.*
- *(iv) For any arc $a = (u, v) \in A$ with $u, v \notin V(C)$ the intersection of all dicycles in $A(V(C) \cup \{u, v\})$ is nonempty.*

Proof. Let $x(C) \leq |C| - 2$ be facet defining and let $\{v_1, \ldots, v_k\}$ be the set of nodes of C. We are going to construct an $|A| \times |A|$ block diagonal matrix

$$
\begin{pmatrix}
M_1 & 0 & 0 \\
* & M_2 & 0 \\
* & * & M_3
\end{pmatrix},
$$

where every entry is a submatrix with appropriate dimensions. M_2 itself is a block diagonal matrix corresponding to arcs incident with exactly one node of $V(C)$ and M_3 is an identity matrix corresponding to arcs not incident with a node of $V(C)$.

The k arc sets that are node induced by $V(C) \setminus \{v_i\}$, $i = 1, \ldots, k$, are acyclic since C has no chord. Restricted to C the corresponding incidence vectors give the $|C| \times |C|$ matrix

$$
M_1 = \begin{pmatrix} 0 & 0 & 1 & 1 & \ldots & 1 & 1 \\ 1 & 0 & 0 & 1 & \ldots & 1 & 1 \\ \vdots & \vdots & \vdots & \vdots & \ddots & \vdots & \vdots \\ 0 & 1 & 1 & 1 & \ldots & 1 & 0 \end{pmatrix} .
$$

M_1 is nonsingular because C is odd.

Now let $v \in V \setminus V(C)$, $A_1 = \emptyset$ (the case $A_2 = \emptyset$ is treated analogously) and $W = \{w \in V(C) \mid (w, v) \in A\}$. For every $w \in W$ the arc set induced by $(V(C) \setminus \{w\}) \cup \{v\}$ is acyclic. The corresponding incidence vectors contribute to M_2 the diagonal $|A_2| \times |A_2|$ block

$$
\begin{pmatrix} 0 & & 1 \\ & \ddots & \\ 1 & & 0 \end{pmatrix} .
$$

Let $A_1 = \{(v, w)\}$ and $A_2 = \{(u, v)\}$. The incidence vectors of $A((V(C) \setminus \{w\}) \cup \{v\})$ and $A((V \setminus \{u\}) \cup \{v\})$ contribute a 2×2 identity matrix to M_2. Next assume that $A_1 = \{(v, u)\}$ and $A_2 = \{(w, v), (z, v)\}$ (the case $A_1 = \{(v, u), (v, w)\}$ and $A_2 = \{(z, v)\}$ is treated in a similar way). W.l.o.g. let w be on the directed path from u to z (within C). Because of (iii)(c) there is a node x on this path between u and w. Therefore the subdigraphs induced by $A((V \setminus \{x\}) \cup \{v\})$, $A((V \setminus \{u\}) \cup \{v\})$, $A((V \setminus \{w\}) \cup \{v\})$, are acyclic and their incidence vectors contribute to M_2 the regular block

$$
\begin{pmatrix} 1 & 1 & 1 \\ 0 & 1 & 1 \\ 1 & 0 & 1 \end{pmatrix} .
$$

Finally, let $a = (u, v)$ be an arc such that $\{u, v\} \cap V(C) = \emptyset$. Due to condition (iv) there exists some node $w \in V(C)$ such that $A((V(C) \setminus \{w\}) \cup \{u, v\})$ is acyclic. The associated incidence vector contributes a diagonal 1 entry to M_3.

We will now show that conditions (i)-(iv) are also necessary. Suppose C has a chord. Together with a part of C this chord forms a dicycle K. If the incidence vector of a node induced acyclic arc set satisfies $x(C) = |C| - 2$ then it also satisfies $x(K) = |K| - 2$, but the converse is not true. The case $|C|$ is even will be discussed below.

Assume now that none of the conditions (a)-(c) holds in case (iii), i.e. $A_1, A_2 \neq \emptyset$, $|A_1| + |A_2| \geq 3$ and there is a (shortest) dicycle K of length three different from C in $A(V(C) \cup \{v\})$ or $|A_1| + |A_2| \geq 4$. Let us start with the first case, i.e. $|A_1| + |A_2| = 3$ and a dicycle K of length three different from C exists. Let e,

g denote the arcs of $K \cap (A_1 \cup A_2)$ and let f be the arc in $(A_1 \cup A_2) \setminus K$. Since there is no possibility to break K without deleting v or exactly one endnode of $\{e, g\}$ on C such that $x(C) = |C| - 2$ holds, this equality implies $x(C) + x_f - x_e - x_g = |C| - 2$. The second case $|A_1| + |A_2| = 4$ is similar. Again, let K denote the shortest dicycle different from C in $A(V(C) \cup \{v\})$. Now, there must exist at least two arcs f and h in $(A_1 \cup A_2) \setminus K$ yielding the implication $x(C) = |C| - 2 \Rightarrow x(C) + x_f - x_h = |C| - 2$. This completes the proof that at least one of (a)-(c) must hold.

Finally, assume for some arc $a = (u, v)$ with $\{u, v\} \not\subseteq V(C)$ the intersection of all dicycles in $A(V(C) \cup \{u, v\})$ is empty. To break these cycles, at least two nodes of $V(C) \cup \{u, v\}$ must be deleted. Therefore $x(C) = |C| - 2$ implies $x(C) + x_a = |C| - 2$ and $x(C) \leq |C| - 2$ is not facet defining. $\qquad \square$

Now consider an even dicycle $C = \{(v_1, v_2), (v_2, v_3), \ldots, (v_{2k-1}, v_{2k}), (v_{2k}, v_1)\}$ and let $C_o := \{(v_1, v_2), (v_3, v_4), (v_5, v_6), \ldots, (v_{2k-1}, v_{2k})\}$ and $C_e := C \setminus C_o$. Then every node induced acyclic subdigraph obviously satisfies $x(C_o) \leq |C_o| - 1$ as well as $x(C_e) \leq |C_e| - 1$. Adding these two inequalities implies $x(C) \leq |C| - 2$ proving that this inequality is not facet defining. Under slightly more complicated prerequisites these *even cycle inequalities* are facet defining. Complications arise from the fact that incidence vectors of both $V(C_o \setminus \{v_i\})$ and $V(C_o \setminus \{v_i, v_{i+1}\})$ satisfy $x(C_o) = |C_o| - 1$ for $i = 1, 3, 5, \ldots, 2k-1$. Therefore $x(C_o) \leq |C_o| - 1$ can also define a facet, if a chord is given. It is not clear, under which conditions a chord prevents $x(C_o) \leq |C_o| - 1$ to be facet defining. Furthermore, in condition (iii) of our last theorem 8 we have now five different subconditions. Even if $|A_1| + |A_2| \geq 5$ and $A_1, A_2 \neq \emptyset$ holds, $x(C_o) \leq |C_o| - 1$ can define a facet if some technical conditions are satisfied. Due to space reasons we give the result here only as far as it can be immediately adopted from the proof of theorem 8.

Theorem 9. *Let $C \subseteq A$ be an even dicycle of length $2k$. Then the even cycle inequality $x(C_o) \leq |C_o| - 1$ defines a facet of $P_{NAC}^A(D)$ if the following conditions are satisfied.*

(i) *C has no chord.*

(ii) *For a node $v \notin V(C)$ define $A_1 := \{(v, w) \in A \mid w \in V(C)\}$ and $A_2 := \{(w, v) \in A \mid w \in V(C)\}$. Then, for every $v \notin V(C)$ one of the following conditions is true.*

    (a) *$A_1 = \emptyset$ or $A_2 = \emptyset$.*

    (b) *$|A_1| = |A_2| = 1$.*

    (c) *$|A_1| + |A_2| = 3$ and $A(V(C) \cup \{v\})$ contains no dicycle of length three different from C.*

(iii) *For an arc $a = (u, v) \in A$ with $u, v \notin V(C)$ there must exist an arc $b \in C_o$ such that $A((V(C) \setminus V(b)) \cup \{u, v\})$ is acyclic.*

The study of even cycle inequalities exhibits some possibilities for generalization. We only indicate possible generalizations of even cycle inequalities obtained by combining dicycles or admitting external and consecutive support arcs.

Dicycles can be combined to a closed walk and external arcs can be added. Furthermore, along a dicycle the support arcs of the inequality do not have to alternate with non-support arcs. As an example, consider Figure 1, composed of two even dicycles, one *external support arc* a_{15}, and *consecutive support arcs* a_2 and a_3. Note that $a_7 = a_{13} = (v_7, v_8)$.

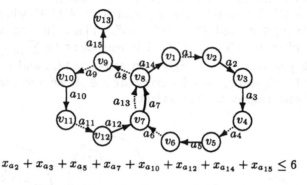

$$x_{a_2} + x_{a_3} + x_{a_5} + x_{a_7} + x_{a_{10}} + x_{a_{12}} + x_{a_{14}} + x_{a_{15}} \leq 6$$

Fig. 1. Generalized even cycle inequality

There are many examples where the corresponding *generalized even cycle inequalities* define facets. But it is an open question under which general conditions they define facets of $P_{NAC}^A(D)$ for a given digraph $D = (V, A)$.

6 Separation

We will now discuss how the valid inequalities derived above can be used in a branch–and–cut framework. For an introduction into branch–and–cut algorithms see e.g. [6]. Throughout this section, let $\bar{x}$ always denote the optimal solution of the current LP relaxation.

6.1 Node Inducing Inequalities

Node inducing inequalities can be separated in linear time. For every node $v \in V$ we compute $d_v = \max\{\bar{x}_a \mid a \text{ incident with } v\}$ and set $a_v = a$ where a is incident with v and $\bar{x}_a = d_v$. If there exists an arc $a = (u, v)$ such that $d_u - \bar{x}_a + d_v > 1$ then the node inducing inequality $x_{a_u} - x_a + x_{a_v} \leq 1$ is violated.

6.2 Odd Cycle Inequalities

The separation of odd cycle inequalities uses shortest paths techniques as described by in [3]. Because $x(C) \leq |C| - 2$ is equivalent to $\sum_{a \in C}(1 - x_a) \geq 2$ separation can be performed by computing shortest odd dicycles with respect to arc weights $1 - \bar{x}_a$. If a dicycle of length < 2 is found, we have a violated dicycle inequality. In our implementation we do not pay attention whether a dicycle C of a violated inequality $x(C) \leq |C| - 2$ is odd or even, because violated inequalities for even dicycles C can also be added. Obviously, this is an exact polynomial separation for odd cycle inequalities.

6.3 Even Cycle Inequalities

Separation of even cycle inequalities has to take into account that we have two kinds of arcs that must appear in a specific order.

We construct the digraph $\overline{D} := (\overline{V}, \overline{A})$ by setting $\overline{V} := V' \cup V''$ where $V' := \{v' \mid v \in V\}$, $V'' := \{v'' \mid v \in V\}$, and $\overline{A} := A' \cup A''$ where $A' := \{a' = (u', v'') \mid a = (u, v) \in A\}$ and $A'' := \{a'' = (u'', v') \mid a = (u, v) \in A\}$.

Like for odd cycles $x(C_o) \leq |C_o| - 1$ is equivalent to $\sum_{a \in C_o} (1 - x_a) \geq 1$ for an even dicycle $C = \{a_1, \ldots, a_{2k}\}$ with associated C_o and C_e. Defining arc weights $c_{a'} = 1 - \overline{x}_a$ and $c_{a''} = 0$ for every arc $a \in A$, we now compute shortest dicycles in $\overline{D}$ with respect to these weights.

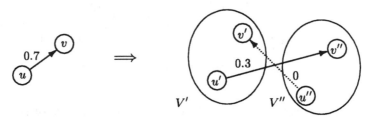

Fig. 2. Transformation to separate even cycle inequalities

An even dicycle $C = \{a_1, \ldots, a_{2k}\}$ in D corresponds to the even dicycles $\{a_1', a_2'', \ldots, a_{2k-1}', a_{2k}''\}$ and $\{a_1'', a_2', \ldots, a_{2k-1}'', a_{2k}'\}$ in $\overline{D}$. If $x(C_o) \leq |C_o| - 1$ is violated, the corresponding dicycle in $\overline{D}$ has length smaller than 1 with respect to the weights defined above.

On the other hand, a dicycle $\overline{C} = \{a_1', a_2'', \ldots, a_{2k-1}', a_{2k}''\}$ in $\overline{D}$ (note that all dicycles in $\overline{D}$ are even because they alternate between V' and V'') corresponds to a closed walk $P = \{a_1, \ldots, a_{2k}\}$ in D. If $\overline{C}$ has length smaller than 1, then we conclude from $1 > \sum_{i=1}^{k} c_{a_{2i-1}''} = \sum_{i=1}^{k} (1 - \overline{x}_{a_{2i-1}})$ that $\sum_{i=1}^{k} x_{a_{2i-1}} \leq k - 1$ is violated by $\overline{x}$. If for an incidence vector x of a subdigraph $\sum_{i=1}^{k} x_{a_{2i-1}} > k - 1$ holds, then all arcs $\{a_1, a_3, \ldots, a_{2k-3}, a_{2k-1}\}$ are present in this subdigraph. If it is node induced, then all arcs $\{a_2, a_4, \ldots, a_{2k-2}, a_{2k}\}$ must be present as well. Since $\{a_1, a_2, \ldots, a_{2k-1}, a_{2k}\}$ is a closed walk, this subdigraph cannot be acyclic. Hence, we find in polynomial time a violated inequality, if an even cycle inequality is violated.

7 Primal Heuristic

For computing feasible solutions in the branch–and–cut algorithm we implemented an LP based primal heuristic. First, nodes are sorted with respect to decreasing average LP values of their incident arcs with positive values. Then we consider the nodes in this sorted sequence.

Starting with the empty set W, a node is added to W with a probability that depends on the average LP values of its incident arcs, provided that $A(W)$ is still acyclic. This reflects the expectation that nodes having incident arcs with high average LP values are likely to be contained in a node induced acyclic subdigraph with high objective function value which is the sum of given arc weights in our case. Having determined this initial W, we use several local exchange heuristics to improve it.

A survey of randomization in combinatorial optimization can be found in [10].

8 Computational Results

We implemented a branch–and–cut algorithm for solving node induced acyclic subdigraph problems with arc variables. The implementation is based on the general framework $ABACUS$ ([16]) and uses CPLEX 2.2 ([1]) as LP solver.

The implementation has three separation procedures, namely for separating node inducing inequalities (*sep1*) and for separating odd and even cycle inequalities (*sep2*, *sep3* resp.), as described in section 6. The primal heuristic is performed after every 15th LP solution.

Computational experiments were performed on unweighted problems (i.e., node induced acyclic subdigraphs with maximum number of arcs have to be determined) in random graphs. For $n = 25, 30, 35$ digraphs with densities 10%, 20% and 30% were generated giving 9 problem instances. The generation of anti-parallel arcs was admitted.

For the computational results presented here, we have used the default settings of $ABACUS$. Our first interest was to get some insight into the difficulties of feedback problems. So far we used no cut generation strategy: All inequalities found by the separation procedures were added to the linear program. Anti-parallel arcs were fixed to zero. Tables 1 and 2 display the respective computational statistics. For table 1 separators *sep1*, *sep2*, and *sep3* were always called, results in table 2 were obtained by invoking *sep3* only if *sep1* and *sep2* failed. The enumeration of the branch–and–cut tree was done by depth–first search.

Columns of the table give problem name (where r_nn_dd is a problem on nn nodes with density dd%), number of arcs, total number of cuts found by *sep1*, total number of cuts found by *sep2*, total number of cuts found by *sep3*, maximum level reached in the branch-and-cut tree, number of nodes of the branch-and-cut tree, lower and upper bound at the root node, value of optimum solution, and CPU time on a SUN SPARCstation 10/20 (in h:mm:ss).

The tables show that the problems are fairly difficult for dense graphs. Only density 10% could be solved at or near the root node. Furthermore, the gap between LP bound at the root node and optimum solution value is large for larger and denser graphs, showing a need for further separation procedures. The primal heuristic often finds feasible solutions of good quality. For example, the optimal solution of r_25_20 was found after 15 seconds. Therefore we decided to use depth first search for these experiments.

Even cycle inequalities are not contained in the trivial integer programming formulation. Hence we were interested to see their impact on the branch–and–

Table 1. Results with *sep1-sep3*

prob-lem	arcs	cuts of sep1	cuts of sep2	cuts of sep3	level	nodes	gap at root		opti-mum	time
r_25_10	57	23	14	19	1	1	41	41.2	41	0:00:01
r_30_10	84	233	27	64	1	1	54	54.0	54	0:00:02
r_35_10	111	2684	263	524	6	19	57	61.0	59	0:00:24
r_25_20	127	26548	2035	4468	37	121	26	46.5	32	0:02:20
r_30_20	154	34420	3064	5771	25	93	51	62.6	51	0:03:41
r_35_20	246	345867	27056	53909	94	621	47	87.5	50	1:03:10
r_25_30	172	40947	2541	5587	50	119	32	52.0	32	0:02:58
r_30_30	238	149224	12393	24145	79	269	39	73.5	42	0:15:11
r_35_30	356	662027	49880	87940	156	873	34	102.5	43	1:52:34

Table 2. Results with *sep3* only when *sep1* and *sep2* failed

prob-lem	arcs	cuts of sep1	cuts of sep2	cuts of sep3	level	nodes	gap at root		opti-mum	time
r_25_10	57	34	21	0	1	1	41	41.5	41	0:00:00
r_30_10	84	336	39	0	1	1	54	54.0	54	0:00:01
r_35_10	111	2530	239	30	2	3	59	61.0	59	0:00:10
r_25_20	127	32517	2367	61	35	131	23	46.5	32	0:01:50
r_30_20	154	49614	4383	161	22	127	48	62.6	51	0:03:25
r_35_20	246	408274	32201	657	94	645	46	87.5	50	1:11:19
r_25_30	172	44103	2619	6	50	119	32	52.0	32	0:02:06
r_30_30	238	166195	13687	71	79	261	37	73.5	42	0:10:53
r_35_30	356	743324	55804	364	157	875	35	102.5	43	1:19:52

cut tree. Tables 1 and 2 show that using *sep3* only if *sep1* and *sep2* did not find a violated inequality is in most cases slightly better.

There is much room for improvement of these results. In the future we will investigate in particular the development of additional cut generation routines and appropriate cut generation strategies and switch from depth–first search to other enumeration strategies. Furthermore, the currently used branching on arcs does not seem to be a good choice, since the branching rule should take into account that we are looking for node induced subdigraphs.

References

1. CPLEX. Using the Cplex callable library and Cplex mixed integer library (1994). Cplex Inc.

2. Grötschel, M., Jünger, M., Reinelt, G.: On the acyclic subgraph polytope. Mathematical Programming, **33** (1985) 28–42
3. Grötschel, M., Pulleyblank, W.: Weakly bipartite graphs and the max–cut problem. Operations Research Letters, **1** (1981) 23–27
4. Hackbusch, W.: On the feedback vertex set problem for a planar graph. Technical report (1994). Institut für Informatik und Praktische Mathematik, Christian–Albrechts–Universität zu Kiel, D-24098 Kiel, Germany
5. Jünger, M.: Polyhedral Combinatorics and the acyclic subdigraph problem., Research and exposition in mathematics **7** (1985). Heldermann Verlag, Berlin
6. Jünger, M., Reinelt, G., Thienel, S.: Practical problem solving with cutting plane algorithms in combinatorial optimization. In Cook, W., Lovász, L., Seymour, P., editors, Combinatorial Optimization, DIMACS series in Discrete Mathematics and Theoretical Computer Science **20** (1995) 111–152
7. Karp, R.: Reducibility among combinatorial problems. In Miller, R., Thatcher, J., editors, Complexity of Computer Computations, (1971) 85–103. Plenum Press, New York
8. Kevorkian, A.: General topological results on the construction of a minimum essential set of a directed graph. IEEE Transactions on Circuits and Systems, **CAS-27(4)** (1980) 293–304
9. Kunzmann, A. Wunderlich, H.: An analytical approach to the partial scan problem. Journal of electronic testing: Theory and Applications, **1** (1990) 163–174
10. Lovàsz,L. Randomized Algorithms in Combinatorial Optimization. In Cook, W., Lovász, L., Seymour, P., editors, Combinatorial Optimization, DIMACS series in Discrete Mathematics and Theoretical Computer Science **20** (1995) 153–179
11. Monien, B. Schulz, R.: Four approximation algorithms for the feedback vertex set problem. In Graphtheoretic concepts in computer science, Proc. 7th Conf. Linz/Austria (1985) 315–326
12. Rosen, B.: Robust linear algorithms for cutsets. Journal of algorithms, **3** (1982) 205–217
13. Smith, W., Walford, R.: The identification of a minimal feedback vertex set of a directed graph. IEEE Transactions on circuits and systems, **CAS-22(1)** (1975) 9–15
14. Speckenmeyer, E.: Untersuchungen zum Feedback Vertex Set Problem in ungerichteten Graphen. Theoretische Informatik Bericht **16** (1983). Universität Paderborn
15. Stamm, H.: On feedback problems in planar digraphs. Graph-theoretic concepts in computer science, Proc. Int. Workshop, Berlin/Germany 1990, Lect. Notes Comput. Sci. **484** (1992) 79–89. Springer Verlag
16. Thienel, S.: ABACUS - A Branch And CUt System. PhD thesis (1995). Angewandte Mathematik und Informatik, Universität zu Köln

Separating over Classes of TSP Inequalities Defined by 0 Node-Lifting in Polynomial Time

Robert Carr

Dept. of Computer Science, University of Ottawa, Ottawa ON K1N-6N5, Canada

Abstract. Many important cutting planes have been discovered for the traveling salesman problem. Until recently (see [5] and [2]), little was known in the way of exact algorithms for separating these inequalities in polynomial time in the size of the fractional point x^* which is being separated. Any facet-defining inequality can be neatly categorized by the simple inequality which it is a 0 node-lifting of. Given a class of inequalities consisting of simple inequalities occuring on a fixed sized graph together with all their 0 node-liftings, an algorithm is presented here that separates over this class of inequalities in polynomial time.

This algorithm uses a relaxation of the TSP which we will call cycle-shrink. Cycle-shrink is a compact description of the subtour polytope, and has some other theoretically interesting properties.

1 Introduction

Given a complete graph $G = (V, E)$ with costs on the edges, the traveling salesman problem (TSP) consists of finding a minimum cost hamilton cycle in G. In an attempt to solve a given instance of the TSP, one may solve a linear programming relaxation of the TSP for this instance. The set of variables used in such a relaxation typically consists of a variable x_e for each edge $e \in E$, just like in the usual integer programming formulation. If this relaxation yields an integral solution, then this solution is the incidence vector of a minimum cost hamilton cycle for this instance (assuming there are no subcycles). In trying to acheive such an integral optimal solution for this relaxation, one uses some of the valid inequalities of the TSP as constraints of the relaxation. Usually, this attempt will fail at first, yielding a fractional optimal solution x^* for this relaxation.

However, if one can find a valid inequality of the TSP that cuts off x^*, then one could add this inequality as a constraint in this linear programming relaxation, and resolve this relaxation. The task of finding such an inequality efficiently goes by the name of the *separation problem*. This inequality which can now be added to the linear programming relaxation is called a *cutting plane*.

In the next section, we introduce an interesting relaxation of the TSP which we call cycle-shrink. In section 3, we define classes of TSP inequalities by 0 node-lifting. Finally, in section 4, we use cycle-shrink as a tool to help us separate over these classes of inequalities. In fact, separating over slightly more generally defined classes of inequalities can be achieved.

2 The Cycle-Shrink Relaxation

Cycle-shrink is a polynomial sized linear programming relaxation developed by the author that implies the validity of all the subtour elimination constraints. There are several other such relaxations that are already known that also imply the validity of these constraints, [8]. But the importance of cycle-shrink lies in the way it will be used in this paper to separate naturally defined classes of TSP inequalities.

To formulate cycle-shrink, additional variables are used as follows. First, rename the edge variables in the usual formulation of the TSP as x_e^0 for each $e \in E$. Arbitrarily label the vertices in V with the integers from 1 through n. Any such labeling imposes a total ordering on the vertices, which is used in the cycle-shrink formulation. Construct the family of graphs $G_i = (V_i, E_i)$ for $i \in V$ such that G_i is the subgraph of the complete graph G which is induced by the set V_i of all those vertices whose labels are greater than i. Then on each graph G_i, create additional variables x_e^i for each $e \in E_i$.

Consider an incidence vector x^0 of a Hamilton cycle $H^0(x^0)$ in G. The values that we want the additional variables of cycle-shrink to have, given the values of x^0, can be determined by considering the following family of Hamilton cycles. Let $H^1(x^0)$ be the Hamilton cycle on G_1 formed by removing vertex 1 from $H^0(x^0)$ and linking the neighbors of 1 in $H^0(x^0)$ with an edge. Let $H^i(x^0)$ be the Hamilton cycle on G_i formed by removing vertex i from $H^{i-1}(x^0)$ and linking the neighbors of i in $H^{i-1}(x^0)$ with an edge.

Then the values that we want the additional variables of cycle-shrink to have in order to represent the Hamilton cycle H^0 are as follows.

$(x_e^0 \,|\, e \in E)$ is the incidence vector of H^0.
$(x_e^i \,|\, e \in E_i)$ is the incidence vector of H^i for all $i \in \{1, \ldots, n-3\}$.

A complete feasible solution x for cycle-shrink can thus be represented by

$$x := (x^0, x^1, \ldots, x^{n-3}) \tag{1}$$

where for each $i \in \{0, 1, \ldots, n-3\}$, x^i is a vector having a component for each edge in G_i.

To this end, we define cycle-shrink to be the following linear program.

$$\text{minimize } \sum_{e \in E} c_e x_e^0$$
$$\text{subject to}$$
$$x_e^0 \geq 0 \quad \forall e \in E.$$

$$\sum_{e \in \delta(\{j\}) \cap E_i} x_e^i = 2 \quad \forall i \in \{0, \ldots, n-3\} \text{ and}$$
$$\forall j \in V_i.$$
$$x_e^i \geq x_e^{i-1} \quad \forall i \in \{1, \ldots, n-3\} \text{ and}$$
$$\forall e \in E_i.$$

Let any Hamilton cycle H^0 of G be given. Then the variable values which represent H^0 as described above are feasible for the linear program cycle-shrink since the incidence vectors of H^i for $i = 0, \ldots, n-3$ satisfy all the constraints. Hence, cycle-shrink is a valid relaxation of the TSP. Call the part of a feasible cycle-shrink solution given by x^k as the k-th *level* of this feasible solution.

2.1 Theorems About Cycle-Shrink

We now have the following theorem.

Theorem 1. *If x is feasible for cycle-shrink, then $x^i(\delta(S) \cap E_i) \leq x^{i-1}(\delta(S) \cap E_{i-1})$ for all $S \subset V$ and all $i \in \{1, \ldots, n-3\}$*

Proof: Let x be feasible for cycle-shrink. Let $S \subset V$ and $i \in \{1, \ldots, n-3\}$ be given. Define $T := V \backslash S$. Assume that $S \cap V_i, T \cap V_i \neq \emptyset$ since otherwise the theorem follows trivially.

Assume that $i \in T$. Then by the last set of constraints for cycle-shrink, it follows that

$$x^i(E(S) \cap E_i) = x^i(E(S) \cap E_{i-1}) \geq x^{i-1}(E(S) \cap E_{i-1}) \tag{2}$$

But the degree constraints (the third set of constraints) ensure that

$$2x^{i-1}(E(S) \cap E_{i-1}) + x^{i-1}(\delta(S) \cap E_{i-1}) = 2|S \cap V_{i-1}| \tag{3}$$

These constraints also insure that

$$2x^i(E(S) \cap E_i) + x^i(\delta(S) \cap E_i) = 2|S \cap V_i| \tag{4}$$

But $|S \cap V_i| = |S \cap V_{i-1}|$, so we can combine equations (3) and (4) to obtain

$$2x^{i-1}(E(S) \cap E_{i-1}) + x^{i-1}(\delta(S) \cap E_{i-1}) = 2x^i(E(S) \cap E_i) + x^i(\delta(S) \cap E_i) \tag{5}$$

But equation (5) implies that the inequality (2) is equivalent to

$$x^i(\delta(S) \cap E_i) \leq x^{i-1}(\delta(S) \cap E_{i-1}) \tag{6}$$

Now assume $i \in S$. Then, the above argument shows that

$$x^i(\delta(T) \cap E_i) \leq x^{i-1}(\delta(T) \cap E_{i-1}) \tag{7}$$

But since $\delta(S) = \delta(T)$, the inequalities (6) and (7) are equivalent. Hence, the theorem follows. □

This theorem roughly states that capacities of cuts decrease as one goes down the levels of cycle-shrink. We now have:

Corollary 2. *If x is feasible for cycle-shrink, then x^0 satisfies all the subtour elimination constraints.*

Proof: Let x be feasible for cycle-shrink. Suppose there is a cut whose shores are S and T such that $x^0(\delta(S)) < 2$. Pick the smallest i such that either $|S \cap V_i| = 1$ or $|T \cap V_i| = 1$. Without loss of generality, assume $|S \cap V_i| = 1$. Then, by Theorem 1, $x^i(\delta(S) \cap E_i) < 2$. But, by the degree constraints, $x^i(\delta(S) \cap E_i) = 2$, which is a contradiction. □

2.2 The Equivalence of the Cycle-Shrink and Subtour Polytopes

The *subtour polytope* is defined to be the set of feasible solutions to the fractional relaxation of the usual integer programming formulation for the traveling salesman problem. We can prove the following:

Theorem 3. *The projection of the cycle-shrink polytope onto the space of original variables is exactly the subtour polytope.*

The proof of this relies on Lovasz's edge splitting theorem, [6].

2.3 Discussion of Yannakakis' Theorem

The cycle-shrink linear program is a compact description of the subtour polytope which violates an important symmetry condition that is defined by Yannakakis, [8].

The results Yannakakis obtained were that neither the matching polytope nor the TSP polytope could be expressed by a polynomial size symmetric linear program, [8].

The violation of this symmetry by cycle-shrink is interesting because it leaves open the possibility that there may be a similar formulation of the matching polytope which could be polynomial in size, but doesn't contradict Yannakakis' theorem because of its asymmetry.

3 A Natural Partitioning of TSP Facets

Separating exactly over the class of all traveling salesman inequalities is known to be NP-hard. Therefore, in order for exact separation to be a useful procedure, one probably needs to find important subclasses of inequalities for which there is a polynomial time exact separation algorithm.

In this section, each facet of the TSP will be put into a class for which, as will be shown later, exact separation procedures are possible.

3.1 The Graphical Traveling Salesman Problem

The graphical traveling salesman problem (GTSP) is to find the cheapest *tour* on a graph of n vertices, where a tour is a multiset of edges on which our salesman can leave the home vertex, visit every other vertex at least once, and go back to the home vertex. Note that our salesman is allowed to traverse an edge more than once in this tour.

3.2 Tight Triangular Form

Any TSP inequality can be put into a unique form, up to a constant multiple, known as *tight triangular form*, [3]. For the tight triangular form, abbreviated TT form, an inequality states that a left hand side is greater than or equal to a right hand side, unlike a closed form of the inequality.

An inequality $fx \geq f_0$ is in tight triangular form if and only if:

i) $f_{uv} \leq f_{uw} + f_{vw}$

ii) For all $w \in V$, there exists $u, v \in V$ such that $f_{uv} = f_{uw} + f_{vw}$.

The theoretical significance of an inequality being in TT form is that if $hx \geq h_0$ is facet-defining for the TSP, then its tight triangular form $fx \geq f_0$ is facet-defining for the GTSP, [7].

3.3 Zero Node-Lifting

An important operation for generating new inequalities from old ones in the TSP is the operation of zero node-lifting, [7]. The operation of zero node-lifting is as follows. Let $hx \geq h_0$ be a facet-defining TSP inequality in TT form on the complete graph $K_n = (V_{(n)}, E_{(n)})$ of n vertices. Add k more vertices to $V = V_{(n)}$, obtaining the vertex set $V_{(n+k)}$. We zero node-lift node u to obtain the inequality $h^* x^* \geq h_0$, where:

i) $h_e^* = h_e$ for all $e \in E_{(n)}$

ii) $h_{ij}^* = h_{uj} \quad \forall i \in V_{(n+k)} \setminus V_{(n)} \quad \forall j \in V_{(n)}$

iii) $h_{ij}^* = 0 \quad \forall i, j \in V_{(n+k)} \setminus V_{(n)}$

We call the vertices in $V_{(n+k)} \setminus V_{(n)}$ *copies* of u. An inequality which has no 0 coefficients when it is expressed in TT form is called a *simple* inequality. It is known that every facet-defining inequality is derivable from a simple inequality through zero node-lifting, [7]. Moreover, for all TSP inequalities known to date, it has been observed by Naddef and Rinaldi, [4] that:

Fact 4 *Zero node-lifted inequalities of any known facet-defining TSP inequality inherit the property of being facet-defining.*

It is these properties of zero node-lifting which make it attractive for partitioning TSP facets into inequality classes.

3.4 Defining Classes of Inequalities

We first partition the subcollection of simple inequalities into classes of inequalities in the most obvious way. Call two simple inequalities *isomorphic* if a bijective mapping of vertices transforms one of the inequalities into the other one. That is, if $\alpha x \geq \alpha_0$ is defined on a complete underlying graph $H = (W, E)$ and $\alpha' x \geq \alpha_0$ is defined on a complete underlying graph $H' = (W', E')$, then these two inequalities are isomorphic if there exists a bijective mapping $\pi : W' \to W$ such that

$$\alpha_{\pi(i)\pi(j)} = \alpha'_{ij} \qquad \text{for all } (i,j) \in E' \tag{8}$$

and hence

$$\sum_{(i,j) \in E'} \alpha_{\pi(i)\pi(j)} x_{\pi(i)\pi(j)} = \sum_{(i,j) \in E'} \alpha'_{ij} x_{ij}^{prime} \tag{9}$$

This partitions the simple inequalities into isomorphism equivalence classes.

For example, suppose one has the following simple comb inequality on an underlying graph $H = (W, E)$, where $W := \{1, 2, 3, 4, 5, 6\}$:

$$x(\delta\{1, 2, 3\}) + x(\delta\{1, 4\}) + x(\delta\{2, 5\} + x(\delta\{3, 6\}) \geq 10 \tag{10}$$

Suppose one also has the simple comb inequality on an underlying graph $H' = (W', E')$, where $W' := \{1, 3, 5, 7, 9, 11\}$:

$$x(\delta\{1, 3, 7\}) + x(\delta\{1, 5\}) + x(\delta\{3, 9\} + x(\delta\{7, 11\}) \geq 10 \tag{11}$$

The mapping $\pi : W' \to W$ defined by

$$\pi(1) = 1 \ \pi(3) = 2 \ \pi(7) = 3$$
$$\pi(5) = 4 \ \pi(9) = 5 \ \pi(11) = 6$$

shows that these inequalities are isomorphic. That is, replacing 3 in (11) with 2, and 7 in (11) with 3, etc., makes inequality (11) identical to inequality (10).

Now, let $fx \geq f_0$ be a facet-defining TT inequality. As stated before, it is then derivable from some simple inequality $hx \geq f_0$ through zero node-lifting.

Call the equivalence class of simple inequalities to which $hx \geq f_0$ belongs S. We then extend the inequalities in the above partition to include all the zero node-liftings of $hx \geq f_0$ and its isomorphisms so that $fx \geq f_0$ is put into the equivalence class S as well.

For example, consider the comb inequality:

$$x(\delta\{1, 1', 2, 2', 2'', 3\}) + x(\delta\{1, 1', 4\}) + x(\delta\{2, 2', 2'', 5\}) + x(\delta\{3, 6\}) \geq 10 \tag{12}$$

Here, $1'$ is a zero node-lifting of node 1, and $2'$ and $2''$ are zero node-liftings of node 2. Hence, inequality (12) belongs in the same equivalence class of inequalities as inequality (10).

4 Separating Inequalities with Cycle-Shrink

Let $G = (V, E)$ be a complete graph on n vertices. Let a fractional point x^* for the TSP on the underlying graph G be given. Let a class S of valid TSP inequalities be given. The solution of the separation problem over this class S consists in finding an inequality $ax \geq b$ in S which is violated by x^*. We will show in this section that we can solve the separation problem over numerous classes S of inequalities. In order to describe those classes S of inequalities which we can separate over, we need to discuss simple inequalities again.

Consider a simple inequality $\hat{h}x \geq b$ defined on the underlying graph G. Form the complete weighted graph $G_{\hat{h}} = (V, E, \hat{h})$ by giving each edge $e \in E$ the weight $\hat{h}_e$. Call $G_{\hat{h}}$ the *inequality graph* of $\hat{h}x \geq b$. Denote the weighted subgraph of $G_{\hat{h}}$ which is induced by a subset $U \subset V$ of vertices by $G_{\hat{h}}[U] = (U, E(U), \hat{h}|_{E(U)})$, where $\hat{h}|_{E(U)}$ is defined to be $(\hat{h}_e \mid e \in E(U))$.

Let $v \in V$, and let $U := V \setminus \{v\}$. We call a node $v \in V$ an *optional node* if $\hat{h}|_{E(U)}x \geq b$ is a valid TSP inequality. Conversely, if for some set $U \subset V$, both

$\hat{h}|_{E(U)}x \geq b$ and $\hat{h}x \geq b$ are both valid supporting TSP inequalities, then we say that $\hat{h}x \geq b$ is an *extension* of $\hat{h}|_{E(U)}x \geq b$.

Define $hx \geq b$ to be a *basic inequality* if $hx \geq b$ is a simple inequality which has no optional nodes. We have the following elementary proposition.

Proposition 5. *Every simple inequality is an extension of a basic inequality.*

In Figure 1, we give the handle and tooth picture of a basic three tooth comb inequality. The inequality is

$$x(\delta\{1,2,3\}) + x(\delta\{1,4\}) + x(\delta\{2,5\}) + x(\delta\{3,6\}) \geq 10 \tag{13}$$

In Figure 2, we show a picture of a three tooth comb inequality which is an extension of the inequality pictured in Figure 1 because of the hollow nodes in Figure 2. This inequality is

$$x(\delta\{1,2,3,7\}) + x(\delta\{1,4\}) + x(\delta\{2,5\}) + x(\delta\{3,6\}) \geq 10 \tag{14}$$

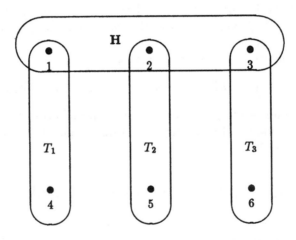

Fig. 1. Picture of a basic three-tooth comb

For each basic inequality, we will define a class S of inequalities which we can separate over with the help of cycle-shrink. Let $\overline{\mu}x \geq \mu_0$ be a basic inequality defined on the graph $G[B]$, where $B \subset V$. Denote by S'_{μ} the set of all zero node-liftings of $\overline{\mu}x \geq \mu_0$ that use the vertices in V. The inequality $\overline{\mu}x \geq \mu_0$ could be the comb inequality defined on the vertices of $B = \{1,3,5,7,9,11\} \subset V = \{1,\ldots,100\}$ shown here:

$$x(\delta\{1,3,5\}) + x(\delta\{1,7\}) + x(\delta\{3,9\}) + x(\delta\{5,11\}) \geq 10 \tag{15}$$

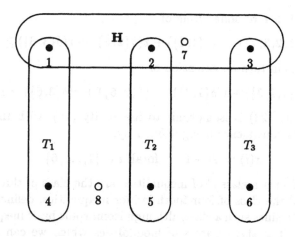

Fig. 2. Picture of a three-tooth comb with optional vertices

An example element of $S'_{\underline{\mu}}$ is:

$$x(\delta\{1,10,3,30,31,5\}) + x(\delta\{1,10,7,70\}) + x(\delta\{3,30,31,9\}) + x(\delta\{5,11\}) \geq 10 \tag{16}$$

For each extension $\hat{\mu}x \geq \mu_0$ of $\overline{\mu}x \geq \mu_0$ that uses the vertices in V, define $S'_{\hat{\mu}}$ to be the set of all zero node-liftings of $\hat{\mu}x \geq \mu_0$ that use the vertices in V. For notational convenience, let $(\hat{\mu}, \mu_0)$ represent the inequality $\hat{\mu}x \geq \mu_0$. Denote by F the set of all simple inequalities $(\hat{\mu}, \mu_0)$ which are extensions of $\overline{\mu}x \geq \mu_0$. We define $S_{\overline{\mu}}$ by

$$S_{\overline{\mu}} = S'_{\underline{\mu}} \cup (\cup_{(\hat{\mu},\mu_0)\in F} S'_{\hat{\mu}}) \tag{17}$$

An example of an inequality in $S_{\overline{\mu}}$ is:

$$x(\delta\{1,10,3,30,5,2,20\}) + x(\delta\{1,10,7,70\}) + x(\delta\{3,30,9\}) + x(\delta\{5,11\}) \geq 10 \tag{18}$$

This inequality is a zero node-lifting of an inequality $\hat{\mu}x \geq \mu_0$ defined as

$$x(\delta\{1,3,5,2\}) + x(\delta\{1,7\}) + x(\delta\{3,9\}) + x(\delta\{5,11\}) \geq 10 \tag{19}$$

But inequality (19) is an extension of inequality (15) because of the presence of node 2.

Denote by F' the set of all basic inequalities $(\mu^{\#}, \mu_0)$ which are defined on a subset of the vertices of V, and which are isomorphic to $\overline{\mu}x \geq \mu_0$. Finally, we define S by:

$$S = \cup_{(\mu^{\#},\mu_0)\in F'} S_{\mu^{\#}} \tag{20}$$

This class S is said to be *generated* by the inequality $\overline{\mu}x \geq \mu_0$. An example of an inequality in S when $\overline{\mu}x \geq \mu_0$ is given by (15) is

$$x(\delta\{1,10,11,2,3,7\}) + x(\delta\{1,10,11,4\}) + x(\delta\{2,5\}) + x(\delta\{3,6\}) \geq 10 \tag{21}$$

This inequality is a zero node-lifting of

$$x(\delta\{1,2,3,7\}) + x(\delta\{1,4\}) + x(\delta\{2,5\}) + x(\delta\{3,6\}) \geq 10 \qquad (22)$$

Inequality (22) is in turn an extension of

$$x(\delta\{1,2,3\}) + x(\delta\{1,4\}) + x(\delta\{2,5\}) + x(\delta\{3,6\}) \geq 10 \qquad (23)$$

Finally, inequality (23) is isomorphic to inequality (15), with the mapping π which shows this equivalence being defined by:

$$\pi(i) := 2i - 1 \qquad \text{for all } i \in \{1, \ldots, 6\}$$

Examples of such a class S of inequalities are the class of three-tooth comb inequalities, and the class of four-tooth ladder inequalities, defined in [1]. Each basic inequality defines such a class, although isomorphic basic inequalities define the same class. It is these classes of inequalities which we can separate over by utilizing the cycle-shrink LP relaxation. The tremendous generality of these classes is revealed by the following observations. As stated earlier, every facet-defining inequality is a zero node-lifting of a simple inequality, [7]. Furthermore, every simple inequality is an extension of some basic inequality. Therefore, every facet-defining inequality belongs to one of these classes. And, as we will see, each one of these classes can individually be separated over in a polynomial amount of time. The key features that each such class S of inequalities have are that the coefficients of the inequalities in S satisfy the triangle inequality, and the right hand side is the same for all of the inequalities in S.

4.1 Main Separation Result

Let a fractional point x^* for the TSP be given. Let S be a class of inequalities generated by a given basic inequality (such as a three-tooth comb or a four-tooth ladder inequality). If there is an inequality $ax \geq \mu_0$ in S which is violated by x^*, we will find this out through the following procedure. Pick a basic inequality $\overline{\mu}x \geq \mu_0$ in S, defined on the graph $G[B]$, where $B \subset V$. Define $k := |B|$. Consider the subproblem of separating over the class $S_{\overline{\mu}}$ of those inequalities which are zero node-liftings of extensions of the inequality $\overline{\mu}x \geq \mu_0$.

First, consider any inequality $\mu x \geq \mu_0$ in $S_{\overline{\mu}}$. Suppose its inequality graph is $H_\mu = (W, E, \mu)$. Consider the vertex induced subgraph $H[W \setminus U]$, where $U \subset W$ and $U \cap B = \emptyset$. Then $H[W \setminus U]$ is an inequality graph of a valid inequality $\mu'x \geq \mu'_0$ in the class $S_{\overline{\mu}}$ and on the underlying graph having vertex set $W \setminus U$.

Now, pick $u \in W \setminus B$ and set $U := \{u\}$. Consider removing the edges (u, v) incident to u from the weighted support graph of the fractional solution x^*. But also add some weight λ_{vw} to each edge (v, w) in the remaining graph so that the degree at each vertex in the weighted support graph remains the same, namely 2. Note that Lovasz's Theorem described in section 2.2 easily shows that one can always do this, but the values of λ_{vw} are not necessarily uniquely determined. Call this new solution x^{u*}. Call this operation a *cycle-shrink removal of node* u. We wish to show that the difference $\mu x^* - \mu' x^{u*}$ must be non-negative given any cycle-shrink removal of node u.

Theorem 6. *If a cycle-shrink removal of a node u is performed, then $\mu x^* - \mu' x^{u*} \geq 0$.*

Proof: Define $\lambda_{vw} := x_{vw}^{u*} - x_{vw}^*$ for all edges (v, w) in the graph without vertex u. We then have the identity:

$$\sum_{w \in W \setminus \{u,v\}} \lambda_{vw} = x_{uv}^* \tag{24}$$

We thus have:

$$\mu x^* - \mu' x^{u*}$$

$$= \left(\sum_{v,w \in W \setminus \{u\}} \mu_{vw} x_{vw}^* + \sum_{v \in W \setminus \{u\}} \mu_{uv} x_{uv}^* \right) - \left(\sum_{v,w \in W \setminus \{u\}} \mu_{vw} x_{vw}^* + \sum_{v,w \in W \setminus \{u\}} \mu_{vw} \lambda_{vw} \right)$$

$$= \sum_{v \in W \setminus \{u\}} \mu_{uv} x_{uv}^* - \sum_{v,w \in W \setminus \{u\}} \mu_{vw} \lambda_{vw}$$

$$= \sum_{v \in W \setminus \{u\}} \mu_{uv} \sum_{w \in W \setminus \{u,v\}} \lambda_{vw} - \sum_{v,w \in W \setminus \{u\}} \mu_{vw} \lambda_{vw}$$

$$= \sum_{v,w \in W \setminus \{u\}} (\mu_{uv} + \mu_{uw}) \lambda_{vw} - \sum_{v,w \in W \setminus \{u\}} \mu_{vw} \lambda_{vw} \tag{25}$$

$$= \sum_{v,w \in W \setminus \{u\}} (\mu_{uv} + \mu_{uw} - \mu_{vw}) \lambda_{vw} \tag{26}$$

The line labeled (25) here is obtained by observing that a λ_{vw} term occurs when v is fixed, and w is chosen from $W \setminus \{u, v\}$; and also when w is fixed, and v is chosen from $W \setminus \{u, w\}$. But the coefficients of μ satisfy the triangle inequality, so the expression (26) is greater than or equal to 0. □

As an example of Theorem 6, consider the ladder inequality $\mu x \geq 20$ on 9 nodes which is shown below.

$$x(\delta(H_1)) + x(\delta(H_2)) + x(\delta(T_1)) + x(\delta(T_2)) \tag{27}$$
$$+ 2x(\delta(T_3)) + 2x(\delta(T_4)) - 2x(T_1 \cap H_1, T_2 \cap H_2) \geq 20$$

Inequality (27) is a valid ladder inequality when the handles and teeth of this ladder inequality are defined as follows:

$$\begin{array}{ll} H_1 = \{2, u, 4, 6\} & H_2 = \{3, 5, 7\} \\ T_1 = \{1, 2, u\} & T_2 = \{8, 3\} \\ T_3 = \{4, 5\} & T_4 = \{6, 7\} \end{array}$$

On the handle-tooth picture of this ladder inequality in Figure 3, we superimpose a fractional solution x^*. If $x_e^* = 1$, then a solid line segment joining the endpoints of e is shown in Figure 3. If $x_e^* = \frac{1}{2}$, then a dotted line segment joining the endpoints of e is shown in Figure 3. We have $\mu x^* = 18$, so this ladder inequality is violated by x^*. Consider any cycle-shrink removal of node u, resulting in the solution x^{u*}. Then, by Theorem 6, we must satisfy $\mu|_{E(\{1,\ldots,8\})} x^{u*} \leq 18$.

Hence, the ladder inequality $\mu|_{E(\{1,...,8\})}x \geq 20$ is violated by x^{u*}. The (in this case unique) example of a vector x^{u*} resulting from a cycle-shrink removal of node u is shown in Figure 4. In this example, $\mu|_{E(\{1,...,8\})}x^{u*} = 18$.

With $B = \{1,...,8\}$ and x^* being any fractional point violating (27), a cycle-shrink removal of node u will not decrease the violation of this ladder inequality.

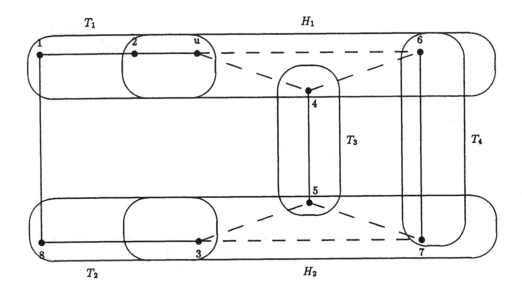

Fig. 3. fractional solution x^*

It is easy to see that the constraints of cycle-shrink force any two consecutive levels x^{i-1} and x^i in the cycle-shrink solution to have the following relationship. Namely, given the vector x^{i-1}, there exists a cycle-shrink removal of node i which yields the vector x^i. Hence, consider the following idea for finding an inequality in $S_{\overline{\mu}}$ which is violated by x^*. For the purposes of formulating a cycle-shrink LP, order the nodes in V arbitrarily, but so that the nodes in B are the last k nodes in the ordering. Denote the variables in the level of the cycle-shrink solution having only these k nodes left as x^{n-k}. By theorem 6, the expression $\mu|_{E(U)}x^i$ decreases as one goes down the levels of a feasible cycle-shrink solution, i.e. as the set $U = \{i+1,...,n\}$ becomes smaller and x^i represents the corresponding level in a feasible cycle-shrink solution. Note that because of the way the class of inequalities $S_{\overline{\mu}}$ was defined, we have $\mu|_{E(B)} = \overline{\mu}$. So, consider minimizing this

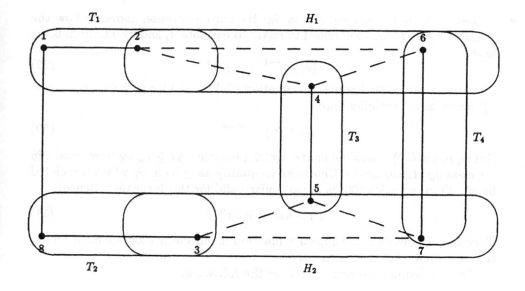

Fig. 4. after cycle-shrink removal

decrease with the following linear program.

$$\text{maximize } \overline{\mu} \cdot x^{n-k}$$
$$\text{subject to}$$
$$x_e^0 = x_e^* \quad \forall e \in E.$$

$$\sum_{e \in \delta(\{j\}) \cap E_i} x_e^i = 2 \quad \forall i \in \{0, \ldots, n-3\} \text{ and} \tag{28}$$
$$\forall j \in V_i.$$
$$x_e^i \geq x_e^{i-1} \quad \forall i \in \{1, \ldots, n-3\} \text{ and}$$
$$\forall e \in E_i.$$

Recall that μ_0 is the right hand side for all the inequalities in $S_{\overline{\mu}}$. Now, we have the following theorem:

Theorem 7. *If there exists an inequality $ax \geq \mu_0$ in $S_{\overline{\mu}}$ which is violated by x^*, then the above linear program will have an optimum objective function value which is less than μ_0.*

Proof: Consider solving the above linear program, and obtaining the optimum solution $\overline{x} = (x^0, x^1, x^2, \ldots)$. Let x^{i-1} be the vector at level $i-1$ in this cycle-shrink solution, and x^i be the vector at the next level after node i is removed. As we observed earlier, the constraints of cycle-shrink impose that the following relationship holds between x^{i-1} and x^i. Given the vector x^{i-1}, there is a cycle-shrink removal of i operation which yields the vector x^i.

Let $\mu x \geq \mu_0$ be an inequality in $S_{\overline{\mu}}$. By our observation above on how the levels x^{i-1} and x^i are constrained to relate to each other, and by Theorem 6, we have:

$$\mu|_{E_{i-1}} x^{i-1} \geq \mu|_{E_i} x^i \tag{29}$$

Since the relationship described in equation (29) is valid for each $i \in \{1, \ldots, n - k\}$, we obtain the relationship:

$$\mu x^* \geq \overline{\mu} \cdot x^{n-k} \tag{30}$$

But equation (30) is valid no matter which inequality $\mu x \geq \mu_0$ we have chosen in the class $S_{\overline{\mu}}$ of inequalities. Choose an inequality $ax \geq \mu_0$ in $S_{\overline{\mu}}$ which is violated by x^*. Then, equation (30) is in particular valid for this inequality. Hence,

$$\mu_0 > ax^* \geq \overline{\mu} \cdot x^{n-k} \tag{31}$$

Therefore, in this case, the objective function in our linear program must be less than μ_0, even when it is maximized. □

Complementing theorem 7, we have the following.

Theorem 8. *If x^* is a convex combination of Hamilton cycles, then the optimum value of $\overline{\mu} \cdot x^{n-k}$ for our linear program is greater than or equal to μ_0.*

Proof: Suppose x^* is a Hamilton cycle. Then x^{n-k} is a Hamilton cycle on the underlying graph whose vertex set is B. Hence, x^{n-k} satisfies all the valid inequalities of the TSP. In particular, this means that $\overline{\mu} \cdot x^{n-k} \geq \mu_0$.

But one can see from the following that the above analysis holds when x^* is a convex combination of Hamilton cycles. Let the incidence vector for a Hamilton cycle H be χ^H. One can perform a sequence of cycle-shrink operations individually on each of these Hamilton cycles H_i for $i \in I$ in a convex combination $x^* = \sum_{i \in I} \lambda_i \chi^{H_i}$ until only the vertices in B remain. Let the resulting Hamilton cycle on the set of vertices B from this sequence of cycle-shrink operations on H_i be H_i'. Define $x^{n-k} = \sum_{i \in I} \lambda_i \chi^{H_i'}$. Then the solution x^{n-k} is a convex combination of Hamilton cycles for the graph whose underlying vertex set is B. Clearly, x^{n-k} is feasible for our linear program, and the analysis of the last paragraph applies as well. □

Recall that for the purposes of formulating the cycle-shrink constraints in linear program (28), the vertices were ordered arbitrarily, but with the provision that the vertices in B were the last k vertices in this ordering. We use functional notation to denote this ordering as follows. Define a function f by:

$$f(B, i) := \text{the } i\text{-th vertex in the above described ordering.}$$

Similarly, make this ordering explicit in LP (28) by naming the variables in LP (28) as:

$$\overline{x}^B := (x^0, x^{B,1}, x^{B,2}, \ldots)$$

Now define the following:

$$V_{B,i} := \{f(B, r) \mid i < r \leq n\}$$
$$E_{B,i} := E \cap (V_{B,i} \times V_{B,i})$$

And consider the following linear program:

$$\text{minimize } c \cdot x$$
$$\text{subject to}$$
$$x_e^0 := x_e \geq 0 \quad \forall e \in E.$$

$$\sum_{e \in \delta(\{j\}) \cap E_{B,i}} x_e^{B,i} = 2 \quad \forall i \in \{0, \dots, n-3\} \text{ and} \quad (32)$$
$$\forall j \in V_{B,i}.$$
$$x_e^{B,i} \geq x_e^{B,i-1} \quad \forall i \in \{1, \dots, n-3\} \text{ and}$$
$$\forall e \in E_{B,i}.$$
$$\overline{\mu} \cdot x^{B,n-k} \geq \mu_0$$

The constraints of LP (32) are essentially those of LP (28) with the constraint $\overline{\mu} \cdot x^{B,n-k} \geq \mu_0$ added in. Note that Theorem 7 states that if x^* violates one of the inequalities in the class $S_{\overline{\mu}}$, then the optimal objective function value of LP (28) is less than μ_0. But this means that the components of x^* are an infeasible set of values for the vector x^0 in the linear program (32) because of the last constraint in LP (32). Theorems 7 and 8 thus yield the following consequence:

Theorem 9. *Fractional solutions x^* that violate one of the inequalities in the class $S_{\overline{\mu}}$ can be separated from the TSP polytope in polynomial time.*

Proof: Suppose the fractional solution x^* violates one of the inequalities in the class $S_{\overline{\mu}}$. Then Theorem 7 implies that the constraints of LP (32) cut-off x^* as discussed above. But Theorem 8 ensures that no points in the actual TSP polytope gets cut-off by the constraints of LP (32).   □

Note that for any feasible point $\underline{x}$ of LP (32), we have that the 0-th level of this solution x^0 satisfies all the inequalities in the class $S_{\overline{\mu}}$. We now have the following corollary:

Corollary 10. *Fractional solutions x^* that violate one of the inequalities in class S can be separated from the TSP polytope in polynomial time.*

Proof: We can separate over the inequalities in $S_{\overline{\mu}}$ for some given basic inequality $\overline{\mu}x \geq \mu_0$ in S and on a graph $G[B]$ by Theorem 9.

But there are order $|V|^k$ basic inequalities in class S, which is polynomial in $|V|$. By running LP (28) for each of these basic inequalities, we can find out for which basic inequality (or inequalities) $\overline{\mu}x \geq \mu_0$ that the constraints of LP (32) can cut-off x^*. Hence, one can separate over the class S of inequalities in polynomial time.   □

Specifically, the time required to achieve the separation stated in this theorem is thus the time needed to solve order $|V|^k$ cycle-shrink linear programs.

References

1. S. Boyd and W. H. Cunningham (1991), Small travelling salesman polytopes, *Mathematics of Operations Research 16 259-271*

2. B. Carr, (1995) Separating clique tree and bipartition inequalities having a fixed number of handles and teeth in polynomial time, *IPCO proceedings*
3. M. Jünger, G. Reinelt, and G. Rinaldi (1994), The traveling salesman problem, *Istituto Di Analisi Dei Sistemi Ed Informatica, R. 375, p. 53*
4. M. Jünger, G. Reinelt, and G. Rinaldi (1994), The traveling salesman problem, *Istituto Di Analisi Dei Sistemi Ed Informatica, R. 375, p. 59*
5. D. Karger, (1994), Random Sampling in Graph Optimization Problems, *PhD Thesis, Department of Computer Science, Stanford University*
6. L. Lovasz (1976), On some connectivity properties of Eulerian graphs, *Acta Math. Acad. Sci. Hungar., Vol. 28, 129-138*
7. D. Naddef, G. Rinaldi (1992), The graphical relaxation: A new framework for the Symmetric Traveling Salesman Polytope, *Mathematical Programming 58, 53-88*
8. M. Yannakakis (1988), Expressing combinatorial optimization problems by linear programs, *Proceedings of the 29th IEEE FOCS, 223-228*

Separating Maximally Violated Comb
Inequalities in Planar Graphs

Lisa Fleischer * and Éva Tardos **

School of Operations Research and Industrial Engineering
Cornell University, Ithaca, NY 14853
lisaf@orie.cornell.edu
eva@cs.cornell.edu

Abstract. The Traveling Salesman Problem (TSP) is a benchmark problem in combinatorial optimization. It was one of the very first problems used for developing and testing approaches to solving large integer programs, including cutting plane algorithms and branch-and-cut algorithms [16]. Much of the research in this area has been focused on finding new classes of facets for the TSP polytope, and much less attention has been paid to algorithms for separating from these classes of facets.

In this paper, we consider the problem of finding violated comb inequalities. If there are no violated subtour constraints in a fractional solution of the TSP, a comb inequality may not be violated by more than 0.5. Given a fractional solution in the subtour elimination polytope whose graph is planar, we either find a violated comb inequality or determine that there are no comb inequalities violated by 0.5. Our algorithm runs in $O(n + \mathcal{MC}(n))$ time, where $\mathcal{MC}(n)$ is the time to compute all minimum cuts of a planar graph.

1 Introduction

The *traveling salesman problem* (TSP) is given by a graph G with node set V, edge set E, where each edge $e \in E$ has a cost c_e. We consider the case when the costs are *symmetric*: for edge with end points a and b, $c_{ab} = c_{ba}$. We are interested in finding a hamiltonian cycle, here called a *tour*, in G of minimum cost.

The TSP is NP-complete, and yet there is interest in finding optimal solutions to instances of this problem. It was one of the very first problems where the cutting plane algorithms and branch-and-cut algorithms were developed and tested. For instance see [1, 4, 8, 9, 19]. The TSP can be formulated as an integer program with variables indexed by edges of G. The basic idea of the cutting plane algorithms is to form a linear programming relaxation of the problem:

$$\text{minimize } c^T x \text{ subject to } Ax \geq b \qquad (1)$$

---
* Supported in part by an NDSEG fellowship.
** Supported in part by a Packard Fellowship, an NSF PYI award, NSF through grant DMS 9505155, and ONR through grant N00014-96-1-0050.

for some set of linear inequalities $Ax \geq b$ that are satisfied by all incidence vectors of tours, x. The simplex method to guarantees that our solution x^* is a vertex of the polyhedron described by $Ax \geq b$. Thus if x^* is not the incidence vector of a tour, then there is a *valid* inequality $\alpha x \leq \beta$, an inequality that is satisfied by all tours x, that is not satisfied by x^*. (See Figure 1.) This inequality is called a *cutting plane* or a *cut* and can be added to (1) to form a tighter relaxation. If we have a method of finding cuts, then we can repeatedly solve LP's and add cuts until we find a relaxation that yields a tour as the optimal solution.

In order to guarantee that we make substantial progress towards finding a solution we would like to add cuts that define facets of the *TSP polytope*, polytope defined by the convex hull of the incidence vectors of all tours. Over the last 20 years a large number of classes of facets have been found for the TSP, and each class has an exponential number of members. Given a class of facets, we would like to have a *separation algorithm* for the class—an algorithm that finds an inequality in the class violated by a fractional solution x^*, or determines that no such inequality exists.

While there has been lots of research in identifying new classes of facets, very few classes have polynomial time separation algorithms. Since a complete, polynomial time separation algorithm in combination with the ellipsoid method would imply a polynomial time algorithm to solve the TSP, it is unlikely that it will be possible to separate efficiently from all facets of the TSP. In fact, if we could even recognize all classes of facets of the TSP, then co-NP would equal NP. In practice, heuristics are often used for finding violated inequalities. The two classes of facets for which polynomial separation algorithms are known are the subtour elimnation inequalities, and the 2-matching inequalities.

If x is a fractional solution to a linear programming relaxation of the TSP, the *graph of x* denoted by G_x, is the graph with vertex set V and edge set $\{e | x_e > 0\}$. The graph of an optimal solution is a hamiltonian cycle. Let A and B be disjoint sets of vertices. We define $x(A) = \sum_{i,j \in A} x_{ij}$ and $x(A, B) = \sum_{i \in A, j \in B} x_{ij}$. Clearly, every incidence vector of a tour satisfies $0 \leq x \leq 1$. Also, since we are looking for a hamiltonian cycle, every vertex obeys the *degree constraint* $x(\{v\}, V \backslash \{v\}) = 2$. These are the standard inequalities included in LP relaxations of the TSP. To eliminate integer solutions that are not tours, we need to add the *subtour elimination constraints* or *subtour constraints*:

$$x(A, \overline{A}) \geq 2, \ \forall A \subset V, A \neq \emptyset. \tag{2}$$

where $\overline{A} = V \backslash A$. Violated subtour elimnation constraints can be found in polynomial time by finding a minimum cut in G_x. The objective achieved by optimizing c over the polytope defined by the nonnegativity constraints, the degree constrains, and the subtour elimination constraints is the Held and Karp lower bound on the length of the TSP tour [13, 18].

In this paper, we consider the next most commonly used class of TSP facets known as comb inequalities. A *comb* consists of a set of vertices H called a *handle* and an odd number of vertex sets T_i, called *teeth*, such that $\forall i, \emptyset \neq H \cap T_i \subset T_i$ and $\forall i, j, \ T_i \cap T_j = \emptyset$. For example, see Figure 1. The corresponding *comb inequality* is

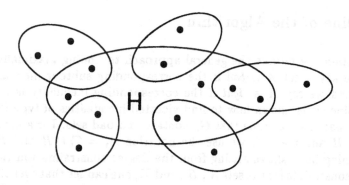

Fig. 1. A comb with 5 teeth

$$x(H) + \sum_{i=1}^{2k+1} x(T_i) \le (|H| - 1) + \sum_{i=1}^{2k+1} (|T_i| - 1) - k. \tag{3}$$

Comb inequalities define facets of the TSP polytope [10, 11], but unlike sub-tour constraints, there is no known polynomial-time separation algorithm for general combs. A slow, but polynomial time separation algorithm is known for the class of 2-matching inequalities, which is a special case of the comb inequali-ties. In [3], Carr describes a separation algorithm that finds a violated comb with t teeth in $O(n^{2t+3})$ time, if one exists. His algorithm can also be generalized to separate over an extesion of comb inequalities, bipartition inequalities, with a fixed number of handles and teeth. Applegate, Bixby, Chvátal and Cook [1] use a nice heuristic to find violated comb inequalities.

In this paper we consider maximally violated comb inequalities in planar graphs. We will assume that there are no violated subtour inequalities. This implies that a comb inequality (or a comb) cannot be violated by more than .5. We call a comb inequality *maximally violated* if it is violated by .5. If G_x is planar, we find a violated comb or provide a guarantee that there are no maximally violated comb inequalities in $O(n^2 \log n)$ time. The main outline of our algorithm is analogous to the heuristic of Applegate, Bixby, Chvátal, and Cook [1]; however, while they enumerate solutions to a set of linear equations, we use just one solution.

Section 2 outlines our general approach, Section 3 provides the details of the algorithm that do not rely on planarity, and Section 4 completes the description and proof of the algorithm. Two matching inequalities are special cases of comb inequalities, combs whose handle consists of two nodes. In the full paper [6] we also give a linear-time algorithm to detect maximally violated 2-matching inequalities and describe how to find all such violated inequalities in linear time per inequality.

2 Outline of the Algorithm

In this section, we outline the general approach to finding maximally violated combs. We say a set A is *bad* if the corresponding subtour inequality (2) is violated, and we say A is *tight* if the corresponding inequality is satisfied at equality. Since it is known how to find violated inequalities of type (2), for the rest of the paper we assume that G_x contains no bad sets. For a comb defined by handle H and teeth T_i we use the notation $A_i = T_i \cap H$ and $B_i = T_i - H$. Combining inequalities arising from the degree constraints and the subtour elimation constraints for the sets A_i, B_i, and T_i, one can see that over the subtour elimination polytope a comb inequality can be violated by most .5. This set of inequalities also implies a structure for the maximally violated combs.

Theorem 1. *A comb inequality is valid for the TSP, and can be violated by at most 0.5 if there are no bad sets. A comb inequality is maximally violated if and only if every T_i, A_i, and B_i is tight and $x(H, \overline{H}) = \sum_{i=1}^{2k+1} x(A_i, B_i)$*

We call pair of vertex sets (A, B) a *domino* if A and B are disjoint, and $T = A \cup B$, A, and B are all tight and non-empty. (See Figure 2a.) Theorem 1 implies that the teeth of maximally violated combs are disjoint dominoes. We call edge set $e(A, B) := \{(a, b) | a \in A, b \in B, x_{ab} > 0\}$ of domino (A, B) a *semicut*. Theorem 1 states that the semicuts of the teeth of a maximally violated comb form a cut in G_x. (Figure 2b) Thus there is an equivalence between a set of teeth of any maximally violated comb and a set of dominoes that satisfy the following conditions:

(i) the cardinality of the set is odd,
(ii) the dominoes are disjoint,
(iii) the semicuts of the dominoes describe a cut in G_x.

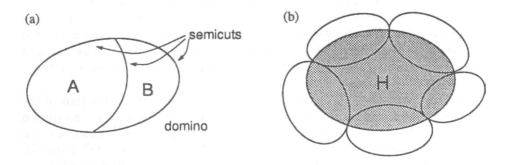

Fig. 2. Properties of a maximally violated comb

Applegate, et al. [1] give details of a heuristic that looks for maximally violated combs by searching for such sets of dominoes. They use a heuristic to find dominoes of G_x and set up a system of equations in GF(2) whose solutions correspond to sets of dominoes that satisfy items (i) and (iii). Given such a solution, they then test, in $O(n)$ time, if it satisfies (ii). To use this method to guarantee that there are no maximally violated combs, it is necessary to find all dominoes, and test all solutions to the system of equations, of which there can be an exponential number. Using the algorithm described in this paper, we find a violated comb, or provide such a guarantee for planar G_x, in $O(n^2 \log n)$ time.

Dominoes and semicuts themselves have nice structure. First note that if (A, B) is a domino, then so are $(A, \overline{T})$ and $(B, \overline{T})$. Also, using the fact that A, B, and T are all tight, we see that the sum of values of edges in a semicut is one, i.e. $x(A, B) = 1$. The semicuts $e(A, B)$ and $e(A, \overline{T})$ together form the tight cut A. It is not hard to show that relation on pairs of semicuts is an equivalence relation. We call the equivalence classes of this relation clusters. A *cluster* is a maximal set of semicuts and dominoes such that any two semicuts in the cluster form a cut, and each domino has its semicut in the cluster. So dominoes (A, B), $(A, \overline{T})$, and $(B, \overline{T})$ are all in the same cluster.

Instead of looking for dominoes we actually look for clusters: Consider the property (*iii*) above. A cut is a non-empty set of edges that intersects every cycle an even number of times. Using the equivalence class definition of the cluster we can see that semicuts in the same cluster intersect any cycle the same number of times *mod* 2. This implies that semicuts in one cluster are equivalent with respect to property (*iii*). If we define a *simple* maximally violated comb to be a maximally violated comb such that no subset of its semicuts form a cut, then the same argument also implies that a simple maximally violated comb contains at most one semicut from each cluster. Also, every maximally violated comb contains a simple maximally violated comb. Thus we can can redefine our search as a search for a set of clusters that satisfy (i) and (iii), and from which we can pick one domino per cluster so that the dominoes are disjoint. This reduces our search considerably, because, while there can be $O(n^2)$ dominoes, in Section 3.1 we show that there are $O(n)$ clusters.

A *pseudo-cut* is a set of edges, some edges possibly represented more than once, that intersects every cycle an even number of times. Note that every cut is a pseudo-cut, but the converse does not hold: e.g., the empty set is a pseudo-cut. We say that a set of clusters forms a pseudo-cut if taking one semicut from each cluster forms a pseudo-cut. By the above argument this is independent of the choice of semicuts. Thus, instead of looking for a set of dominoes that satisfy (i) and (iii), we look for an odd number of clusters whose semicuts form a pseudo-cut. We call such a set a *candidate*. For each cluster in the candidate, we then try to assign one domino in the cluster to *represent* the cluster, such that the representative dominoes are disjoint. If we can do this, the pseudo-cut is a cut, and the set of disjoint dominoes corresponds to the teeth of a maximally violated comb. If a set of disjoint representatives exists for a set of clusters, we call the set *representable*, and we find such a set. Semicuts of disjoint dominoes do not share edgess, which implies the following lemma.

Lemma 2. *If a candidate is representable, then the semicuts of the representatives form a cut in G_x.*

This approach and notion of representability is introduced by Applegate, et al. [1]. Their heuristic repeatedly checks different candidates for representability. Our main contribution is that we check just one candidate, and if G_x is planar and the candidate is not representable, we find a violated comb that uses dominoes from a subset of the clusters in the candidate. A rough outline of our algorithm appears in Figure 3. We discuss the details of this algorithm in the next two sections.

Find and store all clusters of G_x.
Search for a candidate.
If no candidates are found,
    Return guarantee that there are no maximally violated combs in G_x.
Else,
    Check if the candidate is representable.
    If it is,
        Return the corresponding maximally violated comb.
    Else,
        Find a violated comb using dominoes from some of the clusters
        in the candidate.

Fig. 3. Outline of algorithm

3　About Semicuts

In this section, we describe how to find and store all clusters, and how to check a candidate for representability. In doing so, we discuss the general structure of semicuts and clusters – structure that does not assume planarity.

3.1　Finding and Storing Semicuts Using Cactus Trees

In [5], Dinitz, Karzanov, and Lomonosov describe a $O(n)$-sized data structure called a cactus tree which represents all minimum cuts of a graph. In this section, we explain how semicuts and clusters of a graph are also represented in the cactus tree and how they can be easily retrieved.

A *cactus tree* K of a graph $G = (V, E)$ is a tree of cycles that may share vertices but not edges. If λ is the value of the minimum cut, then each edge of K has weight $\lambda/2$. (We assume that there are no cut edges in K by replacing any cut edge by a 2-cycle.) In our case $\lambda = 2$ so each edge of K has weight 1. The vertices of G are mapped to nodes of K by κ so that every minimal cut M of K corresponds to a minimum cut $\kappa^{-1}(M)$ of G, and every minimum cut in G equals $\kappa^{-1}(M)$ for some minimal cut M of K. If $\kappa^{-1}(i) = \emptyset$, we say that i is an *empty* node. Figure 4 contains a graph and its corresponding cactus tree.

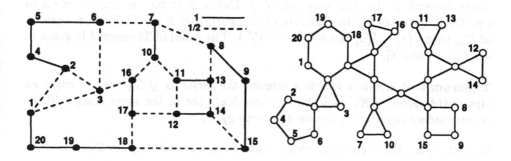

Fig. 4. Cactus tree of a graph

Karzanov and Timofeev [15] describe an $O(nm)$ or $O(\lambda n^2)$ algorithm to construct the cactus tree structure described in [5] from an unweighted graph. Both of these results are nicely summarized in a paper of Naor and Vazirani [17] that describes a parallel cactus algorithm. In [7], Gabow gives an $O(m + \lambda^2 \log(\frac{n}{\lambda}))$ algorithm to construct a cactus tree in unweighted graphs. We are interested in the cactus tree representation of a weighted graph. For this Benczúr [2] describes an $O(n^2 \log^3 n)$ randomized algorithm and also mentions that Hao and Orlin's overall minimum cut algorithm [12] can be used to give a deterministic $O(mn \log n)$ algorithm. Since we are concerned with planar graphs, we can use this last algorithm to compute a cactus tree for G_x in $O(n^2 \log n)$ time. Karger [14] describes a randomized $O(n^2 \log n)$ algorithm to find all minimum cuts of a graph.

Let K_x be the cactus tree of G_x. Since every vertex in G_x obeys the degree constraints and G_x contains no bad sets, each vertex is a minimum cut. Hence $|\kappa^{-1}(i)| \leq 1$ for every node i of K_x and all cut nodes of K_x are empty. As described so far, a graph does not have a unique cactus tree representation. Let i be a node on cycle Y of K_x. Let C_i^Y be the component containing i formed by removing the edges adjacent to i on Y, and let $V_i^Y = \kappa^{-1}(C_i^Y)$. We call i *trivial* if $V_i^Y = \emptyset$. By removing V_i^Y and making the neighbors of i on Y neighbors in Y, we can assume K_x has no trivial nodes. We can also assume that K_x has no empty, 3-way cut-nodes: an empty cut-node whose removal breaks K_x into exactly three components can be replaced by a 3-cycle. If Y is a k-cycle of K_x, we say Y is *significant* if $k \geq 3$. The following lemma is implicit in [5].

Lemma 3. *If Y is a significant cycle of K_x then*

1. *The sum of the costs on edges between any two vertex sets V_i^Y and V_j^Y of adjacent nodes i and j on cycle Y equals 1.*
2. *There are no edges between vertex sets V_i^Y and V_h^Y of nonadjacent nodes i and h on Y.*

It turns out that the significant cycles of the cactus tree correspond to clusters of G_x, and the edges on the cycles correspond to the semicuts. This correspondence depends on the following mapping. Define ϕ to be the map of edges of significant cycles of K_x to edge sets of G_x, such that if $e = (i, j)$ is on cycle Y of K_x, then $\phi(e) = \{(u, v) | u \in V_i^Y, v \in V_j^Y\}$. The proof of Theorem 4 is given in the full version [6].

Theorem 4. *ϕ defines a bijection between the semicuts of G_x and the edges on significant cycles of K_x such that two semicuts are in the same cluster iff the corresponding edges of K_x are on the same cycle.*

Corollary 5. *There are $O(n)$ clusters and semicuts arising from G_x.*

3.2 Representability and the Structure of Semicuts

In this section, we explain how to check if a given set of clusters is representable. First, we describe how dominoes of different clusters can intersect. Note that if the graph of a tight set is not connected, one of the connected components is a bad set. This proves the following lemma.

Lemma 6. *The graph induced by x on the vertex set of any domino is connected and removing the edges in the semicut creates two connected components.*

We say that a semicut *crosses* a tight set if the removal of the edges in the semicut disconnects the subgraph of G_x spanned by the tight set. If the node sets of two dominoes intersect, we say one *overlaps* the other. The following lemma and corollary describe how dominoes in G_x can overlap.

Lemma 7. *If a semicut F crosses a tight set T, all of the edges in F are contained in T; and F and the semicuts that define T are contained in the same cluster.*

Proof. If F crosses T, it divides T into two componenets, A and B. By the no-bad-sets assumption, $4 \leq x(A, \overline{A}) + x(B, \overline{B}) \leq x(T, \overline{T}) + 2\sum_{e \in F} x_e = 4$. Thus all the edges in semicut F are contained in T and F is the semicut of domino (A, B). ∎

A cluster defines a cyclic ordering on its semicuts and on the minimal cuts described by these semicuts. We call these minimal cuts *sections* of the cluster.

Corollary 8. *Given two clusters, all but one of the sections of one cluster are completely contained in one section of the other cluster.*

Proof. Since there are no bad sets, two tight sets whose union is not the entire graph intersect in a tight set. If the intersection of two sections from different clusters is a proper, nonempty subset of each of the sections, then this intersection must be a tight set. Let A be one of these sections, B the other, and $C = A \cap B$. $A \backslash C$ is a tight set since it equals $A \cap \overline{B}$. But then $A \backslash C$ and C are in the same cluster and, by similar reasoning, so is $B \backslash C$. Since any cluster which contains $A \backslash C$ and C also contains A, this contradicts the assumption that A and B are minimal sections from distinct clusters. ∎

If all but one of the sections of cluster S_1 are contained in section s_2 of cluster S_2, we say that s_2 *contains* S_1. Figure 5 shows an arrangement of clusters. If clusters S_1 and S_2 are both in some candidate, and $s_2 \in S_2$ contains S_1, then s_2 cannot be used in a domino to represent S_2 in the candidate. Using this observation, there is an easy $O(n^2)$ algorithm for checking representability: for each pair of clusters in the set, mark the section in each that contains the other. A cluster is *representable* in a set of clusters if it has two neighboring sections that do not contain any clusters in the set. After we have looked at each pair of clusters, the representable clusters will have at least two neighboring sections that are unmarked; these form a domino that represents the cluster. If all the clusters are representable, then the set is representable. A linear time algorithm to check representability is presented by Applegate, et al. in the first section of [1].

Fig. 5. An arrangement of clusters

4   The Planar Case

In this section, we assume G_x is planar with a fixed embedding. Planarity together with Lemma 6 imply that a semicut can be drawn as a line from one face of G_x to another, through the edges it contains. (See Figure 6). Thus, a semicut defines an ordering of its edge set. A *partial semicut* is a proper, nonempty subset of consecutive edges in a semicut. The *value* of a set of edges F in G_x is $v(F) = \sum_{e \in F} x_e$. Thus if F is a semicut, $v(F) = 1$; and if F' is a partial semicut, $0 < v(F') < 1$.

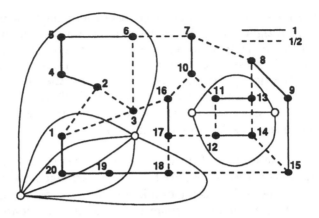

Fig. 6. G_x and part of G_x^S

Let G_x^S be the embedded graph whose vertices correspond to faces of G_x and whose edges correspond to the semicuts of G_x. The endpoints of an edge e_F in G_x^S are the faces of G_x whose boundaries have exactly one edge in the semicut F, and edge e_F is drawn from one end point to another through the edges of G_x contained in F, as in Figure 6. Hence clusters in G_x correspond to maximal sets of parallel edges in G_x^S, and G_x^S contains no loops. The *endpoints* of a cluster are the endpoints of its semicuts. To simplify matters, we will sometimes wish to work with G_x^C, the graph on the same vertex set as G_x^S but with one edge per cluster, i.e. there is one edge in G_x^C for every set of parallel edges in G_x^S. G_x^C is defined independent of the particular embedding of G_x. Note that, although the vertex sets of G_x^S and G_x^C are the same as the vertex set of the planar dual of G_x, neither graph is the planar dual, nor are they necessarily planar.

Lemma 9. *If x maximally violates some comb inequality, G_x^C contains an odd cycle.*

Proof. If x maximally violates some comb inequality, then Theorem 1 implies that the semicuts of the $2k+1$ teeth of the comb form a cut of value $2k+1$ in G_x. Since G_x is planar, the above observations demonstrate that a cut formed by semicuts in G_x corresponds to a cycle in G_x^C. Thus a cut of odd value formed by semicuts in G_x corresponds to an odd cycle in G_x^C. ∎

Recall that a candidate is an odd number of clusters whose semicuts form a pseudo-cut. We now have a method of finding a candidate: construct G_x^C and search for a simple, odd cycle in this graph. The edges of this cycle define the clusters of the candidate. We look for simple cycles since we observed in Section 2 that we can restrict our search to simple maximally violated combs. If the candidate corresponding to the odd cycle is representable, then we have found a maximally violated comb.

The converse of Lemma 9 is not true. The problem, as indicated earlier, is that the clusters that form an odd cycle may not be representable. Instead, we prove something weaker.

Theorem 10. *If G_x^C contains an odd cycle, then x violates some comb inequality, and if we are given a semicut graph of G_x, we can find one such inequality in linear time. The semicut graph can be computed in $O(n^2 \log n)$ time—the best known time to find all minimum cuts.*

Consider the cycle of non-representable clusters as drawn in Figure 7a. This cycle consists of a chain of representable clusters plus one cluster, γ_1 that overlaps some of the others. It is possible to have more than one cluster that overlaps some of the others, but we analyze this simpler case first.

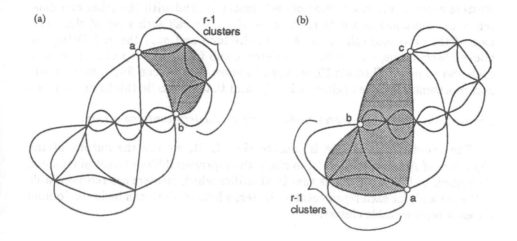

Fig. 7. Violated combs produced by a non-representable cycle

Proof of the simple case. In Figure 7a, there are $r - 1$ clusters in a subchain of the cycle that start at end point a of the overlapping cluster and continue until the first vertex b on a semicut of this cluster. The semicuts of the representable dominoes of these $r - 1$ clusters do not form a cut in G_x, but if we include the partial semicut from a to b of the overlapping domino, then these do form a cut in G_x, namely the shaded region in Figure 7a. Let $0 < \alpha < 1$ be the value of the partial semicut. The value of this cut is $r - 1 + \alpha$. Thus if $r - 1 = 2k + 1$ for some $k > 0$, then the comb that has the $r - 1$ representable dominoes as teeth and the above mentioned cut as the handle produces the following inequality:

$$x(H) + \sum_{i=1}^{2k+1} x(T_i) = |H| + \sum_{i=1}^{2k+1} (|T_i| - 1) - k - .5 - \frac{\alpha}{2}$$

$$> |H| + \sum_{i=1}^{2k+1} (|T_i| - 1) - k - 1$$

Hence this comb violates (3) by $0 < (1-\alpha)/2 < 1/2$. If $r-1 = 2k$ as in Figure 7b, then we include the overlapping domino as the $2k + 1^{st}$ tooth and the handle is the shaded region described by the semicuts of the $2k + 1$ dominoes plus the partial semicut from b to the other end point, c, of the overlapping domino. The resulting comb inequality is also violated, this time by $\alpha/2$. ∎

What happens if a cycle of clusters contains more than one overlapping cluster? More work is required to find a subset that will yield a violated comb. Examine the $r + 1$ clusters in Figure 7 that include the r clusters discussed above, the overlapping cluster γ_1, plus the next cluster after vertex b. This set of clusters is a contiguous subsequence of clusters of the cycle with the middle clusters contained in one section of end cluster γ_1, and with the other end cluster, γ_{r+1}, contained in a different section of γ_1. We call such a set of clusters a *loop* of the cycle and call γ_1 the *head* of the loop and γ_{r+1} the *tail*. Using the above observations, we can find a violated comb from any loop of clusters of a cycle. So to prove Theorem 10, we need to show that we can find a loop in any non-representable cycle of clusters in G_z^C, and that we can do this in linear time.

Lemma 11. *Every non-representable cycle of clusters contains a loop.*

The proof of this lemma is constructive. In it, we use the output of the algorithm of Applegate, et al. [1] to check the representability of a set of clusters. One result of this algorithm is that it identifies which clusters are representable in the set and, for each representable cluster, which sections can be in its domino for each representable cluster.

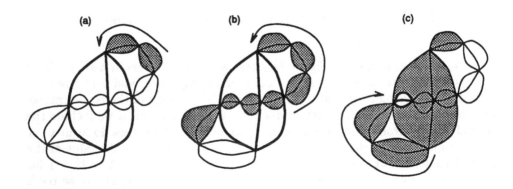

Fig. 8. Finding a loop in a non-representable cycle

Proof. Start with a representable cluster of the cycle. (The structure of clusters described in Corollary 8 implies that one exists in every set of clusters. For example, take a minimal cluster in the embedded arrangement as in Figure 5.). Assign it a domino as indicated by the representability-checking algorithm in [1], add it to the subsequence S, and continue along the cycle, adding clusters with their dominoes to S until a non-representable cluster of the cycle is reached. This cluster γ' may be representable in S, as S may not contain all the clusters of the cycle, e.g. see Figure 8a. Fix three semicuts of γ' and consider the three mega-sections defined by these semicuts. If two of the megasections of γ' do not contain dominoes in S, these two megasections form a domino for this cluster, which can be added to S, before continuing to the next cluster. Otherwise, at least two of these mega-sections contain clusters in S, and γ' is the head of a loop contained in S. (See Figure 8b). Let γ be the previous cluster added to S. The tail of the loop is the last cluster added to S that is not contained in the section of γ' that contains γ. Removing all clusters from S that were added to S before the tail cluster, we are left with a loop.

Until a loop is found, continue checking the next cluster on the cycle. For each non-representable cluster encountered, repeat the above check. For each representable cluster γ'' encountered after the first non-representable cluster, check one of the vertices in the domino identified by the algorithm in [1] to see whether it is contained in a domino in S. If so, γ'' is the tail of a loop and the cluster that contains the already assigned domino is the head. (See Figure 8c). Add γ'' to S, and remove all clusters added to S before the head cluster. The clusters remaining in S form a loop. If the vertex of the domino of γ'' is not contained in an already assigned domino, then this cluster is representable in S. Add the cluster to S and continue to the next cluster. Since the cycle is not representable, at some point we find a loop. ∎

The algorithm in Figure 9 summarizes how we find a violated comb given G_x^S. A naive implementation of the above algorithm might require $O(n)$ time per non-representable cluster, and hence $O(n^2)$ time before we find a non-representable subsequence. With a bit more effort, we can do this in linear time. Details are in the full paper [6].

Proof of Theorem 10. If G_x^C contains an odd cycle, then the cycle is either representable, in which case we find a maximally violated comb, or it is not representable, in which case, Lemma 11 implies that we find a violated comb. Since searching for an odd cycle, finding a loop, and checking representability can all be done in $O(n)$ time, we find a violated comb in $O(n)$ time. ∎

Actually, if a cycle is not representable, we find at least two violated combs: the comb described above and the comb found if we look for loop starting from γ_1 and continue along Γ in the opposite direction.

```
FindViolatedComb($G_x^S$)

Construct $G_x^C$ from $G_x^S$.
Look for an odd cycle, $\Gamma$ in $G_x^C$.
If no odd cycle,
      Return guarantee that there are no maximally violated combs in $G_x$.
Else,
        Check if $\Gamma$ is representable.
        If so,
            Return the corresponding maximally violated comb.
        Else
            Find a loop $\Gamma'$ of $\Gamma$, ordered $\{\gamma_1,\ldots,\gamma_r\}$ (as explained in text)
            If needed, flip order of clusters in $\Gamma'$ so that $\gamma_r$ and $\gamma_{r-1}$ are
                in different sections of $\gamma_1$.
            Remove $\gamma_r$ from $\Gamma'$. ( Note that $\Gamma'\backslash\{\gamma_r\}$ is representable.)
            Let $a$ be the first end point of $\gamma_1$ and $b$ be the second.
            Let $c$ be the last endpoint of $\gamma_{r-1}$.
            Find $\Delta = \{\delta_i\}$, a set of disjoint dominoes for the clusters in $\Gamma''$.
            If $|\Delta|$ is odd,
                Return $\Delta$ and the partial semicut from $a$ to $c$.
            Else
                Return $\Delta\backslash\{\delta_1\}$ and the partial semicut from $b$ to $c$.
```

Fig. 9. Finding a violated comb in G_x^S

References

1. Applegate, D., Bixby, R., Chvátal, V., Cook, W.: Finding Cuts in the TSP (A Preliminary Report). Unpublished manuscript. (1994)
2. Benczúr, A. A.: Augmenting undirected connectivity in RNC and in randomized $\tilde{O}(n^3)$ time. Proc. 26th Annual ACM Symp. on Theory of Comp. (1994) 658–667
3. Carr, R. D.: Separating Clique Tree and Bipartition Inequalities in Polynomial Time. Proc. 4th International IPCO Conference. Editors Balas, E., Clausen J., Springer-Verlag Lecture Notes in Computer Science 920 (1995) 40–49
4. Crowder, H., Padberg, M. W.: Solving large-scale symmetric travelling salesman problems to optimality. Management Sci. 26 (1980) 495–509
5. Dinits, E. A., Karzanov, A. V., Lomonosov, M. V.: On the structure of a family of minimal weighted cuts in a graph. Studies in Discrete Optimization, Editor Fridman, A. A., Moscow Nauka, in Russian (1976) 290–306
6. Fleischer, L., Tardos, É. Separating maximally violated comb inequalities in planar graphs. Cornell University, Department of Operations Research and Industrial Engineering, TR1150 (1996)
7. Gabow, H. N.: Applications of a poset representation to edge connectivity and graph rigidity. Proc. 32nd Annual Symp. on Found. of Comp. Sci. (1991) 812–821
8. Grötschel, M.: On the symmetric travelling salesman problem:solution of a 120-city problem. Math. Programming Study 12 (1980) 61–77

9. Grötschel, M., Holland, O.: Solution of large-scale symmetric travelling salesman problems. Math. Programming **51** (1991) 141–202

10. Grötschel, M., Padberg, M. W.: On the Symmetric Travelling Salesman Problem I: Inequalities. Math. Programming **16** (1979) 265–280

11. Grötschel, M., Padberg, M. W.: On the Symmetric Travelling Salesman Problem II: Lifting Theorems and Facets. Math. Programming **16** (1979) 281–302

12. Hao, J., Orlin, J. B.: A faster algorithm for finding the minimum cut in a graph. Proc. of 3rd ACM-SIAM Symp. on Discrete Algorithms (1992) 165–174

13. Held, M., Karp, R. M.: The traveling-salesman problem and minimum spanning trees. Operations Research **18** (1970) 1138–1162

14. Karger, D. R.: Minimum cuts in near-linear time. To appear in the Proceedings of the 28th ACM Symp. on Theory of Computing (1996)

15. Karzanov, A. V., Timofeev, E. A.: Efficient Algorithms for Finding all Minimal Edge Cuts of a Nonoriented Graph. Cybernetics **22** (1986) 156–162, Translated from Kibernetika **2** (1986) 8–12

16. Lawler, E. L., Lenstra, J. K., Rinnooy Kan, A. H. G., Shmoys, D. B.: The Traveling Salesman Problem: A Guided Tour of Combinatorial Optimization. John Wiley & Sons (1985)

17. Naor, D., Vazirani, V. V.: Representing and enumerating edge connectivity cuts in RNC. Proc. Second Workshop on Algorithms and Data Structures, Springer-Verlag, Lecture Notes in Computer Science **519** (1991) 273–285

18. Nemhauser, G. L., Wolsey, L. A.: Integer and Combinatorial Optimization. John Wiley & Sons (1988)

19. Padberg, M. W., Rinaldi, G.: Optimization of a 532-city symmetric traveling salesman problem by branch and cut. Oper. Res. Letters **6** (1987) 1–7

The Travelling Salesman and the PQ-Tree *

Rainer E. Burkard[1]   Vladimir G. Deĭneko[1]   Gerhard J. Woeginger[2] **

[1] TU Graz, Institut für Mathematik B, Steyrergasse 30, A-8010 Graz, Austria.
[2] Eindhoven University of Technology, Department of Mathematics and Computing Science, P.O. Box 513, NL–5600 MB Eindhoven, The Netherlands.

Abstract. Let $D = (d_{ij})$ be the $n \times n$ distance matrix of a set of n cities $\{1, 2, \ldots, n\}$, and let T be a PQ-tree with node degree bounded by d that represents a set $\Pi(T)$ of permutations over $\{1, 2, \ldots, n\}$. We show how to compute for D in $O(2^d n^3)$ time the shortest travelling salesman tour contained in $\Pi(T)$. Our algorithm may be interpreted as a common generalization of the well-known Held and Karp dynamic programming algorithm for the TSP and of the dynamic programming algorithm for finding the shortest pyramidal TSP tour.

This result has two surprising consequences. The first consequence concerns large sets of permutations, so-called exponential neighborhoods, over which the TSP can be solved efficiently. Up to now, the largest known neighborhoods had cardinality $2^{\Theta(n)}$, whereas our result yields new neighborhoods of cardinality $2^{\Theta(n \log \log n)}$. The second consequence is that the shortcutting phase of the "twice around the tree" heuristic for the Euclidean TSP can be optimally implemented in polynomial time. This contradicts a statement of Papadimitriou and Vazirani as published in 1984.

Keywords. Travelling salesman problem, Polynomial algorithm, Dynamic programming, Combinatorial optimization, Euclidean travelling salesman problem, PQ-tree.

1 Introduction

In the *Travelling Salesman Problem* (TSP), the objective is to find for a given $n \times n$ distance matrix $D = (d_{ij})$ a permutation π of the set $\{1, 2, \ldots, n\}$ that minimizes the function

$$d(\pi) = \sum_{i=1}^{n-1} d_{\pi(i)\pi(i+1)} + d_{\pi(n)\pi(1)} \qquad (1)$$

In other words, the travelling salesman has to visit the cities 1 to n in arbitrary order and in the end he has to return to the city from which he had started. His

* This research has been supported by the Spezialforschungsbereich F 003 "Optimierung und Kontrolle", Projektbereich Diskrete Optimierung.
** Supported by a research fellowship of the Euler Institute for Discrete Mathematics and its Applications.

goal is to minimize the total travel length while doing this. The permutations over $\{1, 2, \ldots, n\}$ will also be called *tours*, the elements of $\{1, 2, \ldots, n\}$ are called *cities*, and two consecutive cities $\pi(i)$ and $\pi(i+1)$ in the tour are said to form an *edge* of the tour. The TSP is one of the fundamental problems in combinatorial optimization. The reader is referred to the book by Lawler, Lenstra, Rinnooy Kan and Shmoys [12] for a guided tour through this problem.

Main results of this paper. In this paper we investigate specially structured sets of permutations that can be represented via PQ-trees, a well-known data structure from Theoretical Computer Science (see Section 2 for an exact definition). As main result, we will prove that over such a set of permutations, the form (1) can be minimized in time $O(n^3 2^d)$ where d is the maximum degree of the underlying PQ-tree. This result has two nice applications: First, it yields a neighborhood of size $2^{\Theta(n \log \log n)}$ for the TSP over which we can optimize in polynomial time. This is currently by far the largest known neighborhood for the TSP that can be searched in polynomial time. Secondly, the result can be used for implementing the shortcutting step of the "twice around the tree" approximation algorithm for the Euclidean TSP optimally in *polynomial time*. Below, we sketch some of the history of these two applications.

Neighborhood	Size	log(Size)	Time	Reference
2-OPT	$\Theta(n^2)$	$\Theta(\log n)$	$O(n^2)$	Croes [6]
k-OPT	$\Theta(n^k)$	$\Theta(\log n)$	$O(n^k)$	Lin [13]
PYRAMIDAL	$\Theta(2^n)$	$\Theta(n)$	$O(n^2)$	Klyaus [11]
BELPERM	$\Theta(n 2^n)$	$\Theta(n)$	$O(n^3)$	Carlier & Villon [4]
EDGE EJECT.	$\Theta(\sqrt[3]{12}^n)$	$\Theta(n)$	$O(n)$	Glover & Punnen [8]
DS-k-OPT	$\Omega(k!^n)$	$\Theta(n)$	$O(n^k)$	Potts & van de Velde [17]
PQ-TREE	$2^{\Theta(n \log \log n)}$	$\Theta(n \log \log n)$	$O(n^3)$	This paper (Theorem 10)
ALL TOURS	$(n-1)!$	$\Theta(n \log n)$	$O(n^2 2^n)$	Held & Karp [9]

Table 1. Some neighborhoods from the literature, sorted by increasing size.

Neighborhoods. By $\mathcal{S}_n$ we denote the set of all permutations over $\{1, 2, \ldots, n\}$. The *canonical* tour is the tour that visits the cities in the order $\langle 1, 2, 3, \ldots, n \rangle$. A *neighborhood* $\mathcal{P}$ is a sequence $\langle \mathcal{P}_n \rangle_{n \geq 1}$ of sets of permutations with $\mathcal{P}_n \subseteq \mathcal{S}_n$. For every n, the set $\mathcal{P}_n$ forms the neighborhood of the canonical tour $\langle 1, 2, 3, \ldots, n \rangle$. The set $\mathcal{P}_n$ also induces a neighborhood for arbitrary tours $\pi \in \mathcal{S}_n$ by renaming the cities in such a way, that the canonical tour in the renamed instance corresponds to π in the original instance. Around each neighborhood $\mathcal{P}$, one can build a local search algorithm that works as follows: The algorithm starts at some initial tour π and then searches the neighborhood of π for a shorter tour. If no such tour exists, the procedure halts with a local optimum with respect to $\mathcal{P}$.

Otherwise, the local search algorithm uses the shorter tour as new starting point and repeats the procedure. This process eventually halts as there is only a finite number of possible tours. For an extensive discussion of local search algorithms, see Papadimitriou and Steiglitz [15].

Probably the oldest and best investigated neighborhoods are the k-OPT neighborhoods that contain all tours that differ in at most k edges from the canonical tour (Croes [6] and Lin [13]). A somewhat larger neighborhood is formed by the *pyramidal tours* (cf. Klyaus [11], and Gilmore, Lawler and Shmoys [7]). A tour is called *pyramidal*, if it starts in city 1, then visits cities in increasing order until it reaches city n, and finally returns through the remaining cities in decreasing order back to city 1. Carlier and Villon [4] designed the heuristic *BelPerm* that is based on a kind of cyclic shift of pyramidal tours. Only very recently, Glover and Punnen [8] introduced the so-called *edge ejection* neighborhood and Potts and van de Velde [17] introduced the *DynaSearch* neighborhood DS-k-OPT. All the neighborhoods that we mentioned till here share two common properties: (i) the shortest tour in these neighborhoods can be found in polynomial time and (ii) their sizes are bounded by $O(\beta^n)$ for appropriate numbers $\beta > 1$. These results are summarized in Table 1, together with our new neighborhood that is based on PQ-trees and together with the well-known Held and Karp dynamic programming algorithm [9] for finding the overall shortest TSP tour. Our paper leaves as an open problem whether there exist neighborhoods of size $\Omega(\gamma^{n \log n})$, for some $\gamma > 1$, over which one can optimize in polynomial time.

Approximation algorithms based on minimum spanning trees. The TSP is well-known to be an NP-hard problem. In fact even the so-called *Euclidean TSP* is NP-hard, i.e. the case where the cities are points in the Euclidean plane and where the distance d_{ij} equals the Euclidean distance measured between city i and city j (Papadimitriou [14]). NP-completeness has stimulated research into the direction of approximation algorithms. Among the most illustrious approximation algorithms for the Euclidean TSP is the "twice around the tree" heuristic which works as follows.

(S1) Compute the minimum spanning tree of the cities.

(S2) Double every edge in the tree to get a Eulerian graph.

(S3) Find an Eulerian trail of this Eulerian graph and then transform the trail into a TSP tour by *shortcutting*: for every city remove all but one of its occurrences in the Eulerian trail.

The resulting tour is guaranteed to have no more than twice the length of the optimal one [10, 15]. Independently from each other, Christofides [5] and Serdyukov [18] suggested an improvement of this heuristic. The Christofides-Serdyukov approximation algorithm performs the same steps (S1) and (S3), but uses another Eulerian graph. It replaces step (S2) by the following step (S2*).

(S2*) Construct a minimum length matching for the odd-degree nodes in the minimum spanning tree and add this matching to the tree. This yields a Eulerian graph.

For this modified approximation algorithm, the resulting tour is at most 50% longer than the optimal one [5, 18, 10, 15]. Papadimitriou and Vazirani [16] concentrated on the optimal implementation of the shortcutting step (S3) for the Christofides-Serdyukov algorithm. They proved that, given a set of points in the Euclidean plane, the minimum spanning tree, and a shortest matching of the odd-degree nodes, it is NP-hard to find the shortest TSP tour that can be obtained from them by shortcuts. In the discussion of their paper Papadimitriou and Vazirani also claimed that their NP-hardness result can be carried over to the shortcutting step of the "twice around the tree" heuristic. However, unless $P = NP$ this claim cannot hold true: We will prove that the set of all tours obtainable from the doubled tree by shortcutting can always be represented by a PQ-tree whose maximum degree is bounded by seven. With this, our main result yields an $O(n^3)$ polynomial time algorithm for finding the optimal shortcutting.

Organization of the paper. Section 2 introduces some notation on permutations and gives all relevant definitions for PQ-trees. Section 3 describes how to find the shortest tour represented by a given PQ-tree with the help of a simple dynamic programming approach. Section 4 presents two modifications of the dynamic program for the case where the PQ-tree possesses a very special, very simple structure. Section 5 demonstrates that the Held-Karp approach and the concept of pyramidal tours both fit into our framework. Moreover, it constructs the large neighborhood of size $2^{\Theta(n \log \log n)}$ over which the TSP can be solved efficiently. Section 6 deals with the shortcutting step of the "twice around the tree" algorithm. Finally, Section 7 gives the conclusion.

2 Definitions

The set of all permutations over $\{1, 2, \ldots, n\}$ is denoted by $\mathcal{S}_n$. For $\pi \in \mathcal{S}_n$, we adopt the notation $\pi = \langle x_1, x_2, \ldots, x_n \rangle$ as abbreviation for "$\pi(i) = x_i$ for $1 \le i \le n$". For two permutations $\pi_1 = \langle x_1, \ldots, x_n \rangle$ and $\pi_2 = \langle y_1, \ldots, y_m \rangle$, their *concatenation* $\pi_1 \star \pi_2$ is the permutation $\langle z_1, \ldots, z_{n+m} \rangle$, where $z_i = x_i$ for $1 \le i \le n$ and $z_{n+j} = y_j$ for $1 \le j \le m$. For two sets Π_1 and Π_2 of permutations, we define
$$\Pi_1 \star \Pi_2 = \{\pi_1 \star \pi_2 \mid \pi_1 \in \Pi_1, \ \pi_2 \in \Pi_2\}.$$
Clearly, '$\star$' is an associative operation on sets of permutations.

A *PQ-tree* T over the universal set $U = \{1, \ldots, n\}$ is a rooted, ordered tree whose *leaves* are pairwise distinct elements of U and whose *internal* nodes are distinguished as either P-nodes or Q-nodes. Every internal node has at least two children. By LEAF(T) we denote the set of leaves in T.

With every PQ-tree T, we associate a set $\Pi(T)$ of permutations of LEAF(T) as follows. If T consists of only a single leaf $u \in U$, then $\Pi(T) = \langle u \rangle$. Otherwise, the root $r(T)$ of T is a P-node or a Q-node. Let $v_1, \ldots, v_m$ denote the sons of $r(T)$, ordered from left to right, and let T_i denote the maximal subtrees rooted at v_i, $1 \le i \le m$. If $r(T)$ is a P-node, then
$$\Pi(T) = \bigcup_{\psi \in \mathcal{S}_m} \Pi(T_{\psi(1)}) \star \Pi(T_{\psi(2)}) \star \cdots \Pi(T_{\psi(m)}),$$

and if $r(T)$ is a Q-node, then

$$\Pi(T) = \Pi(T_1) \star \Pi(T_2) \star \cdots \Pi(T_m) \ \cup \ \Pi(T_m) \star \Pi(T_{m-1}) \star \cdots \Pi(T_2) \star \Pi(T_1).$$

In other words, a P-node allows arbitrary permutations of its sons, whereas a Q-node only allows reversals. For more information on manipulating PQ-trees, for examples and applications, the reader is referred to the original paper by Booth and Lueker [2].

3 A Dynamic Programming Algorithm

In this section we design a fast and simple dynamic programming algorithm for the following problem: Given a PQ-tree T with node degree bounded by d, and a distance matrix $D = (d_{ij})$, find the shortest TSP tour for matrix D that is contained in the set $\Pi(T)$ of permutations.

We start with some more notation. The node set of the PQ-tree T is denoted by V. The set V consists of n *leaves* that correspond to the cities $1, \ldots, n$ and of several *internal* P- and Q-nodes. Note that the number of internal nodes is $O(n)$. By $T(v)$ we denote the maximal subtree rooted at node v. For an internal node v and a subset S of sons of v, we define $T(v, S)$ to be the PQ-tree that results from removing from $T(v)$ all the subtrees $T(w)$ rooted at sons w of v that are not contained in S. For a leaf a in $T(v)$, we denote by $T(v, a)$ the tree $T(w)$ where w is the unique son of v with a in $\text{LEAF}(T(w))$.

For an internal node v, two leaves a and b in $\text{LEAF}(T(v))$ are said to be *separated by* v, if and only if $T(v, a) \neq T(v, b)$. We denote this by $(a, b) \in \text{SEP}(v)$. For the sake of completeness, we moreover define that for every leaf v, $(v, v) \in \text{SEP}(v)$ holds.

Lemma 1. *The overall number of triples (v, a, b) with $(a, b) \in \text{SEP}(v)$ equals n^2. Moreover, we can enumerate and store all the sets $\text{SEP}(v)$, $v \in V$, in $O(n^2)$ overall time.*

Proof. Since every pair of leaves in T is only separated by the top-most node v on the unique path in T that connects a to b, the overall cardinality of all sets $\text{SEP}(v)$ is n^2. To get the claimed time complexity, one first computes for every node v a list that contains the cities in $\text{LEAF}(T(v))$, and afterwards forms cross products of appropriate lists. □

Next, we define three auxiliary arrays $X[*]$, $Y[*]$ and $Z[*]$. The indices of these arrays range over nodes or subsets of nodes, and their values are non-negative numbers. With 'shortest path' in the following we always mean a shortest path with respect to the distance matrix D.

- For two leaves a and b in $T(v)$ with $(a, b) \in \text{SEP}(v)$, we define $X[v; a, b]$ as the length of the shortest Hamiltonian path that starts in city a, then runs through all the other cities in $\text{LEAF}(T(v, a))$ while obeying the PQ-tree restrictions, and finally jumps away from $T(v, a)$ over to city b. If v itself is a leaf, we define $X[v; v, v] = 0$.

- For a subset S of sons of some node v, and for two leaves a and b in $T(v)$ with $(a, b) \in \text{SEP}(v)$, we define $Y[v; S, a, b]$ as the length of the shortest Hamiltonian path through all the cities in $\text{LEAF}(T(v, S))$ that starts in city a and ends in city b while obeying the PQ-tree restrictions. In case no such path exists, $Y[v; S, a, b]$ is set to ∞. This will e.g. be the case if a or b are not contained in the set $\text{LEAF}(T(v, S))$.
- For two leaves a and b in $T(v)$ with $(a, b) \in \text{SEP}(v)$, we will write $Z[v; a, b]$ short for $Y[v; S^*, a, b]$ where S^* is the set of all sons of v. In other words, $Z[v; a, b]$ is the length of the shortest Hamiltonian path from a to b through all cities in $\text{LEAF}(T(v))$ that obeys all the PQ-tree restrictions. If v itself is a leaf, we define $Z[v; v, v] = 0$.

Our goal is to compute all the values $Z[v; *]$. The values $X[v; *]$ and $Y[v; *]$ are auxiliary values: some of these entries are used for certain internal computations that only concern node v. E.g. for Q-nodes v, it will not be necessary to compute the values of *all* entries $Y[v; *]$. All computations are done in a bottom-up fashion, starting at the leaves v of T and moving up towards the root. When we are dealing with a father, all the values $Z[*]$ of all its sons have already been computed.

Next, we describe in detail how to perform these computations. If v is a leaf, we set $Z[v; v, v] = 0$. Otherwise, v is an internal node. We will first show how to deal with P-nodes and then sketch the treatment of Q-nodes.

Lemma 2. *(Computation of the arrays for P-nodes)*
All the values $Z[v; a, b]$ for all P-nodes v and all leaves a and b with $(a, b) \in \text{SEP}(v)$ can be computed in $O(2^d n^3)$ overall time.

Proof. Let v be a P-node with sons $v_1, \ldots, v_s$ and assume that the arrays $Z[*]$ for all of its sons are already known. Define v_i to be the root of the subtree $T(v, a)$. Then it is easy to verify that the equation

$$X[v; a, b] = \min\{ Z[v_i; a, c] + d_{c, b} \mid c \in \text{LEAF}(T(v_i)) \} \tag{2}$$

holds. The values of $Y[v; S, a, b]$ are computed in order of increasing $|S|$. Again, let v_i be the root of $T(v, a)$ and let v_j be the root of $T(v, b)$. If S does not contain both nodes v_i and v_j, the value of $Y[v; S, a, b]$ equals ∞. Otherwise $|S| \geq 2$, and the first non-trivial case occurs for $|S| = 2$ with $S = \{v_i, v_j\}$. In this case, the equation

$$Y[v; \{v_i, v_j\}, a, b] = \min\{ X[v; a, c] + Z[v_j; c, b] \mid c \in \text{LEAF}(T(v_j)) \} \tag{3}$$

holds. For sets $|S| \geq 3$ with $\{v_i, v_j\} \subseteq S$, we have

$$Y[v; S, a, b] = \min\{X[v; a, c] + Y[v; S - \{v_i\}, c, b] \mid c \in \text{LEAF}(T(v, S - \{v_i, v_j\}))\}. \tag{4}$$

Finally, we set $Z[v; a, b] = Y[v; \{v_1, \ldots, v_s\}, a, b]$ for all leaves a and b in $T(v)$ with $(a, b) \in \text{SEP}(v)$.

Next, we analyze the time complexity of the above computations. In a global preprocessing step, we determine for every v the set $\text{SEP}(v)$ as described in

Lemma 1. Since the arrays $X[v; a, b]$, $Y[v; S, a, b]$, and $Z[v; a, b]$ are only defined for triples (v, a, b) with $(a, b) \in \text{SEP}(v)$, Lemma 1 yields that there are only n^2 entries to compute for $X[*]$, only $O(2^d n^2)$ entries for $Y[*]$, and n^2 entries for $Z[*]$. Finally, since all the minima in the equations (2), (3) and (4) are taken over at most $O(n)$ values, the computation of all relevant entries in $X[*]$ can be done in $O(n^3)$ time, the computation of all relevant entries in $Y[*]$ in $O(2^d n^3)$ time, and the computation of all relevant entries in $Z[*]$ in $O(n^2)$ time. Summarizing we arrive at the claimed overall time complexity. □

Lemma 3. *(Computation of the arrays for Q-nodes)*
All the values $Z[v; a, b]$ for all Q-nodes v and all leaves a and b with $(a, b) \in \text{SEP}(v)$ can be computed in $O(n^3)$ overall time.

Proof. Let v be a Q-node with sons $v_1, \ldots, v_s$ ordered from left to right. Again we assume that the arrays $Z[*]$ for all sons are already known. We will only explain the main differences to the handling of P-nodes as described in Lemma 2: Essentially, equation (2) may be applied also in this case for computing $X[v; a, b]$. However, if the roots of $T(v, a)$ and $T(v, b)$ are not neighboring sons of v, then no path that obeys the PQ-tree restrictions may jump directly from $T(v, a)$ to $T(v, b)$ and hence $X[v; a, b]$ has to be set to ∞.

In $Y[v; S, a, b]$, we only compute the value of those entries for which S forms a contiguous interval of sons of v and moreover contains at least one of v_1 or v_s. Let $b \in \text{LEAF}(T(v_s))$, $a \in \text{LEAF}(T(v_i))$ with $i < s$ and let $S = \{v_i, v_{i+1}, \ldots, v_s\}$. Then

$$Y[v; S, a, b] = \min\left\{X[v; a, c] + Y[v; S - \{v_i\}, c, b] \mid c \in \text{LEAF}(T(v, \{v_{i+1}\}))\right\}. \tag{5}$$

This yields all the relevant entries with $v_s \in S$. Deriving the relevant entries with $v_1 \in S$ is done in a symmetric way. Finally, an entry $Z[v; a, b]$ has a finite value only if one of a and b is a leaf in $T(v_1)$ and the other one is a leaf in $T(v_s)$. Observe that hence the computed values of $Y[v; *]$ are indeed sufficient to get all finite entries $Z[v; a, b]$.

The analysis of the time complexity can also be done parallel to Lemma 2. The time complexity for computing $X[*]$ and $Z[*]$ does not change. The main difference consists in the computation of $Y[v; *]$. In fact, formula (5) describes a standard dynamic programming algorithm for finding shortest paths in a kind of s-partite graph. For a vertex with sons $v_1, \ldots, v_s$ and n_i leaves in $T(v_i)$, it takes

$$O(n_1 n_2 + n_2 n_3 + \cdots + n_{s-1} n_{s-2} + n_1(n_2 + \cdots + n_s))$$

operations to compute all the $n_1 n_s$ relevant shortest paths (in this expression, the lefthand part accounts for the number of edges, and the righhand part accounts for the number of shortest paths that start in $T(v_1)$ and end in one of the other subtrees). Hence, the overall number of operations for treating all nodes v is proportional to

$$\sum_{v \in V} (n_1 n_2 + n_2 n_3 + \cdots + n_{s-1} n_{s-2} + n_1(n_2 + \cdots + n_s))$$

$$\leq \sum_{v \in V} n(n_1 + n_2 + \cdots + n_s) \leq n^3,$$

and thus the overall time complexity drops down to $O(n^3)$. □

Theorem 4. *For a PQ-tree T with node degree bounded by d, and a distance matrix $D = (d_{ij})$, the shortest TSP tour for matrix D contained in $\Pi(T)$ can be computed in $O(2^d n^3)$ overall time.*

Proof. Let r be the root of T and note that $\Pi(T) = \Pi(T(r))$. The shortest TSP tour in $\Pi(T(r))$ decomposes into a single edge $(1, x)$ and a shortest Hamiltonian path through $\{1, \ldots, n\}$ that starts in city x and ends in city 1, for an appropriate city x. Hence, its length equals

$$\min \{ d_{1,x} + Z[r; x, 1] \mid x \in \text{LEAF}(T) \}. \tag{6}$$

Now the global strategy is obvious: Compute all relevant values $Z[*]$ according to Lemmas 2 and 3 in $O(2^d n^3)$ overall time, and then evaluate (6). □

4 Modifications of the Dynamic Program

In this section we describe two modifications of the dynamic programming procedure developed in the preceding section that take advantage of cases where the PQ-tree possesses a very simple, special structure.

Throughout this section, β will denote a small positive integer that is considered to be constant and not part of the input. Let v be an interior node in some PQ-tree T, and let $v_1, \ldots, v_s$ denote the sons of v. We say that v has a *β-small son*, if for some son v_i the inequality $|\text{LEAF}(T(v_i))| \leq \beta$ holds. We say that v is *β-unbalanced*, if for some son v_i the inequality $|\text{LEAF}(T(v_i))| \geq |\text{LEAF}(T(v))| - \beta$ holds (in other words, in all subtrees rooted at sons v_j with $j \neq i$ there are altogether at most β nodes). All results in this section are centered around the following two definitions.

(SSC) A PQ-tree T fulfills the *β-small son condition*, if every interior node in T has a β-small son.
(UBC) A PQ-tree T is called *β-unbalanced*, if every interior node in T is β-unbalanced.

In the following, we will use the same notation and concepts as introduced in the preceding section.

Lemma 5. *For a PQ-tree T that fulfills (SSC) and that has node degree bounded by d, and for a distance matrix $D = (d_{ij})$, the shortest TSP tour for matrix D in $\Pi(T)$ can be computed in $O(2^d n^2 + n^3)$ overall time.*

Proof. Most parts of the dynamic programming algorithm developed in Section 3 run in $O(n^3)$ time. The expensive bottleneck is the computation of the relevant entries in $Y[*]$: this takes $O(2^d n^3)$ overall time for all P-nodes and $O(dn^3)$ overall time for all Q-nodes. The main idea for speeding up these computations under condition (SSC) is to exploit the subtrees rooted at β-small sons as central point for the computations in equations (3) and (4), and to compute only those entries of $Y[*]$ that correspond to paths starting or ending in such a subtree. To keep the notation clean, we introduce another auxiliary array $W^+[*]$ that is defined as follows. For every interior node v, we fix a unique β-small son v^β.

– For an interior node v with β-small son v^β, for a subset S of sons of v with $v^\beta \notin S$, and for a leaf a in $\text{LEAF}(T(v, S))$ and a leaf $b \in T(v^\beta)$, we define $W^+[v; S, a, b]$ as the length of the shortest Hamiltonian path through all the cities in $\text{LEAF}(T(v, S)) \cup \{b\}$ that behaves as follows: The Hamiltonian path starts in city a, then runs through all leaves in $T(v, S)$ while obeying the PQ-tree restrictions, and finally jumps away from $T(v, S)$ over to city b.

Let v be a P-node and let v_i be the root of $T(v, a)$. Similarly as in equations (3) and (4), the values of $W^+[v; S, a, b]$ are computed in order of increasing $|S|$. If $S = \{v_i\}$, then $W^+[v; S, a, b] = X[v; a, b]$. Otherwise $|S| \geq 2$, and

$$W^+[v; S, a, b] = \min \left\{ X[v; a, c] + W^+[v; S\text{-}\{v_i\}, c, b] \mid c \in \text{LEAF}(T(v, S\text{-}\{v_i\})) \right\}$$
(7)

Note that in (7), the relation $(a, c) \in \text{SEP}(v)$ holds. Since every pair $(a, c) \in \text{SEP}(v)$ is handled at most once per set S, Lemma 1 yields an $O(2^d n^2)$ upper bound on the overall cost of computing all relevant entries $W^+[v; S, a, b]$ for all v. By symmetric arguments and symmetric computations, we compute in $O(2^d n^2)$ overall time all relevant entries of the symmetrically defined array $W^-[v; S, a, b]$ (here the corresponding Hamiltonian path runs in the reverse direction and starts in b and ends in a).

Next, we explain how to compute all relevant entries $Z[v; a, b]$ with $(a, b) \in \text{SEP}(v)$ in $O(2^d)$ time per entry. Once more, let v_i be the root of $T(v, a)$ and v_j the root of $T(v, b)$. A Hamiltonian path corresponding to entry $Z[v; a, b]$ starts in city a, then runs through the cities in $T(v, S_1)$ for an appropriate subset S_1 of the sons of v with $v_i \in S_1$, then jumps to some $c_1 \in T(v^\beta)$, runs through all of $\text{LEAF}(T(v^\beta))$, leaves $T(v^\beta)$ by leaving a city $c_2 \in T(v^\beta)$, jumps over to $T(v, S_2)$ where $S_2 = S - (S_1 \cup \{v^\beta\})$, runs through all cities in $T(v, S_2)$ and finally stops in b. Hence, its length equals

$$W^+[v; S_1, a, c_1] + Z[v^\beta; c_1, c_2] + W^-[v; S_2, c_2, b].$$

Since there are only $O(2^d)$ choices for S_1 and since there are only a constant number of choices for c_1 and c_2, the value of $Z[v; a, b]$ is indeed determined in $O(2^d)$ time.

The computations for Q-nodes can be performed in $O(n^2)$ overall time in a similar but simpler fashion. Similarly as in the proof of Lemma 3, it is not necessary to compute all the entries of $Y[v; S, a, b]$, but only an appropriate small subset of them. This yields the desired speedup and completes the proof.
$\square$

Lemma 6. *For a PQ-tree T that fulfills (UBC) and that has node degree bounded by d, and for a distance matrix $D = (d_{ij})$, the shortest TSP tour for matrix D in $\Pi(T)$ can be computed in $O(2^d n^2)$ overall time.*

Proof. Since (UBC) implies (SSC), the results of Lemma 5 apply, and most parts of the dynamic program can be implemented in $O(2^d n^2)$ time. This time, the bottleneck is the computation of the relevant entries in $X[*]$ which takes $O(n^3)$ overall time.

Let v be a P-node in T, let v_k be the son of v that maximizes $|\text{LEAF}(T(v_k))|$ among all sons of v, and let w_j be the son of v_k that maximizes $|\text{LEAF}(T(w_j))|$ among all sons of v_k. Note that this implies that v has at most β sons outside of $T(v_k)$ and that v_k has at most β sons outside of $T(w_j)$. We analyze the computation of the entries $X[v; *]$ according to equation (2). Let $(a, b) \in \text{SEP}(v)$. If a is not a leaf in $T(v_k)$, then only the $O(\beta)$ cities in $\text{LEAF}(T(v)) - \text{LEAF}(T(w_j))$ are candidates for c, and thus the expression in (2) can be evaluated in constant time. Similarly, in case a is a leaf in $T(w_j)$, only the cities in $\text{LEAF}(T(v_k)) - \text{LEAF}(T(w_j))$ are candidates for c and again expression (2) can be evaluated in constant time. Hence, it remains to consider the cases where a is in $\text{LEAF}(T(v_k)) - \text{LEAF}(T(w_j))$ and b is in $\text{LEAF}(T(v)) - \text{LEAF}(T(v_k))$. In these cases, the evaluation of expression (2) may cost $O(n)$ time; however, there are only $O(\beta)$ choices for a and only $O(\beta)$ choices for b.

Summarizing, for every P-node v the computation of an entry in the array $X[v; *]$ can be done in constant time, with a constant number of exceptions for which the computation takes $O(n)$ time. Clearly, this yields $O(n^2)$ overall time for all P-nodes. For Q-nodes, once more it is not necessary to compute all the entries of $Y[v; S, a, b]$, but only an appropriate small subset of them (cf. the proof of Lemma 3). This completes the proof. □

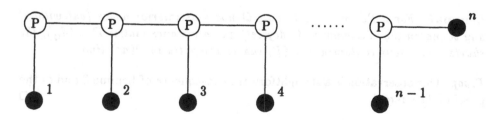

Fig. 1. A PQ-tree that represents the set of pyramidal tours. The root is the leftmost P-node.

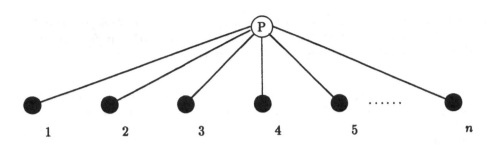

Fig. 2. A PQ-tree that represents the set of all tours.

5 Application: Special Neighborhoods

This section describes several applications of our main Theorem 4. It reveals the connections between PQ-trees and pyramidal tours, it constructs a large neighborhood of size $2^{\Theta(n \log \log n)}$ over which the TSP can be solved in polynomial time, and it poses two open problems.

Lemma 7.
(i) The set of pyramidal tours can be represented by a PQ-tree with maximum degree two that fulfills (UBC) with $\beta = 1$. Hence, there is an $O(n^2)$ algorithm for computing the shortest pyramidal TSP tour.

(ii) The set S_n of all permutations can be represented by a PQ-tree with maximum degree n that fulfills (SSC) with $\beta = 1$. Hence, there is an $O(n^2 2^n)$ algorithm for computing the shortest TSP tour. $\qquad\Box$

These statements can be proved by looking at the two pictures in Figures 1 and 2. With this, Lemma 5 and Lemma 6 yield the algorithms with the claimed time complexities. Note that the time complexities of these algorithms exactly match the time complexities of the corresponding algorithms by Klyaus [11] and by Held and Karp [9] as shown in Table 1.

Lemma 8. *For a PQ-tree T with only Q-nodes as interior nodes (but without any bound on the maximum node degree!) and a distance matrix $D = (d_{ij})$, the shortest TSP tour contained in $\Pi(T)$ can be computed in $O(n^3)$ time.*

Proof. This observation is a straightforward consequence of Lemma 3 and of the proof of Theorem 4. $\qquad\Box$

Aurenhammer [1] calls a permutation π a *twisted sequence* if there exists a PQ-tree T with only Q-nodes as interior nodes that fulfills $\pi \in \Pi(T)$ and $\langle 1, 2, 3, \ldots, n \rangle \in \Pi(T)$. Note that e.g. $\langle 1, 4, 3, 2, 5 \rangle$ is a twisted sequence, whereas $\langle 1, 3, 5, 4, 2 \rangle$ is not. It is easy to verify [1] that for every n, there are at least $\Omega(2^n)$ distinct twisted sequences contained in S_n.

Question 9. What is the computational complexity of finding for a given distance matrix $D = (d_{ij})$ the shortest TSP tour that is a twisted sequence ?

Theorem 10. *There exists a sequence of sets $Q_n \subset S_n$, $n \geq 1$, of permutations such that*

(i) $\log |Q_n| = \Theta(n \log \log n)$.
(ii) The shortest tour in Q_n can be computed in $O(n^3)$ time.

Proof. Define $d_n = \log n$ (throughout this proof, by $\log n$ we always mean the base two logarithm $\log_2 n$) and define $p_n = n/(d_n - 1)$. Construct a PQ-tree T_n with maximum node degree d_n as follows: All interior nodes are P-nodes. These P-nodes form a path $w_1, w_2, \ldots, w_{p_n}$ where w_1 is the root and where w_i is the father of w_{i+1} for $1 \leq i \leq p_n - 1$. The cities $\{1, 2, \ldots, n\}$ are partitioned into p_n groups, where every group consists of $d_n - 1$ cities. For every i, $1 \leq i \leq p_n$, the cities in the i-th group are the children of the P-node w_i. Clearly, the resulting tree T_n has its node degree bounded by d_n.

We define $Q_n = \Pi(T_n)$. Since for every one of the p_n interior vertices, there are $d_n!$ possible permutations of its children, the total number of permutations in Q_n is at least

$$\prod_{i=1}^{n/d_n} d_n! \geq \prod_{i=1}^{n/d_n} \left(\frac{d_n}{e}\right)^{d_n} = 2^{n[\log(\log n) - \log(e)]} = 2^{\Theta(n \log \log n)}.$$

Here we applied Stirling's formula. Since T_n fulfills (SSC) with $\beta = 1$, by applying Lemma 5 we can optimize over Q_n in $O(n^2 2^{d_n} + n^3)$ time, which in this case boils down to $O(n^3)$. $\qquad\square$

Question 11. Does there exist a sequence of sets $R_n \subset S_n$, $n \geq 1$, of permutations such that (i) $\log |R_n| = \Theta(n \log n)$ holds and such that (ii) the shortest tour in R_n can be computed in polynomial time ?

6 Application: The Shortcutting Step of the Tree Heuristic

In this section, we finally prove that the shortcutting step of the "twice around the tree" heuristic can be optimally implemented in polynomial time.

Hence, let $\{1, 2, \ldots, n\}$ be a set of cities in the Euclidean plane, let MST denote their minimum spanning tree and let MST$\times 2$ denote the Eulerian multigraph that results from doubling every edge in MST. We say that a tour in S_n is *embedded* in MST$\times 2$, if it is a subsequence of some Eulerian trail of MST$\times 2$. Clearly, a tour is embedded in MST$\times 2$ if and only if it can be obtained from shortcutting a Eulerian cycle in MST$\times 2$ as described in Section 1. By T_1 we denote the set of all tours that are embedded in MST$\times 2$.

Next, we root MST at an arbitrary city r. By MST(i) we denote the maximal subtree rooted at city i. We call a tour π *compatible* with the subtree MST(i)

if π visits the set of cities contained in MST(i) consecutively. By $\mathcal{T}_2$ we denote the set of tours that are compatible with every subtree MST(i), for $1 \leq i \leq n$. Note that for the definition of $\mathcal{T}_2$, it does not play a role which specific city r has been taken to be the root.

Lemma 12. $\mathcal{T}_1 \subseteq \mathcal{T}_2$.

Proof. Let π be an arbitrary tour in $\mathcal{T}_1$ and let τ denote the Eulerian trail in MST×2 from which π was obtained by shortcuts. Consider some city i together with its subtree MST(i).

The only possibilities for τ to enter and to leave MST(i) is via the two edges that connect i to its father. Obviously, τ must behave as follows: At some time it enters MST(i) through one of these two edges, then runs through all cities in MST(i), and finally leaves MST(i) through the other, remaining edge. Hence, τ visits the cities in MST(i) consecutively. Since π is a subsequence of τ that results from removing duplicate cities from τ, the cities in MST(i) are also visited consecutively by π. Consequently, π is compatible with MST(i). Since this argument holds for every i, tour π is in $\mathcal{T}_2$. □

Lemma 13. $\mathcal{T}_2 \subseteq \mathcal{T}_1$.

Proof. For any two cities a and b in MST, there is a unique simple path in MST going from a to b. By IC(a, b) we denote the ordered sequence of intermediate cities along this simple path, not including a and b. Next, let $\pi = \langle x_1, x_2, \ldots, x_n \rangle$ be an arbitrary tour in $\mathcal{T}_2$. Consider the cyclic trail τ in MST that visits the nodes in the ordering

$$x_1 - \text{IC}(x_1, x_2) - x_2 - \text{IC}(x_2, x_3) \cdots - \text{IC}(x_{n-1}, x_n) - x_n - \text{IC}(x_n, x_1) - x_1.$$

We claim that this trail τ is a Eulerian trail for MST×2. Observe that τ visits all cities, that it uses only edges available in MST, and that it returns to its starting city in the end. Consequently, τ visits every edge in MST, and it does so an even number of times. It remains to show that τ cannot pass through the same edge (f, s) in MST four or more times. Suppose otherwise, and let f be the father of s in the rooted MST: The assumption yields that at least four of the paths IC(x_i, x_{i+1}) pass through (f, s), which in turn implies that π enters and leaves MST(s) at least twice. Hence, π is not compatible with MST(s), a contradiction.

Summarizing, τ is a Eulerian trail of MST×2 and obviously π is a subsequence of τ. This gives $\pi \in \mathcal{T}_1$. □

Theorem 14. *For a given set of n points in the Euclidean plane and their minimum spanning tree MST, the shortest TSP tour embedded in MST×2 can be computed in $O(n^3)$ time. In other words, the shortcutting step of the "twice around the tree" heuristic can be optimally implemented in $O(n^3)$ time.*

Proof. We root the minimum spanning tree MST at an arbitrary node and then transform this rooted tree into a corresponding PQ-tree T^*. Essentially, the tree T^* has the same combinatorial structure as the rooted MST, but it does contain some additional leaves. The exact construction is performed as follows: First, we replace every non-leaf node in MST (that corresponds, say, to city i) by a corresponding interior P-node i'. All connections of this node i' to its father and its sons are the same as in the original rooted MST. To this P-node i' we attach a new leaf labeled by city i. Moreover, every leaf in the rooted MST remains a leaf in T^* corresponding to the same city.

We claim that the shortest tour π^* in $\Pi(T^*)$ constitutes the desired solution and that π^* can be computed in $O(n^3)$ time. By construction, $\Pi(T^*)$ contains exactly the tours that are compatible with every subtree MST(i), for $1 \le i \le n$. Consequently, by Lemmas 12 and 13 the set $\Pi(T^*)$ contains exactly those tours that are obtainable by shortcutting a Eulerian cycle in the "twice around the tree" heuristic. The algorithm in Theorem 4 computes the shortest tour in $\Pi(T^*)$ in $O(2^d n^3)$ time, where d is the maximum node degree in T^*. It is a folklore result that the minimum spanning tree of points in the Euclidean plane does not have nodes of degree seven or more, as no angle in it can be smaller than 60 degrees. The maximum degree d of T^* equals the maximum degree of MST plus one. Hence, $d \le 7$ and the running time of the dynamic programming algorithm becomes $O(n^3)$. □

7   Conclusion

In this paper we investigated a special case of the travelling salesman problem where optimization is done over a specially structured set of permutations: The only feasible permutations are those that are represented by a given PQ-tree. As a main result we showed that for PQ-trees whose maximum degree is bounded by a constant, this special case of the TSP is solvable in polynomial time. Our results can easily be carried over to the Bottleneck version of the TSP (where the goal is to minimize the length of the longest travelled edge) and to the general algebraic traveling salesman problem (cf. Burkard and van der Veen [3]).

Up to now, research on well-solved special cases of the TSP mainly concentrated on certain combinatorial structures in the *underlying distance matrix* or on combinatorial structures in a certain *underlying graph* (see Gilmore, Lawler and Shmoys [7]). Maybe now is the time to start a systematic investigation of the effect of specially structured *sets of permutations* on the computational complexity of the TSP.

Acknowledgement. We would like to thank Bettina Klinz for several discussions and for helpful comments on an earlier version of this paper.

References

1. F. Aurenhammer, On-line sorting of twisted sequences in linear time, *BIT* **28**, 1988, 194–204.

2. K.S. Booth and G.S. Lueker, Testing for the consecutive ones property, interval graphs and graph planarity using PQ-tree algorithms, *Journal of Computer and System Sciences* **13**, 1976, 335–379.

3. R.E. Burkard and J.A.A. Van der Veen, Universal conditions for algebraic traveling salesman problems to be efficiently solvable, *Optimization* **22**, 1991, 787–814.

4. J. Carlier and P. Villon, A new heuristic for the travelling salesman problem, *RAIRO – Operations Research* **24**, 1990, 245–253.

5. N. Christofides, Worst-Case Analysis of a new heuristic for the travelling salesman problem, Technical Report CMU, 1976.

6. G.A. Croes, A method for solving travelling-salesman problems, *Operations Research* **6**, 1958, 791–812.

7. P.C. Gilmore, E.L. Lawler and D.B. Shmoys, Well-solved special cases, Chapter 4 in [12], 87–143.

8. F. Glover and A.P. Punnen, The travelling salesman problem: New solvable cases and linkages with the development of approximation algorithms, Technical report, University of Colorado, Boulder, 1995.

9. M. Held and R.M. Karp, A dynamic programming approach to sequencing problems, *J. SIAM* **10**, 1962, 196–210.

10. D.S. Johnson and C.H. Papadimitriou, Performance guarantees for heuristics, Chapter 5 in [12], 145–180.

11. P.S. Klyaus, The structure of the optimal solution of certain classes of travelling salesman problems, (in Russian), *Vestsi Akad. Nauk BSSR, Physics and Math. Sci.*, Minsk, 1976, 95–98.

12. E.L. Lawler, J.K. Lenstra, A.H.G. Rinnooy Kan and D.B. Shmoys, *The travelling salesman problem*, Wiley, Chichester, 1985.

13. S. Lin, Computer solutions to the travelling salesman problem, *Bell System Tech. J.* **44**, 1965, 2245–2269.

14. C.H. Papadimitriou, The Euclidean travelling salesman problem is NP-complete, *Theoretical Computer Science* **4**, 1977, 237–244.

15. C.H. Papadimitriou and K. Steiglitz, Combinatorial optimization: algorithms and complexity, Prentice Hall, New Jersey, 1982.

16. C.H. Papadimitriou and U.V. Vazirani, On two geometric problems related to the travelling salesman problem, *J. Algorithms* **5**, 1984, 231–246.

17. C.N. Potts and S.L. van de Velde, Dynasearch – Iterative local improvement by dynamic programming: Part I, The travelling salesman problem, Technical Report, University of Twente, The Netherlands, 1995.

18. A. Serdyukov, On some extremal walks in graphs, *Upravliaemye systemy* **17**, Institute of Mathematics of the Siberian Academy of Sciences, USSR, 1978, 76–79.

Author Index

Springer-Verlag
and the Environment

We at Springer-Verlag firmly believe that an international science publisher has a special obligation to the environment, and our corporate policies consistently reflect this conviction.

We also expect our business partners – paper mills, printers, packaging manufacturers, etc. – to commit themselves to using environmentally friendly materials and production processes.

The paper in this book is made from low- or no-chlorine pulp and is acid free, in conformance with international standards for paper permanency.

Lecture Notes in Computer Science

For information about Vols. 1–1013

please contact your bookseller or Springer-Verlag